MULTILAYER FLEXIBLE PACKAGING

PLASTICS DESIGN LIBRARY (PDL)
PDL HANDBOOK SERIES
Series Editor: Sina Ebnesajjad, PhD (sina@FluoroConsultants.com)
President, FluoroConsultants Group, LLC
Chadds Ford, PA, USA
www.FluoroConsultants.com

The **PDL Handbook Series** is aimed at a wide range of engineers and other professionals working in the plastics industry, and related sectors using plastics and adhesives.

PDL is a series of data books, reference works and practical guides covering plastics engineering, applications, processing, and manufacturing, and applied aspects of polymer science, elastomers and adhesives.

Recent titles in the series

Biopolymers: Processing and Products, Michael Niaounakis (ISBN: 9780323266987)
Biopolymers: Reuse, Recycling, and Disposal, Michael Niaounakis (ISBN: 9781455731459)
Extrusion, 2e, John Wagner & Eldridge Mount (ISBN: 9781437734812)
Fluorinated Coatings and Finishes Handbook, 2e, Laurence McKeen (ISBN: 9780323371261)
Fluoroelastomers Handbook, 2e, Jiri George Drobny (ISBN: 9780323394802)
Fluoroplastics, Volume 1, 2e, Sina Ebnesajjad (ISBN: 9781455731992)
Fatigue and Tribological Properties of Plastics and Elastomers, 3e, Laurence McKeen (ISBN: 9780323442015)
Future Trends in Medical Plastic Engineering and Manufacturing, Markus Schönberger (ISBN: 9780323370233)
Handbook of Biopolymers and Biodegradable Plastics, Sina Ebnesajjad (ISBN: 9781455728343)
Handbook of Polymer Applications in Medicine and Medical Devices, Kayvon Modjarrad & Sina Ebnesajjad (ISBN: 9780323228053)
Handbook of Thermoplastic Elastomers, Jiri G Drobny (ISBN: 9780323221368)
Handbook of Thermoset Plastics, 2e, Hanna Dodiuk & Sidney Goodman (ISBN: 9781455731077)
High Performance Polymers, 2e, Johannes Karl Fink (ISBN: 9780323312226)
Introduction to Bioplastics Engineering, Syed Ali Ashter (ISBN: 9780323393966)
Introduction to Fluoropolymers, Sina Ebnesajjad (ISBN: 9781455774425)
Manufacturing Flexible Packaging, Thomas Dunn (ISBN: 9780323264365)
Manufacturing and Novel Applications of Multilayer Polymer Films, Deepak Langhe (ISBN: 9780323371254)
Multilayer Flexible Packaging, 2e, John R. Wagner, Jr. (ISBN: 9780323371001)
Plastic Films in Food Packaging, Sina Ebnesajjad (ISBN: 9781455731121)
Plastics in Medical Devices, 2e, Vinny Sastri (ISBN: 9781455732012)
Printing on Polymers, Joanna Izdebska (ISBN: 9780323374682)
Reactive Polymers, 2e, Johannes Karl Fink (ISBN: 9781455731497)
Specialized Injection Molding Techniques, Hans-Peter Heim (ISBN: 9780323341004)
The Effect of Creep and Other Time Related Factors on Plastics and Elastomers, 3e, Laurence McKeen (ISBN: 9780323353137)
The Effect of Long Term Thermal Exposure on Plastics and Elastomers, Laurence McKeen (ISBN: 9780323221085)
The Effect of Sterilization on Plastics and Elastomers, 3e, Laurence McKeen (ISBN: 9781455725984)
The Effect of Temperature and Other Factors on Plastics and Elastomers, 3e, Laurence McKeen (ISBN: 9780323310161)
Thermoforming of Single and Multilayer Laminates, Ali Ashter (ISBN: 9781455731725)
Thermoplastics and Thermoplastic Composites, 2e, Michel Biron (ISBN: 9781455778980)
Thermosets and Composites, 2e, Michel Biron (ISBN: 9781455731244)
UHMWPE Biomaterials Handbook, 3e, Steven Kurtz (ISBN: 9780323354011)

To submit a new book proposal for the series, or place an order, please contact David Jackson, Acquisitions Editor
david.jackson@elsevier.com

MULTILAYER FLEXIBLE PACKAGING

Second Edition

Edited by

John R. Wagner, Jr.

ELSEVIER

AMSTERDAM • BOSTON • HEIDELBERG • LONDON
NEW YORK • OXFORD • PARIS • SAN DIEGO
SAN FRANCISCO • SINGAPORE • SYDNEY • TOKYO
William Andrew is an imprint of Elsevier

William Andrew is an imprint of Elsevier
The Boulevard, Langford Lane, Kidlington, Oxford, OX5 1GB, UK
50 Hampshire Street, 5th Floor, Cambridge, MA 02139, USA

British Library Cataloguing-in-Publication Data
A catalogue record for this book is available from the British Library.

Library of Congress Cataloging-in-Publication Data
A catalog record for this book is available from the Library of Congress.

ISBN: 978-0-323-37100-1

For Information on all William Andrew publications visit our website at http://elsevier.com/

Publisher: Matthew Deans
Acquisition Editor: David Jackson
Editorial Project Manager: Zara Preston
Production Project Manager: Caroline Johnson
Designer: Mark Rogers

Typeset by MPS Limited, Chennai, India

Contents

Preface to the First Edition

In the beginning of this project, I had expected to complete it in less than a year. Now, four years later, it has finally come to completion. Along the road, the pressures of limited resources and conflicting objectives changed the timeline and contributors.

This book had ambitious goals set. What has resulted is an incomplete description of multilayer films and the technology to produce them. It is a snapshot that gives a comprehensive view of multilayer film designs and technology used. As in any evolving system, things change and what we can describe today may be somewhat different tomorrow. The value this book brings is that it will help interested parties to understand what is being made and how it is produced. This will help them deal with current problems and issues and move forward to invent new products and processes that will meet future challenges.

The authors who have contributed their time and extensive knowledge are industry experts and respected educators from around the world. They have shared their thoughts and ideas with you so that you can better understand this important technology and improve your contributions in your area of expertise.

The book is organized as follows:

Introduction

Raw materials

- Resins
- Additives
- Rheology

Die design

- Flat
- Blown

Process considerations

Technologies

- Blown film, cast film, and lamination processes
- Machine direction orientation process
- Biaxial orientation
- Blending
- Coating technologies
- Vacuum-deposited coating

Multilayer film designs

- PE-based film
- Oriented films

Appendix: Writing guide for describing multilayer film structures.

As the technology of multilayer films is constantly developing, it is virtually impossible to have an up-to-the minute description of every multilayer film and technology used. So, this book, by definition, is only a snapshot of what is available. By reading this book, whether you are a manager, purchasing agent, user, engineer or technician, you can improve your knowledge and insight into this important technology that provides safety, freshness, and visual appeal for point of sale awareness.

The introduction starts by presenting a historical perspective of the materials used to produce flexible plastic films. Then a look at how the markets for flexible plastic films have grown along with an extrapolation to 2020. The main body of this book is divided into three sections:

- materials
- hardware and processes
- multilayer film designs with applications

In the materials section, there are chapters about polyethylene and polypropylene, the two major materials used for flexible film, a comprehensive chapter on additives used to make polymers functional and then a chapter on rheology which presents important concepts needed to understand non-Newtonian viscoelastic flow.

The hardware and process section begins with a chapter that describes the dies used to produce multilayer film. Annular dies are used for blown processes. The several ways multilayer dies can be designed are presented. Flat dies are used for cast and oriented films. Multilayer structures can be produced with multicavity dies, feedblocks and by combining feedblocks with multicavity dies.

There is a special chapter on process engineering and how important this function is to producing multilayer films that meet specifications and cost targets.

Then there follow chapters about the technologies used to produce multilayer films. The areas covered are production of multilayer films, laminating films to produce a structure that cannot be produced in a single step and coating processes. The chapter on blending is especially useful as blending technology is widely used and not always easy to understand.

The last section presents PE based and oriented film-based multilayer structures. In these two chapters, you will learn about many different structures and their applications.

The appendix presents a nomenclature or shorthand language that can be used to describe multilayer films.

Read, Learn, Enjoy.

Preface to the Second Edition

It is now six years after the first edition went to press and we have listened to reviewers of this book and added three new chapters. While this book is still an incomplete description of multilayer films and the technology to produce them, it is an updated and more comprehensive edition.

The new chapters are on:

- Bio Polymers
- Winding
- Food regulations

The new table of contents is:

- Introduction
- PE [new author]
- PP
- Biopolymers [new author]
- Rheology
- Blown Film Dies
- Process Engineering [new author]
- Polymer blends
- Coating Technology
- Vacuum Metallizing
- Winding [new author]
- PE Films
- Regulatory Aspects of Food Packaging — A Global Matter [new author]
- Appendix

The updates include new information and improved grammar for easier reading.

I trust you will find these additions, updates, and improvements useful in your work.

Read, Learn, Enjoy.

Acknowledgments

It is unfortunate in today's economy that everyone is very busy and such was the case with this second edition. I am extremely grateful to the authors who have contributed their time and shared their knowledge so that you, our readers, can do your jobs more effectively and efficiently.

While plastic overall is a positive contributor to our health and well-being, plastic has a negative connotation due to the pollution that occurs in our environment. We need to be painfully aware of this issue and engineer products that are more sustainable, nontoxic, and economically viable.

We also need to educate people not to pollute and to establish programs to clean our oceans and water ways of plastic.

We all need to be part of the solution, not part of the problem.

Best to you,
John R. Wagner, Jr

List of Contributors

Joachim Bayer Songwon International, Frauenfeld, Switzerland

Charles A. Bishop C.A. Bishop Consulting Ltd, United Kingdom

Jürgen Breil Brückner Maschinenbau GmbH & Co. KG, Siegsdorf, Germany

Thomas I. Butler Blown Film Technology, LLC, Lake Jackson, TX, United States

Allison Calhoun Whitman College, Walla Walla, WA, United States

Edward D. Cohen Consulting Services Inc., Fountain Hills, AZ, United States

Sina Ebnesajjad FluoroConsultants Group, LLC, Chadds Ford, PA, United States

Edgar B. Gutoff Consulting Chemical Engineer, Brookline, MA, United States

Eric Hatfield Clopay, Mason, OH, United States

Duk-Suk Jung Songwon Industrial, Ulsan, South Korea

Sung-Leal Jung Songwon Industrial, Ulsan, South Korea

Klaus Keck-Antoine Songwon Industrial, Ulsan, South Korea

Els Lievens Agfa Gevaert, Mortsel, Belgium

John Mara Songwon International Americas, Friendswood, TX, United States

Scott B. Marks E.I. du Pont de Nemours & Company, Wilmington, DE, United States

Antonio Meraldo Lux Research, Inc., Singapore

Barry A. Morris DuPont Performance Materials, Wilmington, DE, United States

Eldridge M. Mount III EMMOUNT Technologies, Canandaigua, NY, United States

Rajen M. Patel The Dow Chemical Company, Freeport, TX, United States

John Perdikoulias Compuplast International Inc., Zlín, Czech Republic

David R. Roisum Finishing Technologies, Inc., Neenah, WI, United States

David Strutt Tencate Performance Composites, Baycomp Company, Burlington, ON, Canada

Vincent Vezza Independent Consultant, Rochester, NY, United States

John Vlachopoulos Department of Chemical Engineering, McMaster University, Hamilton, ON, Canada

John R. Wagner, Jr. Crescent Associates, Inc., Rochester, NY, United States

Contributors

Joachim Bayer

Joachim Bayer joined Songwon in June 2007 as director of technical sales for Europe and Middle East. Joachim has 13 years of experience in antioxidants, UV additives, dispersing agents, and antistats. Joachim acquired key competences in physical forms and polymer conversion processes. He held various positions in R&D, TS, Marketing and Sales. Before Songwon, he worked for Hoechst (D), Clariant (D), Great Lakes (CH), and Chemtura (CH)

Charles A. Bishop

Charles left school at 16 and completed an engineering apprenticeship as a toolmaker. He took a degree in Materials Technology at Loughborough University. He then moved to the Physics Department and completed his Masters and Doctorate degrees by research into vacuum deposition onto polymer webs. After a brief time as a post doctoral researcher and consultant, he moved into industry working for ICI and DuPont for 15 years before setting up his own consultancy business in 1998.

He authored the book *Vacuum deposition onto webs, films and foils* and authored or co-authored over 60 papers and 5 patents. He jointly runs a vacuum training website (www.VacuumCoatingTraining.com) and edits the AIMCAL Blog www.vacuumcoatingblog.com and is on their panel of "Experts" to answer problems members send in. He received the SVC Mentor Award 2008.

Jürgen Breil, PhD

Dr Jürgen Breil graduated from the Rheinisch-Westfälische Technische Hochschule (RWTH) Aachen in 1980. He remained at RWTH Aachen and was employed in the field of science and industrial projects in the faculty of Mechanical Engineering. He focused on automating extrusion lines and received his PhD from the Institute for Plastic Processing (IKV).

At Krauss Maffei Kunststofftechnik, Munich, he worked on automating injection-molding machines. He joined Brückner Maschinenbau GmbH in 1986 and worked on several automation projects. He was promoted to department leader for technology and automation and, in 1993, he became responsible for the Research and Development department. Since 2002, he has been the head of division Process Technology and R&D.

Dr Breil is the author of 14 papers and 9 patents on orientation technologies.

Thomas Butler

Thomas I. (Tom) Butler received his BS in Chemical Engineering from New Mexico State University. He has over 39 years of experience in the development of application and products for the plastics and chemical industry. Tom now provides consulting services with his company Blown Film Technology, LLC. Tom retired from The Dow Chemical Company in 2004 where he made significant contributions in the development of polyethylene polymers and processing technology while in the Polyethylene Technical Service and Development Group. His interests include expertise in extrusion process design and control, coextrusion technology and blown film process technology. He has been involved in film application development for film liners, stretch cling, grocery sacks, adhesive resins, sealant resins, skin packaging, snack food applications, processed meat packaging, hygiene applications, and produce packaging.

Allison Calhoun, PhD

Dr Allison Calhoun is Associate Professor of Chemistry and teaches courses in general and physical chemistry and the applications of physical chemistry in environmental and polymeric systems at Whitman College in Washington state. Prior to joining the faculty at Whitman, she worked for five years in the polymer industry and holds three patents for the use of mineral additives in polymeric thin films. She is the co-author of a polymer science textbook and her current research projects examine the oxidation of polyolefins and the impact of this oxidation on the interface of minerals in polymers. She received her PhD from the University of Georgia with a focus in physical chemistry. She sits on the Board of Directors for the Flexible Packaging Division of the Society of Plastics Engineering and is a member of the Physical Chemistry Test Writing Committee through the American Chemical Society.

Edward D. Cohen
Dr Cohen is a technical consultant in all aspects of the web coating process. He received a PhD from the University of Delaware and a BS in Chemical Engineering from Tufts University. His expertise is in the coating and drying of thin films, coating process development and scale-up, coating defects and new product development.

He has over 45 years' experience in coating research and manufacturing technology with the DuPont Company and as an independent consultant. He has extensive publications in coating drying technology. Currently, he is Technical Consultant for AIMCAL.

Sina Ebnesajjad
Dr. Sina Ebnesajjad is the President and Founder of FLUOROCONSULTANTS GROUP, LLC. He had worked for nearly 24 years for the DuPont Fluoroproducts before retiring at the end of 2005. He has extensive experience with fluoropolymer products and technologies using granular, fine powder and dispersion PTFE (Teflon®), additives and melt processable fluoropolymers including PFA, FEP, ETFE,PVDF, ECTFE, PVF, PCTFE, and films.

Dr. Ebnesajjad authored over a dozen books including the incumbent handbooks of fluoropolymer technologies all published by Elsevier.

Edgar B. Gutoff
Edgar B. Gutoff, ScD, PE, has been a consultant on coating and drying since 1988. He organized the first of the 14 International Coating Symposia at the AIChE Spring 1982 meeting and was the founding Secretary of the International Society for Coating Science and Technology, which now sponsors them. He has organized a number of coating courses and seminars and has co-authored *Coating and Drying Defects*, *Modern Coating and Drying Technology* and *The Application of Statistical Process Control to Roll Products*. In 1994, he was awarded the AIChE John A. Tallmadge Award for Contributions to Coating Technology. He is a fellow of the AIChE and IS&T and is on AIMCAL's Technical Advisory Panel. He worked for Polaroid for 28 years and now is an Adjunct Professor at Northeastern University. He received an ScD and an SM from MIT and a BChE from CCNY.

Eric Hatfield
Eric Hatfield is co-owner of MDO Engineering which specializes in Machine Direction Orientors (MDO). He just recently was a Partner and the Managing Director of Operations Technology at FlexTech Packaging Ltd, an MDO & high barrier based film company where he was responsible for Engineering and Process Technologies. Formally, Eric was the manager of Process and Equipment Development for James River's Flexible Packaging Group. Duties there included design of combining adapters, flat and round dies, MDOs and the creation and commercialization of process equipment and technologies. Eric has a BS in Chemical Engineering from the Illinois Institute of Technology and an MBA from Keller Graduate School. He also has several patents.

Duk-Suk Jung
Duk-Suk Jung received an education as chemist and process engineer in South Korea. Duk-Suk is team leader and the manager of the technical department in Songwon's homebase Ulsan (South Korea). His responsibility involves the process engineering of all synthesis plants. He has over 24 years' experience in technical service in Asian countries in antioxidants, UV absorbers, HALS, and PVC stabilizers.

Sung-Leal Jung
Sung-Leal Jung received an education as chemist and process engineer in South Korea. Sung-Leal currently holds the position of manager of the Asian technical service group for PVC stabilizers, based in Ulsan (South Korea). For this product group, he has more than 15 years of experience. Before that, he was involved in the engineering of Songwon's stabilizer plants.

Klaus Keck-Antoine
Klaus Keck-Antoine studied Polymer Chemistry at FH Reutlingen in Germany until 1991. He joined Songwon in September 2007 after an extensive career in plastic stabilizers and additives. In the past 16 years, he worked on antioxidants, UV additives, inhibitors, dispersing agents and antistats for polyolefins, engineering resins and non-polymeric substrates. Before Songwon, he worked for Hoechst (Germany), Hoechst Celanese (USA), Clariant (USA), Great Lakes (USA and Belgium), and shortly for Chemtura (Belgium). Currently, he is in charge of implementing Songwon's Global Application and Technical Service Community (GATSC). GATSC is a global forum of and network for industry experts and scientific laboratories to enhance the systematic understanding of polymer additives and provide this service and knowledge to Songwon's customers. He holds various patents and has published frequently on above-mentioned topics.

Els Lievens
Els Lievens is a professional bachelor of chemistry from the Karel De Grote Hogeschool in Hoboken (B). She started her career in 1992 with Shell Chemicals in the application research group for polypropylene (PP) in Louvain La Neuve (B), more particular PP film applications. In 1995, she started doing application development work on additives for PP throughout all applications. Later, after the merger of Shell into respectively Montell and Basell, she became responsible for the upscaling of new PP additives in the northern European plants and technical qualification of new additives. In 2002, she joint Great Lakes — later on Chemtura — to be project leader for application development of polymer stabilizers. Since 2007, she has held the position of chemical sourcing manager in the chemistry research department of Agfa Graphics in Mortsel (B).

John Mara
John Mara is the Technical Sales Manager for Songwon International — Americas, a position he has held since 2007. During this time, his primary responsibility has been to grow Songwon's polymer stabilizer business in the Americas providing commercial and technical support to a broad range of customers. Prior to joining Songwon, John managed the Advanced Polymer Technologies Group at Excel Polymers where his research focused on Rubber Nanocomposites. Prior to Excel, John provided technical support to BASF's Light Stabilizer Business, managed commercial and technical activities involving polymer stabilizers with Great Lakes Chemical Company and managed the Polymer Science Analytical Group at Chevron Chemical Company. John received his PhD in Polymer Science/Plastics Engineering from the University of Massachusetts-Lowell (1992) where he studied the synthesis and characterization of novel liquid crystalline polymers. John also holds a Six-Sigma Black belt from the University of Akron (2007) and has authored and co-authored over a dozen technical papers on polymer stabilization.

Scott B. Marks
Scott Marks was educated at Rutgers University where he earned degrees in Mechanical Engineering and Business Administration. He is employed by E.I. DuPont de Nemours & Company, starting initially as a Construction Engineer and then, in 1984, transferred into the Polymers business unit working in the Flexible Packaging industry. He has held Market Development and Technical Service positions concentrating on the Asia-Pacific and North American regions the majority of his career, but also has supported Latin American and Europe.

Scott's process expertise is in multilayer blown film, cast film, extrusion coating and sheet. He has also worked with adhesive lamination and thermal lamination. He has a broad packaging knowledge including meat, cheese, seafood, pharmaceutical and healthcare, personal care, and industrial and consumer goods markets. In addition, he has worked with non-packaging applications such as composite building panels, cable shielding, cable sheathing, and surf and boogie boards.

Scott's work is mainly with polyolefins, specializing in modified ethylene copolymers for sealants, adhesives and bulk layers in extrusions and laminations. In addition, he has worked with Nylon and ethylene vinyl alcohol coextruded with polyolefins and polyesters, polystyrenes, and newer materials such as thermoplastic starch and polylactic acid.

Scott has presented for TAPPI and SPE in North America and via other organizations in Singapore, Malaysia, Thailand, Taiwan, Australia, and New Zealand.

He is active in TAPPI's PLACE division, having served in several officer positions including

Division Chairman and division council advisor. He is an active member of the Flexible Packaging, Extrusion Coating, and Film Extrusion working committees and co-chairs the International Committee.

Barry A. Morris

Barry A. Morris is a senior technology associate with DuPont Packaging and Industrial Polymers. He received a BS in chemical engineering from the University of Virginia and a PhD from Princeton University. He also has an MBA from the University of Delaware. His research interests include polymer adhesion, blends, compounding, rheology, coextrusion and film and coating converting processes. He has over 20 years of experience in packaging technology and is widely published.

Eldridge M. Mount III

Dr Mount is an independent consultant in the extrusion, metallization and plastic film industries. Originally a chemist, he obtained ME and PhD degrees in chemical engineering at Rensselear Polytechnic in 1979 with the experimental and theoretical study of melting mechanisms and the modeling of polymer melting in single-screw extruders. Since graduation, Eldridge has worked for ICI Americas and ExxonMobil in the areas of biaxially oriented PET and OPP films. His work encompassed process and product design where he was responsible for new screw designs, increased output, coextrusion technology and the development of new metallized and other OPP film products. Eldridge has nine US patents and many publications in the area of metallized OPP films and extrusion. A Fellow and Honored Service Member of SPE, he has been active on the Executive Committee and the Extrusion Division Board and has been a speaker at ANTEC, TAPPI and AIMCAL as well as other professional societies in the areas of metallized films, extrusion and film processing. Since June 2000, Eldridge has been an independent consultant with EMMOUNT Technologies. In his spare time he is a long standing adult Boy Scout Leader and Silver Beaver recipient. He enjoys hunting, hiking, and camping.

Rajen M. Patel

Rajen Patel joined Polyolefins R&D of the Dow chemical company in June 1991 in the materials science group and transferred to Polyolefins TS&D of the Dow chemical company in November 2007. He has worked in various research and applications development areas including polyolefins characterization (thermal & rheological), structure-processing-properties relationships in polyolefins, and product development in variety of applications such as hygiene & medical, oriented shrink films, collation shrink film, sealants, cast stretch films, blown films, extrusion coating, tie-layers, barrier, elastic films and fibers. He has researched extensively on materials science and applications development of single-site (metallocene) catalyzed Polyolefins. He received 2 Dow Inventor of the Year Awards and The Dow Excellence in Science Award. He wrote a chapter on structure-properties and applications of Polyolefins produced by single-site catalyst technology in encyclopedia of chemical processing and design. He co-authored 22 technical peer reviewed journal publications and 4 book chapters. He is also a co-inventor of 47 granted US patents. In 2009, he was elected as Fellow of Society of Plastics Engineers (SPE) in recognition of his technical and commercial accomplishments in Polyolefins and especially in single-site catalyzed (metallocene) Polyolefins. He is currently a board member of SPE's Engineering Properties and Structure division (ESPDIV). He is currently an Associate Research Fellow in the Packaging and Specialty Plastics technical service and development (TS&D) group in the Dow Chemical Company leading packaging applications development.

Rajen Patel obtained his BS in Chemical Engineering from the University of Bombay in 1984. He joined university of Tennessee, Knoxville, USA, in September 1985 and obtained MS in December 1987 and PhD in May 1991, both in Polymer Science & Engineering.

David R Roisum

Dr. Roisum is a well-known authority in the area of web handling and converting. He has authored nine books that include Winding, Rollers and Web-Handling with his most recent being the Web Machine Buying Guide. He has coauthored or edited several others. Additionally, he has written 200 articles, conference papers and other publications. He has also written hundreds of blog posts and made one hundred YouTube instructional clips. He has been a technical editor for Converting Magazine with a monthly column entitled Web

Works since 1994. Dave has been honored by TAPPI with their Finishing & Converting Division Award, Thomas W. Busch Prize and Finest Faculty awards. He was elected TAPPI Fellow in 2003. He received AIMCAL's President's Award in 2012.

Dave received his BS in Mechanical Engineering and MS in Engineering Mechanics from the University of Wisconsin at Madison, and his PhD from the Web Handling Research Center where he later became an Industrial Advisory Board member.

Dave has worked for the Beloit Corporation as a designer of high-speed winding machinery and later as a manager of research, and for Kimberly-Clark as an internal converting expert serving all business units. He is now a principal of Finishing Technologies Inc., providing consulting services to more than 300 clients who convert or manufacture: paper, film, foil, nonwovens and many other materials. He has accumulated much practical experience working in nearly 1000 plants over the course of more than three decades. He has shared his experience with nearly 5000 students.

David Strutt

Dr David Strutt earned his PhD in chemical engineering from the University of Waterloo in 1998, in the area of simulation of mixing in reactive flows of polypropylene in co-rotating twin-screw extruders. He has held several industrial and consulting positions related to plastics processing and currently works in process development for thermoplastic composite tapes at Baycomp Ltd. He is the author of 10 papers on single- and twin-screw extrusion and a co-author of chapters in two SPE books.

Vincent Vezza

Vincent Vezza is an independent consultant with over 35 years industrial experience. He graduated with a MS degree in applied and mathematical statistics and a BS degree in chemistry from Rochester Institute of Technology. He has held positions as Statistician, Quality Engineer and Process Engineer at Eastman Kodak Company and Ortho Clinical Diagnostics a Johnson and Johnson Company. He retired from Ortho Clinical Diagnostics, a Johnson and Johnson Company.

John Vlachopoulos

John Vlachopoulos was educated at NTU, Athens, Greece (Dipl Ing) and at Washington University, St Louis, MO (MS, DSc). He subsequently started teaching and doing research at McMaster University, Hamilton, Canada, on polymer processing, applied rheology and computer-aided methods for the solution of flow problems. He is the author of more than 250 publications, frequent lecturer at conferences and consultant to the polymer processing industry worldwide. He is currently Professor Emeritus of Chemical Engineering and President of Polydynamics, Inc. a company specialized in process simulation software.

John R. Wagner, Jr.

John R. Wagner Jr is President of Crescent Associates, Inc., a consulting firm that specializes in plastic films and flexible packaging. He graduated from the University of Notre Dame with a BS and MS in chemical engineering. His career began with B.F. Goodrich Chemical Co. working on PVC manufacturing from monomer production through polymerization and compounding. He then worked for Mobil Chemical Co. Films Division doing research and development on oriented polypropylene films (OPP) for the food packaging industry. His areas of expertise included product and process development and optimization, processing equipment troubleshooting/improvement and applying statistical methods to improve the efficiency and effectiveness of experimental investigations. He currently holds 23 patents on flexible packaging films and coating technology.

John has presented several papers at local, national and international technical and scientific meetings. He has authored book chapters and co-edited three books: *The SPE Guide on Extrusion Technology and Troubleshooting*; *The Plastics Technician's Toolbox – Extrusion*; *Extrusion – The Definitive Processing Guide and Handbook*. He is also editor-in-chief of the *Journal of Plastic Film and Sheeting*.

An SPE member since 1969, he has been a member of the Extrusion Division Board of Directors since 1984, ANTEC Technical Program Chair 1995, Board Chair 1996–97 and elected Fellow of the Society in 1998. He has served as treasurer for the SPE Extrusion Division Board of Directors since 2001. He was Technical Program Co-Chair for Flexible Packaging Conference 2004 and was on the technical program committee for Flexible Packaging Conference 2005. He is on the Flexible Plastics Packaging Division Board of Directors and has served as Treasurer since 2007.

PART I

INTRODUCTION

1 Introduction

John R. Wagner, Jr. [1] ***and Scott B. Marks*** [2]

[1]Crescent Associates, Inc., Rochester, NY, United States; [2]E.I. du Pont de Nemours & Co., Wilmington, DE, United States

1.1 Materials: A Historical Perspective

Today we have many clear plastic packaging films. Jacques E. Brandenberger invented the first common clear film, "cellophane" in 1908. He was a Swiss textile engineer whom first thought of the idea for a clear, protective, packaging layer in 1900. He was seated at a restaurant when a customer spilled wine onto the tablecloth. As the waiter replaced the cloth, Brandenberger decided that he would invent a clear flexible film that could be applied to cloth, making it waterproof.

Brandenberger experimented with different materials and tried applying liquid viscose (a cellulose product known as rayon) to cloth, but the viscose made the cloth too stiff. His idea failed but he noted that the coating peeled off in a transparent film. Like so many inventions, the original use was abandoned and new and better uses were found. By 1908 he developed the first machine to manufacture transparent sheets of regenerated cellulose. By 1912 he was making a saleable thin flexible film used in gas masks. He obtained patents to cover the machinery and the essential process ideas [1]. The rights to produce cellophane were then obtained by E.I. du Pont de Nemours & Co, who began to produce and further refine the process for producing cellophane. Breakthrough improvements included adding a moisture barrier layer to the cellophane, in the form of a nitrocellulose coating. This allowed for better stiffness retention in the cellophane and facilitated using the film as a food overwrap film. This coating was then refined to make it heat sealable as well, creating the first readily sealable packaging film. Later on, other coatings were applied to cellophane including PVDC (polyvinylidene chloride), which added oxygen barrier and moisture barrier to the cellophane, resulting in the original nonmetal barrier film for food packaging.

Today we have many polymers such as PE (polyethylene), PP (polypropylene), PET (polyester), and PS (polystyrene) that are used to produce clear films for packaging. While these aforementioned polymers are commonly used in monolayer format, they are also used in multilayer films produced by coextrusion and/or lamination processes. There also exist many specialty polymers and compounds, which sometimes are used monolithic, but are more commonly used in coextrusions.

Polyethylene was first synthesized by the German chemist Hans von Pechmann who prepared it by accident in 1898 while heating diazomethane. When his colleagues Eugen Bamberger and Friedrich Tschirner characterized the white, waxy, substance that he had created, they recognized that it contained long $-CH_2-$ chains and termed it "polymethylene" [2].

The first industrially practical polyethylene synthesis was discovered (again by accident) in 1933 by Eric Fawcett and Reginald Gibson at the ICI works in Northwich, England. Upon applying extremely high pressure (several hundred atmospheres) to an ethylene and benzaldehyde mixture, they again produced a white, waxy, material. Because the reaction had been initiated by trace oxygen contamination in their apparatus, the experiment was, at first, difficult to reproduce. It was not until 1935 that another ICI chemist, Michael Perrin, developed this accident into a reproducible high-pressure synthesis for polyethylene that became the basis for industrial low-density polyethylene (LDPE) production beginning in 1939 [2].

Subsequent landmarks in polyethylene synthesis have revolved around developing several catalysts that promote ethylene polymerization at more mild temperatures and pressures. The first was a chromium trioxide–based catalyst discovered in 1951 by Robert Banks and J. Paul Hogan at Phillips Petroleum. In 1953 the German chemist Karl Ziegler developed a catalytic system based on

Multilayer Flexible Packaging. DOI: http://dx.doi.org/10.1016/B978-0-323-37100-1.00001-6

titanium halides and organoaluminum compounds that worked at even milder conditions than the Phillips catalyst. The Phillips catalyst is less expensive and easier to work with; however, both methods are used in industrial practice.

By the end of the 1950s, both the Phillips- and Ziegler-type catalysts were being used to produce high-density polyethylene (HDPE). Phillips initially had difficulties producing a uniform quality HDPE product and filled warehouses with off-specification plastic. However, financial ruin was unexpectedly averted in 1957, when the hula hoop, a toy consisting of a circular polyethylene tube, became a fad among youth in the United States.

A third catalytic system, one based on metallocenes, was discovered in 1976 in Germany by Walter Kaminsky and Hansjörg Sinn. The Ziegler and metallocene catalyst families have since proven to be very flexible at copolymerizing ethylene with other olefins and have become the basis for the many polyethylene resins available today, including very low-density polyethylene and linear low-density polyethylene. These new catalysts have allowed creation of some very special polyethylene resins. One such example is the polymer used to produce DSM's "Dyneema" fiber, an ultra high-strength material that competes with aramid-based fibers in applications where high strength is needed, but not where high temperature resistance is also needed. As it has always been, new materials find their useful niches in the market.

Until recently, the metallocene systems were the most active single-site catalysts for ethylene polymerization known. New catalysts are typically compared to zirconocene dichloride. Much effort is currently exerted on developing new, single-site (so-called postmetallocene) catalysts that may allow greater tuning the polymer structure than is possible with metallocenes. Recently the work by Fujita at the Mitsui Corporation (among others) has demonstrated that certain salicylaldimine complexes of Group 4 metals show substantially higher activity than the metallocenes [2].

Dr. Karl Rehn at Hoechst AG in Germany first polymerized polypropylene in 1951 and did not recognize the importance of his discovery. It was then rediscovered on March 11, 1954, by Giulio Natta [3]. Biaxially oriented polypropylene (BOPP) was introduced in the 1960s. Initially, it was developed in a blown film process, and the first commercial production was started by DuPont in IA, USA. Eventually other companies started to produce BOPP in a cast sheet/orientation process which made the film production more economical. This change was instrumental in replacing cellophane, as it was lower cost and more environmentally friendly to produce. As BOPP usage gained in the industry, there was a corresponding decline in cellophane usage and production. Today cellophane production is a mere fraction of what it once was.

British chemists, John Rex Whinfield and James Tennant Dickson, employed by the Calico Printer's Association of Manchester, patented "polyethylene terephthalate" (also called PET or PETE) in 1941, after advancing the early research conducted by Wallace Carothers of DuPont. They saw that Carothers's research had not investigated the polyester formed from ethylene glycol and terephthalic acid. Polyethylene terephthalate is the basis of synthetic fibers such as generic polyester fiber and specialty fibers such as Invista's "Dacron." Whinfield and Dickson along with inventors W.K. Birtwhistle and C.G. Ritchie also created the first polyester fiber called "Terylene" in 1941 (first manufactured by Imperial Chemical Industries or ICI). The second polyester fiber was DuPont's "Dacron" (now an Invista Corporation product) [4].

Polystyrene was discovered in 1839 by Eduard Simon [5], an apothecary in Berlin. From storax, the resin of the Turkish sweet gum tree (*Liquidambar orientalis*), he distilled an oily substance, a monomer which he named styrol. Several days later, Simon found that the styrol had thickened, presumably from oxidation; into a jelly he dubbed styrol oxide (Styroloxyd). By 1845 English chemist John Blyth and German chemist August Wilhelm von Hofmann showed that the same styrol transformation took place in the absence of oxygen. They called their substance metastyrol. Analysis later showed that it was chemically identical to Styroloxyd. In 1866 Marcelin Berthelot correctly identified that forming metastyrol from styrol was a polymerization process. About 80 years went by, before it was realized that heating styrol starts a chain reaction which produces macromolecules, following the thesis of German organic chemist Hermann Staudinger (1881–1965). This eventually led to the substance receiving its present name, polystyrene.

The company I.G. Farben began manufacturing polystyrene in Ludwigshafen, Germany, about 1931, hoping it would be a suitable replacement

Table 1.1 Flexible Plastic Packaging Films Market in the United States: Breakdown of Volume Sales by Material Type for the Years 2002, 2007, and 2012 (in Million Pounds)

Flexible Plastic Packaging Film Type	2002	2007	2012
Polyethylenes	7676.6	9584.3	10,788
Polypropylene	1160.3	1449.2	1218
Thermoplastic polyesters	131.2	170.0	1566
Total	8968.1	11,203.5	13,572

for die-cast zinc in many applications. Success was achieved, when they developed a reactor vessel that extruded polystyrene through a heated tube and cutter, producing polystyrene in pellet form.

In 1959 the Koppers Company in Pittsburgh, PA, developed expanded polystyrene (EPS) foam [5].

Other polymers used in the packaging industry have had similar invention stories ranging from purely accidental discovery to hard-wrought research.

1.2 Markets: A Global Economy

The dynamics of the industrialized nations post–World War II have led to a plethora of applications for many polymers from the mid-20th century continuing into the 21st century. The flexible packaging industry has been one significant part of this growth engine, and adoption for use in many applications has expanded into a global economy, as many newly industrialized nations have become global players in trade.

Table 1.1 [6] shows the flexible plastic packaging films market (in million pounds) in the United States for three packaging films for the years 2002, 2007, and 2012.

Fig. 1.1 plots this data and shows a trend line for the total film market. This exponential fit trend line calculates a 4.1% growth rate from the 2002 to 2012 data. At this growth rate, the US flexible packaging film market could be 20 billion pounds in 2022. This is a slower growth rate than predicted in the first edition with the 2002 and 2007 data. The business recession of 2008–10 likely explains the lag in growth from 2007 to 2012. Another difference is the drop in the PP volume and the dramatic increase in the thermoplastic polyester volume. These numbers have not been confirmed, and like any statistical data they may get revised. It is possible that the PET data is correct as Mitsubishi has announced that they will spend $100 million to expand its Greer, S.C., facility with the installation of a new BOPET stretch film line due to start running in mid-2017. This new line will add

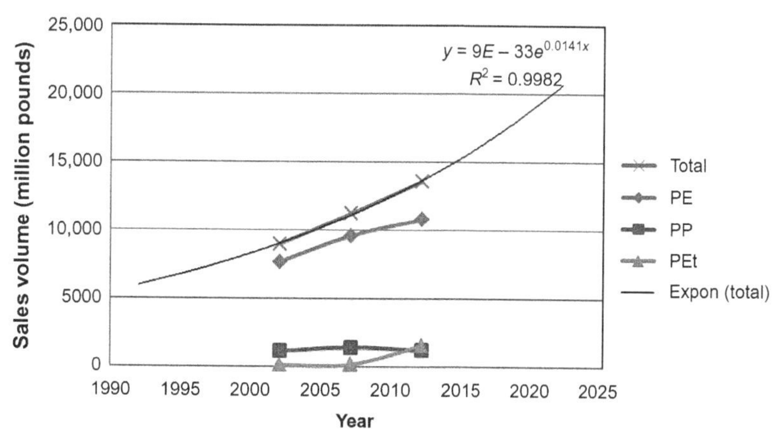

Figure 1.1 Flexible Packaging Film Market in the United States - Sales Volume versus Year.

Table 1.2 Selected Data from the US Census Office for the NAICS Code 326112 (Unsupported Plastics Packaging Film & Sheet Mfg.)

Year	Companies	Employees	Total Value Shipments ($1000)	Total 1997 Value Shipments ($1000)
1997	131	15,428	3,638,911	3,638,911
1998		15,288	3,998,678	3,937,645
1999		16,447	4,813,330	4,638,284
2000		17,327	4,823,210	4,495,845
2001		16,875	4,782,978	4,335,645
2002	207	24,301	6,456,000	5,760,602
2007	351	25,137	9,722,000	7,524,801
2012	401	30,792	12,771,000	8,924,067
2013		30,810	12,342,053	8,500,226

55 million pounds per year of capacity. No further project details are available at this writing. Toray Films in Rhode Island has also expanded production of BOPET in the United States which could explain part of the uptick for the United States.

Selected data from the US Census Office for the NAICS code 326112 (unsupported plastics packaging film & sheet mfg.) is shown in Table 1.2 [7]. In addition to this selected data, the total value shipments have been adjusted to 1997 dollars using data from the Misery index website [8].

Fig. 1.2 plots the data in Table 1.2 for employees, total value shipments, and the adjusted total value shipments to 1997 dollars versus year. The exponential curve fits calculate the total value shipment growth rate at 7.9% and the employee growth rate at 4.6%. When the total value shipments are adjusted to 1997 dollars, the calculated growth rate is 5.5%. These projections are below what was calculated in the first edition which were for total growth 9.16%, employees 6.96%, and adjusted to 1997 dollars 6.73%. This suggests that the market growth rate is down about 1%, employee growth down 2.4%, and adjusted to 1997 growth down 1.2%. This could be due to many factors such as increased market penetration, inflation, and lower raw material costs. The lower employee growth versus sales suggests higher productivity. The drop between 2012 and 2013 may be a trend or statistical variation. One would not know without a better understanding of how the data was taken and reported. Time will also provide an answer.

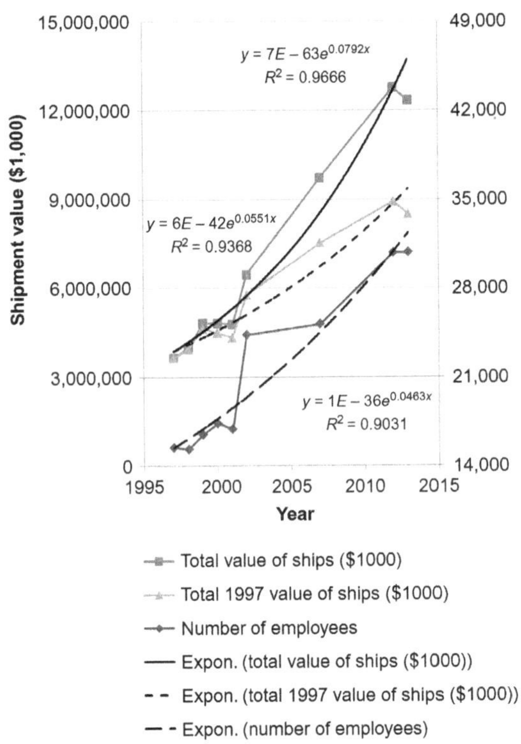

Figure 1.2 Shipment Value and Number of Employees versus Year

If one were to do a more rigorous future growth estimate, one would have to have good historical sales and population data. Market growth is a function of market penetration, that is, pounds of material used per person and number of people. Looking at global data one would find that there are countries with minimal market penetration, some with average market penetration, and some where the flexible plastic packaging value add is highly recognized and are leaders in taking advantage of the benefits flexible films give.

A market research report on companiesandmarkets.com is predicting that the plastic film market in the United States will grow at 16.65% CAGR during 2015–19 [9]. From this report, they offer some new insights for the US Plastic Film Market[1]:

> *Plastic films are used for packaging chemicals, consumer goods, F&B, and pharmaceutical products. These films are lightweight, non-corrosive, relatively cheap, and can be molded into various shapes, making them highly applicable for packaging. Their inert nature and ability to preserve products for a long period make them a popular choice in the F&B industry. They are also used by other industries for packaging of products, such as automotive parts, industrial goods, pharmaceutical products, and other consumer goods. The various materials used in the manufacturing of packaging films are BoPET, BOPP, cellulosic, CPP, EVOH, HDPE, LDPE, LLDPE, PVC, PVDC, PVOH, and polyamide.*
>
> *Analysts recognize the following companies as the key players in the Plastic Films Market in the US: AEP industries, Amcor, Bemis, Berry Plastics, Jindal Poly Films, Printpack and* Sealed Air.
>
> *Other Prominent Vendors in the market are: Allied Plastics, ALPLA-Werke, Ampac, APPE, Britton Group, CharterNex Films,* Coveris (formerly Exopack)*, Daibochi Plastic, DS Smith, GP Plastics, Graphic Packaging, International Precision Components, King Plastics, Northeast Packaging Company,* Sigma Plastics, *and Tegrant Alloyd Brands*
>
> *Biodegradable plastic films are easily decomposed by microorganisms. They have a significantly less harmful effect on the environment than conventional plastics. They are not necessarily entirely made from renewable materials; many are made from petrochemicals with biodegradable additives. The adoption of these materials, such as polyvinyl alcohol, polyanhydrides, and polylactic acid, by plastic film manufacturers will reduce the market shares of conventional plastic manufacturers. With environment compliance regulations worldwide becoming more stringent, vendors will be forced to in order to avoid paying significant penalties.*
>
> *According to analysts, demand for blister packs has been increasing in the pharmaceutical industry in the US. The growing incidence of diseases, the aging population, concern for health and wellness, and changing consumer lifestyles have increased the need for pharmaceuticals, which has in turn led to high demand for blister packaging. The implementation of FDA regulations, which require that all prescribed pharmaceuticals dispensed in hospitals and nursing homes be packaged in unit-dose formats with barcodes to reduce dispensing errors, has increased the sales of blister packs over the years. Demand for contact lenses in blister packs has increased tremendously in recent years.*
>
> *Further, analysts state that stringent rules and rising environment issues will lead to a decline in the usage of the plastic films during the forecast period.*
>
> *Key Vendors:*
>
> - *AEP industries*
> - *Amcor*
> - *Bemis*
> - *Berry Plastics*
> - *Jindal Poly Films*
> - *Printpack*
>
> *Other prominent vendors*
>
> - *Allied Plastics*
> - *ALPLA-Werke*
> - *Ampac*
> - *APPE*
> - *Britton Group*
> - *Charter Nex Films*
> - *Daibochi Plastic*
> - *DS Smith*
> - *Exopack*

[1]Text in italics represent updated information from what was presented in this report

- *GP Plastics*
- *Graphic Packaging*
- *International Precision Components*
- *King Plastics*
- *Northeast Packaging Company*
- *Tegrant Alloyd Brands*

Key market driver

- *Growing demand for blister packaging in pharmaceutical industry*

Key market challenge

- *Environmental concerns*

Key market trend

- *High Demand for biodegradable plastic packaging films*

1.3 Processes, Materials, Needs

Coextrusion is used in the following processes:

- Blown film
- Cast film (<10 mils or <250 μm)
- Cast sheet (> 10 mils or >250 μm)
- Blow molding
- Tubing and sheathing
- Extrusion coating and laminating
- Profile

Fig. 1.3 is a coextrusion feedblock flow diagram where three extruders are employed to produce a three-layer asymmetrical structure. This particular design has the barrier on the outside with a tie layer to provide adhesion to the core resin.

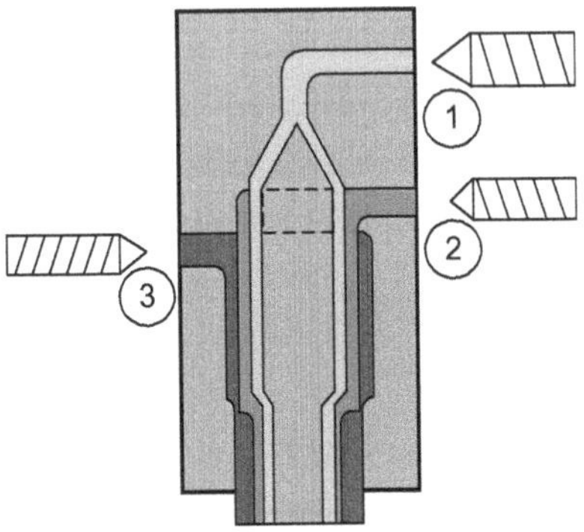

Figure 1.3 Flow diagram of a coextrusion feed block where three extruders are employed to produce a three-layer asymmetrical structure.

While much polymer extrusion is done in monolithic form, coextruding polymers has been growing now for more than half a century. The polymer extrusion concept grew from the already existing technology for extruding processed foods ranging from cereals to meats. Likewise, coextrusion was practiced in the food industry, before it was practiced with molten polymers. Without realizing it, almost every reader has likely consumed some coextruded food in the form of candy, bakery item, cereals, or processed meats.

Some considerations as to why we may wish to coextrude polymers are:

- to bring desirable properties of different materials into one structure
- to reduce emissions from solvents used in adhesive lamination processes
- to process multiple materials in one pass into a structure to save process steps, cost, and time.

Coextruding polymers is no simple matter. In the food industry they have their compatibility issues, and in the polymer industry we take the complexity to a higher level. We must consider the three main scientific property arenas related to moving a polymer from solid form to liquid form, and back to differently shaped solid form, which are:

- thermal
- fluid
- mechanical [10].

Thermal considerations are important because the processes are nonisothermal. The science of fluid flow is important, because the fluids are non-Newtonian viscoelastic with nonlinear temperature-dependent viscosities. These fluids have interfacial boundaries that depend on the fluids' viscosity and normal stress difference. Mechanical considerations are important, because the metal that contains the fluid flow is subjected to internal pressure forces, which distort the metal. The metal has to be shaped properly and has thermal mass and chemical resistance characteristics, which interact with the flowing fluids.

However, the benefits of combining materials in coextrusion far outweigh the challenges associated with such an endeavor. Coextrusion provides solutions to meet product functionality. There are many film properties that need to be taken into consideration when designing a packaging film. Some properties relate to the surface characteristics, while others relate to the entire thickness/body of the film. Surface characteristics include: coefficient of friction (COF), gloss, haze (surface induced), and sealability, as well as surface tension and chemical receptivity, which are important for materials to be printed. "Body" characteristics include: tensile properties, elongation, haze (internally induced), transparency, color, impact strength, and several other parameters.

Additionally, packaging requirements may also include gas and aroma barrier, chemical resistance, puncture resistance, formability, and shrink properties. In a world where energy prices keep rising, the biggest challenge is to have a structure that includes all the desirable properties in the most economical way. This requires that processors properly understand the material properties, as well as grasp polymer rheology to achieve an optimal formulation especially when attempting to coextrude [11].

In designing a coextruded film, one has to consider the fitness for use which translates into the following substitute quality characteristics:

- Physical properties
  - tensile, elongation, flexure, stiffness,
  - hardness, toughness, puncture, COF,
  - sealability, peelability, etc.
- Barrier properties
  - oxygen, moisture, oil, chemical,
  - aroma/fragrance, carbon dioxide, etc.
- Additives
  - colors, mineral fillers, COF modifiers, melt fracture inhibitors, scavengers, etc. [12]

Coextruded structures normally have a bulk layer, a barrier layer, a sealant layer, and often an adhesive layer to join incompatible layers. In some applications where using reclaimed materials are acceptable, the "Reclaim" also has to be considered. You can put it back into its own layer or as a combined blend into the bulk layer with virgin resin [13].

Common bulk layer choices are usually from the following groups [13]:

- PE (polyethylene)
  - HD (high density), LLD (linear low density), LD (low density), VLD (very low density)
- PP (polypropylene)
  - HoPP (homopolymer PP), CoPP (ethylene-propylene copolymer), Ter-PP (ethylene-propylene-butene terpolymer), and impact CoPP
- Acrylates
  - EMA (ethylene methyl acrylate)
  - EBA (ethylene butyl acrylate)
  - EEA (ethylene ethyl acrylate)
  - PMMA (poly(methyl methacrylate))
  - EMMA (ethylene/methyl methacrylate)
- EVA (ethylene vinyl acetate)
  - Low %, Mid %, High % VA
- PS (polystyrene)
  - HIPS (high-impact PS), GPPS (general purpose PS)

Important criteria for consideration are:

- cost
- adhesion
- flexibility
- softness
- stiffness, etc.

Resin selection criteria are:

- melt index—MW & MWD (physical properties tensile strength, organoleptics)
- density—crystallinity (opticals, permeability, modulus—ie, LDPE)
- molecular branching—melt strength, processing parameters
- comonomer content—all properties
- melt point—crystallinity (seal initiation temperature—thermoforming).

Following are generic names for extrudable resins. When the polymer brand/grade is unknown, these names help communicate about the material [13].

- Polyethylenes
  - HDPE—high-density polyethylene
  - HMW-HDPE—high molecular weight HDPE
  - MDPE—medium-density polyethylene
  - LDPE—low-density polyethylene
  - LLDPE—linear low-density polyethylene
  - VLDPE—very low-density (linear) polyethylene (also VLLDPE)
  - ULDPE—ultra low-density polyethylene (also ULLDPE)
  - mPE—metallocene polyethylene, generic.
  - mLMDPE—metallocene linear medium-density polyethylene
  - mLLDPE—metallocene linear low-density polyethylene
  - mVLDPE—metallocene very low (linear) density (sometimes referred to as a "plastomer")
- Polypropylenes
  - PP—polypropylene (generic indication when minimal information is known).
  - CoPP—copolymer polypropylene (usually random copolymer)

- HoPP—homopolymer polypropylene
- Ter-PP—terpolymer polypropylene
- Impact-CoPP—impact copolymer polypropylene
- Acid copolymer resins
  - ACR—acid copolymer resin, generic name for EAA and EMAA resins
  - EAA—ethylene acrylic acid copolymer, such as DuPont Nucrel, Dow Primacor, and ExxonMobil Escor.
  - EMAA—ethylene methacrylic acid copolymer, such as DuPont Nucrel
  - Ionomer—generic name for ionomeric copolymer resins, such as DuPont Surlyn
- Acrylates
  - Acrylate—generic name for various acrylate copolymers, such as DuPont Elvaloy AC, Arkema Lotryl, Westlake EMAC, and ExxonMobil Optema
  - EBA—ethylene butyl acrylate
  - EEA—ethylene ethyl acrylate
  - EMA—ethylene methyl acrylate
  - EMMA—ethylene methyl methacrylate
  - EiBA—ethylene isobutyl acrylate
  - EnBA—ethylene normal-butyl acrylate
- Ethylene vinyl acetates
  - EVA—ethylene vinyl acetate; examples: DuPont Elvax, Celanese, Ateva, LyondellBasell Ultrathene, ExxonMobil Escorene Ultra
- Polystyrenes
  - PS—polystyrene
  - EPS—expanded (or foamed) polystyrene
  - HIPS—high-impact polystyrene
  - GPPS—general purpose polystyrene
- Polyvinyl chlorides
  - PVC—polyvinyl chloride
  - PVDC—polyvinylidene chloride. Most commonly seen as a coating on a film, but there are also extrudable grades, such as Dow Saran and SolVin Ixan
- Terpolymers
  - EVACO—terpolymer of ethylene, vinyl acetate, and carbon monoxide
  - EnBACO—terpolymer of ethylene, normal-butyl acrylate, and carbon monoxide
  - EiBAMAA—terpolymer of ethylene, isobutyl acrylate, and methacrylic acid

  (***There are various other terpolymers in the industry.)
- Grafted resins
  - EVA-gMAh—ethylene vinyl acetate with a graft of maleic anhydride
  - LLDPE-gMAh—LLDPE with a graft of maleic anhydride

  (***There are various other grafted resins in the industry.)
- Others
  - Peel Seal or Easy Peel—generic name for a peelable sealant resin of unknown brand and grade. Example brand names include: DuPont Appeel, Mitsui-DuPont CMPS, Yasuhara Hirodyne, or Toyo Topco
  - TPS—generic name for "thermoplastic starch" resin

1.4 Materials and Barrier

When one talks about "barrier," please note that there are differences in interpretation of what this relates to. For some it may relate to oxygen, while for others moisture barrier may be more important. Barrier to other parameters may also be of greater or lesser importance, including carbon dioxide, oils, acids, flavors, aromas, fragrances, solvents, and other chemicals.

Each polymer has its own unique characteristics and properties, and thus will have differing "barrier" performance. Which polymer or combination of polymers you may wish to coextrude will depend upon whether you need a barrier for oxygen, water, oil, aroma, or something else and to what degree for each property. Some common resins that can be used for certain barriers are [13]:

- Nylon, polyamide (PA)
  - oxygen barrier, aroma barrier, and some oil barrier
- EVOH
  - oxygen barrier and flavor/aroma barrier
- PET
  - moisture barrier, some flavor/aroma barriers, and some chemical barrier needs
- PVDC
  - moisture, oxygen, flavor, aroma, and some chemical barriers
- HDPE
  - moisture barrier
- PP
  - moisture barrier
- Ionomer
  - oil barrier and some chemical barrier

When choosing one or more of these resins to coextrude in addition to the barrier properties, you must also understand the criteria for flex durability, clarity, and other physical parameters related to each material, and how it adds or diminishes from the needs of the overall coextrusion.

Following here is some general information on nylon. This material is used broadly in coextrusions and has some unique properties related to processing, so cannot be dropped into any extruder at will [13,14].

Nylon or polyamide common types:

- Nylon 6: MP = 220°C (430°F)
  - The most common grade used in flexible packaging coextrusions.
- Nylon 66: MP = 250°C (480°F)
- Nylon 6 or 66: MP = 210°C (410°F)
- Amorphous Nylon
  - Used mainly in blends with Nylon 6 or 66, but occasionally is used pure.
- Nylon 11 and 12: MP = 180°C (360°F)

It is important to note that extrusion screws for nylon are generally very different in design versus PP, PE, and ethylene copolymer screws. Thus do not just try extruding nylon resin through a screw not known to be designed for nylon. Please consult with the equipment manufacturer for advice. Without doing so, you risk damaging equipment including but not limited to the screw, gearbox, and drive motor on the extruder.

There are many extrusion grade nylon resin producers who supply globally:

- Honeywell
- BASF
- Lanxess
- DuPont
- EMS
- Mitsubishi
- Ube
- Evonik
- Radici, etc.

The features that one can obtain with nylon are:

- good gas barrier
- good thermoformability
- good toughness, abrasion resistance, tensiles
- good optics
- good oil resistance.

The extrusion issues for consideration with nylon are:

- crystalline melt point
- flow properties change rapidly from solid to melt and back to solid. SAFETY!
- moisture sensitivity
- various types, so need to choose extrusion temperature profiles carefully
- "high" extrusion temperature needed
- almost always processed in coextrusion.

Ethylene vinyl alcohol, commonly known as "EVOH," has grown into a very useful barrier resin in the flexible packaging industry. EVOH began its life in the 1950s as an offshoot of research into polyvinyl alcohol adhesive polymers. EVOH was originally developed and patented by the DuPont Company, but the polymer did not have valuable adhesive properties. In the 1950s there was no barrier film coextrusion market, so looking at the barrier properties was not even a consideration. The EVOH technology was eventually sold to the Monsanto Company. Monsanto in turn later sold it to a Japanese corporation, and by this timeline, coextrusion was in swing, and barrier films containing nylon were becoming prevalent. The oxygen barrier properties of EVOH were studied and found to be very interesting if you keep the polymer dry, protecting it from moisture by other polymers in coextruded films [13,14]. Today there are two major EVOH suppliers and one newer minor supplier.

EVOH or ethylene vinyl alcohol suppliers are:

- Kuraray Eval (Japan and United States)
- Nippon-Gohsei Soarnol (Japan and United States)
- Arkema & Chang Chun Petrochemicals (Taiwan) Evasin—(new market entrant).

EVOH or ethylene vinyl alcohol characteristics are:

- good oxygen barrier when dry
- almost always run in coextrusion
- needs tie layers (except to Nylon)
- moisture sensitivity
- thermal sensitivity
  - needs short, smooth flow paths when designing extrusions systems
- good optical properties
- high stiffness
- flex crack sensitivity.

Polyester has been previously discussed in this introduction, with regard to its history. There are many polyesters, and the most important one for films and sheet is PET (polyethylene terephthalate). PET is typically oriented to provide strength, clarity, and barrier properties. PET for extrusion coating is more challenging to process compared to common polyolefins such as PE and PP and requires special equipment. Coextrusion makes it easier to run, but will require in-line dryers and proper PET screw designs [13,14]. PET resin suppliers for film, sheet, and extrusion coating resins in the United States would be:

- Indorama Ventures
- DAK Americas
- M&G.

Polyester polymers found in the packaging industry include:

- PET—generic extruded polyester. Could be monolayer or in a coextrusion.
- PET ext.ctg.—generic extrusion coating of polyester
- APET—amorphous polyester
- CPET—crystalline polyester
- PET or PETE—polyethylene terephthalate
- PETG—polyester copolymer with glycol
- PBT—polybutylene terephthalate
- PPT—polypropylene terephthalate
- PTT—polytrimethylene terephthalate
- PEN—polyethylene naphthalate
- PLA—polylactic acid
- PGA—polyglycolic acid
- PHA—polyhydroxyalkanote.

Other resins that can be considered to provide a barrier of some type are:

- PAN—polyacrylonitrile, sometimes written as "ACN"
- SAN—styrene acrylonitrile copolymer
- AN-MA—acrylonitrile methyl acrylate copolymer (such as Innovene and Barex)
- ABS—acrylonitrile butadiene styrene copolymer
- LCP—liquid crystal polymer
- COC—cyclic olefin copolymer, such as TAP Topas or Mitsui Apel
- PUR—polyurethane, extrudable type when seen in a coextrusion for example
- SBC—styrene butadiene copolymer (such as Chevron Philipps K-Resin).

1.5 Materials and Sealing

When assembling a coextruded film for the flexible packaging industry, most films require that they are sealable. The most common seal method is thermal, using a constant temperature sealer or a variable temperature impulse sealer. Other seal methods include:

- radio frequency
- high frequency
- ultrasonic
- pressure sensitive.

Given that thermally created seals are the most common, following is an overview of what to consider when considering a sealing layer material.

The sealant layer choice should be made considering the application needs, since each parameter does not need to be maximized in order to perform in the given end use. What is important is finding the appropriate balance with properties such as:

- hot tack strength
- heat seal strength
- seal initiation temperature
- sealing speed
- COF
- ability to seal through a given contaminant
- chemical resistance to the material to be contained
- economics.

The most common sealant layer resins range from high performance to commodity performance. Inherent property differences in the polymers will affect the sealing performance, and since there are so many polymer grades in each family, it is best to consult with polymer suppliers to match your given needs to the performance of a given resin such as:

- ionomers of acid copolymers
- acid copolymer (EAA or EMAA)
- mVLDPE (very low density)
- EVA or EMA blends with LLDPE
- CoPP/Ter-PP, EVA, EMA, LLDPE, mLLDPE
- LDPE or PP.

1.6 Bringing Things Together

When assembling coextrusion structures, some materials will bond to other materials; however, the adhesion level may be insufficient for the durability needs of the given application. In order to create a more robust structure, one must use a material designed specifically to bond one material to another with improved interlayer adhesion. The polymers that will work as a "facilitator of adhesion" in this case are commonly called "tie resins or coex tie resins." They act as glue and provide adhesion between the incompatible layers. These "coex tie" resins provide both mechanical and chemical bondings to other polymers when in a molten state. The adhesive strength obtained will be a function of the tie resin formulation and the coextrusion process parameters. It should be recognized that the extrusion process can have a significant impact in generating interlayer adhesion. For example, taking the same structure from a blown film process to a cast film process using the same resins usually will not translate into obtaining the same interlayer adhesion. This is a complex topic for another discussion [13].

To achieve higher bond strengths than attainable with a base polymer, resins are formulated with special materials, including grafted functional groups, and/or are compounded with modifiers. The correct "tie" resin choice is based on many variables, including extrusion processing parameters and material end use parameters. Your "tie" resin supplier can provide helpful recommendations [13,14].

Base resins typically used for tie layers are:

- LDPE, LLDPE, HDPE
- EVA, PP
- acrylate copolymers
- acid copolymers.

Common modifiers used are:

- maleic anhydride
- olefinic tougheners
- tackifiers
- rubbers.

Some tie resin suppliers include:

- DuPont
- Mitsui
- LyondellBasell
- Arkema
- Dow
- Westlake, etc.

When selecting a coextrudable adhesive resin, it is always useful to choose the correct resin for a given process and/or end use application. If an EVA resin is selected, care must be paid to temperature limitations of the extrusion process, since EVA should not be processed over 235°C (455°F). For all resins, the tie layer rheology must match with adjacent materials to assure good layer uniformity with no "interfacial instability." Some materials have a minimum temperature that is necessary to generate adhesion via chemical and/or mechanical interaction. Extrusion "process time" is also an important factor and related to the time between the polymers exiting the die, and solidification. This relates to the "cooling" and "drawing" of the polymers and internally induced stresses during production. Coextrusion feedblock and die design are also influential in how a "tie resin" may be able to generate interlayer adhesion [13].

This book is an attempt to provide a snapshot of multilayer flexible films and where they are used. Like any still picture, this book cannot describe the future and the innovative uses, applications, and film designs that will drive these products to fill customer needs.

References

[1] <http://en.wikipedia.org/wiki/Cellophane>.

[2] <http://en.wikipedia.org/wiki/Polyethylene>.

[3] <http://en.wikipedia.org/wiki/Polypropylene>.

[4] <http://inventors.about.com/library/inventors/blpolyester.htm>.

[5] <http://en.wikipedia.org/wiki/Polystyrene>.

[6] Flexible Plastic Packaging Films Market in the United States. Breakdown of volume sales by material type for the years 2002 & 2007. Converted flexible packaging market—Global Strategic Reports, Table 5. Global Industry Analysts Inc.; March 1, 2008.

[7] <http://thedataweb.rm.census.gov/TheDataWeb_HotReport2/econsnapshot/2012/snapshot.hrml?NAICS=326112>.

[8] <http://www.miseryindex.us/indexbyyear.aspx?type=IR>.

[9] <http://www.4spe.myindustrytracker.com/en/article/68142?utm_source=Mandrill-4spe&utm_medium=newsletter&utm_campaign=4spe-284-s-en-021015>.

[10] Cloeren Jr P. Application driven coextrusion. In: SPE Extrusion Division TopCon 2008, Cincinnati, OH; October 21–22, 2008.

[11] Xiao K. Blown film coextrusion. In: SPE Extrusion Division TopCon 2008, Cincinnati, OH; October 21–22, 2008.

[12] Marks SB. Material selection for coextrusion. In: SPE Extrusion Division TopCon 2008, Cincinnati, OH; October 21–22, 2008.

[13] Marks SB. Specialty resins for film & extrusion coating. In: Tappi PLACE Short Course 2015, Concord, NC; August 18–20, 2015.

[14] Marks SB. Barrier in flexible packaging. In: Tappi PLACE Flexible Packaging Summit 2009, Columbus, OH; April 28–29, 2009.

PART II

RESINS

2 Polyethylene

Rajen M. Patel
The Dow Chemical Company, Freeport, TX, United States

2.1 Introduction

Polyethylene is the highest volume global plastic available today, produced from ethylene gas. By utilizing different manufacturing processes, varying catalyst and comonomer types, a broad range of ethylene homopolymers and copolymers can be produced. The ability to manufacture so many variations allows producers to tailor resins for specific applications such as packaging films, rigid containers, drums, and pipes. When we talk about polyethylene (PE) today, we have to specify whether it is a high density or low density or linear low density or polar ethylene copolymer, what catalyst was used, what is the density and melt index (MI), what comonomer was used to control density, what manufacturing process was used, and what molecular structure resulted such as molecular weight and composition distribution.

Polyethylene resins are typically classified by their MI and density. While the MI indicates the resin molecular weight and its processability, density measures product crystallinity, which influences stiffness and impact performance of articles produced from such PE resins. This chapter will cover:

- low-density polyethylene (LDPE)
- polar ethylene copolymers
- high-density polyethylene (HDPE)
- linear low-density polyethylene (LLDPE)
- very low-density polyethylene (VLDPE)
- single-site or metallocene-catalyzed homogeneous polyethylene such as polyolefin plastomers (POPs), polyolefin elastomers (POEs), and linear low-density polyethylene (m-LLDPE)

and the manufacturing processes to make them.

2.2 Polyethylene History and Basic Characterization

The history of PE began with the accidental discovery of LDPE by Imperial Chemical Company (ICI) in 1933 while researching ethylene reactions at high temperatures and pressures [1]. The first high-pressure LDPE commercial plant was built in 1939 by ICI using an autoclave reactor. Subsequently, BASF developed a tubular reactor for the high-pressure LDPE process to obtain higher ethylene conversion per pass and therefore higher production rates. Today LDPE is still produced using similar chemistry with autoclave or tubular reactors.

The basic raw material for LDPE is the ethylene monomer. Ethylene (Fig. 2.1) is a colorless gas with a slightly sweet smell and is obtained by cracking ethane or naphtha at high temperatures (about 850°C) in a steam cracker. Ethane is found in natural gas and naphtha is an oil refining byproduct. The ethylene is compressed to very high pressures, and free radical polymerization is initiated using a peroxide to produce LDPE.

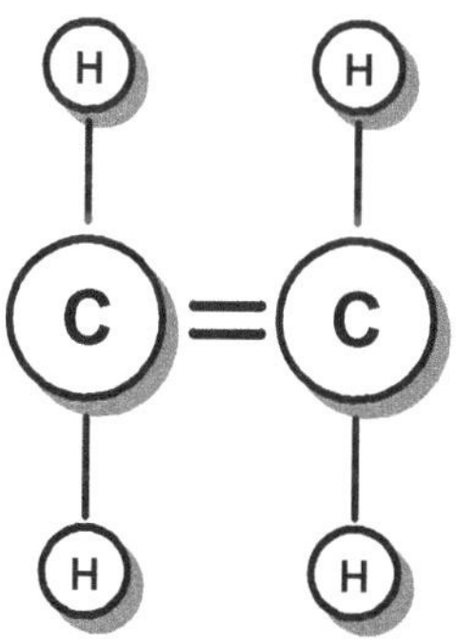

Figure 2.1 Ethylene molecule (C_2H_4: molecular weight = 28).

In 1953 Carl Ziegler developed a new catalyst system, known today as a Ziegler–Natta (Z-N) catalyst, using transition metal chemistry that allowed PE production at lower temperatures and pressures. Around the same time frame, Hogan and Banks at Phillips Petroleum synthesized HDPE using a silica-/alumina-supported chromium oxide catalyst at relatively low pressures [2].

Multilayer Flexible Packaging. DOI: http://dx.doi.org/10.1016/B978-0-323-37100-1.00002-8

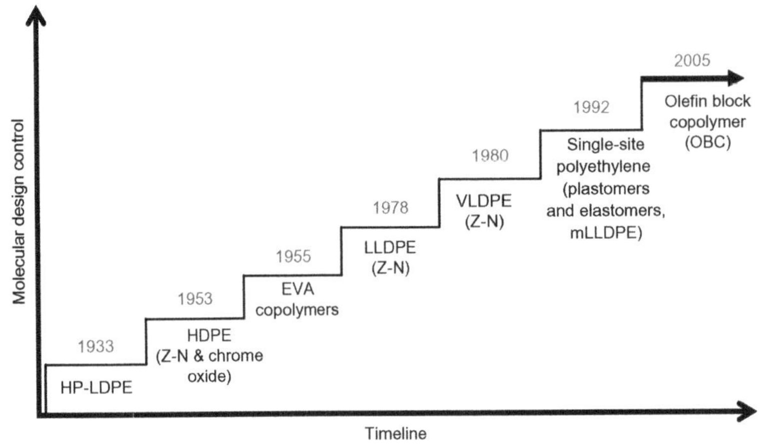

Figure 2.2 Polyethylene innovations versus time.

HDPE polymers were introduced with these new catalysts and processes [3]. In 1973 the Z-N catalyst efficiency was dramatically improved using a supported catalyst technology. In 1978 LLDPE polymers were introduced using the Z-N catalyst technology and incorporating an alpha-olefin comonomer (eg, 1-butene, 1-hexene, or 1-octene). The latest major new technology, introduced in early 1990s, was the single-site catalyzed or metallocene-catalyzed polyethylene resins. This resulted in the commercial introduction of several new polyethylenes. Fig. 2.2 shows major polyethylene innovations versus time.

Polyethylene resin selection is in general based on the physical properties required for the product for the end-use application. PE is typically characterized by MI and density.

2.3 Melt Index

MI or melt flow index refers to the extruded polymer flow in grams per 10 minutes from a die as described in ASTM D-1238, under a specified load as shown in Fig. 2.3. Condition E is typically used for polyethylene grades with a 2.16-kg piston load and the die dimensions are $D = 0.0825$ in and $L = 0.315$ in ($D = 2.095$ mm and $L = 8$ mm). The test is carried out at 190°C.

Under the ASTM D-1238 conditions of the MI measurement with the 2.16 kg load, the wall shear stress can be calculated to be $\tau_w = 1.932 \times 10^4$ Pa ($=2.802$ psi) and the wall shear rate is $\dot{\gamma} = 2.48 \times \text{MI s}^{-1}$. Shear viscosity at the calculated shear rate is about 7804/MI Pa s.

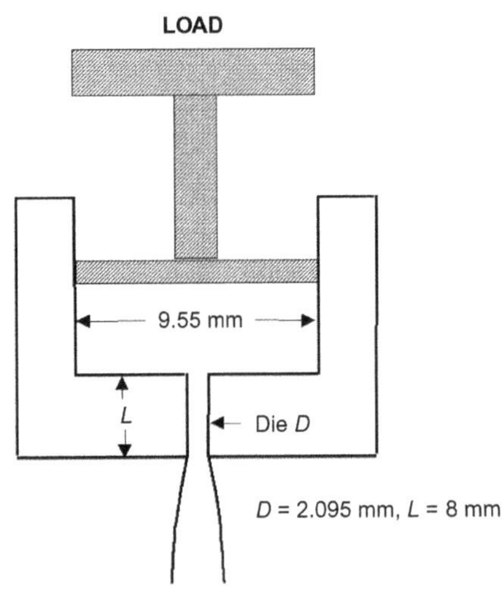

Figure 2.3 Schematic of a melt indexer.

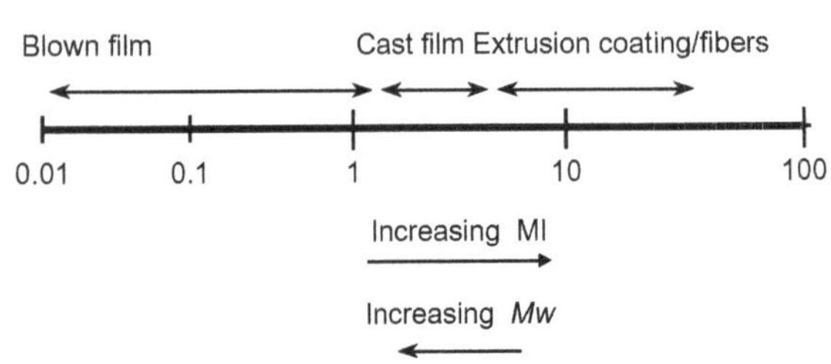

Figure 2.4 PE melt index range suitable for various extrusion processes.

Polymer MI is reported as grams/10 minutes or (dg/min). PE blown film resin MI is typically between 0.01 and about 2 dg/min (Fig. 2.4). MI for cast film PE grades is typically between 2 and 6 dg/min.

Fig. 2.5 shows the inverse relationship between the MI and the weight average molecular weight (M_w) (as measured by GPC) for LLDPE polymers

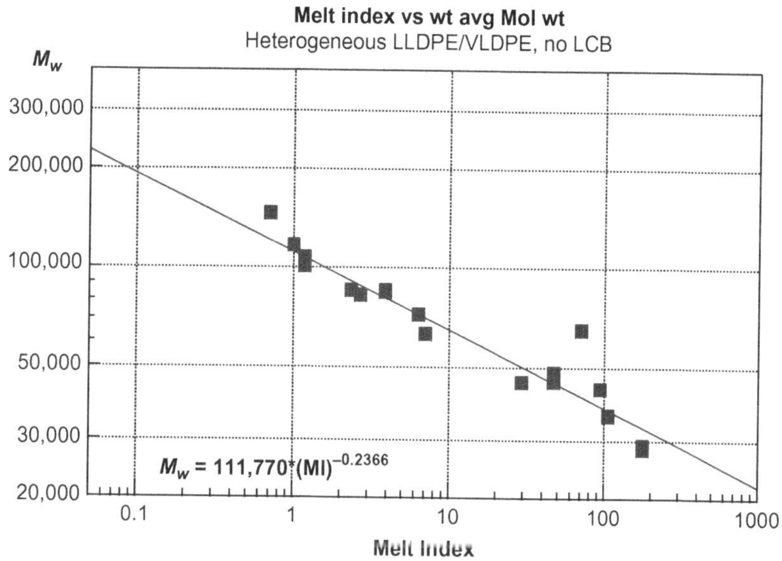

Figure 2.5 Weight average molecular weight (M_w) versus melt index.

(made using the Z-N catalyst). Lower MI polymers have higher weight average molecular weight. Long-chain branching (LCB) effects are not shown, but they will lower the M_w at the same MI.

2.4 Density

Many polymers contain some crystalline structures when they solidify. These polymers are semicrystalline as not all polymer chains incorporate into the crystalline structure. For these polymers, the chain segments that are not crystallized make up a second phase known as the amorphous phase. Molecules in the amorphous phase are packed in a random manner (no order). Molecules in the crystalline phase are packed tightly together in a regular three-dimension order. Hence, the crystalline phase has a higher density than the amorphous phase. Polyethylene is a semicrystalline polymer, and has a solid state density between 0.857 and 0.967 g/cm^3 depending upon the crystallinity. Resins with higher solid densities contain higher crystallinity.

Short-chain branching (SCB) level either made in situ (eg, LDPE) or due to comonomer incorporation (eg, LLDPE) controls resin density/crystallinity. Alpha-olefin comonomer level determines the SCB level in ethylene/alpha-olefin copolymers (eg, LLDPE, VLDPE, POPs, and POEs). Fig. 2.6 shows the resin density versus octene comonomer for homogeneous polyethylene. Note that LCB, present in LDPE and in some single-site catalyzed resins, also affects resin density/crystallinity.

Polyethylene density, measured using ASTM D792, is a measure of crystallinity at 23°C.

The relationship between density and weight% crystallinity ($\%X_w$) for polyethylene is expressed in Eq. (2.1). The volume% crystallinity ($\%X_v$) is calculated using Eq. (2.2).

$$\%X_w = \frac{\rho - \rho_a}{\rho_c - \rho_a} \cdot \frac{\rho_c}{\rho} \cdot 100 \tag{2.1}$$

$$\%X_v = \frac{\rho - \rho_a}{\rho_c - \rho_a} \cdot 100 \tag{2.2}$$

where ρ is the nonpolar polymer density (LDPE, HDPE, LLDPE, etc.), ρ_a is the amorphous phase density (0.855 g/cc) for nonpolar PE, and ρ_c is the crystalline phase density (1.000 g/cc).

For polar ethylene copolymers, as the comonomer level increases, the crystallinity (and hence, stiffness) decreases but density increases. This is due to bulky polar side groups which increase amorphous phase density and therefore the overall density even though crystallinity is lower.

PE density is directly related to the polymer crystallinity measured using differential scanning calorimetry (DSC) (Fig. 2.7).

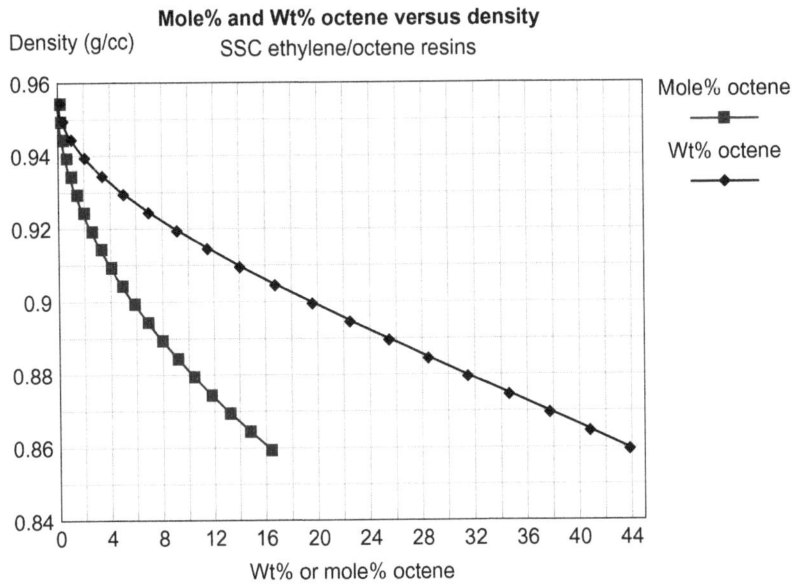

Figure 2.6 Single-site catalyzed polyethylene density versus comonomer level.

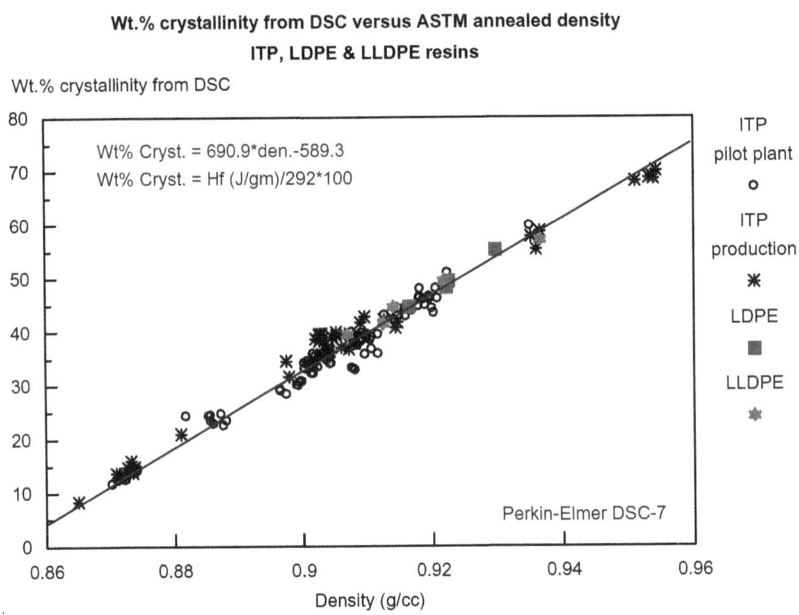

Figure 2.7 Density versus crystallinity for PE polymers.

2.5 Gel Permeation Chromatography

Polymerization gives rise to a molecular weight distribution (MWD) as individual polymer chains end up with a repeat unit distribution. GPC technology measures the MWD in a polymer. The polymer molecules are dissolved in a solvent and processed through columns packed with porous beads that separate the molecules based on their hydrodynamic volume, which are detected as they emerge from the column (Fig. 2.8).

Due to this chain length distribution and thus MWD, it is not possible to describe the resin using a single molecular weight. Numerous methods have been developed with the two most common being the number-average molecular weight (M_n) (Eq. 2.3) and the weight-average molecular weight (M_w) (Eq. 2.4).

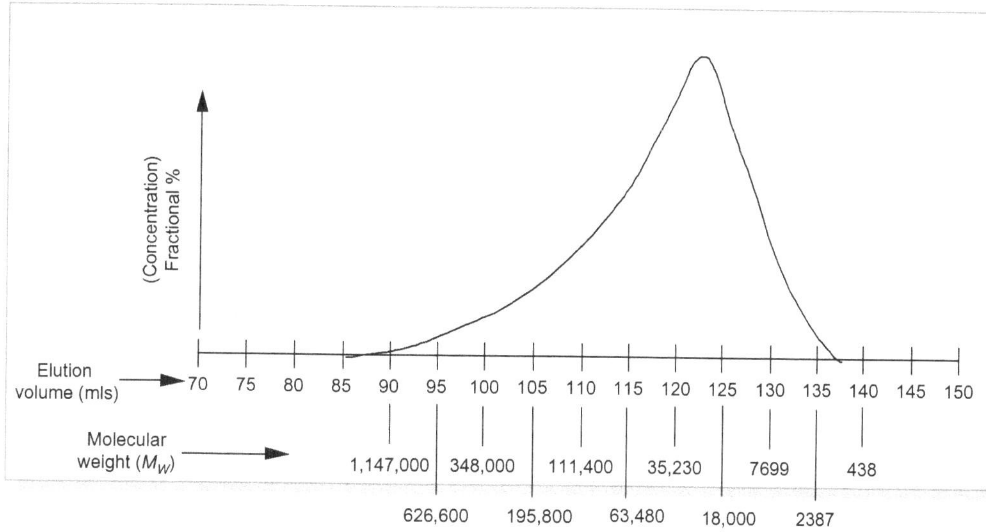

Figure 2.8 GPC curve showing concentration of various molecular weights of a polymer.

$$M_n = \frac{\sum n_i M_i}{\sum n_i} \qquad (2.3)$$

$$M_w = \frac{\sum n_i M_i M_i}{\sum n_i M_i} = \sum w_i M_i \qquad (2.4)$$

where n_i is the number of molecules with a molecular weight of M_i and w_i is the weight fraction of molecules with a molecular weight of M_i.

Molecular weights can range from several thousand to several million. Another common descriptor for a resin is the polydispersity or MWD. The MWD is the ratio of M_w to M_n (Eq. 2.5).

$$\text{MWD} = \frac{M_w}{M_n} \qquad (2.5)$$

The resin MWD depends on how it is produced; that is, the catalyst type, reactor conditions, etc. PE MWD is typically between 2 and 30, and Z-N LLDPE MWD is between 4 and 4.5 for single reactor polymers. Bimodal polymers (produced using multiple reactors) have up to 20 MWD. MWD of single-site catalyzed polyethylenes is typically from 2 to 3.

Polyethylene resins have molecular structures shown in Fig. 2.9. The most common polymer structure is linear (Fig. 2.9a), where the repeating units are connected one after another. Common linear polymers include HDPE, polystyrene (PS), polycarbonate (PC), and polyamide (PA) (although

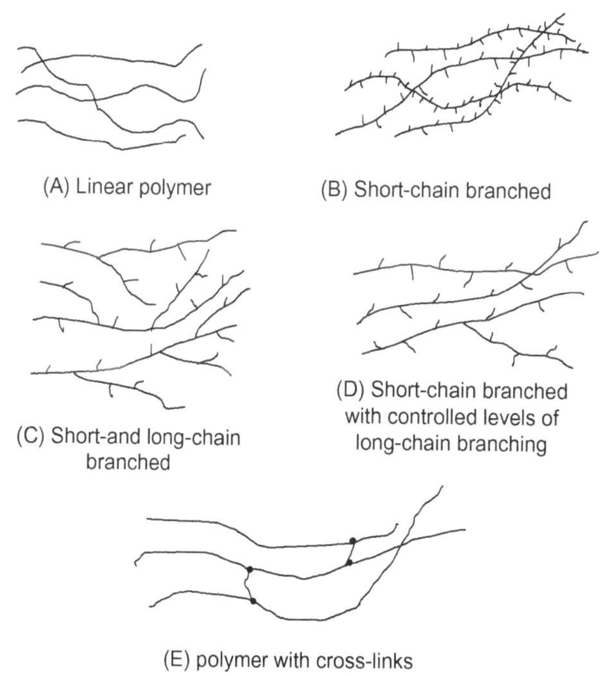

Figure 2.9 Schematic of common polymer chain structures.

branching can be designed into these materials). Some polymers have a comonomer that results in short-chain branches (Fig. 2.9b). The most common polymer with short-chain branches is LLDPE. This polymer is produced by the incorporating an alpha-olefin like 1-butene, 1-hexene, or 1-octene into an ethylene backbone. Patented single-site catalysts systems control both long- and short-chain branches

(Fig. 2.9d), tailoring both the physical and processing properties to the desired process and end-use application [4–6].

2.6 Low-Density Polyethylene

The high-pressure LDPE process produces both short-chain branches and long-chain branches (typically 2 to 3 LCB/1000 carbon atoms). LDPE has ratio of about 10 short-chain branches to every long-chain branch (Fig. 2.9c). As will be discussed later, the polymer structure and weight average molecular weight highly influence the polymer rheological properties. When two or more polymer chains are connected through a chemical bond, the resin is said to contain cross-linked segments. See schematic for a cross-linked polymer chain in Fig. 2.9e.

2.7 Free Radical Polymerization Processes

Fig. 2.10 shows the free radical chain propagation mechanism. A free radical is generated with an initiator (either a peroxide or an oxygen is injected into the reactor). The free radical then reacts with an ethylene molecule being inserted into the polymer chain. The free radical is transferred to the end of the just inserted ethylene molecule, and the reaction continues to grow the polymer chain.

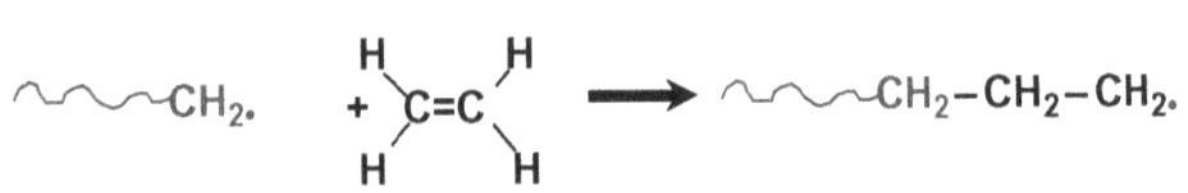

Figure 2.10 Chain propagation mechanism for free radical polymerization.

Fig. 2.11 shows that the free radical molecule can also transfer the radical to a carbon atom on another polymer molecule. Then the free radical will start another polymer chain (branch) growing at the free radical site. Typically when the branch is greater than 350 carbon atoms (>5000 g/mole) it will be designated as a long-chain branch (LCB).

Figure 2.11 Mechanism for insertion of LCB in free radical polymerization.

This is because the branch length needs to be at least four to six times the entanglement molecular weight to see the entanglement effect.

Fig. 2.12 shows a third important free radical reaction for the formation of short-chain branches. The active center is transferred through a "back-biting" mechanism. This example shows a butyl branch (4 carbons SCB) created when the new free radical continues to grow the polymer chain.

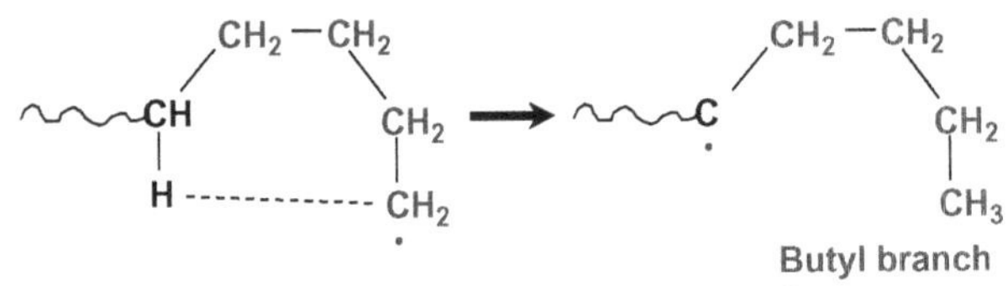

Figure 2.12 Mechanism for SCB formation in free radical polymerization.

2.8 Autoclave Reactor

The autoclave is a stirred cylindrical reactor (Fig. 2.13). These reactors are typically about 6.1 m (20 ft) long and 0.38 to 0.91 m (15 to 36 in) in diameter. Ethylene is introduced into the primary compressor along with the low-pressure recycled gas from the recovery units. The compressed gas is then further compressed in the hyper compressor up to the target pressures of the reactor (1200–2400 bars/120–240 MPa/17,400–34,800 psi) and is cooled to remove the heat of polymerization. The peroxide initiator is injected with the fresh ethylene. Ethylene polymerization is an exothermic reaction. The heat of reaction must be removed by the

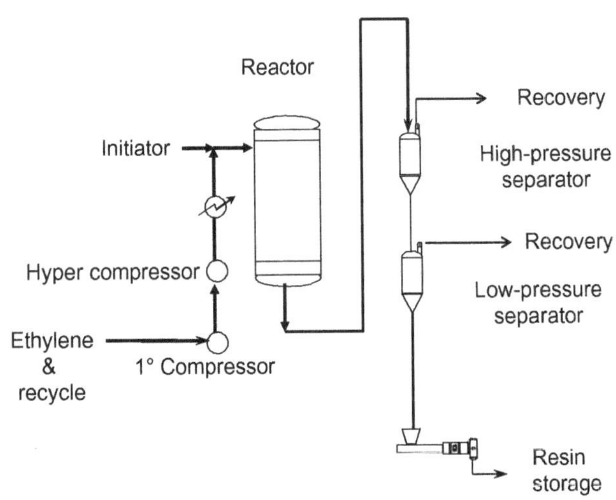

Figure 2.13 Autoclave reactor (simplified).

incoming cool gas to maintain the target reactor temperature. Proprietary baffle designs partition the reactor into discrete zones enabling better molecular reaction control. This will control the polymer's M_w, MWD, and density.

The autoclave reactor efficiently back-mixes the gas/polymer. The reactor walls are very thick to accommodate the high pressures. Typical ethylene conversion to polyethylene is about 14–20% per reactor pass. When the polymer/gas exits the reactor, it passes through separators where the unreacted ethylene is recycled, and the molten polymer is prepared for additive injection and then pelletized. Typical production rates for autoclave reactors are 165–495 MM lb/year (75–225 kT/year).

2.9 Tubular Reactor

The tubular reactor is a long pipe reactor (Fig. 2.14). These reactors are about 1.25 km (0.78 mile) long and 2.5 to 7.6 cm (1 to 3 in) inside diameter. Ethylene is introduced into the primary compressor along with the low-pressure recycled gas from the recovery units. The compressed gas is then further compressed in the hyper compressor up to 2000–3200 bars (200–320 MPa or 29,000–46,400 psi). The oxygen or peroxide initiator is injected with the fresh ethylene, sometimes at multiple points along the tube. The heat of reaction is removed with cooling jackets to maintain the reactor temperature profile. This will control the polymer's M_w, MWD, and density.

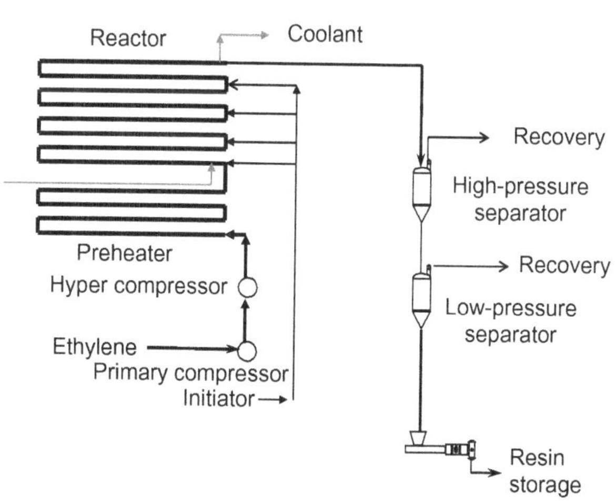

Figure 2.14 Tubular reactor (simplified). *Source: From Univation Technologies.*

The tube reactor has poor back-mixing. Typical ethylene conversion to polyethylene is 21–38% per reactor pass. When the polymer/gas exits the reactor, it passes through separators where the unreacted ethylene is recycled, and the molten polymer is prepared for additive injection and then pelletized. Tubular reactor production rates are between 330 and 880 MM lb/year (150–400 kT/year).

LDPE is a homopolymer, because it is produced using only one monomer (ethylene). The branching controls the polymer density. The free radical polymerization process incorporates both LCB and SCB into the polymer molecule (Fig. 2.15). The SCB (and LCB junction) creates a defect when the polymer molecule crystallizes, forcing smaller lamella to form (which lowers crystallinity and density).

Fig. 2.16 shows some subtle differences in the branching distributions for autoclave and tubular

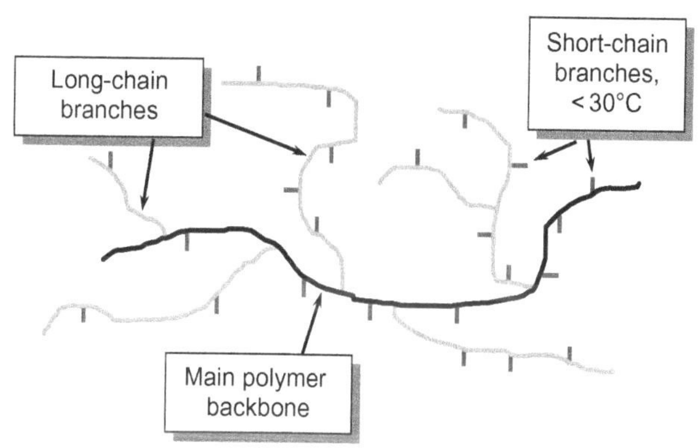

Figure 2.15 LDPE homopolymer with LCB and SCB in the molecule.

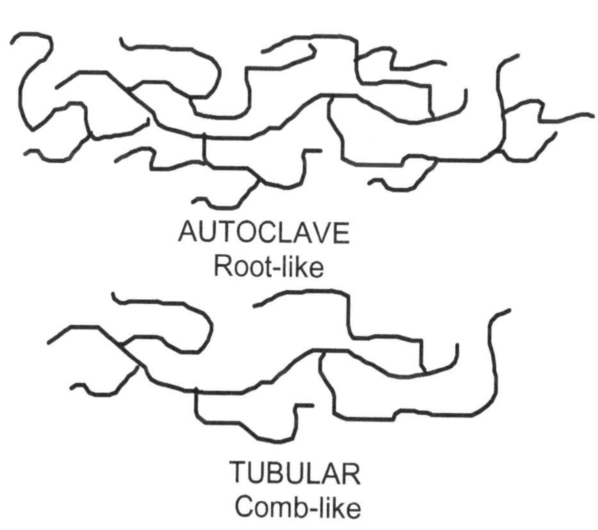

Figure 2.16 Branching for autoclave and tubular LDPE polymers.

polymers. The autoclave process produces a branching distribution that is more root like, while the tubular process produces a distribution that is more comb-like.

2.10 Polar Ethylene Copolymers

There are several polar ethylene copolymers that are commercially significant.

Ethylene-vinyl acetate	EVA
Ethylene-methyl acrylate	EMA
Ethylene-ethyl acrylate	EEA
Ethylene-butyl acrylate	EBA
Ethylene-acrylic acid	EAA
Ethylene-methacrylic acid	EMAA
Ethylene-carbon monoxide	ECO
Ionomers	ION

The largest volume of these is the ethylene-vinyl acetate (EVA) copolymer. Similar free radical polymerization processes used for LDPE homopolymer are used to produce EVA polymers. The vinyl acetate (VA) comonomer is randomly distributed along the polymer backbone and long-chain branches (Fig. 2.17). Both VA and methyl acrylate copolymers exhibit polar functionality. Low crystallinity and high polarity gives the polymer adhesive characteristics as well as low seal initiation temperatures. However, EVA copolymers are thermally less stable compared to acrylate copolymer (eg, EMA) due to weaker carbon–oxygen bond which leads to acetic acid (vinegar) generation during process and cross-linked gel formation during

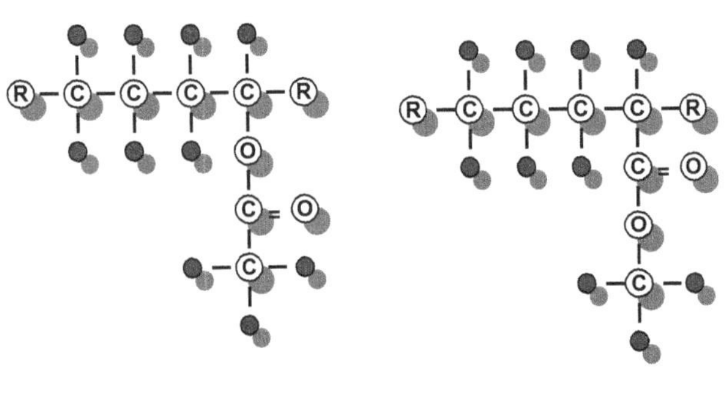

Figure 2.17 Polar copolymers of ethylene EVA and EMA.

high-temperature melt processing. Therefore, EVA resins are processed below 230°C, whereas acrylate copolymers can be processed at much higher melt temperatures.

VA has a higher density than the ethylene monomer and it disrupts the polymer crystallinity such that as VA levels increase the density increases, and crystallinity decreases. Fig. 2.18 shows EVA crystallinity versus wt% VA.

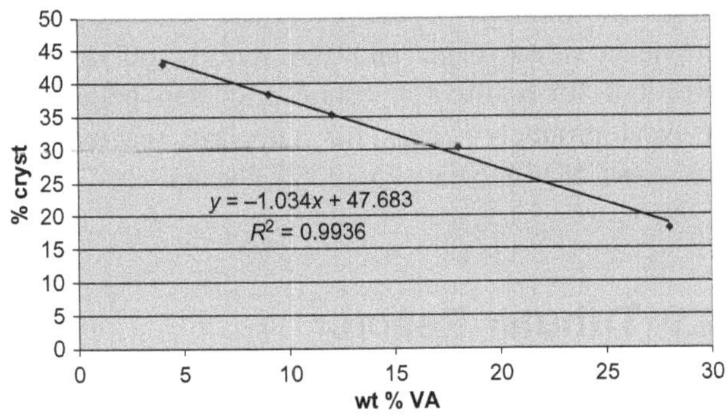

Figure 2.18 EVA crystallinity versus wt% VA. *Courtesy: Dr. Ken Anderson, Celanese.*

Ethylene-acrylic acid (EAA) and ethylene-methacrylic acid (EMAA) copolymers (Fig. 2.19) hydrogen bond due to their acid functionality. Therefore, the acid copolymers are good adhesive polymers for bonding aluminum foil, metalized substrates, and polyamides to other layers in multilayer packaging structures. The adhesion strength depends on the acid comonomer content.

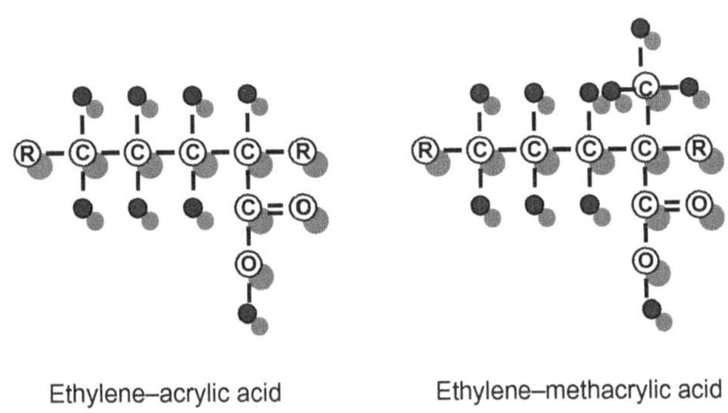

Figure 2.19 Acid copolymers of ethylene (EAA and EMAA).

Ionomers are partially neutralized acid copolymers (usually EMAA) using sodium or zinc bases. Fig. 2.20 shows a sodium ionomer made from EAA. Neutralization creates ionic clusters that act

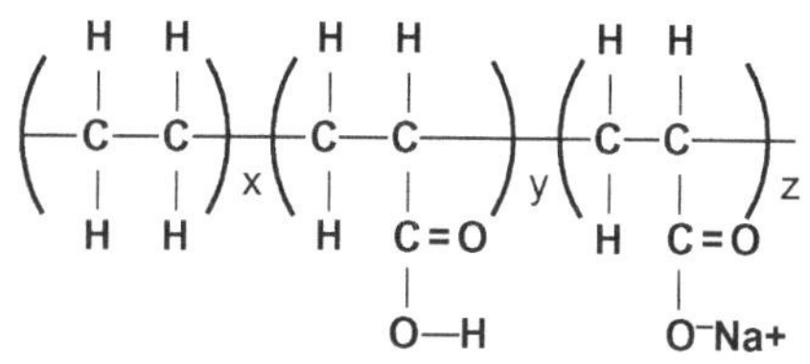

Figure 2.20 Sodium ionomer from EAA.

like cross-linked networks. The melt index (MI) decreases as neutralization increases. This results in a polymer that has a very high melt strength and provides a polymer that has excellent heat seal and hot tack strength.

2.11 Low-Pressure Polymerization Processes

The transition metal catalyst chemistry developed by Ziegler/Natta & Hogan/Banks led to the development of several low pressure processes that produce linear polyethylene (no LCB) or polyethylene having a very small level of LCB (eg, when made using chrome oxide catalyst). The processes are characterized by the low pressures (compared to the free radical high pressure processes) where they operate. These processes are as follows:

- Gas phase
- Solution
- Slurry

The first PE made with Z/N catalyst was HDPE in 1953. Phillips Petroleum commercialized chrome-catalyzed HDPE resins using a loop slurry process in the late 1950s. The Z/N catalyst produces a linear molecule when ethylene is the only monomer used (a carbon chain than has no short or long branches). In the late 1970s LLDPEs were introduced by copolymerizing with alpha olefins to create SCB to reduce density.

2.12 Gas-Phase Process

The gas-phase process operates in the gaseous ethylene state and as the PE molecule is polymerized it precipitates as a solid (powder) dispersed as a fluid bed in the reactor (Fig. 2.21) [7]. Raw materials include: ethylene, hydrogen, comonomer, catalyst, and cocatalyst. The gas circulates through the polymerizing fluidized bed. The reactor temperature must be kept <115°C to prevent the powder melting. The pressure is typically 350 psi (2.4 MPa). The gas recycles through the reactor, while the polymer exits the reactor bed, into a separator to recover the gas. Subsequently, additives and catalyst kill are added, and the polymer is pelletized.

There are three main catalysts used in gas-phase reactors and come in many variations:

- Ziegler–Natta
- Chrome oxide based
- Metallocene

The exothermic heat of polymerization is removed by injecting a cooled gas stream, and inert monomer such as isopentane or propane (condensed or super-condensed mode of operation). The alpha-olefin comonomer content controls the polymer density. The comonomers used are either 1-butene or 1-hexene. Increasing the comonomer content lowers the polymer density.

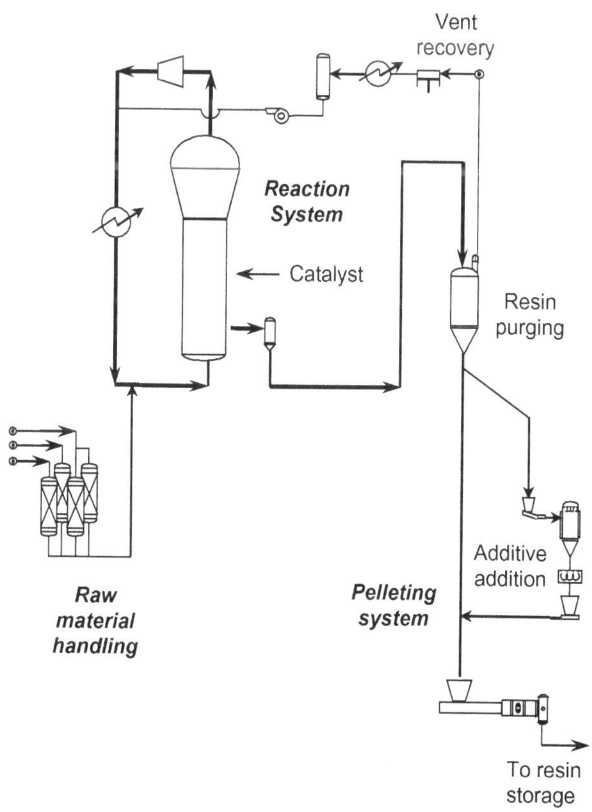

Figure 2.21 Gas-phase process (simplified).
Source: Univation Technologies.

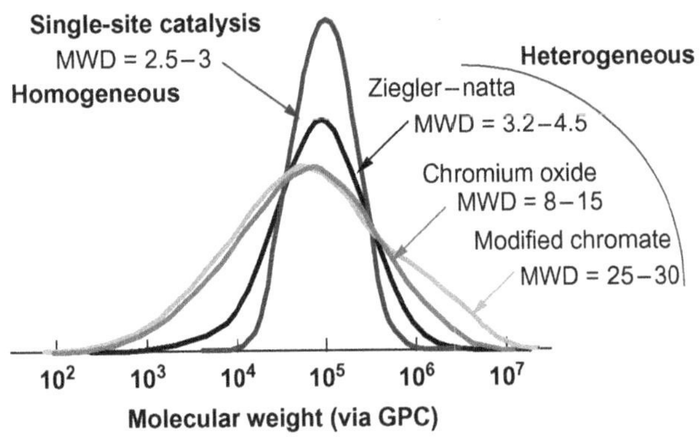

Figure 2.22 Gas-phase process MWD versus catalyst.

The reactor conditions are controlled to produce a target MI, density, and MWD. Single reactor or product from multiple reactors is formulated to produce a resin. Fig. 2.22 shows the MWDs available via different catalysts.

The gas-phase process produces many different PE products from very high M_w to moderately low M_w, densities from 0.91 to 0.96 g/cc, and MWD from 2.5 to 30.

2.13 Solution Process

The solution process operates with the ethylene and polymer dissolved in an inert solvent (Fig. 2.23). Raw materials include: ethylene, hydrogen, comonomer, and catalyst. The ethylene, hydrogen, and a comonomer is dissolved in an inert hydrocarbon fluid which is used as the diluent. The reactor temperature must be kept greater than 150°C to keep the polymer in solution. The pressure is about 500 psi (3.5 MPa). The heat of reaction is removed by the cool fresh ethylene and recycled cool solvent injected into the reactor. The polymer solution from the reactor is transferred into a series of separators to recover the unreacted monomers and solvent, and to isolate the molten polymer. Then catalyst kill agents, neutralizers, and other additives are added, and the polymer is pelletized. The unreacted monomers and solvent are recycled.

The catalysts systems used in solution reactors and which come in many variations are:

- Ziegler–Natta
- Single-site catalyst

The alpha-olefin comonomer content controls the polymer density. This is typically 1-octene, but also could be either 1-hexene or 1-butene. Higher comonomer content lowers the polymer density. The reactor conditions control the MI, density, and MWD. Single reactor or product from multiple reactors is formulated to produce a resin. Single catalyst or multiple catalysts are used to produce many different MWD products.

The solution process also produces moderate M_w to very low M_w, densities from 0.857 to 0.967 g/cc, and MWD from 2.2 to 20.

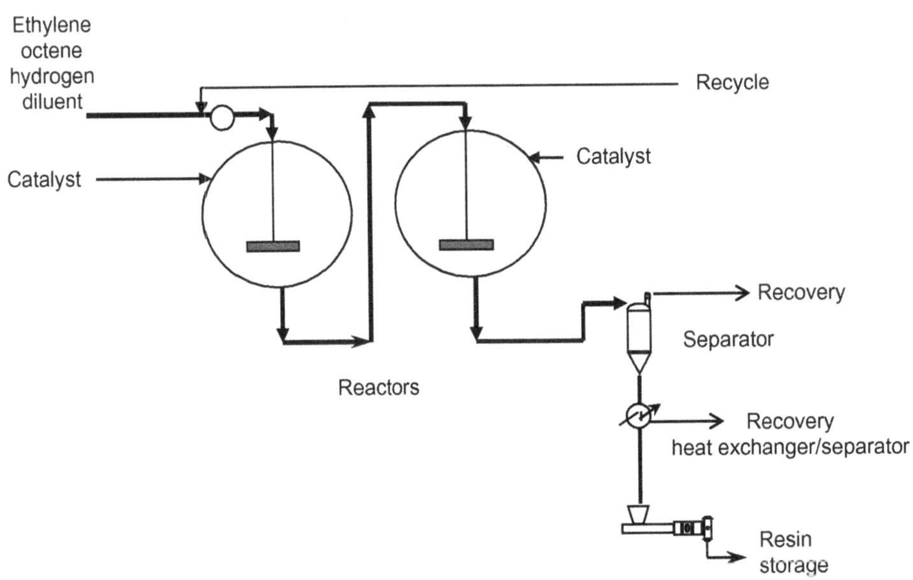

Figure 2.23 Solution process (simplified).

2.14 Slurry Processes

The slurry polymerization is the oldest and widely used polyethylene production process. Slurry processes utilize either continuous stirred-tank reactors (CSTRs) or loop reactors. Fig. 2.24 shows the loop slurry process. The slurry process operates with two phases. The liquid phase contains the ethylene and the diluent fluid and the solid phase is the polymer [8]. Raw materials include ethylene, hydrogen, comonomer, catalyst, and cocatalyst. The ethylene, hydrogen, and a comonomer are dissolved in an organic hydrocarbon fluid which is used as the diluent. The reactor temperature remains below the polymer melting point. The polymer crystallizes upon formation, creating a solid particle in the solvent (slurry). The reactor temperature must be kept below the polymer melting point (typically <100°C) to keep the polymer solid. The pressure is typically <300 psi (2 MPa). The unreacted monomers and diluent recycles through the reactor. The heat of reaction is removed by jacket cooling the loop. The polymer slurry is removed from the reactor into a flash tank to recover the ethylene. Then catalyst kill agents and other additives are injected, and the polymer is pelletized.

The main catalysts used in slurry reactors and which come in many variations are as follows:

- Ziegler–Natta
- Chrome oxide based
- Metallocene

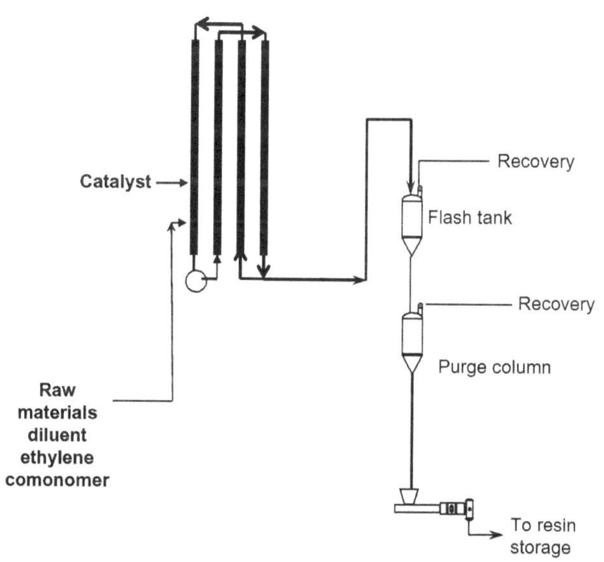

Figure 2.24 Slurry loop process (simplified). *Source: Chevron-Phillips Chemical.*

The alpha-olefin comonomer content controls the polymer density. This could include several alpha-olefin comonomers. The more comonomer incorporated in the polymer the lower the polymer density. The reactor conditions controlled the MI, density, and MWD. Single catalyst in single reactor or multiple catalysts in a single or multiple reactors are used. Advanced CSTR processes may have more than two reactors. Slurry process can produce products; from ultra high M_w to moderate low M_w, densities from 0.914 to 0.965 g/cc and MWD from 2.2 to 30 (depending on the catalyst and reactor configuration used).

2.15 Catalyst Chemistry

2.15.1 Ziegler–Natta

The early Z/N catalysts yielded about 1000 lbs PE per lb of Titanium (Fig. 2.25). This is low compared to today's catalysts.

In 1973 the catalyst technology advanced to a supported catalyst system (Fig. 2.26). This increased the efficiency 1000× to 1,000,000 lbs polymer/lb of Ti.

The supported Z-N catalyst has multiple active sites that produce slightly different polymers (Fig. 2.27). Site A on the catalyst support produces low M_w polymers with high comonomer incorporation. Site B produces a medium M_w polymer with a medium short-chain branches on the backbone. Site C produces high M_w polymers with low SCB due to low copolymer incorporation.

The original Z-N and also the supported Z-N catalysts control both the MWD and the branching distribution (Fig. 2.28).

$$Al(C_2H_5)_3 + 2TiCl_4 \longrightarrow [2TiCl_3 + Al(C_2H_5)_2Cl] + C_2H_5Cl$$

"Catalyst"
Yield ~ 1000 lbs. Polymer/lb of Ti (1964–73)

Figure 2.25 Typical Z/N catalyst (low efficiency).

"Supported catalyst" $MgCl_2 + Ti(OR)_4 + Al(C_2H_5)Cl_2$

1,000,000 lbs. Polymer/lb. Ti (1973–present)

Figure 2.26 Supported Z/N catalyst technology (1973 to present).

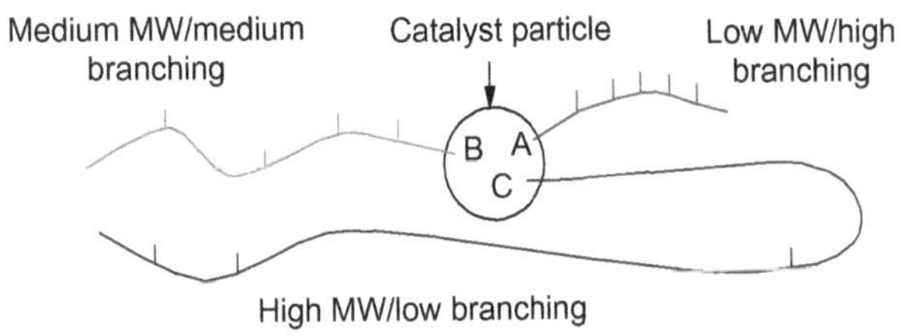

Figure 2.27 Schematic of a multiple site Z-N catalyst.

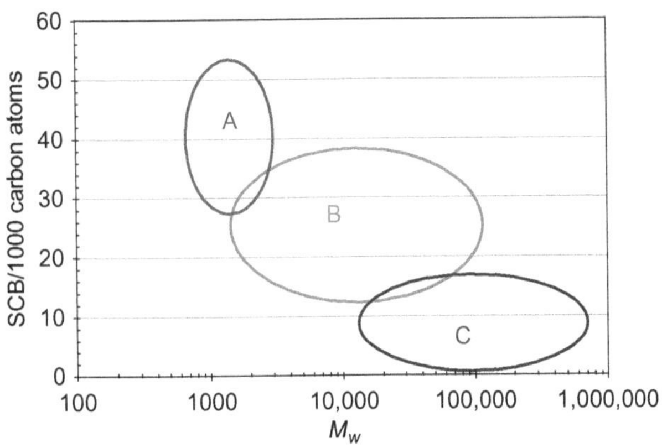

Figure 2.28 Branching distribution and M_w for a multiple site Z/N-supported catalyst producing a heterogeneous polymer.

When semicrystalline polymers are heated in a DSC instrument, the melt point (T_m) is taken as the temperature at which the highest crystal melting energy rate is recorded. Fig. 2.29 shows the DSC melting curve for an octene-based Z-N LLDPE (heterogeneous polymer) at 1.0 MI, 0.9207 g/cc. Fig. 2.29 shows that the wt% crystallinity at 23°C calculated from DSC and the measured quick density are in close agreement. The LLDPE melting point is 122.6°C. The melting point is directly related to the crystalline lamella thickness. Using the DSC melting endotherm, the heat of fusion (ΔH_f = area under the DSC curve) is determined and the wt% crystallinity can be calculated (100% crystallinity = 292 J/g). This heterogeneous LLDPE has a broad melting distribution due to broad lamellae thickness distribution. It should be pointed out that Z/N catalyst chemistry variations and different process conditions will produce different heterogeneous characteristics.

The broad melting distribution of heterogeneous LLDPE polymer reflects a broad comonomer distribution incorporated into the polymer molecules. The higher density crystalline fraction melts at a higher temperature, while the lower density fraction melts at lower temperatures.

Table 2.1 shows how different alpha-olefin comonomers can affect Z-N catalyzed LLDPE resin film properties. The butene and hexene LLDPE's were made using a gas-phase process, whereas the octene LLDPE was made using a solution process. The SCB length is directly related to the comonomer used. The higher alpha olefins have longer branches and this leads to improved film properties such as dart impact tear strength, and tensile strength [9].

2.16 Chrome Oxide–Based Catalyst

Chrome oxide–based catalysts are used for both slurry and gas-phase processes. Chrome-based catalyst produce very high molecular weight–polymers with very broad MWD. The high melt strength applications for these polymers are as follows:

- blow molding
- pipe
- thermoforming
- high stalk blown film.

In some cases, the chrome-based catalyst technology incorporates some LCB into the polymer structure further improving melt strength and processability.

2.17 Single-Site (Metallocene) catalyzed Polyethylene

Single-site catalyzed (SSC) polyethylenes have evolved.. Some fall into a subset technology called metallocene catalyzed and a metallocene catalyst is illustrated in Fig. 2.30.

All these catalyst technologies are characterized by a uniform reactive site that very accurately controls the ethylene and the comonomer insertion into the polymer backbone as shown in Fig. 2.31. The polymers produced using SSC technology will have very narrow MWD and composition or comonomer distribution (CD).

Fig. 2.32 compares single-site catalyzed polyethylene MWD to a typical Z-N catalyzed LLDPE at similar MI and density. Typically, single-site

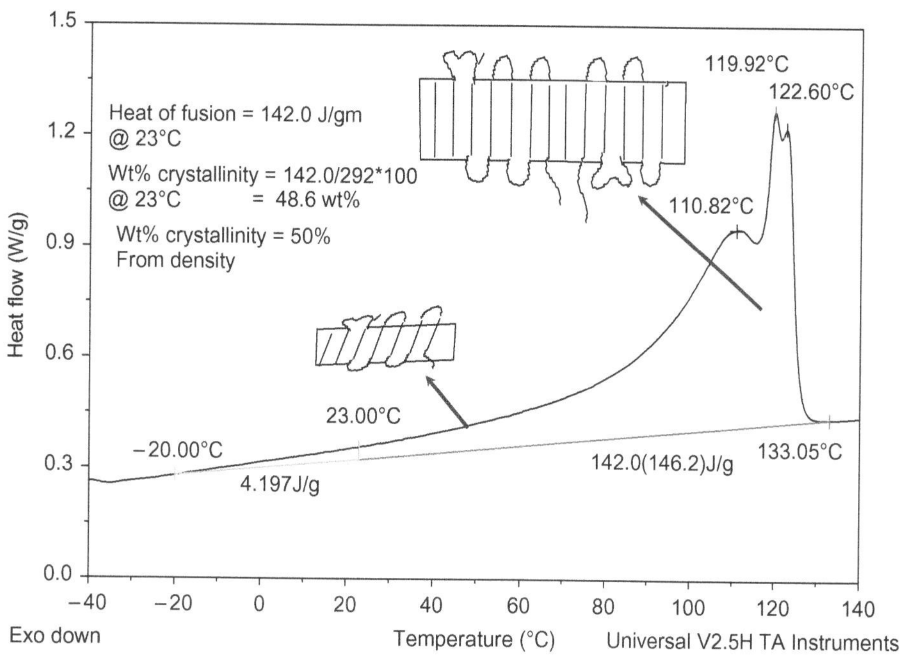

Figure 2.29 DSC second heat melting curve for 0.9207 g/cc ethylene/1-octene Z-N solution LLDPE resin.

Table 2.1 Comonomer Influence on Film Properties

Melt index, dg/min	1.0	1.0	1.0
Density, g/cc	0.919	0.919	0.920
Comonomer	Butene (C4)	Hexene (C6)	Octene (C8)
Dart impact, g/mil	100	200	335
MD Elmendorf tear, g/mil	100	300	400
CD Elmendorf tear, g/mil	300	650	710
MD tensile, MPa	33.1	36.5	46.9
CD tensile, MPa	25.5	31.0	44.1

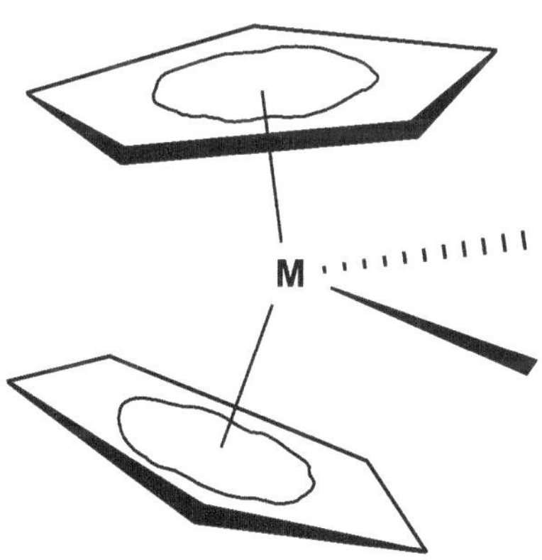

Figure 2.30 Metallocene catalyst contains a metal atom (zirconium) stabilized with in this case two cyclopentyldienenyl rings.

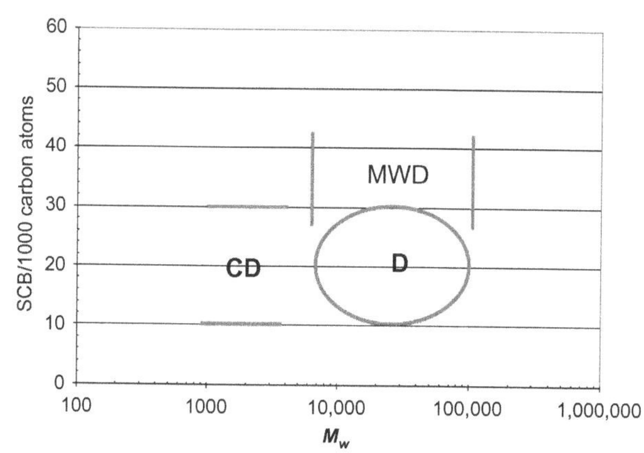

Figure 2.31 Comonomer distribution (CD) and molecular weight distribution (MWD) for single-site catalyzed (SSC) polyethylenes.

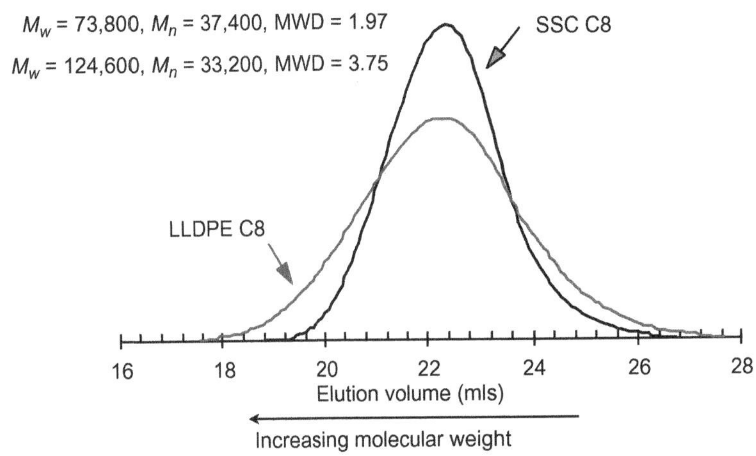

Figure 2.32 MWD comparison between Z-N LLDPE and SSC octene copolymers at similar MI and density (1.0 MI, 0.920 g/cc).

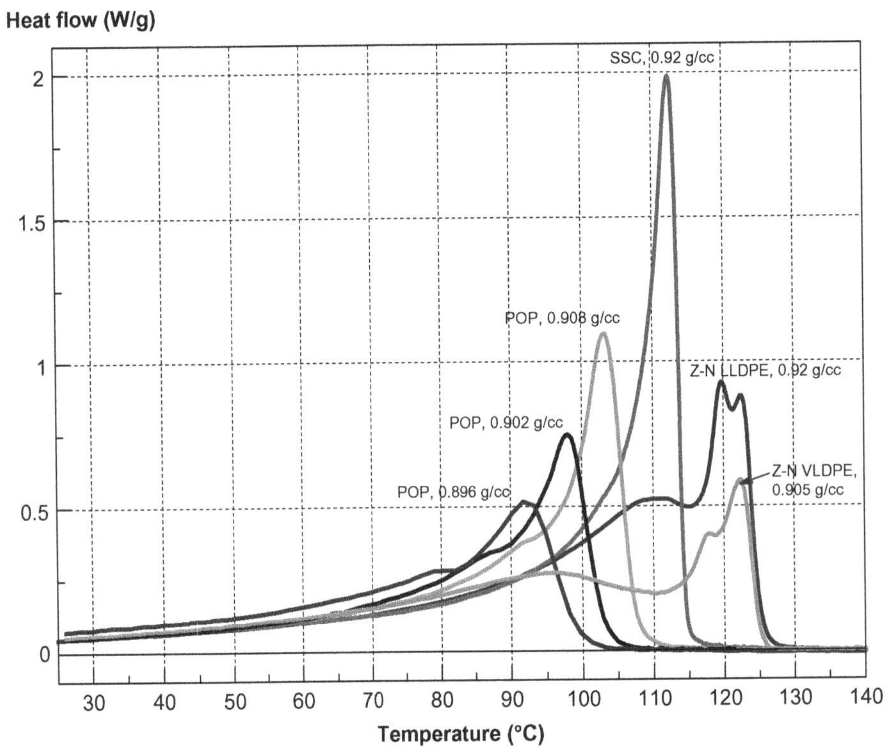

Figure 2.33 DSC curves comparing Z/N octene LLDPE and SSC octene copolymers.

catalyst will produce a 2.0–2.5 MWD (M_w/M_n). The narrow MWD distribution provides:

- increased film strength
- lower extractables
- better optical properties.

However, the narrow MWD will make these polymers more difficult to process, in that they require higher amps and generate higher melt temperatures and pressure.

The composition or comonomer distribution (CD) for single-site catalyzed (SSC) polymers is also very narrow compared to Z-N polymers. Fig. 2.33 show that the solution process octene LLDPE and VLDPE copolymers with different density (0.92 and 0.905 g/cc, respectively) have very similar DSC curves. The HD peaks differ only in the relative amounts, while the lower density peak for the 0.905 g/cc polymer is at a lower temperature. The Z-N catalyst used for these polymers produces a very heterogeneous composition distribution (CD) as shown by the DSC curves. The SSC polyethylenes have narrower DSC curves and the peak melting temperature is proportional to the density.

Fig. 2.34 shows the differences in the peak melting temperature as a function of density for single-site catalyzed, Z-N catalyzed solution process octene copolymers, LDPE and EVA resins. The heterogeneous curve will shift slightly depending on the specific Z-N catalyst characteristics and comonomer employed. LDPE resins have a slightly lower melting peak temperature compared to octene-based SSC resins at a given density. The EVA resin density increases as the VA content increases even though the crystallinity and melting peak temperature decrease. This is due to the bulky nature of the acetoxy group increasing the amorphous phase density as mentioned earlier. Hence, the EVA resins density cannot be compared with the ethylene/alpha-olefin copolymer resin density.

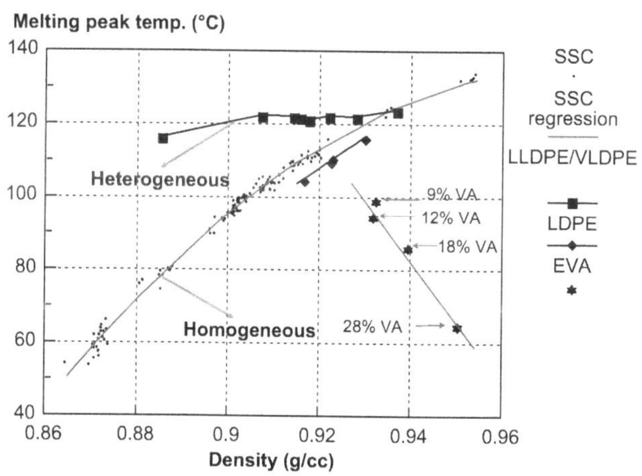

Figure 2.34 Peak melting temperature versus density for single-site catalyzed and Z-N catalyzed solution process octene copolymers.

The SSC polymers in Fig. 2.34 have LCB incorporated into the molecular structure. The reactor conditions can be varied to increase or decrease the LCB. Table 2.2 shows the relationship between polymers with increased LCB and the I_{10}/I_2 ratio for 1.0 MI HDPE polymers produced using INSITE technology from the Dow Chemical Company. Higher I_{10}/I_2 indicates that the shear sensitivity is greater which makes these polymers easier to process.

Fig. 2.35 plots M_w/M_n versus I_{10}/I_2 for Z/N LLDPE octene polymers with no LCB and SSC copolymers with and without LCB. For the Z/N

Table 2.2 I_{10}/I_2 Relationship to LCB for Four SSC Polymers

Polymer #	NMR LCB/ Chain	I_{10}/I_i
1	0.2	8.89
2	0.44	10.63
3	0.53	11.85
4	0.66	16.07

LLDPE, the shear sensitivity (processability) is significantly increased as the MWD is increased. SSC polymers with no LCB fall on lower extrapolation part of the LLDPE curve. Notice that the SSC polymers with LCB show significantly improved shear sensitivity (improved processability) even though they have a very narrow M_w/M_n.

Homogeneous polyethylene random copolymers produced by single-site catalysts (eg, metallocene or constrained geometry catalyst) with a density between 0.885 and 0.910 g/cc are known as POPs and with a density below 0.885 g/cc are known as POEs. Several SSC technology–based polyethylene random copolymer families have been commercialized since the 1990s. These include:

- POEs such as ENGAGE from the Dow Chemical Company
- TAFMER from Mitsui Chemicals
- POPs such as AFFINITY from The Dow Chemical Company
- EXACT from ExxonMobil Chemicals
- Queo from Borealis
- KERNEL from Japan Polychem Corporation.

Single-site catalysts (including metallocene catalysts) also make LLDPE and VLDPE resins via the gas-phase, slurry, and solution processes. They are generically called m-LLDPE and m-VLDPE resins. Examples include:

- Gas-phase m-LLDPE/m-VLDPE resins such as:
  - EXCEED and ENABLE from ExxonMobil Chemical
  - EVOLUE from Mitsui Chemical
  - HARMOREX from The Japan Polychem Corporation
- Slurry m-LLDPE/VLDPE resins
  - mPACT from Chevron Phillips Chemical Company
  - Lumicene from Total Petrochemicals

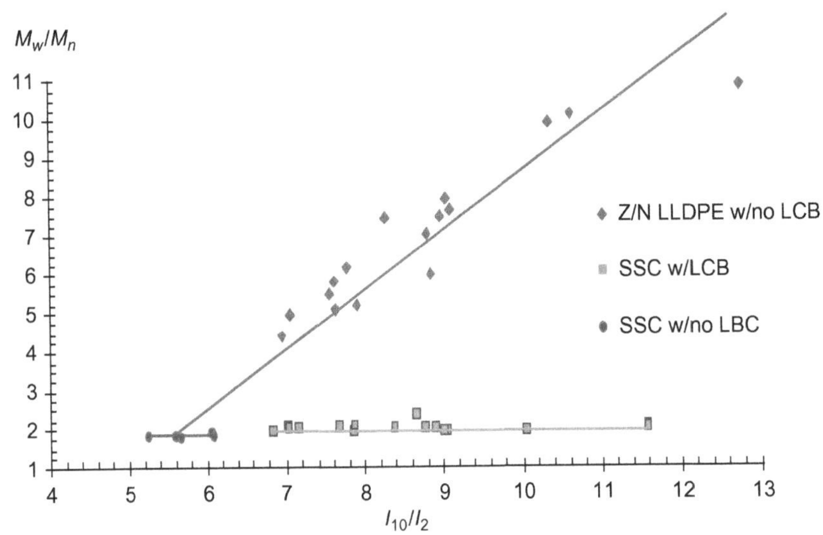

Figure 2.35 MWD versus I_{10}/I_2 for LLDPE, SSC w/o LCB, and SSC w/ LCB.

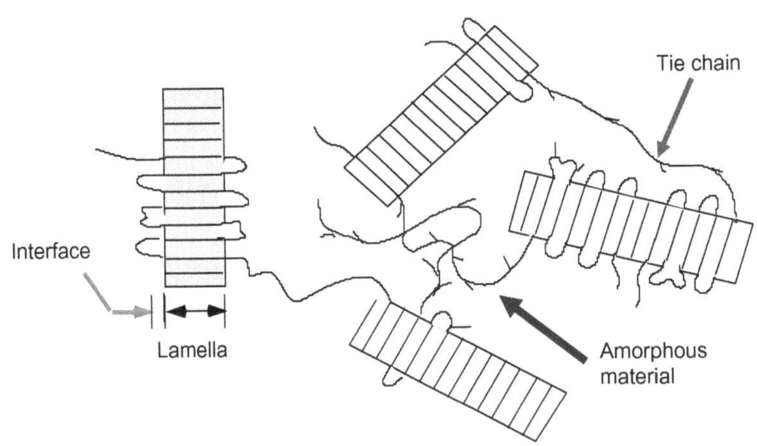

Figure 2.36 Schematic of crystalline and amorphous regions of polyethylene.

- Solution single-site catalyzed resins such as enhanced polyethylenes
  - ELITE and ELITE AT from the Dow Chemical Company
  - SURPASS from Nova Chemicals
  - NEXLENE from SK Chemicals.

2.18 Tie Chain Molecules

Molecular architecture modification enhances PE properties. Film solid-state properties can be optimized by controlling the MWD and comonomer/composition distribution (CD). Fig. 2.36 illustrates the crystalline and amorphous components of semicrystalline polyethylene. The comonomer distribution determines the crystalline lamella size distribution. Larger lamella thickness indicates higher density components (few comonomer insertions along the ethylene backbone).

A tie molecule is an amorphous molecule that is restrained by having its ends embedded in the lamella. Such tie molecules significantly improve toughness and tear properties by providing improved energy dissipation capability when the film is subjected to high stress by puncture or impact.

Tie chain concentration is difficult to measure in polyethylene [10]. Fig. 2.37 shows the calculated relative tie-molecule concentration in homogeneous (single-site catalyzed) and monodisperse polyethylene versus density [11]. Fig. 2.37 shows that an

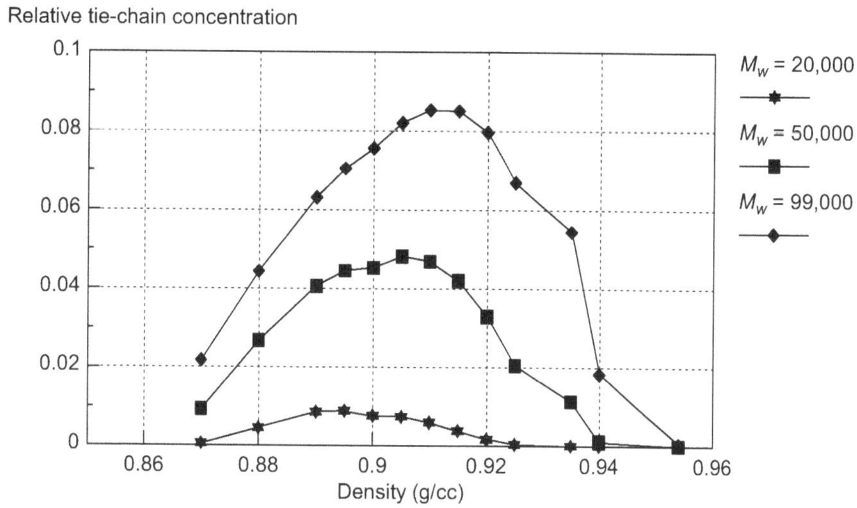

Figure 2.07 Calculated relative tie-chain concentration in homogeneous and monodisperse (MWD = 1) polyethylene copolymers versus density.

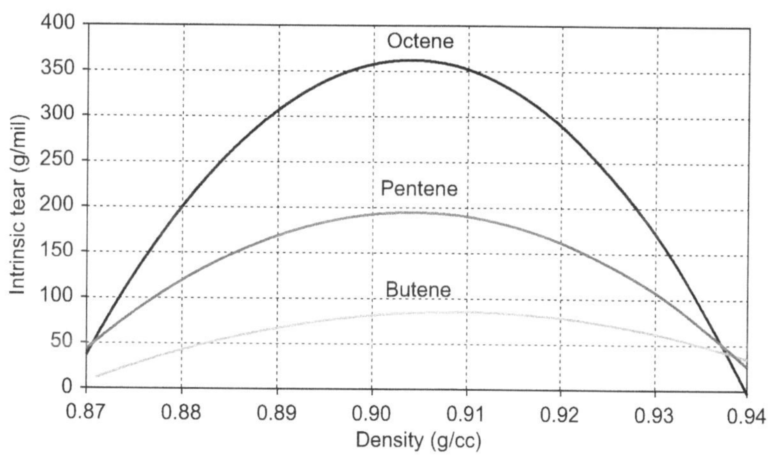

Figure 2.38 Intrinsic tear versus density for octene, pentene, and butene copolymers.

optimum density exists that provides the best properties such as shown in Fig. 2.38 with intrinsic tear versus density for octene, pentene, and butene copolymers [12]. All polymers have a similar MI (1.0 dg/min).

2.19 Summary

Polyethylene has a rich history of product, manufacturing process, and fabrication innovations that have met the growing market needs since 1939. Many different polyethylene homopolymers and copolymers have been developed with broad performance capability, to meet the diverse performance requirements in many applications, and as a result, polyethylene is the highest volume plastic sold today.

Acknowledgment

The author would like to thank his colleagues Herbert Bongartz, Shaun Parkinson, Hrishikesh Munj, and Sanjib Biswas for their critical review and suggestions.

References

[1] Patel RM, et al. Polyethylene: an account of scientific discovery and industrial innovations. ACS Symp Ser 2009;1000 (Innovations in Industrial and Engineering Chemistry):71–102.

[2] McDaniel MP. A review of the Phillips supported chromium catalyst and its commercial use for ethylene polymerization. Adv Catal 2010;53:123–606.

[3] Benham E, McDaniel M. Ethylene polymers, HDPE. Encycl Polym Sci Technol (4th ed.), 2014;5:484–515.

[4] Chum PS, Swogger KW. Olefin polymer technologies–history and recent progress at The Dow Chemical Company. Prog Polym Sci 2008;33(8):797–819.

[5] Torres A, et al. Molecular architecture: a new application design process made possible by single-site catalysts. In: Scheirs J, Kaminsky W, editors. Metallocene-based polyolefins, vol. 1. Chichester (England): John Wiley & Sons, Ltd.; 2000. p. 143–53.

[6] Swogger KW, et al. Improving polymer processability utilizing constrained geometry single site catalyst technology. J Plast Film Sheeting 1995;11(2):102–12.

[7] Xie T, et al. Gas phase ethylene polymerization: production processes, polymer properties, and reactor modeling. Ind Eng Chem Res 1994;33(3):449–79.

[8] Hogan JP, Norwood DD, Ayres CA. Phillips petroleum company loop reactor polyethylene technology. J Appl Polym Sci Appl Polym Symp 1981;36:49–60.

[9] Gupta P, et al. Does the length of the short chain branch affect the mechanical properties of linear low density polyethylenes? An investigation based on films of copolymers of ethylene/1-butene, ethylene/1-hexene and ethylene/1-octene synthesized by a single site metallocene catalyst. Polymer 2005;46(20):8819–37.

[10] Lustiger A, Ishikawa N. An analytical technique for measuring relative tie-molecule concentrations in polyethylene. J Polym Sci B Polym Phys 1991;29(9):1047–55.

[11] Patel RM, et al. Theoretical prediction of tie-chain concentration and its characterization using postyield response. J Appl Polym Sci 1996;60(5):749–58.

[12] Kale LT, et al. Structure-property relationships of ethylene/1-octene and ethylene/1-butene copolymers made using INSITE technology. J Plast Film Sheeting 1996;12(1):27–40.

3 Polypropylene

Allison Calhoun
Whitman College, Walla Walla, WA, United States

3.1 Managerial Overview

Polypropylene is found in many films and multilayer film structures due to its strength, toughness, and high melting point [1–3]. When formulated with the proper additive packages, polypropylene resins can exhibit excellent processability. The homopolymer chemical structure is rather simple, a methyl group located on every other backbone carbon unit as shown in Fig. 3.1.

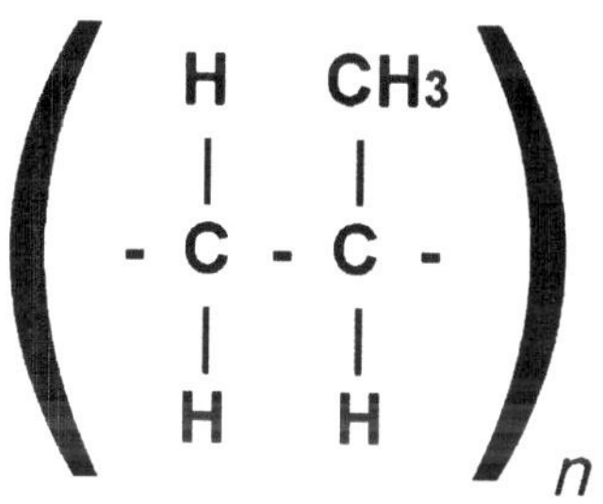

Figure 3.1 The polypropylene monomer unit. The subscript *n* indicates that this is a repeating unit that is connected to *n* number of monomeric units.

The placement of this methyl group around the stereocenter during polymerization creates different tacticities as shown in Fig. 3.2.

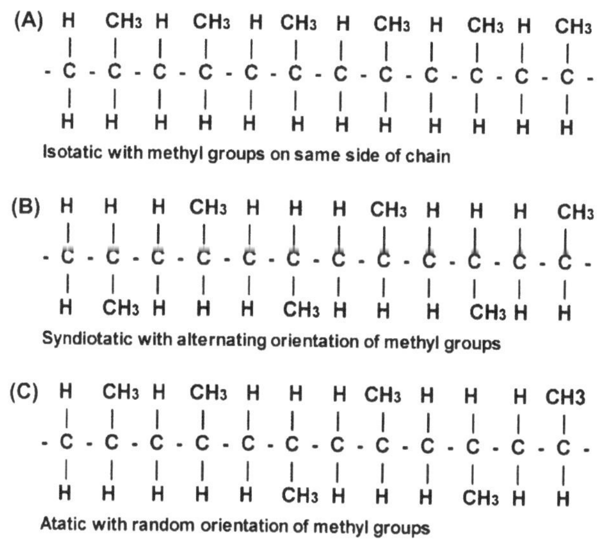

Figure 3.2 Polypropylene tacticities: (A) isotactic with all methyl groups on the same side of the chain, (B) syndiotactic with methyl groups alternating above and below the chain, and (C) atactic with methyl groups in a random orientation.

Methyl groups placed consistently on one side of the polymer chain generate isotactic polypropylene. When these methyl groups alternate from one side to the other the polymer is known as syndiotactic. A random methyl group ordering creates an atactic polymer. The polymer tacticity partially defines the polymer's properties. Molecular weight distribution, degree of crystallinity, and the polymerization method also define the polymer properties. Further chemical alterations via copolymerization with monomeric units such as ethylene and butylene create an even more comprehensive physical property library for specific applications.

3.1.1 Describing Polymers

3.1.1.1 Polypropylene Molecular Weight Characterization

Polymer molecular weight is defined as a distribution rather than a specific number because polymerization occurs in such a way to produce different chain lengths. Weight average molecular weight (M_W) and number average molecular weight (M_N) are two ways we can characterize the polymer molecular weight. Polymer M_W is defined as follows.

$$M_W = \sum_{i=1}^{N} w_i \mathrm{MW}_i.$$

Here, w_i is the weight fraction of polymer chains having a molecular weight of MW_i. The M_W is

Multilayer Flexible Packaging. DOI: http://dx.doi.org/10.1016/B978-0-323-37100-1.00003-X

typically measured by light scattering experiments. When light passes through a liquid polymer-solvent solution, it is scattered by the individual molecules suspended in the solution. The degree of scattering arises from the molecule size and, thus, a molecular weight distribution can be mathematically resolved from the total scattering created by the sample.

M_N is an average based on the number of polymer chains in a sample and is defined as follows.

$$M_N = \sum_{i=1}^{N} X_i \mathrm{MW}_i.$$

In this equation, MW_i refers to the molecular weight of each individual polymer chain and X_i refers to the mole fraction of each chain length included in the summation. This number average means that each chain, regardless of its molecular weight, contributes equally to the reported M_N.

Gel permeation chromatography (GPC), also known as size exclusion chromatography, is used to determine this property. In GPC, a dissolved polymer passes over a material held in a cylindrical column. The material, the substrate, contains holes that temporarily trap low molecular weight molecules. The lower the individual molecule's molecular weight, the more time they spend trapped and, thus, the longer they spend on the column. The polymer molecule concentration in the solvent escaping the column is measured, and residence time on the column is related to the inverse of the molecular weight.

The ratio of (M_w/M_n) allows us to define the polymer polydispersity. Fig. 3.3 illustrates M_w and M_n for a high- and low-polydispersity polymer.

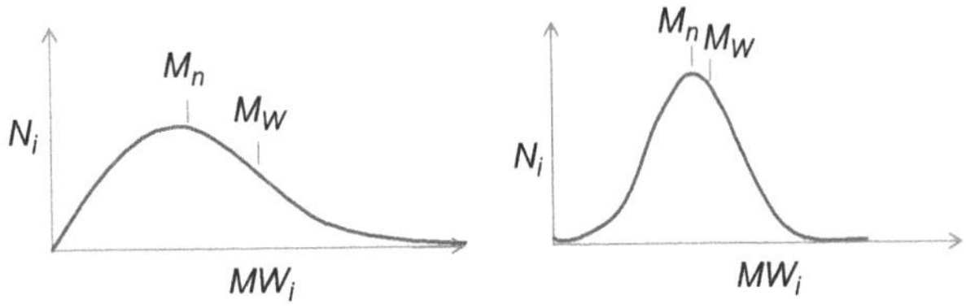

Figure 3.3 Molecular weight distributions for a high- and low-polydispersity index polypropylene. In both materials, $M_w > M_n$. The polydispersity for the left figure is higher because of the broader molecular weight distribution which means that the high molecular weight materials influence M_W more than they influence M_n.

For all real polymers, $M_n < M_w$. The smaller the polydispersity index, the more closely the M_n and M_w match one another because the distribution curves are becoming more symmetric. Polydispersity for most [4,5] isotactic commodity-grade polypropylene used in film manufacture today approximately equals 3.5.

3.1.1.2 Characterizing Polypropylene Tacticity

While it is convenient to think that polypropylene is simply isotactic, syndiotactic, or atactic, most polymeric materials do not exhibit this kind of regularity. Tacticity approximates the behavior of many polymer molecules. Tacticity, typically analyzed through nuclear magnetic resonance spectrometry, is often defined as the fraction of five adjacent units (called pentads) with a specific methyl group orientation (isotactic is all on the same side, syndiotactic is on opposite sides, and atactic is random orientation) relative to the total pentads in the polymer. Tacticity is then reported as a fraction or percentage.

3.1.2 Polymerization

Initial research investigating polypropylene polymerization relied on high pressures and radical initiators [2]. This technology produced low molecular weight oligomers with little commercial value. The Ziegler–Natta catalyst technology allowed the manufacture of commercially viable polypropylene. The resulting product was predominantly a high molecular weight isotactic polypropylene. Metallocene catalysts, the next-generation catalysts, provide a mechanism to control polymerization forming either isotactic or syndiotactic high polymer. Additionally, these catalysts can create specific molecular weight distributions.

3.1.2.1 Ziegler–Natta Catalysis

Ziegler–Natta catalyzed polypropylenes exhibit a relatively broad molecular weight distribution. The Ziegler–Natta catalysts developed for propylene polymerization generally are a mixed catalyst system containing $TiCl_3$ and trialkyl aluminum. The high weight average molecular weights achieved with this chemistry ($\sim 4 \times 10^6$ g/mol) create a polymeric material that can be processed into cohesive films. To better control the molecular weight distribution, polymer tacticity, and to improve the catalyst efficiency, chemists have modified the $TiCl_3$ crystal structure, purified the catalysts, introduced transition metal compounds, and/or a Lewis base to act as an electron-donating species.

In general, a Ziegler–Natta catalyst acts to destabilize the double bond between adjacent alkene carbons. The catalyst itself is a crystalline solid. It catalyzes polymerization on the edge sites of the crystal metal (Ti or Zr) centered complex by forming a bond between the metal atom and the monomer. The electrons from the double bond delocalize making them available for subsequent monomer addition. This process, called a chain migratory insertion, allows for chain growth [4]. Fig. 3.4 shows this process.

3.1.2.2 Slurry Process Polymerization

In slurry polymerization, chain growth occurs as monomers dissolved in a hydrocarbon solvent add to growing chains also dissolved in the solvent. Two fractions result from polymerization: one fraction is soluble in the solvent, whereas the other is not. The two fractions are separated by centrifugation, thereby creating two product streams. The insoluble fraction creates the slurry phase and is predominantly isotactic material. The soluble components are either low molecular weight oligomers or highly atactic polypropylene that precipitates from solution when the solvent is boiled off at high temperatures. Unreacted monomer is also released as a gas when the solvent boils off. This monomer is then recycled back into the monomer stream (Fig. 3.5).

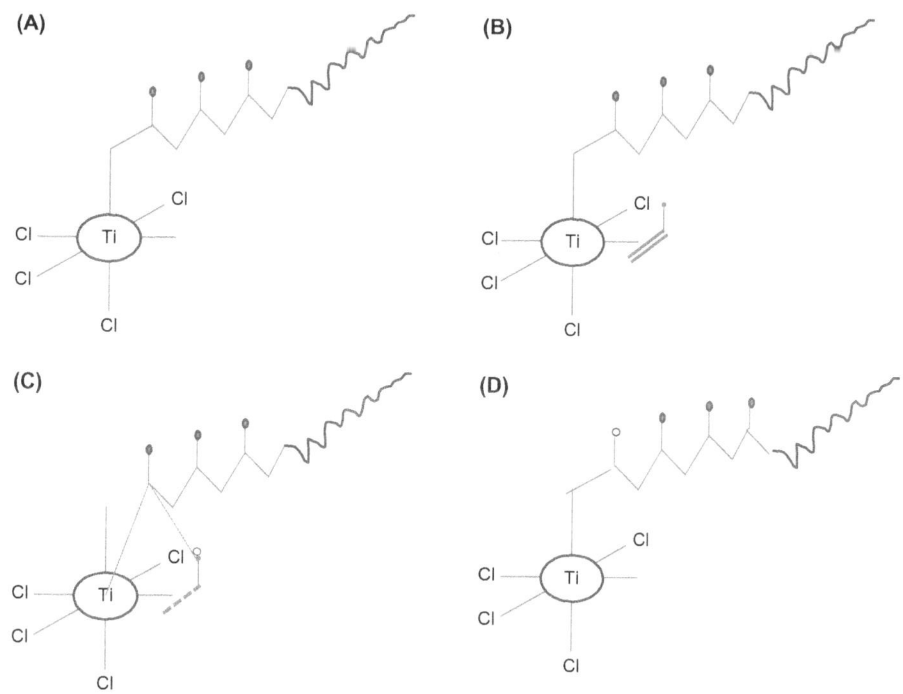

Figure 3.4 Illustrative mechanism of Ziegler–Natta catalysis. (A) The catalyst ready to receive the monomer, (B) the catalyst with a monomer interacting with the alkene bond, (C) the reaction intermediate in which the electron density shifts from the double bond to form a single bond with the polymer held by the catalyst, and (D) the catalyst is now ready to add the next monomer.

These catalysts are often "supported" with added $MgCl_2$ nanoparticles.

The $MgCl_2$ provides a surface to adsorb the Ziegler–Natta catalysts (the $TiCl_4$/trialkyl aluminum). This improves stereoselectivity and process control.

Ziegler–Natta catalyst chemistry allows polymerization in four unique polymerization systems: slurry, liquid propylene, solution, and gas phase.

3.1.2.3 Liquid Propylene Polymerization

Another polymerization process utilizes the liquid monomer as the polymerization solvent. The advantage this process has over the slurry method

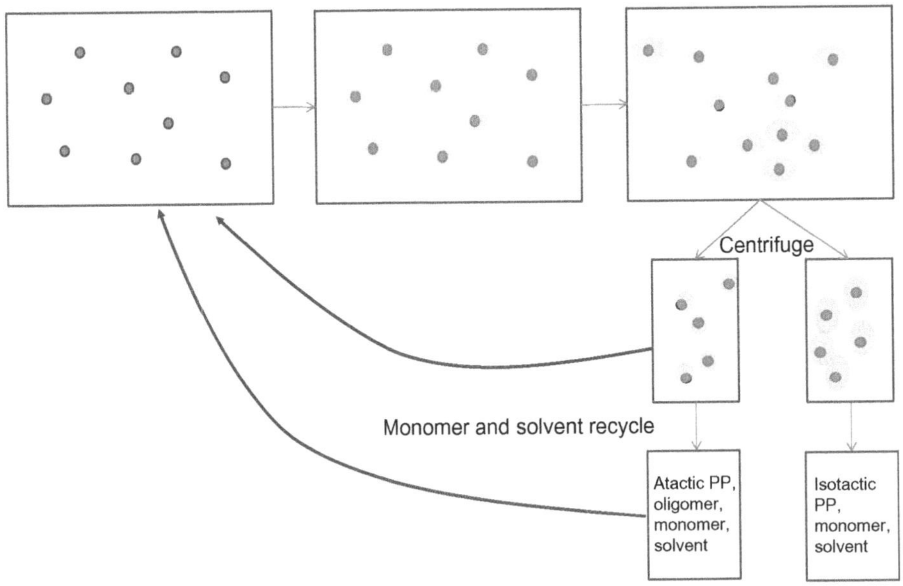

Figure 3.5 Slurry process for polymerizing polypropylene. Catalysts, represented as the solid dots, act to promote the formation of high polymer over time, indicated as lighter halos around the catalysts. Centrifugation removes the high and low weight samples which are then either recycled or purified.

arises from the fact that the high monomer concentration increases the reaction rate. Runaway reactions, uncontrolled reactions that arise from hot spots in the reaction medium, are prevented because the monomer itself can absorb excess heat as it vaporizes. The gas-phase monomer then can be condensed and reintroduced to the reactor. As in the slurry process, high-molecular-weight polypropylene forms an insoluble phase in the condensed monomer. This insoluble phase contains both the atactic and isotactic fractions which are isolated from the liquid propylene at the same time. A second separation step removes the solvent-soluble atactic fraction by using an appropriate organic solvent resulting in a pure isotactic polypropylene product (Fig. 3.6).

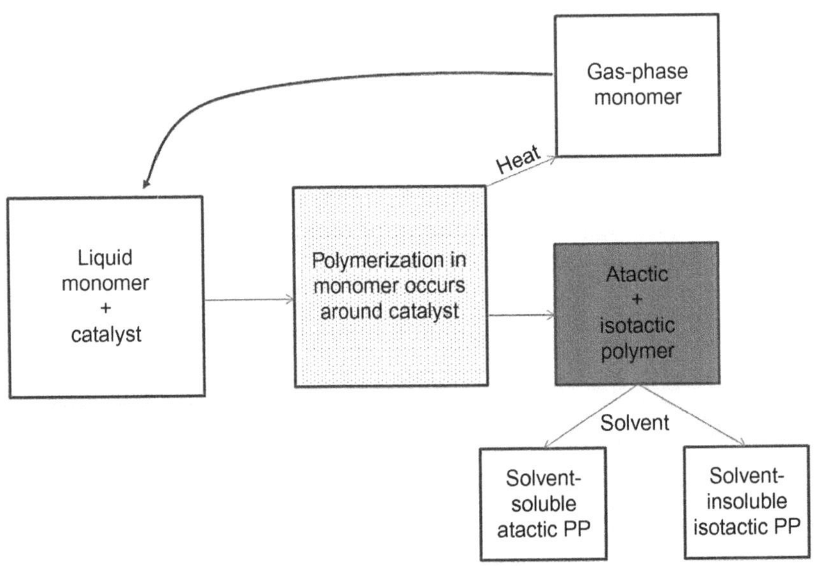

Figure 3.6 A solution containing the liquid monomer and catalyst allows polymerization to occur around the dispersed catalyst. The solution is heated to remove the gas-phase monomer from the polymer. The polymer is then treated with a solvent to separate out the atactic PP.

3.1.2.4 Solution Polymerization

During solution polymerization the monomer, catalyst, and solvent enter a reactor held at a temperature between 175°C and 250°C. Polymerization in the solvent creates a viscous polymer solution. A pump removes the whole solution from the reactor, and the solution is filtered to remove any residual catalyst, if necessary. Vacuum drying removes the solvent from the polymer, leaving behind isotactic and atactic polymer (Fig. 3.7).

3.1.2.5 Gas-Phase Polymerization

Gas-phase polymerization passes gaseous monomer through a fluidized bed containing polymer granules and an inert media–supported catalyst. Some monomer reacts within the bed to create polymer, and the remainder passes through where it is collected and recycled. The high-polymer granules are removed from the fluidized bed bottom along with some inert media–supported catalyst. The catalyst is replenished in the reactor at the same time (Fig. 3.8).

3.1.2.6 Metallocene Catalyzed Polymerization

Once manufacturers recognized the benefits of polypropylene in composite film structures, there was an increased demand for tightly controlled polydispersity and tacticity. The Ziegler–Natta catalysts were unable to meet these market needs. Additionally there was a need for improved catalyst performance to reduce residual catalyst levels in the final polymer. Entrained catalysts act as prodegradants thereby reducing the polymer's stability. The most effective way found to achieve these goals is to use metallocene catalysts.

Metallocene catalysts control the polymerization by permitting only those monomers that approach with a specific orientation to attach to the growing chain. The catalyst chemical structure has two distinct regions. The first region is the active site, the actual locus of polymerization. The second region has bulky molecular structures that hinder access to the region in which polymerization occurs (Fig. 3.9).

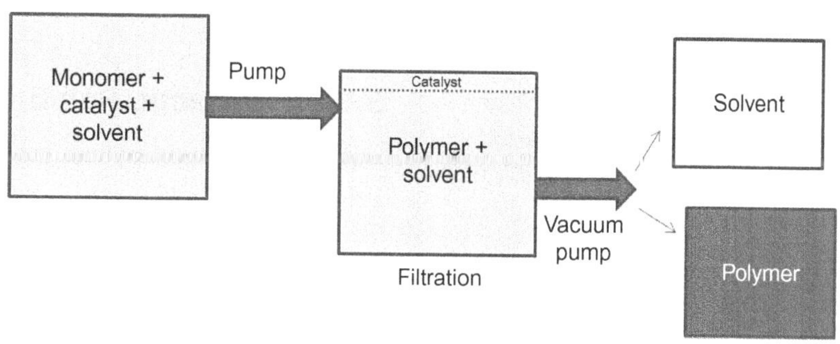

Figure 3.7 A solution containing the liquid monomer and catalyst allows polymerization to occur around the dispersed catalyst. The solution is heated to remove the gas-phase monomer from the polymer. The polymer is then treated with a solvent to separate out the atactic PP.

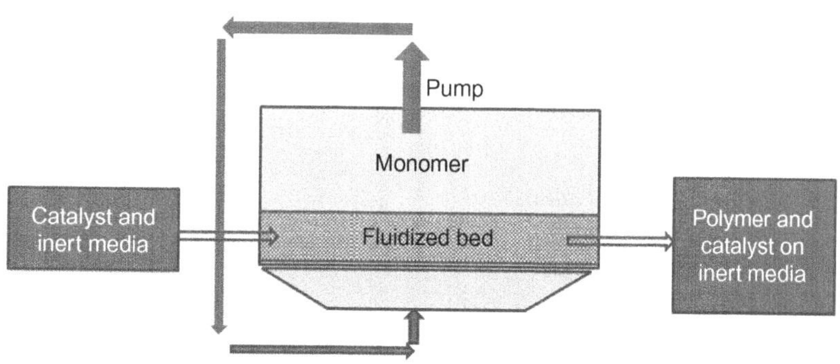

Figure 3.8 Gas-phase polymerization. The monomer stream and its recycle are indicated as *solid arrows* leading into the fluidized bed reactor. The *open arrows* indicate the introduction and removal of the catalyst and inert media. Upon leaving the reactor, the catalyst and inert media contain polymer that can then be extracted by a solvent.

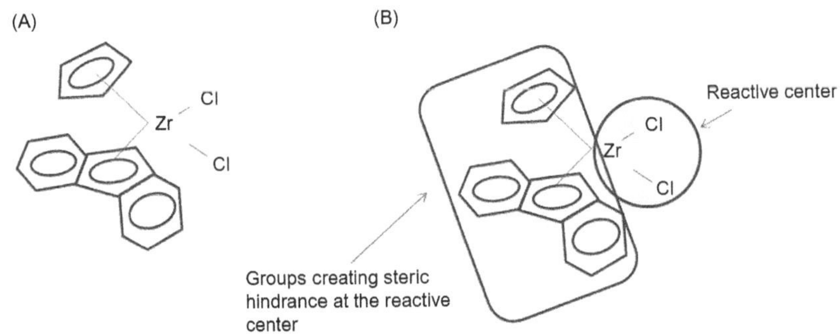

Figure 3.9 (A) A Zr centered metallocene catalyst. (B) The same catalyst with the reactive center and groups creating the steric hindrance highlighted.

Both the active site and the bulky structures can be designed to alter catalyst efficiency (generally by changing metal center composition) and tacticity (generally by altering the bulky structures). In addition, the metallocene catalysts have only one active polymerization site; this produces a narrower molecular weight distribution than is achieved by the Ziegler–Natta catalysts ($M_w/M_n \approx 2.0$). In general, metallocene catalysts process polymers with a slightly lower melting temperature than can be achieved with the best Ziegler–Natta catalysts.

Metallocenes are homogeneous catalysts, meaning they exist in the same phase as the reaction they are catalyzing and are often soluble in organic solvents. Therefore, polymerization can occur via a solution process with a nonpolar diluent dissolving the propylene gas, the catalyst, and the cocatalyst system. They can also be adsorbed onto an inert substrate which acts as part of the fluidized bed for gas-phase polymerization processes (Fig. 3.8).

3.1.3 General Polypropylene Resin Classes

3.1.3.1 Atactic Polypropylene

Atactic polypropylene, though the first manufactured polypropylene, finds little commercial use in the film industry. The random methyl group placement creates a polymer with low crystallinity. It is a soft and tacky material with low strength. It has high vapor and gas permeability and little solvent resistance. Commercial applications are very limited; however, it can be found as a component in adhesives and roofing systems.

3.1.3.2 Isotactic Polypropylene

Isotactic polypropylene produced through current catalyst technology can be more than 95% isotactic as defined by the percent isotactic pentads in the polymer chain. The advantages of isotactic polymers over the syndio- and atactic polypropylene arise from their high crystallinity. Typical isotactic polypropylenes are between 40% and 70% crystallinity. The resulting polymer has a higher melting point (160°C–180°C), higher density, and higher strength relative to the syndiotactic and atactic forms.

In general, isotactic homopolymer polypropylene has high crystallinity, thereby creating a strong material with low permeability to vapor or solvents, and high chemical resistance. Isotactic polypropylene, with the exception of the mesomorphic crystalline form, is typically opaque due to the high crystallinity.

Isotactic polypropylene exhibits four crystalline structures: alpha, beta, gamma, and mesomorphic [3,5]. Each structure forms under specific processing conditions and defines the polypropylene properties. In polypropylene containing the alpha, beta, and gamma structures, the final material is typically opaque due to the light scattering by spherulites with sizes similar to the wavelength of visible light.

The crystallinity type is generally analyzed via X-ray diffraction (XRD) techniques. Light passing through a sample scatters in specific patterns based on the repeating structures within the polymer. The crystallinity can be calculated based on the peak intensity associated with specific repeating structures. A typical XRD pattern can be found below with representative peaks labeled (Fig. 3.10). The degree of crystallinity can be measured based on the relative polymer fraction solubility as a function of temperature (the higher the dissolution temperature, the more crystalline the material) or

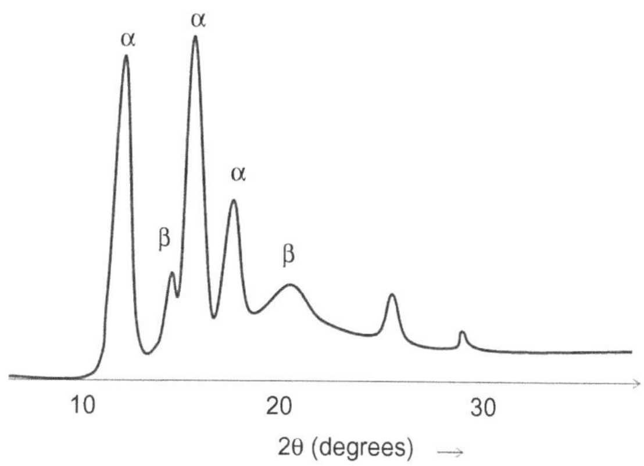

Figure 3.10 Typical X-ray diffraction pattern for isotactic PP showing alpha and beta peak locations.

through differential scanning calorimetry (a way to investigate the polymer melting temperature range).

3.1.3.2.1 Alpha Crystallinity

Alpha crystallinity arises from a helical conformation of the polymer chains described by a monoclinic unit cell. This structure represents the most thermodynamically stable structure and forms readily as the polypropylene cools below its crystallization temperature. The helices align to form crystallites which then associate to form sheet-like lamellae. Crosshatching microstructure arises as secondary lamellae grow perpendicularly off the primary lamellae. These secondary lamellae create a supermolecular structure by connecting the neighboring primary lamellae. The secondary lamellae can further associate to create spherulites as shown in Fig. 3.11.

Another way to look at the spherulites and the lamella arrangement in the spherulite is shown in Fig. 3.12.

Fig. 3.13 [6] is a picture of a PP spherulite grown between glass slides on a microscope hot stage.

Polypropylene crystallites can be the size of the wavelength of light and therefore tend to refract light and be translucent with high haze and low transparency. There are several additives that can promote nucleation, reduce the spherulite size below the wavelength of light and make the quenched PP more transparent. Fig. 3.14 [7] is an optical micrograph that shows the effect of a nucleating agent on spherulite size.

3.1.3.2.2 Beta Crystallinity

Beta crystallinity forms when the polymer is cooled quickly to a temperature below its melting

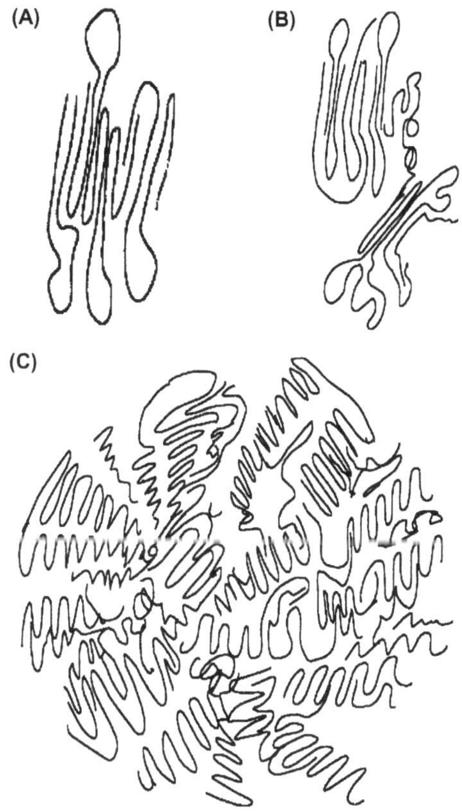

Figure 3.11 Polypropylene semicrystalline structures. (A) Crystallite formed from polymer chains associating; (B) secondary crystallization associating crystallites to form lamellae; (C) spherulite formation from long range structuring of lamellae.

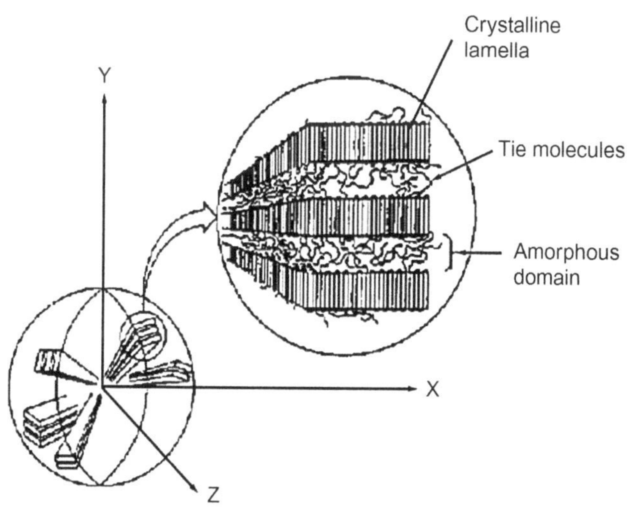

Figure 3.12 Spherulite crystal morphology showing the lamella, the molecules, and amorphous domain.

Figure 3.13 A Maltese cross pattern of birefringence obtained using optical microscopy under crossed polarizers. Polypropylene was isothermally crystallized at a pressure of 100 MPa.

(Approximate magnification × 120)

General purpose polypropylene | Nucleated polypropylene

Figure 3.14 Two optical micrographs show the effect of a nucleating agent on spherulite size. In nucleated polypropylene (*right*) spherulites are smaller and more numerous than the unnucleated polypropylene (*left*).

point. High processing speeds and fast cooling times created by cooling and quenching equipment create these structures in films. Beta crystallinity arises because polymer chains do not have the time required to form the thermodynamically stable alpha crystallinity. Because of the reduced crystallinity, beta crystalline polymer exhibits lower density and lower melting point than a polymer containing the alpha crystalline form. Once the polymer is heated or placed under strain, the beta crystallinity reverts to the alpha structure. Annealing processes take advantage of this transition to increase the polymer strength via increased alpha crystallinity.

3.1.3.2.3 Gamma Crystallinity

Isotactic polypropylene also forms gamma crystallinity, though this form rarely arises from processing under standard conditions. The gamma crystallites form a crosshatched structure just as seen in alpha crystallinity. However, the crystallites grow in two directions at the same time resulting in a high-crystallinity, high-density material. The mechanism for forming gamma crystallinity is believed to arise from shearing an alpha crystalline polymer to create a more uniform crystal structure.

3.1.3.2.4 Smectic or Mesomorphic Crystallinity

Rapidly quenched isotactic polypropylene has a fourth crystalline form. The rapid cooling prevents large crystalline domains from forming. Therefore the crystalline regions are small with poor alignment between the individual crystallites. The smectic crystalline polypropylene properties lie between alpha crystalline and atactic polypropylene. The polymer will be transparent and will revert to an alpha crystalline structure upon heating.

The crystallinity and crystallite orientation in a mesomorphic polypropylene can be affected both by the cooling rate and by subsequent annealing. Fast cooling creates small crystalline domains with limited crystallite orientation, whereas slowing the cooling rate will allow for larger domains and greater long range structure. Annealing allows the polymer to retain the dispersed structured domains while converting to the alpha crystalline phase. This process creates a ductile film because the crystalline domain distribution is also strong because of the alpha crystallinity [5].

3.1.3.3 Syndiotactic Polypropylene

With the introduction of single-site catalysts, syndiotactic polypropylene became commercially available. Its use, however, is not widespread in multilayer film manufacture. Until recently, syndiotactic polypropylene was not a viable commercial product as its manufacture was not feasible with the prior catalyst technologies [3,5]. Polymer produced with single-site catalyst rarely exceeds 75% syndiotacticity as determined from pentad sequences. The polymer, relative to isotactic polypropylene, has a lower melting point (approximately 138°C versus 155–160°C) and a lower density (0.89 g/cm^3 versus 0.93 g/cm^3). The crystallites that form in syndiotactic polypropylene are more complex than those formed in the isotactic polymer. This complexity reduces the rate at which the crystallites can form meaning that the polymer crystallizes very slowly [6,7]. Syndiotactic polypropylene has found use as a core layer in multilayer films to provide high tear strength and clarity.

Its slow crystallization rate limits its use as a monolayer film as that leads to material that can distort or stick to processing equipment. Syndiotactic polypropylene can also be incorporated into other polypropylene materials to improve strength.

3.1.3.4 Polypropylene Copolymers

Propylene monomers can be polymerized with other monomers to create copolymers. Random incorporation of nonpropylene monomers creates a random copolymer while structures with alternating chains of the two monomers can create block copolymers. In general, incorporating other monomers interferes with the pure polypropylene crystalline structure. Thus, copolymers are more flexible than homopolymers while retaining good strength. Introducing a second monomer creates an opportunity for different chemistry to alter the polymer reactivity, surface chemistry, or rheology.

Ethylene–propylene copolymers are used as sealant layers in multilayer films. The ethylene content changes the melting point and thus the sealing properties. Ethylene–propylene–butene terpolymers are also used as seal layers in multilayer films.

3.1.3.5 EPR or EPDM Elastomers

A common polypropylene copolymer produced with ethylene is an elastomer. The Polymerizing ethylene into the propylene chain generates a flexible and extensible polymer that regains its shape. Thus, like rubbers, they can be used to create structures that recover after strain and are known as elastomers. These copolymers are often referred to as EPR or EPDM rubbers and elastomers. They are used in automotive weather-stripping and seals, glass-run channel, radiator, garden and appliance hose, tubing, belts, electrical insulation, roofing membrane, rubber mechanical goods, plastic impact modification, thermoplastic vulcanizates, and motor oil additive applications. They are not normally used for multilayer film application.

3.1.3.5.1 Processing

Polypropylene's excellent strength, low surface energy, low gas and liquid permeability, and relative ease of processing make it an attractive option for use in multilayer films. However, its opacity may be a detriment for packaging applications.

Orientation breaks up the spherulitic structure, thereby resulting in a transparent film. Additionally, orientation improves tensile properties, impacts strength, and reduces gas permeability.

Polypropylene may be used to manufacture single layer films or as a component in multilayer films via both cast and blown film processing. In multilayer films, polypropylene can serve as the major layer to provide structural integrity to the film or it may be used as a minor layer for specific functionality. In either blown or cast film processing of multilayer films, coextrusion generates the individual layers. Processors accomplish adhesion between the layers by using pressure to create intimate contact between the layers at a time when they are still hot. In blown film processing, this occurs through a double bubble method where the pressure is created by air on the inside of the bubble. In cast film, the layers are pressed together in nip rollers at the outlet of the die.

3.1.3.6 Polypropylene Use Challenges

The challenges associated with using polypropylene in film applications are that it is an intrinsically unstable polymer that degrades via oxidative, thermo-oxidative, and ultraviolet degradation processes [8,9,10,11]. Resin manufacturers recognize this issue so they stabilize them well before sending them to their customers. However, stability may be a considerable issue if the stabilizer is ineffective for the processing conditions or is used up during processing or use.

The most common degradation during manufacturing is thermooxidative degradation. In this process, oxygen and high temperatures create an autocatalytic reaction which promotes the double bond formation in the polymer chain, creates chain scissions, and generates low molecular weight volatile products. The double bonds end up conjugated on the chains which lead to off-color resin with a yellow tint. The low molecular weight volatile materials can lead to holes in the film (under very extreme conditions) or thin spots where the volatiles flashed off at the die. Another issue associated with degradation is that the polymer molecular weight reduction induced by chain scission reduces the polymer melt viscosity. This leads to ineffective polymer mixing and pumping thereby changing the final film properties.

To prevent this degradation from occurring, processors need to confirm that the stabilizer works under the processing conditions including such things as the residence time, shear conditions, and temperature regime that the polymer will experience. This becomes especially critical when recycled material is introduced into the process. The stabilizer will already be depleted in this recycled material which can lead to additional problems during its reprocessing.

Another processing concern is that polypropylene's melt strength may be inadequate to sustain a stable bubble during blown film processing. This means that the bubble may sag or dance under its own weight resulting in uneven film thicknesses across the bubble. A recent improvement to this resin family includes a polypropylene which contains long chain branching. This branching increases entanglements among the polymer chains promoting greater polymer cohesion. This, therefore, increases the melt strength.

3.2 Summary

Polypropylene provides the film manufacturer with strength, toughness, high melt temperature, and processing ease that makes it a star among the polyolefin resins. Films produced with this polymer find many applications in many different arenas. Consumable and medical packaging, labels, stickers, personal hygiene, and construction films all have been made with this versatile resin.

Table of Selected Polypropylene Manufacturers and Example Products

Suppliers	Example Products
BASF	Neopolen
Borealis	HC101BF, TD310BF
BP Amoco	Accpro
Braskem	Symbios
Carmel Olefins	Capilene
Chevron Phillips Chemical Company LP	Marlex Copylene
Dow Chemical	STRANDFOAM
DSM Engineering Plastics	Electrafil

(*Continued*)

Table of Selected Polypropylene Manufacturers and Example Products—*Cont'd*

Suppliers	Example Products
DSM Engineering Plastics	Fiberfil
DuPont	Bynel
Exxon Mobil	Exxon Mobil PP6262
Formosa Plastics Corporation, USA	Formolene
INEOS Technologies	H03G-05, H07G-00
LyondellBasell	Moplen, Profax, Clyrell, Adstif, Hostalen
Mitsui Chemicals	ADMER
Pinnacle Polymers	PP5135
RTP Company	RTP 100
SABIC Innovative Plastics (GE Plastics)	
Sunoco Chemicals (formerly Aristech)	Ultraclear
The Dow Chemical Company	INSPIRE
Total Petrochemicals	PPR1251, PPH3271, PPH 3274, PPC4170

References

[1] Charrier J-M. Polymeric materials and processing: plastics, elastomers and composites. New York, NY: Hanser Publishers; 1990.

[2] Malpass DB, Band E. Introduction to industrial polypropylene: properties, catalysts, processes. Hoboken, NJ: John Wiley & Sons Inc. and Salem, MA, Scrivener Publishing; 2012.

[3] Karian HG. Handbook of polypropylene and polypropylene composites. CRC Publisher; 1999.

[4] Moore EP, The rebirth of polypropylene: supported catalysts. New York, NY: Marcel Dekker, Inc.; 2003.

[5] Pasquini N, editor. Polypropylene handbook, products, technologies and markets. Cincinnati, OH: Hanser/Gardner Publications; 2005.

[6] Odian G. Principles of polymerization. 3rd ed. New York, NY: Wiley Interscience Publications; 1991.
[7] Moore Jr. EP, editor. Polypropylene handbook. 1st ed. Cincinnati, OH: Hanser-Gardner Publications; 1996.
[8] Painter PC, Coleman MM. Fundamentals of polymer science. Lancaster, PA: Technomic Publications; 1997.
[9] Ehrenstein GW. Polymeric materials: structure-properties-applications. Cincinnati, OH: Hanser/Gardner Publications; 2001.
[10] Schnabel W. Polymer degradation: principles and practical applications. New York, NY: Hanser Publishers; 1992.
[11] Pielichowski K, Njugana J. Thermal degradation of polymeric materials. Shrewsbury, United Kingdom: Rapra Technology; 2005.

4 Introduction to Bio-Based Polymers

Antonio Meraldo
Lux Research, Inc., Singapore

Bio-based polymers are defined as polymers made from biological sources. Some of these polymers are formed directly in the polymeric form within the producing organisms (eg, microorganisms, algae, or plants), while others are manufactured ex vivo from bio-based monomers. The simplest bio-based polymers, such as cellulose and starch, have been known and widely used for centuries. However, the use of these pioneering compounds was restricted to a few applications such as packaging, textiles, and construction. Since 1990, advances in chemical processing and biotechnology have allowed the production of novel bio-based polymers such as polylactic acid (PLA) and polyhydroxyalkanoate (PHA) that have received much attention from many industrial players. Many people have opined that these new bio-based polymers have the potential to replace petroleum-based polymers and help solve some of the most urgent problems caused by the overuse of petroleum-based polymers, such as water and soil pollution, deleterious influence to human health, and overdependence on petroleum. Yet others believe that the shortcomings of bio-based polymers, both in the economic and technical aspects, preclude their rapid adoption at least in the near future.

This chapter provides an overview of some novel bio-based polymers. It will not cover bio-based polymers that are chemically equivalent to their petroleum-based counterparts (eg, bio-based polyethylene), nor will it cover first-generation bio-based polymers (eg, starch or cellulose). An important part of this chapter is the discussion about how these bio-based polymers perform, which is inextricably linked to their current and potential future applications. To aid readers' understanding, these bio-based polymers are compared with two representative petroleum-based plastics, polyethylene (PE), and polyethylene terephthalate (PET). The evaluation is performed on six parameters (cost, recyclability, mechanical properties, ease of processing, barrier properties, and biodegradability), and a scoring system for this evaluation was developed, where the score of 5 denotes the best performance and 1 denotes the worst.

4.1 Commercially Important Novel Biopolymer

4.1.1 Polylactic Acid

4.1.1.1 Introduction

PLA is a biodegradable thermoplastic made from renewable resources. Despite carrying the word "acid" in its name, PLA is actually not a polyacid per se, as it does not contain any carboxylic acid side groups except in its termini. Rather, the compound is a polyester consisting of joined lactate residues. Because of the ester bond that it contains, sometimes, the term "polylactide" is used to refer to PLA.

Although it may appear to be the most straightforward method, directly synthesizing high molecular weight PLA from lactic acid is challenging. Direct condensation of lactic acid only yields short-chain oligomers, which are not used industrially due to their inferior thermomechanical properties. A widely used industrial method of synthesizing the desirable long-chain PLA is the ring-opening polymerization of lactide (the lactic acid dimer) in the presence of metal catalyst. The lactide itself is obtained by the depolymerization of the short-chain oligomers of lactic acid.

The raw material for PLA, lactic acid, is a common industrial chemical with an annual production exceeding 400,000 metric tonnes. In the industry, lactic acid is produced by lactic acid bacteria (mostly belonging to the genus *Lactobacilli*), which convert simple carbohydrates such as glucose, sucrose, or galactose to lactic acid. Commercial fermentation is often performed in a batch reactor and usually takes 3–6 days to complete.

Multilayer Flexible Packaging. DOI: http://dx.doi.org/10.1016/B978-0-323-37100-1.00004-1

Lactic acid is chiral and has two optical isomers: the L(+)-lactic acid or (*S*)-lactic acid and the D(−)-lactic acid or (*R*)-lactic acid. Since lactic acid produced through bacterial fermentation comes almost exclusively in the L/*S* configuration, the corresponding PLA, termed poly-L-lactic acid (PLLA), is the most commonly produced PLA. Today, efforts are ongoing to industrially produce the other lactic acid enantiomer, the D/*R* lactic acid and the corresponding PLA, poly-D-lactic acid (PDLA), as it has been discovered that combining PLLA and PDLA (the so-called PLA stereocomplex) results in improved thermomechanical properties compared to pure PLLA.

4.1.1.2 Performance

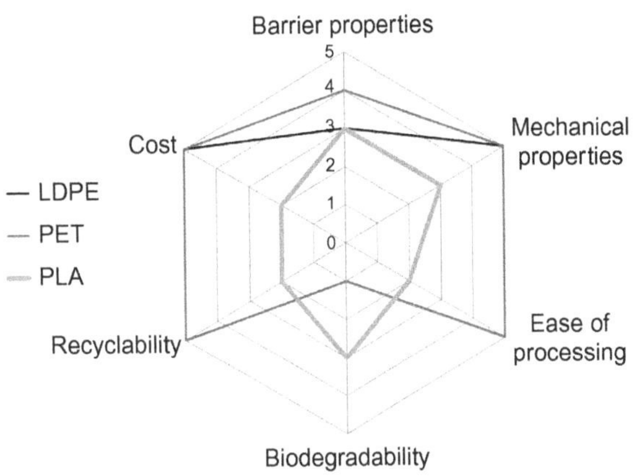

PLA's properties depend considerably on the ratio of the D/*R* and L/*S* enantiomers of lactic acid used in its production and its configuration in the chain. Generally speaking, PLA whose individual chains contain both D/*R* and L/*S* enantiomers exhibits low heat stability, as the presence of both enantiomers in the chains prevent efficient packing of the strands. For example, PLA made from a 80:20 L/*S* and D/*R* lactic acids ratio exhibits a glass transition temperature (T_g) and a melting temperature (T_m) of only 56°C and 125°C, respectively. PLA made from only one enantiomer exhibits higher transition temperatures; for example, PLLA has a T_g and T_m of 63°C and 178°C, respectively. It has been reported that a 1:1 mixture of pure PLLA and PDLA yields the so-called PLA stereocomplex that exhibits a set of thermomechanical properties better than either pure polymers, even resembling PE in some metrics.

Most common processes for thermoplastics can be applied for PLA. However, the processing of PLA presents several hurdles. Due to its hydrophilicity, PLA resin rapidly absorbs ambient moisture. When the resin is not adequately dried, downstream processing issue might arise. For example, due to the fact that PLA is a condensation polymer, the presence of water during melt processing can lead to the hydrolysis of polymer chains. Additionally, just like PET, PLA requires crystallization so that the PLA particles will not stick together.

4.1.1.3 Application

Today, PLA is widely used in food packaging. Some products made from PLA include trays, tablewares, cups, and cutleries. The drive behind the adoption of PLA for packaging is its biodegradability, which resonates well with end users. Additionally, most PLA types exhibit attractive glossy appearance that is important for some packaging applications.

In addition, owing to its biodegradability and balanced mechanical properties, PLA finds promising application in agriculture. Its transparency renders it suitable for the fabrication of mulching films or delayed release materials for pesticide and fertilizer. When used as mulch film, at the end of the cultivation season, PLA-based mulch can be left in the field, where it gradually biodegrades down into small pieces that can pass through harvesting machinery without difficulty.

PLA has also been used in the medical industry as implants, stents, and bone support splints. The mechanical properties of PLA can be tailored to mimic those of the surrounding tissue, facilitating easy integration. The use of PLA as implant is relatively safe, since the degradation of PLA by hydrolysis leads to the formation of lactic acid, a natural product easily decomposed by the body's normal metabolism.

There have also been some efforts to use PLA in 3D printing. The company NatureWorks has announced a collaboration with 3DOM to develop several PLA grades for 3D printing and aims to launch these products in 2015.

4.1.1.4 Players

Today, NatureWorks is the world's largest producer of PLA. The company has received investments by Cargill and PTT Global Chemical. NatureWorks owns a manufacturing facility in Blair, NE, with nameplate capacity of 140,000 TPA of Ingeo biopolymer, its PLA brand. The company recently announced Thailand as a preferred location for a second manufacturing plant.

Taghleef Industries offers a portfolio of films for food, labeling, and nonfood application. One of its products is bioaxially oriented PLA that is sourced from NatureWorks' Ingeo. The company has a total production capacity (mainly non-PLA plastics) of 360,000 TPA.

Corbion Purac is a food and chemicals company based in the Netherlands. One of its products is PLA polymer. The company produces a variety of PLA and PLA derivatives, including PLLA homopolymer resins, PDLA homopolymer resins, and PLA compounded blends. The company claims that its technology allows a wide customization of PLA and that its PLA products can go into a broad range of markets and applications that include packaging, consumer goods, automotive, and textiles. In the end of 2014, the company announced the intention to construct a 75,000-TPA PLA facility in Thailand.

The Canadian company Solegear Bioplastic produces PLA-based resin with the brand name Polysole. The company targets Fortune 500 brands, partnering with plastics molders and processors that make consumer products, such as toys, cosmetics, personal care, office, and home and housewares market applications. The company also intends to eventually grow into automotive and medical applications. The company currently has a 200,000 tonnes per year of resin-manufacturing capacity through contract manufacturing.

Other companies that have PLA capacity include Sulzer, Zhejiang Hisun Biomaterials, Henan Piaoan, Toray, and FkuR.

4.1.2 Polybutylene Succinate

4.1.2.1 Introduction

Similar to PLA, polybutylene succinate (PBS) is an aliphatic polyester. The most common way to produce PBS is through direct esterification of succinic acid with 1,4-butanediol. This process consists of two steps. In the first step, 1,4-butanediol is reacted with succinic acid to form PBS oligomers. Afterward, these oligomers are catalytically *trans*-esterified to form a high molecular weight PBS.

The two monomers that make up PBS can be produced either using petrochemical feedstocks or through fermentation route. The conventional petrochemical route to PBS consists of the production of 1,4-butanediol using feedstocks such as acetylene and formaldehyde, the inorganic oxidation of 1,4-butanediol to succinic acid, and the condensation of the two petrochemically derived monomers. On the other hand, the key to renewable PBS is bio-based succinic acid produced by the anaerobic fermentation of bacteria or yeast. Some microorganisms used to produce succinic acid include *Anaerobiospirillum*, *Propionibacterium*, *Escherichia coli*, and *Saccharomyces cerevisiae*. The bio-based succinic acid can be easily reduced to form 1,4-butanediol, which is then reacted with the succinic acid to form 100% bio-based PBS.

4.1.2.2 Performance

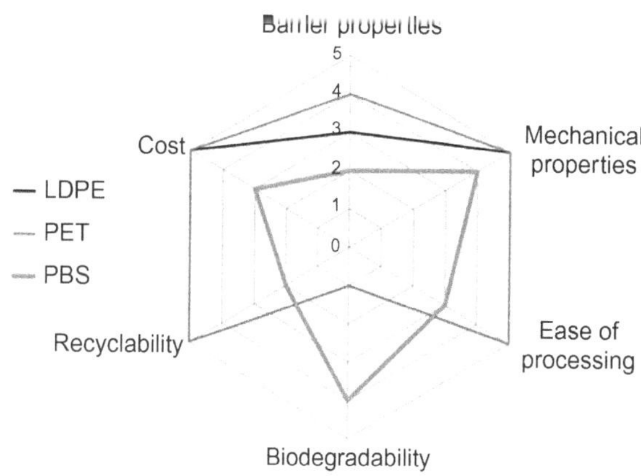

PBS is a semicrystalline polyester. Its thermomechanical properties depend on the degree of crystallinity of the polymer. Generally speaking, PBS has mechanical properties comparable to that of polypropylene (PP). It possesses good tensile and impact strength and good rigidity and hardness. However, some of its mechanical properties, such as tensile strength, show high dependency on the molar mass of the polymer, which declines as the polymer degrades. This variability is the reason why PBS is given a score of 4 for the *mechanical properties* metrics, lower than PE and PET. Despite properties comparable to PP, the cost of PBS, which today is double that of PP, prevents it from substituting PP in most of its intended applications, such as the construction or automotive industries.

4.1.2.3 Application

PBS finds application in the packaging industry in subsegments where biodegradation is of importance and oxygen permeability is of lower priority. Such application includes compostable waste bags, film packaging for foods with a short shelf life, and compostable pouches, plates, and cutlery. PBS' good degradability also renders it appropriate for deployment in the agriculture and forestry industry as mulch film, nets, and trappings.

PBS holds high potential to be deployed into the medical industry. The plastic presents better processability than PLA, better toxicity profile than PLA, and exhibits mechanical properties comparable to PP and PE. The use of PBS in several load-bearing applications within the medical industry, such as bone and cartilage repair, is especially attractive. However, because of PBS' poor flexibility, the blending of PBS with PLA might be required.

4.1.2.4 Players

As can be expected, the main players in the PBS space are the producers of succinic acid. There are currently four major producers of bio-based succinic acid: BioAmber, Myriant, Reverdia, and Succinity.

BioAmber produces succinic acid from genetically modified organisms using one of two platforms: a first-generation *E. coli* or a second-generation yeast. The first-generation *E. coli* is currently used in the 3000 TPA demonstration-scale facility in Pomacle, France. The company is partnered with PTTMCC Biochem (PTTMCC), a joint venture (JV) between PTT PLC and Mitsubishi Chemical, to construct a PBS plant in Thailand. The construction started in 2012, and upon completion, BioAmber will supply PTTMCC with succinic acid on an exclusive basis.

Myriant uses metabolically engineered *E. coli* to ferment sugar to produce succinic acid. PBS is among the applications of succinic acid that Myriant has identified. Capacity-wise, the company is currently producing at its 15,000 TPA facility in Lake Providence, LA. Myriant has signed a JV with PTT Chemical for production of bio-based succinic acid to serve China and the Southeast Asian markets. Additionally, the company also supplies its succinic acid to Showa Denko, which intends to convert the chemical to PBS.

Reverdia develops succinic acid using a genetically modified strain of *S. cerevisiae*. Reverdia's succinic acid plays in different end applications such as PBS, with specific end application including utensils, disposable cups, packaging, and agricultural film. Today, Reverdia is producing succinic acid at its 10,000 TPA commercial facility in Cassano Spinola, Italy.

Succinity is a 50:50 JV between Carbion Purac and BASF. The company uses a proprietary bacteria, *Basfia succiniciproducens*, to convert a range of sugar feedstocks to succinic acid. In 2014, the company started the operation of its 10,000 tpy commercial plant that is located at the Corbion Purac site in Montmeló, Spain.

4.1.3 Polyhydroxyalkanoate

4.1.3.1 Introduction

PHAs are linear polyesters produced by bacterial fermentation. Unlike PLA and PBS, each of which refers to one specific type of compound, PHA is a big family of polymers. More than 150 different monomers of hydroxyalkanoate have been discovered, which can be combined to yield polymers with diverse properties. Within the PHA family, poly-3-hydroxybutyrate (PHB) was one of the first to be characterized. Today, PHB is the PHA that has received the most industrial attention.

A variety of PHA-producing bacteria have been identified. These bacteria produce PHA using renewable waste feedstock, including lignocellulosics, gas, vegetable oils, and fatty acids. A common industrial process to produce PHA consists of fermentation, isolation, and purification. In the first step, the feedstock is introduced into the reactor until bacterial growth and internal PHA accumulation are complete. Subsequently, the cells are concentrated, and the PHA is extracted using organic solvent. The cellular debris that are also produced during this step are removed by solid–liquid separation. Finally, the dissolved PHA is precipitated by the addition of an alcohol.

4.1.3.2 Performance

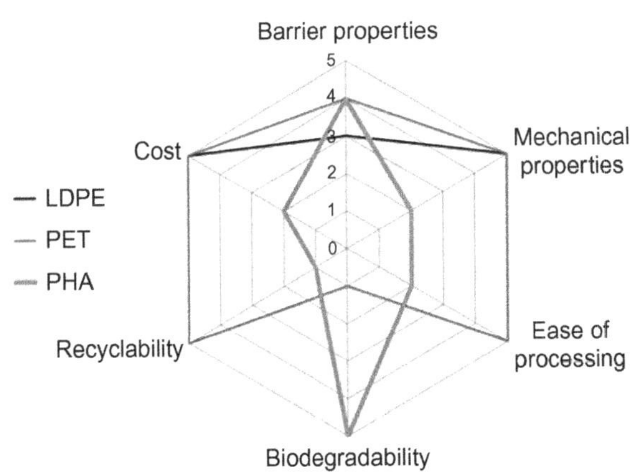

PHA's thermomechanical properties vary and largely depend on the constituent monomers. However, in general, pure PHA suffers from poor mechanical performance. This drawback prevents its widespread use in applications where structural strength is of importance, such as most weight-bearing applications in the construction or automotive industry.

PHB, the most commercially important PHA, has a good resistance to moisture and aroma barrier properties. However, the polymer exhibits relatively low melting points that range from 40°C to 180°C, thus excluding the polymer from going into applications such as retail packaging, where heat-based sterilization is often necessary.

4.1.3.3 Application

Due to its poor thermomechanical properties, for most of its target applications, PHA is blended with other polymers. This is the strategy executed by Metabolix, for example, that collaborates with Samsung Fine Chemicals to create PHA/PBS/PBAT (polybutyrate adipate terephthalate) blends. Most of the PHA producers today target compostable films/bags as their principal applications. PHA's biodegradability and low-to-nonexistent toxicity opens up many potential applications for PHA within the medical and the pharmaceutical industries. These applications include sutures, bone plates, grafts, and a wide variety of implants.

4.1.3.4 Players

Metabolix produces PHA using undisclosed genetically modified microbes that use PHA as energy reserves. Whereas pure PHA (Mirel) was Metabolix's former primary focus, the company now concentrates on developing composites that combine PHA with other polymers, such as starch, PBAT, and polybutylene terephthalate. The company's JV with ADM, dubbed Telles, was dissolved in 2010; today, the company is still scouting for commercial production facilities.

Meredian uses canola oil to produce PHA. The company extracts canola oil from the seeds using its solvent-free Agrocrush technology and then feeds the oil to its proprietary microorganisms that ferment it to produce varying types of PHA. The company's business model is to sell its PHA pellets to companies within the single use consumables and packaging space, as well as hygienic articles (eg, diaper components), fibers, and adhesives. The company's current facility in Bainbridge, GA, has a capacity of 3000 TPA.

Newlight Technologies develops a technology to produce PHA from methane-consuming bacteria (methanotrophs). Newlight focuses on providing PHA resins at specification. Newlight's resins are currently used in a range of applications including items like consumer goods, containers, caps, and bags. In early 2014, the company announced that it has scaled its technology to a small commercial facility with a nameplate capacity in the single-digit million pounds per year range (hundreds of TPA).

4.1.4 Polyethylene Furanoate

4.1.4.1 Introduction

Polyethylene furanoate (PEF) is a polymer made from the condensation polymerization between monoethylene glycol (MEG) and 2,5-furandicarboxylic acid (FDCA). MEG is a common industrial chemical produced mainly from ethylene. FDCA, on the other hand, is a relatively novel chemical made from fructose using multistep catalysis process involving dehydration and oxidation steps.

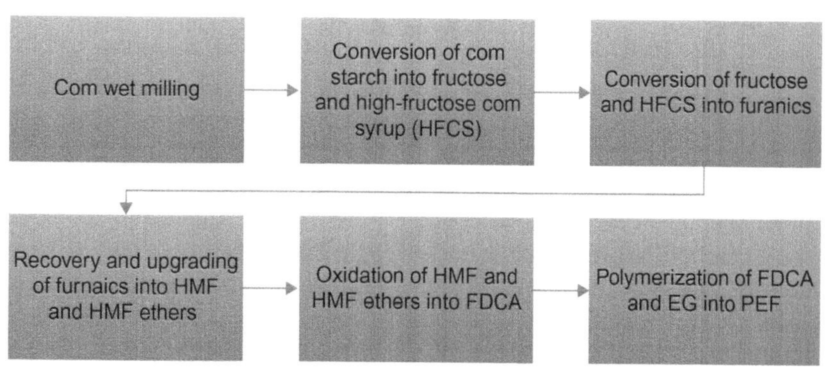

The formation of PEF closely resembles that of polyethylene terephthalate (PET), which is formed from the reaction between MEG and terephthalic acid. In fact, PEF itself is structurally similar to PEF and, as we will see later, is specifically designed to be a replacement for PET.

4.1.4.2 Performance

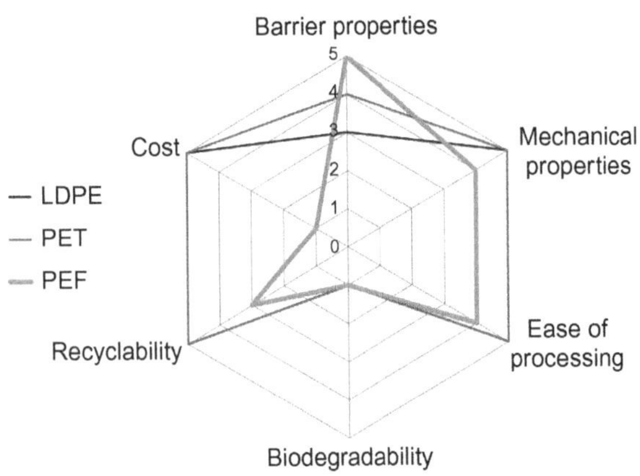

PEF's performance should be compared directly to PET, as the latter is the polymer that PEF is intended to substitute, especially for the packaging market. PEF outperforms PET mainly in barrier properties to oxygen (O_2) and water. Preliminary data from Avantium, a startup focused on scaling and commercializing PEF, states that PEF has an O_2 barrier up to 10 × better than that of PET and water barrier twice of that of PET. These very low permeability values against oxygen and water are the reason behind the perfect score of PEF in the *barrier properties* metric and render PEF a strong contender for the packaging market.

In addition to its strong barrier properties, PEF possesses favorable thermal properties that potentially allow it to be more easily processed compared to PET. PEF's melting point is 230°C, around 30°C lower than that of PET, allowing its melting and extrusion to occur at lower temperature, potentially resulting in reduction in process cost. Additionally, theoretical, early-stage PEF modeling suggests that PEF can have a net benefit in terms of environmental emissions and nonrenewable energy use.

Despite the attractive potential energy saving, however, efforts are still ongoing to verify PEF's compatibility with the existing PET machineries. The yet unresolved questions surrounding the possibility of direct processing of PEF in PET machineries is the reason why we gave PEF a score of 4 in its *ease of processing* metric.

4.1.4.3 Application

The barrier properties and relative easy processing give PEF a great potential to be PET substitute in the packaging industry. In fact, the most prominent company in PEF production, Avantium, explicitly positions its PEF as PET substitute in the packaging industry. Additionally, based on PEF's relatively good tensile strength, Avantium also aims to sell its PEF in the PET film and fiber markets in the future (with applications such as textiles).

4.1.4.4 Players

The main player in the PEF space is Avantium. Avantium's technology takes carbohydrate feedstock through two catalytic steps to form FDCA. The company then copolymerizes FDCA with MEG, forming PEF. Avantium partners with ALPLA, The Coca-Cola Company, and Danone to develop and commercialize PEF. Today, Avantium operates a 40 tonnes per year FDCA demonstration plant. The company aims to have its first commercial plant complete in 2017.

Another player in the space is Industrial Technology Research Institute (ITRI). ITRI is a Taiwan-based, nonprofit research and development (R&D) institution focused on applied technologies. One of the technologies that ITRI is working on is the development of hydroxymethylfurfural (HMF) and FDCA. The technology is similar to that developed by Avantium, with fructose as a feedstock and HMF as an intermediate. Today, the technology is still at laboratory scale.

References

[1] Resin prices, <http://plasticker.de/preise/pms_en.php?show=ok&make=ok&aog=A&kat=Mahlgut>.

[2] Bio-based plastics, <http://cen.acs.org/articles/92/i43/Biobased-Polymers.html>.

[3] Bio-based plastics, <http://www.progressbiomaterials.com/content/pdf/2194-0517-2-8.pdf>.

[4] PLA performance, <http://www.natureworksllc. com/~/media/Technical_Resources/Technical_Data _Sheets/TechnicalDataSheet_2003D_FFP-FSW_ pdf.pdf>.

[5] PBS performance, <http://digilib.lib.unipi.gr/dspace/bitstream/unipi/5575/1/Adamopoulou.pdf>.

[6] Lux Research Reports.

5 Additives to Design and Improve the Performance of Multilayer Flexible Packaging

Klaus Keck-Antoine[1], *Els Lievens*[2], *Joachim Bayer*[3], *John Mara*[4], *Duk-Suk Jung*[1] *and Sung-Leal Jung*[1]

[1]Songwon Industrial, Ulsan, South Korea; [2]Agfa Gevaert, Mortsel, Belgium; [3]Songwon International, Frauenfeld, Switzerland; [4]Songwon International Americas, Friendswood, TX, United States

5.1 Introduction

In order to design tailormade properties for an end application, the characteristics of the principal substrate must be modified. The properties of a polymeric substrate can be modified in a variety of different ways. The two principal approaches are:

- structural modification of the polymer
- incorporation of additives.

Structural modification of the polymer includes different polymerization processes for the same polymer (see chapters: Polyethylene and Polypropylene): copolymerization, polymer alloys, and the entire substitution of one polymer by another.

Additives are chemical substances that are incorporated into the polymeric substrate in order to stabilize, modify, or enhance the initial performance of the substrate. The general term "additive" can be subdivided into:

- stabilizer
- modifier
- filler.

Stabilizers are added to maintain the substrate's original molecular architecture (mechanical, optical, and organolephtic properties) during storage, conversion, and service life.

Modifiers, such as slip agents, antistats, antiblocking agents, lubricants, processing aids, etc., improve or alter polymeric substrate's performance and properties.

Fillers, having a different surface-to-volume ratio and being added at significantly higher amounts than the additives mentioned earlier, improve the physical properties or cost-dilute the polymeric substrate (eg, talc, chalk). In most cases, additive combinations are used to modify the polymeric substrate for a particular end-use.

5.2 Overview

Organic substrates can be modified significantly by incorporating additives. In order to achieve an optimum effect, the incorporation method and the additive chemistry must be considered.

Additive incorporation has two general possibilities for multilayer films. First, the additives can be incorporated into the organic substrate after synthesis (Appendix 5.2, Tables A5.2.1–5.2.4). Second, the additives can be incorporated into the layers of a multilayer film (Appendix 5.1, Table A5.1.1).

The different effects and how to influence those are summarized below. The general comments on additives should also be considered.

5.3 Additive Handling, Addition, and Dosing

Modifying polymer properties via additives requires incorporating them into the polymer. Additives can be incorporated at various points during the lifetime of a polymer.

The specific additive and its most appropriate physical form have an impact on the most appropriate addition point during a substrate's life cycle. Fig. 5.1 is a generic representation that shows the principal addition points in the life cycle of a polymeric substrate from the monomer stage through to the end article (eg, film) [1,2].

Multilayer Flexible Packaging. DOI: http://dx.doi.org/10.1016/B978-0-323-37100-1.00005-3

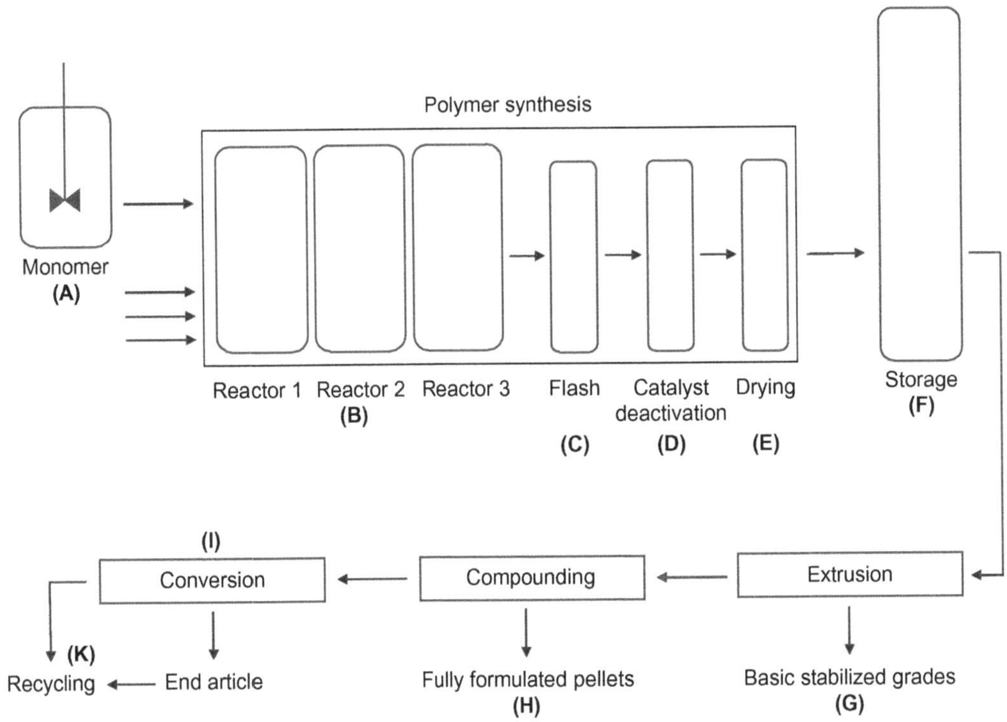

Figure 5.1 Addition points for additives during the lifetime of a substrate.

Depending on the substrate, additives can or already have been incorporated into the monomer (addition point (A)). Selected additives (eg, inhibitors for vinyl chloride monomer, the monomer for polyvinylchloride (PVC)) are designed to be depleted or physically lost during further operations. Depending on the polymerization technology and the substrate, additives can be incorporated at various points during the synthesis, either into the reactor (B) or during immediate postreactor operations, for example, immediately into the reactor discharge, solvent flash-off, or polymer drying (C–E). Only occasionally, additives are incorporated into the storage silo (F). Later on, additives are incorporated via extrusion to yield basic grades (G). Compounds equipped for specific end-use applications or conversion processes are manufactured during a compounding step (H). This compounding step can either be integrated into the polymer manufacture or can be independent. In the first case, addition points (G) and (H) can be identical. The last possibility to incorporate additives is at the actual conversion step, for example, the film manufacturing step (I). If the end article is recycled, further additives can or might have to be added (K).

Especially for polyolefins, it is possible either to use fully formulated grades or, alternatively, to use basic additive grades and top-up additives during the conversion step. In the latter cases, it should be noted that knowledge about additives that are already present in the organic substrate is imperative in order to avoid overdosing or undesirable interactions between different additives or additive classes.

Any process to handle, add, or dose additives must be designed to ensure accurate dosing and accurate dispersion of, in most cases, numerous additives within the organic substrate. Insufficient dispersion typically results in a significant additive performance loss, especially on an industrial scale. Insufficient performance often is corrected by overdosing the corresponding additive in order to improve its performance. It should be noted that the physical form and chemical characteristics of individual additives can span a wide range. In many cases, the additive's basic physical form is not very suitable for an optimum homogenization in the organic substrate.

It is important to match the additive's physical form that will be used in handling and dosing to the industrial plant requirements. Various different physical forms are suitable for addition during synthesis of the organic substrate (depending on which addition point (A) through (F) [3,4]) is used. After synthesis, addition is predominantly via extrusion (G–K). The dominant physical forms used are:

- masterbatch (20–50% active ingredients in a carrier compatible with the organic substrate)
- one pack (100% active ingredient without a carrier).

When manufacturing multilayer films, it should be highlighted that not every additive needs to be added to every layer of the construct. Some additives, like UV additives, need only to be added to the skin layer (Fig. 5.2). Appendix 5.1 provides an overview of the necessity to equip the individual layers of a multilayer film with which type of additive.

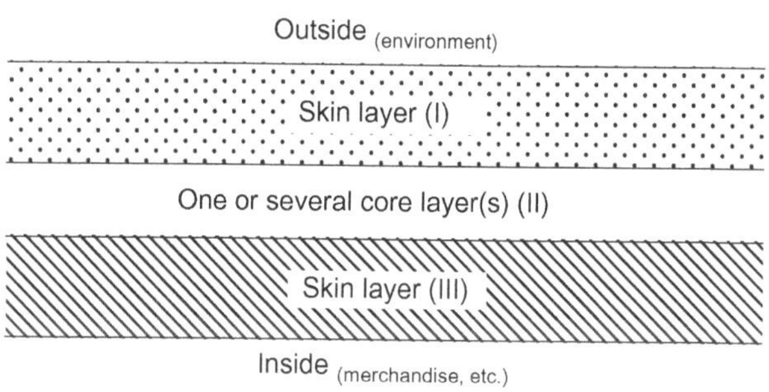

Figure 5.2 Generic multilayer film.

In the case when additives are added to selected layers only, migration effects must be considered. Like in any heterophasic (co-) polymer system, the additive solubility in the different phases results in a partition coefficient which, together with the concentration difference, determines the (potential) additive migration from one phase to the other. Thus, although additives might only be added to one layer, migration into neighboring layers may occur. This effect will decrease the additive concentration in the initial layer.

5.4 Additive Types and Principal Mode of Action

For practical reasons, additives are mainly classified according to the technical effect that should be achieved. This technical effect is generally referred to as *primary effect*. While a simple and practical classification, the approach struggles with additives that can be used for various effects. Nevertheless, this approach is used here. As mentioned earlier, additives can be subdivided into:

- stabilizers
- modifiers
- fillers.

Stabilizers can be further subdivided into:

- antioxidants
- UV stabilizers.

The specific *modifier* classes discussed in this chapter are:

- optical brighteners
- slip additives
- antiblock additives
- antistats
- others.

Additives will be selected based on the primary effect desired. It is advisable to modify this initial ranking based on secondary effects. The obtained ranking should be further "thinned out" according to the permanence in the substrate and industrial suitability needed. Last, but not least, the regulatory guidelines or "approval" status should be considered. General requirements for additive selection are listed below. For any given additive class and substrate, these requirements can be described in more detail. The relative importance of any of the listed requirements can vary case by case. Although this is a technical discussion, the requirement *cost* has been added because, in industrial practice, it is an important selection criterion.

- Activity (primary effect)
  - efficiency
  - molar activity (selected additive classes)
  - concentration
  - "process window"
- Interactions (secondary effect)
  - side effects at excessive concentrations
  - discoloration
  - chemical reactions with other additives
  - physical–chemical effects with other additives
- Permanence in the substrate
  - solubility
  - diffusion rate

- thermal stability (eg, at processing temperature)
- volatility
- chemical stability
- Industrial suitability
  - addition point
  - physical form
  - physical–chemical characteristics (eg, melt point)
  - storage stability
  - end application
- Cost
- Regulatory status
  - registration (local and global)
  - EHSA, FDA
  - indirect food contact.

The effect of an additive can be influenced by its concentration. Most additives exhibit a concentration dependency (Fig. 5.3). This behavior is characterized as a *saturation curve*. In phase (B), the primary effect increases with increasing concentration. In phase (C), the primary effect "flattens-out." Then, in phase (D), there is no further improvement. Within phase (B), the primary effect can be predicted semiquantitatively as a function of additive concentration. Within phase (D), the primary effect is not further improved despite a continued increase in the additive concentration. One might think that incorporating excess additive (a lot helps a lot) is an acceptable approach. However, in phase (D), secondary effects have a negative impact on the overall additive performance. One secondary effect is exceeding the additive solubility limit, which results in plate-out and blooming. Additionally, the cost increases without any beneficial effect.

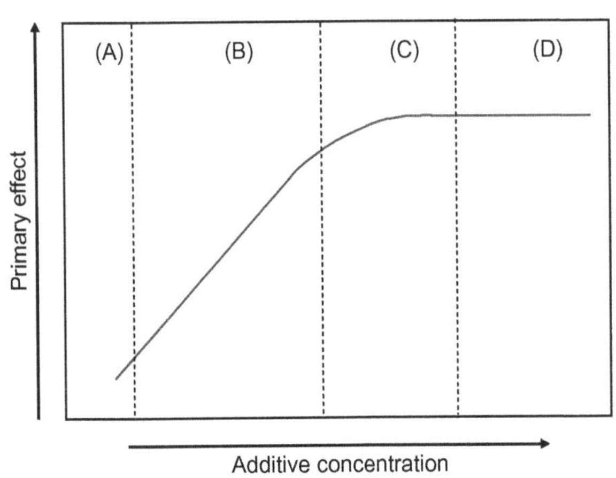

Figure 5.3 Concentration dependency of the primary effect of additives.

The actual concentration should be situated in phases (B) or (C) with a particular emphasis on the required safety margin relative to the (in-) accuracy of additive dosing on an industrial scale.

Phase (A) describes an often experienced phenomenon of a lower critical additive concentration, below which no measurable effect is detected. Phase (A) can be due to real scientific considerations (a real lower critical concentration limit) and/or dosing and dispersion limits on an industrial scale plant operation.

The additive classes will now be discussed by first outlining the principal mode of action, followed by substrate-specific considerations.

5.4.1 Antioxidants and Other Stabilizers

5.4.1.1 Effect

For the purpose of this discussion, the term "antioxidant" is used to describe all stabilizer classes that are utilized to protect organic substrates from thermal and thermooxidative degradation during storage, processing, or conversion and service life in the absence of ultraviolet light (sunlight) and maintain its initial properties (eg, molecular architecture, appearance, and organolephtic properties). The complex degradation chemistry is generally described in Ref. [5]. Substrate-specific degradation patterns can be found in Ref. [6].

The effect of antioxidants is essentially a bulk effect although, during service life, degradation will start where the organic substrate contacts the environment (oxygen diffusion).

5.4.1.2 Principle

On one hand, for any given organic substrate, the requirements for antioxidants used during processing or conversion are quite different and sometimes contrary to the requirements during service life. On the other hand, the degradation pattern strongly depends on the organic substrate. Therefore, the antioxidant (or stabilizer classes) needed vary considerably for polyolefins (polypropylene, polyethylene), polyester, PVC, and polyamide (to mention only the major substrates for flexible packaging).

A very generic principle is outlined in Fig. 5.4. During processing and conversion, the initial properties of the synthesized organic substrate start to deteriorate immediately if the organic substrate is neither intrinsically stable nor stabilized (case (A)). An ideal stabilizer retains the properties of the organic substrate throughout the entire processing or conversion step (B). In reality, most stabilizers or stabilizer combinations are only able to delay the onset of measurable degradation and slow down the degradation rate in order to maintain the organic substrate within specifications throughout the processing or conversion step (C).

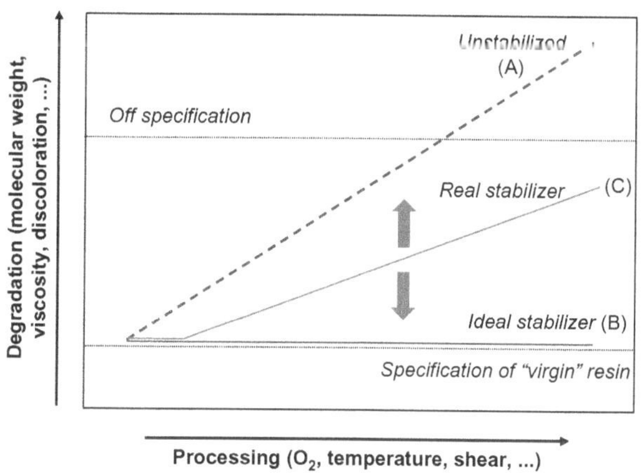

Figure 5.4 Principle of the utilization of antioxidants (and other stabilizers) during processing or conversion.

5.4.1.3 Influencing the Substrate Stability During Processing or Conversion

Adjusting the processing stability is necessary when the organic substrate degrades at the processing conditions in the selected equipment (Table 5.1). If no changes occur during processing or conversion, a stepwise reduction of the antioxidant and stabilizer concentration might be possible within the limitations outlined in Fig. 5.3. It is important to understand the complete chain of use for the material and have a product that will perform properly during its projected life span.

Protecting the organic substrate during service life is illustrated in Fig. 5.5. The gray area represents the service life in general. The area is limited by time (expressed in days, weeks, and eventually months) and a defined acceptable deterioration of properties. Both time and properties are mainly determined by the end application. Stabilizer addition (case (C)) typically results in an induction period during which no or only minor changes in the organic substrate are detectable. As the stabilizer is depleted, changes occur until the specification limit of the end article is reached. The slope of these changes can, in selected cases, be lower compared to the unstabilized case (A).

Table 5.1 "Quick Reference Guide" to Influence Processing Stability

Issue/ Observation	Corrective Action
High change in molecular weight	• Lower processing temperature and shear • Eliminate/reduce oxygen • Use an organic substrate with different initial molecular weight • Increase stabilizer concentration (until saturation) • More efficient stabilizers (substrate specific) • Synergistic stabilizer mixtures (substrate specific)
High discoloration	• Substrate specific • Use of optical brightener

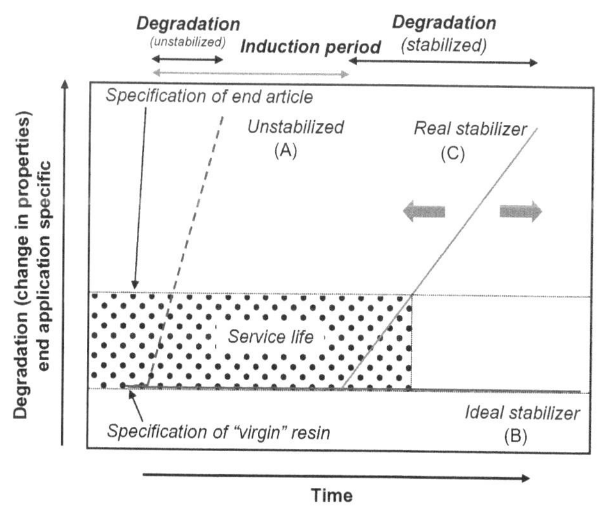

Figure 5.5 Principle of the utilization of antioxidants (and other stabilizers) during service life.

5.4.1.4 Influencing the Substrate Stability During Service Life

The lifetime of the organic substrate should match (including a safety margin) the desired service life of the plastic article. Table 5.2 highlights several alternatives, involving stabilizers or other option, to adjust the lifetime to the requirements.

Table 5.2 "Quick Reference Guide" to Influence the Service Life of Plastic Articles

Issue/Observation	Corrective Action
Time to mechanical property deterioration insufficient	• Increase thickness of plastic article • Surface coating • Substitute with intrinsically more stable substrate • Increase stabilizer concentration (until saturation) • More efficient stabilizers (substrate specific) • Synergistic stabilizer mixtures (substrate specific)
High discoloration	• Substrate specific

The detailed stabilization strategies applied during processing or conversion need to be discussed substrate by substrate.

5.4.1.5 Polyolefins (Polypropylene and Polyethylene)

The term polyolefin (PO) comprises polypropylene (PP, mainly homopolymers and also random copolymers) and the different polyethylenes (PE), namely high density (HDPE), linear low density (LLDPE), and low density (LDPE). Although all polyolefins degrade in a thermooxidative pattern only, the intrinsic stability varies [5]. The general ranking is as follows.

It should be noted that PP and some PE types degrade exclusively via a molecular weight reduction, whereas other PE grades show both molecular weight reduction and cross-linking (molecular weight increase). In industrial practice, the intrinsic stability depends on the specific polymerization process and catalyst system used.

Polyolefin degradation is counteracted by using antioxidants. The fundamentals of antioxidants are summarized in Ref. [7].

5.4.1.5.1 Requirements

Apart from the general requirements, low discoloration of both substrate and antioxidant is desirable. For film or tape manufacturing processes that use a water bath to quench the film with water rather than chill rolls, antioxidants with low water-carry-over (w-c-o) effect are required. The stabilizer's thermal stability can vary depending on the conversion conditions which are defined by the polyolefin type and the conversion process (blown vs cast film).

5.4.1.5.2 Antioxidant Chemistry

Extrusion grade polyolefins are stabilized using the synergistic effect between a phenolic antioxidant (AO) and a phosphite (P). The antioxidant's main purpose is to protect the organic substrate during processing or conversion including the stretching step. The phenolic antioxidant concentration present for processing or conversion usually is enough to extend the organic substrate lifetime beyond the service lifetime requirement. The physical form of all antioxidants recommended in Table 5.3 is solid unless otherwise stated.

Depending on the catalyst system used, acid neutralizers must be added in order to avoid long-term corrosion effects in the processing equipment and improve stability during service life.

5.4.1.5.3 Influencing the Antioxidant Effect

Case (A) in Table 5.4 can be considered as typical multipurpose film stabilization, for example, biaxially oriented polypropylene (BOPP) double bubble and cast film. In the BOPP tenter process, organic substrates with high molecular weight are used, requiring a better stabilization package. This is achieved by increasing the overall concentration and shifting the phenolic antioxidant to phosphite ratio from 1:1 to 1:2. Better color is achieved by choosing a low discoloring phenolic antioxidant (AO-3) combined with reducing the stabilizer responsible for the discoloration (phenolic antioxidant; (C)). It must be pointed out that (C) can only be effective instead (and not on top of) (A) and/or (B). Therefore, knowledge of antioxidants already present in the

Table 5.3 Stabilizer Selection for Polyolefin Extrusion Grades

Antioxidant	PP	HDPE	LLDPE	LDPE	Remark
AO-1	•	○			Multipurpose
AO-2		○	•	•	Multipurpose, good solubility
AO-3	•	•			Reduced color
AO-4	•				Low water-carry-over
AO-37				•	Multipurpose
HYA-1	○	○	•		Excellent color
P-1	•	•	•		Multipurpose
P-2			•		Multipurpose, liquid
P-3	○	○	○	○	Excellent color, related stability P-4 & P-5
Others					

Filled circles (•) indicate a recommended use or application; open circles (○) indicate limited, partial, or conditional use.
Note: For the antioxidant codes used in this chapter, Table 5.19 lists their chemical name, commercial name, and a selection of suppliers.

Table 5.4 Recommendations for Polypropylene Extrusion Grades

Case	Effect	Additive	Load Level (ppm)	Remark
A	General purpose	AO-1	750	• Robust multipurpose blend
		P-1	750	
B	Severe processing	AO-1	750	• Increased stabilizer concentration • Balance shifted to processing
		P-1	1500	
C	Reduced color	AO-3	500	• Low discoloring phenolic antioxidant • Balance shifted to processing
		P-1	1000	
D	Low water-carry-over	AO-4	1000	• Balance shifted towards antioxidant • with lowest water-carry-over
		P-1	500	
E	Severe processing (Top-up during conversion)	AO-1	1000	
		P-3	1000	

Indicative starting point formulations from Ref. [8].

organic substrate is crucial. Low water-carry-over can be achieved by utilizing as much as possible a stabilizer not causing w-c-o (AO-4), which needs to be balanced with an increase in color (D). Case (E) is designed as top-up formulation at the conversion step in case an insufficient stability of the organic substrate is encountered at the film manufacturer level.

The recommendations for HDPE are essentially identical.

PE, in particular LLDPE, is stabilized differently for blown film (F) versus cast film (G). The increased stabilizer concentration is a result of the higher cast film processing temperature. The low discoloring phenolic antioxidant AO-3 has a very high melting point and is not very soluble in LLDPE.

Table 5.5 Recommendations for LLDPE Extrusion Grades

Case	Effect	Additive	Load Level (ppm)	Remark
F	General purpose	AO-2	300	• Robust multipurpose blend
		P-1/P-2	1200	• Blown film
G	Severe processing	AO-2	750	• Cast film
		P-1/P-2	1500	
H	Excellent color	HYA-1	750	• Nondiscoloring stabilizer
		P-1	750	

Indicative starting point formulations from Ref. [8].

Table 5.6 Recommendations for LDPE Extrusion Grades

Case	Effect	Additive	Load Level (ppm)	Remark
I	General purpose	AO-2	500	• Multipurpose and high solubility
K	Severe processing	AO-37	500	

Indicative starting point formulations from Ref. [8].

As a consequence, excellent color can only be achieved by using an intrinsically nondiscoloring stabilizer (HYA-1, (H)). Table 5.5 gives an overview.

As LDPE is the most stable polyolefin type, lower antioxidant levels are suitable (Table 5.6).

5.4.1.5.4 Incorporation

Antioxidants are typically incorporated during the extrusion step directly after polymerization (addition point (G) in Fig. 5.1). As a major exception, formulation (E) from Table 5.4 should be used at the conversion step (addition point (I)) in masterbatch form or to stabilize the masterbatch carrier itself (Appendix 5.2).

5.4.1.6 Polyester

The polyester group comprises polyethylene terephthalate (PET), polybutylene terephthalate (PBT), and, more recently (although scientifically not entirely correct), the biodegradable polylactide (PLA, polylactic acid). In general, PET is considered a bit more stable against degradation during conversion and service life than PBT. Both show the following degradation pattern in descending order of importance:

- hydrolysis
- thermal (mainly during conversion)
- thermooxidative (conversion and service life in particular at high temperatures).

During processing, the hydrolysis reaction kinetics are significantly faster than the thermal and thermooxidative degradation kinetics. Hydrolysis is defined as the ester bond decomposition due to the moisture and water which leads to a significant (melt) viscosity and molecular weight decrease. Hence, the primary stabilization strategy for polyester during processing is to suppress hydrolysis. Hydrolysis can be retarded by adding chemicals such as:

- epoxide (a cyclic ether with only three ring atoms)
- carbodiimide (functional group consisting of the formula $N=C=N$)
- phosphorus acid ester.

However, the most efficient way to prevent hydrolysis does not involve additives but simply good drying.

Thermal decomposition during conversion also results in lower molecular weight, lower viscosity, and more end-standing carboxyl groups. End-standing carboxyl groups have a strong negative impact on hydrolytic stability. Thermal degradation is strongly influenced by metal compounds used as catalysts during the synthesis of the organic substrate. Hence, organic phosphorus compounds are used as complexing agents. Selected molecules include among others:

- bis-stearyl-pentaerythritol-diphosphite (additive code P-7)
- triphenylphosphate (P-10)
- trimethylphenylphosphate (P-11).

Discoloration during processing or conversion is very difficult to control and often only possible with the help of optical brighteners (see Table 5.10 below).

PET is sufficiently stable against *thermooxidative* degradation during processing, so that no antioxidants (as outlined earlier) need to be added. In contrast, PBT benefits from the addition of phenolic antioxidants (eg, additive code AO-2) and phosphite (P-1).

Top-up stabilization during processing or conversion is recommended (Table 5.7).

Table 5.7 Recommendations for Polyester Grades

Substrate	Stabilizer	Load Level (ppm)	Remark
PET	P-3	3000	• Phenol free
	AO-1	600	
	P-1	2400	
PBT	AO-2	800	
	P-1	3200	

Indicative starting point formulations from Ref. [8].

5.4.1.6.1 Requirements

No specific requirements other than mentioned earlier need to be considered.

5.4.1.6.2 Incorporation

Chemicals to retard hydrolysis are added during compounding. Complexing agents to guard against thermal decomposition during processing and conversion are most efficiently added during synthesis. Appendix 5.2 compiles an overview.

5.4.1.7 Polyvinylchloride

Flexible PVC is, of all the substrates that are discussed in detail in this chapter, the most complex and the most difficult to stabilize against degradation mainly during processing or conversion and, to a certain degree, during service life. Even nonindustrial, artificially pure PVC must be considered intrinsically unstable. Polyvinylchloride as an organic substrate degrades via various mechanisms. For flexible PVC, the plasticizer stability must be considered as well. On top of that, (flexible) PVC stabilizer packages tend to be more customer and equipment specific than stabilizer packages for other organic substrates.

(Flexible) PVC degrades according to the following mechanism:

- dehydrochlorination in the absence of oxygen
- thermooxidative degradation (resulting in dehydrochlorination) in the presence of oxygen
- secondary degradation reactions.

The degradation mechanisms of (flexible) PVC are described extensively in Ref. [9]. An effective stabilizer needs to either react with one of the key intermediates in the dehydrochlorination (allylic chloride) and/or scavenge liberated hydrochloric acid (HCl). The following chemical families are available:

- lead based
- tin based
- mixed metal–based
- metal-free.

Mixed metal–based heat stabilizers are the most used in flexible packaging film. They can be further broken down into:

- barium/zinc stabilizers
- barium/cadmium stabilizers
- calcium/zinc stabilizers.

While barium/zinc-based stabilizers are a valid alternative to cadmium-based stabilizer systems, they can be further replaced by calcium/zinc-based systems in order to be completely heavy-metal free. These systems are often liquid. The performance of a barium/zinc-based stabilizer package is dependent on the proper choice of costabilizer (synergist). These synergists are:

- phosphites
- epoxy compounds
- antioxidants.

Lubricants play an important role in the stabilization of PVC during processing or conversion. Due to the complexity of the formulations, Table 5.8 can only give a very rough guidance for a flexible PVC stabilizer system. Other formulations are possible.

Table 5.8 Generic Recommendations for Flexible PVC

Effect	Additive	Load level (ppm)	Remark
Heat stabilizer	BZ	35000	• Liquid, formulated system
Costabilizer	ESBO	30000	
Lubricant	Stearic acid	5000	

Plasticizer content up to 60% (indicative starting point formulations from Ref. [8]).

Plasticizers are incorporated into PVC in order to make it flexible. The plasticization mechanism is described in Ref. [10]. There are many chemical families available to achieve this effect, such as:

- phthalates (most common)
- trimelliates
- aliphatic dibasic esters
- polyesters
- epoxides.

A degrading plasticizer will impact the overall properties of flexible PVC and hence needs to be stabilized. The stabilization strategies employed depend on the plasticizer chemistry. The most common plasticizers are listed in Table 5.9 together with suggested starting point formulations for both processing or conversion and storage stabilization.

5.4.1.7.1 Incorporation

Due to the complexity of the formulations, the potential interactions between the components and the trend for customer specific formulations, all ingredients related to stabilization must be added at the same point, eg, during compounding (addition point (G) in Fig. 5.1).

Table 5.9 Recommendations for PVC Plasticizers

Plasticizer		Stabilizer		
Code	**Family**	**Type**	**Load Level (ppm)**	**Remark**
DOP	Phthalate	AO-2	1500	• General purpose
DINP	Phthalate	AO-8	1500	• Improved regulatory status
DINA	Adipate	AO-2	2000	• Improved low temperature brittleness
DOA	Adipate	AO-2	2000	• Food packaging

Indicative starting point formulations from Ref. [8].

5.4.1.8 Polyamide

Polyamide (PA) can be divided into aromatic and aliphatic polyamide and the latter further into:

- PA 6
- PA 6,6
- PA 4,6
- PA 11
- PA 12.

These notes on preventing degradation during processing or conversion and service life are limited to PA 6 and PA 6,6. Similar to polyester, PA exhibits the following degradation pattern:

- hydrolysis
- thermal
- thermooxidative.

Concerning hydrolysis, the same recommendations apply as for polyester. Polyamide used for film applications does not need to be specifically stabilized during processing. The intrinsic stability of PA against thermal and thermooxidative degradation is mainly influenced by the amount of carboxylic acid present in place of primary amino end groups. A further improvement can be achieved with lubricants. To a certain degree, yellowing must be controlled during processing via phosphites or optical brighteners (Table 5.10).

Table 5.10 "Quick Reference Guide" to Improve Processing Stability of PA

Issue/ Observation	Corrective Action
High change in molecular weight	• Decrease carboxylic end groups • Increase amino end groups • Incorporate lubricants
High discoloration	• Incorporate phosphite (P-1 or P-10) • Incorporate optical brightener
Hydrolysis	• Improved drying • Incorporate hydrolysis inhibitor (eg, polymeric carbodiimide)

Stability during service life can, in principle, be improved by three alternative stabilizer and/or antioxidant classes:

- copper salts (combined with halogen ions or phosphorus compounds)
- aromatic amines
- phenolic antioxidants (eg, AO-21).

While the first two stabilizer classes are very effective (primary effects), both impart negative secondary effects which prohibit their use in packaging applications. For the modest service life stability needed in packaging film applications, phenolic antioxidant AO-21 offers a good balance of properties. A detailed differentiation of the three stabilizer classes is given in Refs. [5,6]. Table 5.11 summarizes the recommendations.

5.4.1.8.1 Incorporation

Phosphites are typically added during compounding (addition point (H) in Fig. 5.1). The phenolic antioxidant is best added during polycondensation (B) or as additional top-up stabilization during compounding (H).

5.4.2 UV Stabilizers

5.4.2.1 Effect

Organic substrates tend to degrade during service life due to various mechanisms. Along with thermal, thermooxidative (induced by oxygen and heat), and hydrolysis initiated degradation (discussed earlier), UV light can initiate oxidation. Organic substrates which are used outdoors or exposed to lamps emitting UV radiation are subject to photooxidative degradation. An overview over photooxidation is provided in Ref. [11]. UV stabilizers are one of several tools to prevent and retard photooxidation.

Photooxidative degradation is a nonhomogeneous effect that always starts at the surface (at the side exposed to UV light) and propagates throughout the sample.

Using antioxidants can lead to an improved resistance against photooxidation. It must be further noted that a suitable stabilization during processing or conversion is mandatory for optimum stability against photooxidation.

5.4.2.2 Principle

Although four principal classes of UV stabilizers exist, only two are suitable for flexible packaging applications:

- *UV absorption*
- *free radical scavenging*

Table 5.11 Recommendations for Polyamide Extrusion Grades

Case	Effect	Additive	Load Level (ppm)	Remark
A	Color suppression (processing)	P-1	2000	• Multipurpose
		P-10	2000	• Multipurpose, low volatility
B	Good balance processing & service life	AO-21	1000	
		P-1	1000	

Indicative starting point formulations from Ref. [8].

- *excited state quenching*
- *hydroperoxide decomposition.*

Ultraviolet absorption (UVA) stabilizers follow a physical principle and essentially depend on the substrate thickness, not the substrate type. As the name implies, UV absorbers absorb UV light and dissipate it as less harmful longer wavelength radiation (heat). UVA can be used to stabilize the packaging material and the merchandise that it protects.

the ability to manufacture noncolored and/or transparent packaging material. Another way to absorb UV energy is with *h*indered *a*mine (*l*ight) *s*tabilizers (HA(L)S), hydroxybenzoates and selected antioxidants to provide free radical scavenging.

Not every additive class can be used in every substrate discussed in this chapter. Table 5.12 gives an overview.

Table 5.12 Use and Purpose of UV Stabilizers

Class	Subclass	PP	PE	PVC	PA	PET	Substrate Protection	Content Protection
UV absorption	Pigment	•	•	•	•	○	○	•
	Dye	○	○	•	•	○		
	UVA	•	•	•	•	•	•	•
Radical scavenger	HA(L)S	•	•	○	•		•	
	Hydroxybenzoate	•	•	○		○	•	
	Antioxidant	○	○		○		•	

Filled circles (•) indicate a recommended use or application; open circles (○) indicate limited, partial, or conditional use.

Free radical scavenging follows a chemical principle that is related to antioxidants. It is substrate dependent and, in particular, suitable for substrates that degrade by a free radical autooxidative mechanism. This is the only mechanism that can be used to stabilize the organic substrate used as the packaging material.

The principle, illustrated in Fig. 5.5, also applies to UV stabilizer utilization during the service life of an organic substrate.

5.4.2.3 Requirements

No specific requirements other than mentioned earlier need to be considered.

5.4.2.4 Chemistry

UV absorber additives are either pigments/dyes, which absorb in the visible and the UV light spectrum or organic UV absorbers that absorb mainly in the UV range. The specific pigment or dye can have a positive or negative effect on the photooxidative stability of an organic substrate. A particular pigment worth mentioning is carbon black, which imparts high UV stability and limits

5.4.2.5 Influencing the Stability Against Photooxidation

Apart from increasing the UV stabilizer concentration up to the saturation limit, stability against photooxidative degradation can be increased by:

- thicker sample (film)
- best possible protection during processing or conversion
- surface coating with a protective layer (skin layer (I) in Fig. 5.2).

Table 5.13 summarizes starting point formulations for various substrates. It should be noted that stability requirements against photooxidation in packaging applications are often low.

5.4.2.6 Incorporation

UV stabilizers are typically added during the compounding step as top-up stabilization of a base grade (addition point (H) in Fig. 5.1) or even later during the conversion step (I). UV stabilizers can be added evenly into all layers of a multilayer film. Alternatively, a skin layer (I) can be heavily loaded with UV absorber, thus filtering out UV radiation in the top layer, thus lower layers do not need to be

Table 5.13 UV Recommendations for Various Substrates

Substrate	Expected Service Life (months)	UV Stabilizer			Remark
		Type	Molecule	Load Level (ppm)	
PP	≤12	HA(L)S	HAS-3	1000	• Multipurpose
	≤12	HA(L)S	HAS-3	600	• Improved UV & reduced secondary effects
		HB	HB-1	400	
	≤12	HA(L)S	HAS-5	800	• Protection of content
		UVA	HBP-1	1200	
PE	≤12	HA(L)S	HAS-5	1000	• Multipurpose
	≤12	HA(L)S	HAS-5	1000	• Protection of content
		UVA	HBZ-1	1000	
PET	≤12	UVA	HBZ-7	3000	• Protection of content
PVC	≤12	HA(L)S	HAS-7	1000	• Multipurpose
		UVA	HBZ-5	1000	
PA	≤12	HA(L)S	HAS-3	1000	• Multipurpose
		UVA	HBZ-7	1000	

Indicated starting point formulations from Ref. [8] based on geographical regions with an annual radiation of 70–90 kLy/a).

stabilized. This second approach does not work well for very thin films.

Appendices 5.1 and 5.2 give an overview over the addition of UV stabilizers (excluding pigments and dyes).

5.4.3 Optical Brighteners

5.4.3.1 Effect and Principle

Organic substrates that impart a slight yellowish color can be modified to appear whiter and brighter by increasing reflected bluish light (in the range of 400–600 nm). One of several options is to utilize additives that absorb in the UV range and reemit the energy at higher wavelength. This effect is called fluorescence and the additives achieving this effect are called optical brighteners or fluorescent whitening agents.

Fluorescent whitening of organic substrate is a surface effect. Optical brighteners are a defensive tool to mask unpreventable discoloration of the organic substrate (or any additive component).

5.4.3.2 Requirements

Optical brighteners are partially efficient at rather low load levels (≤10 ppm). Requirements for homogenization and dispersion are very high. Therefore, optical brighteners tend to be added along with other materials if technically feasible. Limited compatibility and limited photostability lead to a limited service life of the additive and the achieved effect.

Due to the difficult handling and limited stability, optical brighteners should be considered only after approaches described earlier are not successful in reducing discoloration.

5.4.3.3 Chemistry

Various chemical classes are available.

5.4.3.4 Influencing the Effect

The effect of optical brighteners is concentration dependent and the saturation level is reached at low load levels. The performance depends on the processing or conversion conditions and is substrate specific. Additives that absorb UV light (UV absorbers and selected pigments) will decrease the efficiency of optical brighteners.

Table 5.14 proposes some starting point formulations.

Table 5.14 Recommendations for Optical Brighteners

Substrate	Pigmentation	Optical Brightener Type	Optical Brightener Load Level	Remark
PP/PE	Transparent	FWA-1	10	• Compatibility is critical
PET	Transparent	FWA-1	200	
PVC	Transparent	Various	75	• Solubility and migration in plasticizer to be considered
	TiO_2	Various	≤1000	
PA	Transparent	FWA-1	200	

Indicative starting point formulations from Ref. [8].

5.4.3.5 Incorporation

Due to the (very) low concentrations used, the techniques used to incorporate optical brighteners vary considerably. Adding the additive alone should be avoided wherever possible. For polyester, selected optical brighteners are incorporated into the monomer, while for flexible PVC, optical brightener is added preferably predispersed in the plasticizer which is added during compounding.

5.4.4 Slip Additives

Slip is when polymeric films slide parallel over each other. Slip is a surface effect.

Slip is quantified via the coefficient of friction (COF). If films have a high COF, individual film layers have a high surface friction and tend to stick together instead of sliding over one another. The films do not "slip." This phenomenon typically makes the handling, use and conversion of films rather difficult. It can result in decreased line speeds (lower productivity) and/or wrinkled film (higher waste). In order to overcome this issue, slip agents are added.

5.4.4.1 Principle

Slip additives can be divided in two fundamental classes:

- migrating
- nonmigrating.

Migrating slip additives are by far the most common class and must be used above their solubility limit in the polymeric substrate. Slip agents have a part that is soluble in the organic substrate and a part that is insoluble. During processing (in the molten phase), slip additives (as overall effect) are soluble and homogeneously dispersed in the organic substrate. Upon crystallization, the solubility limit is exceeded and the slip additive migrates from the matrix toward the surface.

Initially, the slip additive is homogeneously dispersed in the film and, consequently, its surface concentration is low (Fig. 5.6; Phase A) and the COF is high. Due to the limited compatibility with the organic substrate, the slip additive migrates to the surface, thus the concentration at/on the film surface increases (Phase B) and the COF decreases. At equilibrium (Phase C), a continuous coating of at least one layer of slip additive molecules is formed at/on the film surface and the COF reaches its minimum value. The two major performance criteria are the migration speed to the surface (time until slip effect occurs) and the slip effect achievable (minimum COF). These two criteria are determined by:

- the polymeric substrate (crystallinity)
- slip additive (in-) compatibility (aliphatic chain lengths for fatty acid)
- slip agent concentration
- temperature.

Nonmigrating slip additives are only used in very specific cases. These slip additives are applied externally. An immediate slip effect is achieved, reducing the principle outlined in Fig. 5.6 to "Phase C." Typical applications are very tacky substrates and the outer layer of multilayer films. Due to its limited applicability, this principle is not discussed further.

5.4.4.2 Requirements

Apart from the general additive requirements, migrating slip additives must have a designed low

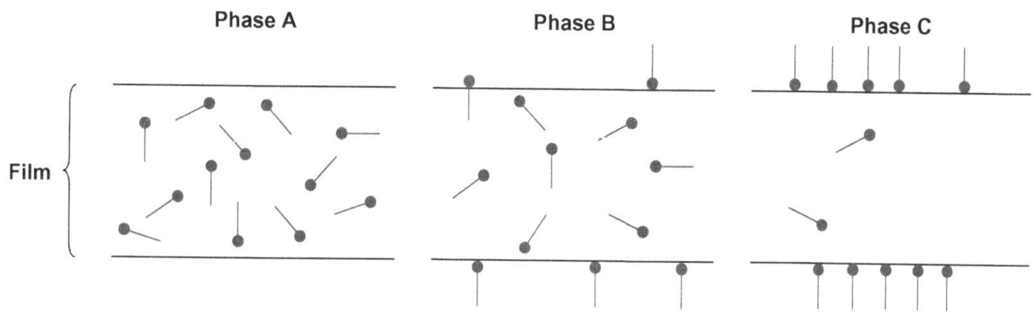

Figure 5.6 Principal mode of action of slip additives.

Table 5.15 Classification of Different Fatty Acid Amides

	Amide Type	Migration Speed	Low COF Achievable	Thermal Stability	Other Effects/Notes
Erucamide (E)	P	++	+++	++	AB (limited)
Oleamide (O)	P	+++	++	+	
Ethylene bis-oleamide (EBO)	S, B	+	+	+++	AB in polar polyolefin copolymers; multilayer film
Stearyl erucamide (SE)	S	+	+	+++	AB Multilayer film
Oleyl palmitamide (OP)	S	+	+	+++	Multilayer film

P, primary amide; S, secondary amide; B, bisamide; AB, antiblock effect. +++, very good; ++, medium; +, poor but established.

compatibility with the polymeric substrate. Chemically, migrating slip additives are mainly fatty acid amides that are derived from natural sources (vegetable or animal based). As with every additive derived from natural sources, purity (manifested as odor, flavor, and color) needs to be defined.

5.4.4.3 Chemistry of Slip Additives

Most commonly, migrating slip agents are fatty acid amides. Various structural modifications are available, differing mainly in the aliphatic chain length, (un-)saturation, and the amide (primary, secondary, or bisamide). Oleamide (O) and erucamide (E) are most often used. However, for multilayer films, fatty acid amides with longer aliphatic chains and correspondingly slower migration speed are favored, for example, ethylene bis-oleamide (EBO), stearyl erucamide (SE), and oleyl palmitamide (OP). Table 5.15 shows a comparison.

Chemically, migrating slip additives are high-molecular-weight organic materials such as polysiloxane. While not migrating, upon prolonged contact, the additive can nevertheless transfer onto a neighboring layer.

5.4.4.4 Influencing the Slip Effect

In general, slip additives should always be tested in the presence of the entire additive package in the film. Various coadditives influence the performance of slip agents. When combining slip additives and antiblock additives, there can be an interaction where increasing one additive actually enhances the effect of the other additive.

The slip effect is influenced by the slip additive concentration. Increasing the concentration will reduce the COF until an equilibrium state is obtained. At (too) high concentrations, slip additives plate onto process equipment and negative interactions with adhesion and lamination operations might be observed. Further, an increased film haze will likely be observed.

The slip properties can also be improved by:

- decrease in film thickness
- antiblocking agents (low concentration)
- fast migrating substances (eg, antifogging agents and selected antistats provide a carrier effect).

The slip properties can be negatively affected by several interactions. Interactions affecting slip properties can occur through:

- polar additives (chemical interaction)
- other surface modifying agents (eg, stearates or antifogging agents; competition for space on the film surface)
- too high slip additive concentration (deposits on processing equipment, printing issues)
- changing the substrate polarity
- increase in film thickness
- insufficient thermal stability of slip additive at processing temperature.

Starting point formulations are given in Table 5.16.

5.4.4.5 Incorporation

Slip additives can alternatively be added during the extrusion, compounding, or conversion step (Fig. 5.1; addition points (G), (H), and (I)). The most suitable physical form depends on the selected addition point. Migrating slip additives should be added mainly to the skin layer but sometimes as well to the core layer, while nonmigrating slip additives are added to the skin layer only (Appendix 5.1).

5.4.5 Antiblock Additives

5.4.5.1 Effect

Blocking is when adjacent film layers (made of organic substrate) stick to one another. Blocking is a surface effect.

The effect is quantified via the force needed to separate two film layers under controlled conditions. A high force results in more difficult opening of blown film tubes after extrusion (lower productivity) and layer separation after storage (reblocking) [12]. In order to overcome this issue, antiblock additives are added.

5.4.5.2 Principle

Antiblock additives can be divided into two fundamental classes:

- inorganic
- organic.

Chemically inert, inorganic antiblock additives are most commonly used. Inorganic antiblock additives migrate to the film surface, partially stick out and create a microroughness of the film surface. Fig. 5.7 illustrates this principle.

The detailed mechanism of how organic antiblock additives work is not yet understood. It is

Table 5.16 Recommendations for Slip Additives

Substrate	Effect	Additive Type	Load Level	Remark
PP	Multipurpose	Erucamide	900	
	Metallized film	–	–	• Interaction with surface treatment
PE	Multipurpose	Erucamide	1000	
PET	Multipurpose	Various	2000	
PVC	Multipurpose	Erucamide	2000	

Indicative starting point formulations from Ref. [8].

thought that a barrier layer is formed on the plastic film surface, thus inhibiting the two adjacent plastic film layers' adhesion. Their usage is limited. Organic antiblock additives were partially discussed earlier and will not be further mentioned here.

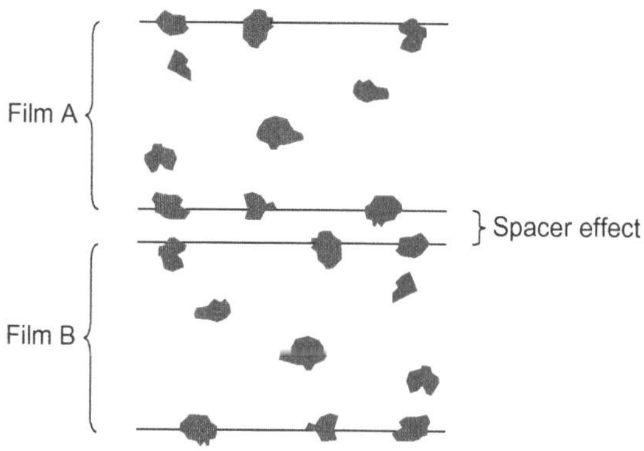

Figure 5.7 Principal mode of action of inert, inorganic antiblock additives.

5.4.5.3 Requirements

Apart from the general additive requirements, inorganic antiblock additives must be chemically inert with the organic substrate. Particle size and homogeneous dispersion has a major influence on the antiblock effect. The right balance between particle size and optical properties (low haze) is crucial. Abrasion effects to the surface of processing or conversion equipment need to be considered as well.

5.4.5.4 Chemistry

Several chemical classes of inorganic antiblock additives are available. Every class has its own performance profile. Table 5.17 selects the antiblock additives for the major organics substrates used in packaging. Table 5.18 recommends load levels for selected antiblock additives.

Table 5.17 Selection Guide for Antiblock Additives Used in Packaging

Substrate	Silica		Talc, chalk china clay, $CaCO_3$	Organic Antiblock	Others
	Synthetic	Natural			
PP	•			•	• Spherical silicon
					• Spherical MMA
PE	○	•	•		
PET	•				• Zeolithes
PVC	•				• Alumosilicates
PA					

•Filled circles indicate a recommended use or application; ○ open circles indicate limited, partial, or conditional use.

Table 5.18 Recommendations for Antiblock Additives

Substrate	Effect	Additive Type	Load Level (ppm)	Remark
PP	Multipurpose	Synthetic silica	2000	
	High quality	Spherical MMA	3000	• Multilayer
PE	Multipurpose	Talc	5000	
PET	Multipurpose	Synthetic silica	2000	
PVC	Multipurpose	Alumosilicates	4000	

Indicative starting point formulations from Ref. [8].

5.4.5.5 Influencing the Antiblock Effect

Apart from the concentration, the antiblock properties will additionally be improved by:

- best possible dispersion
- larger particle size (although too large particles will cause film defects)
- addition of selected slip additives.

The antiblock properties can be negatively affected by several interactions. Interactions related to antiblock properties can occur through:

- additive agglomeration
- increased film haze at too high load level of antiblock additive
- static charges
- high temperatures (during film storage)
- high contact pressure.

5.4.5.6 Incorporation

Antiblock additives are typically added during the extrusion and/or compounding step or sometimes even at the conversion step (Fig. 5.1; addition points (G), (H) and (I)). However, for PET, the additive is added to the monomer (addition point (A)). (See Appendix 5.1 for an overview.)

Antiblock agents must be very well dispersed in order to obtain a good effect.

5.4.6 Antistats

5.4.6.1 Effect

When two (organic) substrates are in contact with each other such that friction occurs, electrostatic charges can built up. Electrostatic charges can impact plastic parts in several ways; one of the most annoying being the attraction of dust particles. One way to counter this effect is to use antistats (or antistatic additives).

This effect is principally a surface effect, although one potential counter measure (conductive fillers) converts it into a bulk effect.

5.4.6.2 Principle

Tools that decrease electrostatic charges and hence increase the conductivity of an organic substrate can be classified as:

- external antistat (surface effect)
- conductive filler (bulk and surface effect)
- internal antistat (surface effect).

An external antistat is applied via a carrier medium to the surface of the plastic part. The same considerations and limitations apply as with nonmigrating slip additives. A conductive filler is incorporated into the organic substrates and builds up a conductive network on a molecular level. While both approaches are used in organic substrates, they are not the most common.

An internal antistat is compounded into the organic substrate and migrates to the plastic part surface. The same principle considerations apply as for migrating slip additives (Fig. 5.6). However, it has to be mentioned that the molecule's orientation is inversed relative to migrating slip additives, meaning that the polar part of the antistat sticks out. It is moisture attracted by the polar functionalities that leads to the buildup of a very fine water layer on the plastic part surface, which is capable of dissipating electrostatic charges. In the following, the focus will be only on internal antistats.

5.4.6.3 Requirement

The same requirements as for migrating slip additives apply. Based on the mechanism for static dissipation, typical antistats only function at a high enough relative humidity. For low humidity applications, special chemistries need to be applied.

5.4.6.4 Chemistry

Internal antistats can be divided into chemical families of which only two are of importance for the scope of this section [13]:

- nonionic
- cationic
- anionic
- amphoteric.

The subclass of nonionic antistats can further be subdivided into:

- fatty acid ester
- ethoxylated alkylamine
- diethanolamine
- others.

Fatty acid esters migrate quickly (in their major application area polyolefins) and are also quickly depleted. Ethoxylated alkylamines migrate slower and the effect remains longer. For optimum performance, mixtures are employed.

5.4.6.5 Influencing the Antistatic Effect

A concentration increase will (initially) increase the antistatic effect. However, certain fatty acid esters do not follow the previously described saturation curve behavior. In those cases, the effect declines beyond an optimum concentration. The antistatic properties will additionally be improved by:

- higher relative humidity
- use of more fatty acid ester versus ethoxylated alkylamine in blends (initial effect).

The antistatic properties can be negatively affected by interactions. Interactions related to antistats can occur through:

- lower relative humidity
- incorporation of filler (lower migration speed)
- recrystallization of (surface of) organic substrate
- volatility
- migration inside the plastic part
- increased surface orientation.

5.4.6.6 Incorporation

Antistats are mainly incorporated during the compounding step (addition points (G) and/or (H) in Fig. 5.1). However, as the chemicals are mainly low melting or liquid, antistats are often added separate from solid additives (eg, antioxidants, UV stabilizers).

5.4.7 Others

Various other additives, like nucleating agents, oxygen scavengers, and fillers, can be used in multilayer packaging film.

Neutralizers and acid scavengers will not be covered systematically in this section. However, it should be mentioned that polyolefins synthesized with Ziegler–Natta catalysts need acid scavengers due to acidic residues. In general, metal soaps (mainly calcium stearate) are preferred for cost reasons. For metallizable and printed film, metal soaps interfere antagonistically with the surface treatment and are therefore substituted by hydrotalcite. Hydrotalcite, being a superior acid scavenger, is used at lower load levels than metal soaps.

Lubricants will also not be covered systematically. However, their importance for the processing and/or conversion of PVC and PA should be mentioned.

Adhesive compounds are not classified as additives and should be considered separately.

5.5 Suppliers and Contacts

Several additives mentioned in this chapter are described with additive codes. Table 5.19 gives their chemical name, commercial name, and a selection of suppliers.

Table 5.19 Additives Discussed in Chapter "Polypropylene"

Additive Code	Chemical Name	Commercial Name	Supplier
AO-1	Tetrakis (methylene-3-(3,5-di-*tert*-butyl-4-hydroxyphenyl)propionate) methane	Irganox 1010	Ciba
		Songnox 1010	Songwon
AO-2	Octadecyl-3-(3,5-di-*tert*-butyl-4-hydroxyphenyl) propionate	Irganox 1076	Ciba
		Songnox 1076	Songwon
AO-3	Tris (3,5-di-*tert*-butyl-4-hydroxybenzyl)iso-cyanurate	Irganox 3114	Ciba
		Songnox 3114	Songwon
AO-4	1,3,5-Trimethyl-2,4,6-tris (3,5-di-*tert*-butyl-4-hydroxybenzyl) benzene	Ethanox 330	Albemarle
AO-8	Tri(butylcresyl) butane	Topanol CA	Vertellus
AO-21	*N,N*-Hexamethylenebis(3,5-di-*tert*-butyl-4- hydroxyhydrocinnamamide)	Irganox 1098	Ciba
		Songnox 1098	Songwon
AO-37	4-Ethyl-2,6-di-*tert*-butylphenol	BHEB	Chevron Phillips chemical Nanging Datang chemical Clariant
HYA-1	*N,N*-distearylhydroxylamine	Irgastab FS 042	Ciba
P-1	Tris(2,4-di-*tert*-butylphenyl) phosphite	Irgafos 168	Ciba
		Songnox 1680	Songwon
P-2	Trisnoylphenyl phosphite	Weston 399	Chemtura
P-3	Bis(2,4-di-*tert*-butylphenyl) pentaerythritol diphosphite	Ultranox 626	Chemtura
		Songnox 6260	Songwon
P-7	Distearyl pentaerythritol diphosphite	Weston 618	Chemtura
		Songnox 6180	Songwon
P-10	Triphenylphosphate	TPF	Bayer
P-11	Trimethylphenylphosphate		
DOP	Di (2-ethylhexyl-)phthalate	DEHP	BASF, LG chem.
DINP	Di (isononyl-)phthalate	DINP	BASF, LG chem.
DINA	Di (isononyl-)adipate	DINA	BASF
DOA	Di (2-ethylhexyl-)adipate	DEHA	BASF, Songwon
BZ	Barium/zinc-mixed metal stabilizer system		Songwon

ESBO	Ethoxidzed soybean oil	ESBO/ESO	Chemtura Arkema
HAS-3	Poly[(6-[(1,1,3,3-tetramethyl butyl)amino]-*s*-triazine-2,4-dinyl][(2,2,6,6-tetramethyl-4-piperidyl) imino] hexamethylene [2,2,6,6-tetramethyl-4-piperidyl] iminol]	Chimassorb 944	Ciba
		Songlight 9440	Songwon
HAS-5	Polymer of dimethyl succinate and 4-hydroxy-2,2,6,6-tetramethyl-1-piperidine ethanol	Tinuvin 622	Ciba
		Songlight 6220	Songwon
HAS-7	Bis(2,2,6,6-tetramethyl-4-piperidinyl) sebacate	Tinuvin 770	Ciba
		Songlight 7700	Songwon
HB-1	Hexadecyl-3,5-di-*tert*-butyl-4-hydroxybenzoate	Cyasorb UV 2908	Cytec
		Songsorb 2908	Songwon
HBP-1	2-Hydroxy-4-*n*-octoxybenzophenone	Cyasorb UV 531	Cytec
		Chimassorb 81	Ciba
HBZ-5	2-(2′-Hydroxy-5′-methylphenyl) benzotriazole	Tinuvin P	Ciba
		Songsorb 1000	Songwon
HBZ-7	2-[2′-Hydroxy-3′,5′-di(1,1-dimethylbenzyl) phenyl]-2*H*-benzotriazole	Tinuvin 234	Ciba
		Songsorb 2340	Songwon
FWA-1	2,5-Thiophenediylbis(5-*tert*-butyl-1,3-benz-oxazole)	Uvitex OB	Ciba

Appendix 5.1

Table A5.1.1 Incorporation of Additives into Various Layers of Multilayer Film Constructs

	Generic Film Construct		
	Skin Layer (I)	**Core Layer (II)**	**Skin Layer (III)**
Antioxidants and heat stabilizers	•	•	•
UV stabilizers			
UV absorber	•	○	○
HA(L)S	•	○	○
Optical brighteners	•		
Slip additive			
Migrating	•	○	•
Nonmigrating	•		•
Antiblock additives	•		•
Antistats	•		•
Neutralizers	•	•	•
Others	○	○	○

• Filled circles indicate a recommended use or application; ○ open circles indicate limited, partial or conditional use

Appendix 5.2

Table A5.2.1 Addition Points of Additives for Polyolefins

	Addition Point										
	A	**B**	**C**	**D**	**E**	**F**	**G**	**H**	**I**	**K**	**Other**
Phenolic antioxidant							•		•	○	
Phosphite							•		•	○	
UV absorber							•		•	○	
HA(L)S							•		•	○	
Optical brightener							•				
Slip additive							•		•		
Antiblock additive							•		•		
Antistats							•	•			
Neutralizers							•				

• Filled circles indicate a recommended use or application; ○ open circles indicate limited, partial or conditional use

Table A5.2.2 Addition Points of Additives for Polyester

	Addition Point									
	A	B	C	D	E	F	G	H	I	K	Other
Antihydrolysis agent							•	○			
Complexing agent		•		○							
Phenolic antioxidant							•				
Phosphite							•				
UV absorber							•		•	○	
Optical brightener	○						•				
Slip additive		•									
Antiblock additive	•										
Antistat							•				

• *Filled circles indicate a recommended use or application;* ○ *open circles indicate limited, partial or conditional use*

Table A5.2.3 Addition Points of Additives for Polyvinylchloride

	Addition Point										
	A	B	C	D	E	F	G	H	I	K	Other
Heat stabilizer							•	•			
Costabilizer							•	•			
UV absorber							•		•	○	
HA(L)S							•		•	○	
Optical brightener							•	•			
Slip additive							•		•		
Antiblock additive							•		•		
Antistat							•				
Lubricant							•	•			

• *Filled circles indicate a recommended use or application;* ○ *open circles indicate limited, partial or conditional use*

Table A5.2.4 Addition Points of Additives for Polyamide

	Addition Point										
	A	B	C	D	E	F	G	H	I	K	Other
Phosphite								•		○	
Phenolic antioxidant		•						•		○	
UV absorber							•		•	○	
HA(L)S							•		•	○	
Optical brightener							•				
Slip additive							•		•		
Antiblock additive							•		•		
Lubricant		•						•	○	○	

• *Filled circles indicate a recommended use or application;* ○ *open circles indicate limited, partial or conditional use*

References

[1] Keck-Antoine K, et al. Considerations and advantages of employing additive blends in polyolefin manufacturing and compounding. In: Dubai Plast Pro 2006 (conference proceedings); 2006.
[2] Keck-Antoine K, Bayer J. The right physical form of stabilisers and additives. In: Songwon customer presentation; 2008.
[3] Keck-Antoine K, et al. Improving polyolefins by adding the right stabilizer type in the right physical form. In: Workshop (on) polyolefin thermoplastics: advancements and innovations (conference proceedings). King Fahd University of Petroleum and Minerals; 2008.
[4] Keck-Antoine K, Kim J. Additive handling during the manufacturing of engineering resins. In: Songwon customer presentation; 2008.
[5] Kramer E, et al. Antioxidants. Plastics additive handbook. 5th ed. New York, NY: Hanser Publishers; 2001.
[6] Gugumus F. Antioxidants. Plastic additives. 3rd ed. New York, NY: Hanser Publishers; 1990.
[7] Kramer E. Fundamentals of polymer degradation and stabilization, 10th Post-doctoral course. Aargau, Switzerland: University of Applied Sciences; 2001.
[8] Songwon Chronicles. Unpublished recommendation guide; 2008.
[9] Kaufhol J, et al. PVC stabilisers. Plastics additive handbook. 5th ed. New York, NY: Hanser Publishers; 2001.
[10] Krauskopf L. Plasticizers. PVC handbook. New York, NY: Hanser Publishers; 2005.
[11] Gugumus F. Light stabilizers for thermoplastics. Gächter/Müller plastic additives. 2nd ed. New York, NY: Hanser Publishers; 1987.
[12] ExxonMobil. Tip from technology—slip and anti-block additives in polyethylene film; 2003.
[13] Pfahler G. Antistatische Ausrüstung von Kunststoffen. In: Hoechst customer presentation; 1994.

6 Rheology of Molten Polymers

John Vlachopoulos[1] *and David Strutt*[2]
[1]Department of Chemical Engineering, McMaster University, Hamilton, ON, Canada; [2]Tencate Performance Composites, Baycomp Company, Burlington, ON, Canada

6.1 Introduction

The term "rheology" was coined by E.C. Bingham in 1929, from the Greek verb *rheo* (=flow) and was defined as the study of deformation and flow of matter. The Society of Rheology was formed in 1929 by Bingham and other prominent scientists and engineers from both Europe and the United States, with its motto "*panta rei*" (all things flow, Heracleitus 6th century BC). The birth of rheology as a separate branch of engineering science coincides with the time when the flow behavior of materials like rubber, plastics, paints, clays, human blood, and other biological fluids started to attract considerable attention.

A central concept in rheology is that force applied to materials results in deformations for solids or flow for fluids. The relative deformation is called *strain* and the applied force per unit area is called *stress*. The main objective of rheology is measurement, examination and development of relations between stresses and the corresponding strains, and the change of strains per unit time (strain rates). Such knowledge is absolutely necessary for problem solving in polymer processing.

Although all materials are composed of molecules, in rheology, we rely heavily on the *continuum hypothesis*, according to which, a material, solid or fluid is considered as an infinitely divisible substance. For molten polymers, we first try to measure the flow properties making use of the continuum concept and, subsequently, we try to find relations between the flow properties and the molecular structure.

In the study of flow, we assume that a fluid adjacent to a solid surface cannot slip relative to the surface. In other words, all fluids at a point of contact with a solid take on the velocity of the solid surface ($V_{\mathrm{fluid}} = V_{\mathrm{wall}}$). This assumption is one of the cornerstones of fluid mechanics and has been shown to apply for air and water over very broad flow conditions. We will see later on in this chapter under what conditions molten polymers can actually slip on the solid walls of process equipment.

The flow properties of polymers depend on their molecular weight and molecular architecture. In subsequent sections, reference will be made to *linear polymers* (with monomeric units linked linearly, like high-density polyethylene (HDPE), polypropylene (PP), and polystyrene (PS)) and *branched polymers* (with side chains attached to the backbone chain like low-density polyethylene (LDPE)). Since the polymer chains in a given polymer sample are not the same molecular weight or same length, we define average molecular weights. The *number average molecular weight*, M_n, is the sum of the individual molecular weights divided by their number. The *weight average molecular weight*, M_w, is the sum of the squares of the weights divided by the sum of the molecular weights. Because of the squares involved in calculating M_w, its value is shifted toward the higher molecular weights. The *polydispersity index* (PDI) M_w/M_n (weight average/number average) would be 1.0 if all chains had exactly the same length (only theoretically possible). Usual grades of polymers have PDI values from 1.5 to 30.

With the advent of metallocene catalysts, it is possible to make controlled molecular weight and molecular architecture polyolefins. Generally, metallocene-catalyzed polyolefins have more narrow molecular weight distributions (lower PDI values) than those catalyzed with Ziegler–Natta catalysts. It will be shown later on in this chapter

Multilayer Flexible Packaging. DOI: http://dx.doi.org/10.1016/B978-0-323-37100-1.00006-5

how rheological properties relate to average molecular weight and molecular weight distributions.

6.2 Viscosity and Melt Flow Index

Viscosity is the most important flow property [1]. It represents the resistance to flow. Strictly speaking, it is the resistance to shearing, that is, flow of imaginary fluid slices like the motion of a deck of cards. Referring to Fig. 6.1, we can define viscosity as the ratio of the imposed *shear stress* (force F, applied tangentially, divided by the area A) and the *shear rate* (velocity U, divided by the gap h)

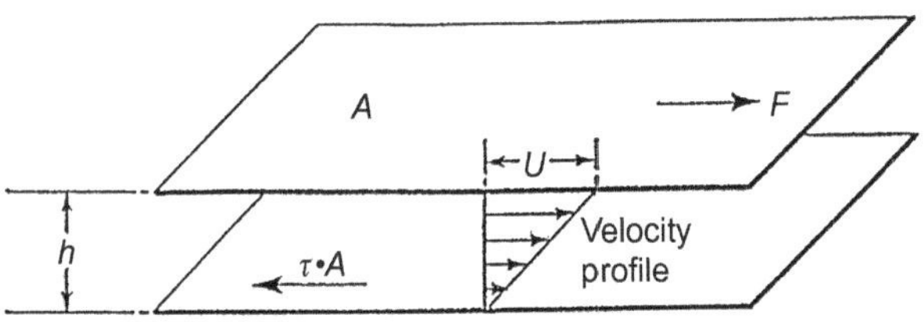

Figure 6.1 Simple shear flow.

$$\eta = \frac{\text{Shear stress}}{\text{Shear rate}} = \frac{F/A}{U/h} = \frac{\tau}{\dot{\gamma}}$$

The Greek letters τ (tau) and $\dot{\gamma}$ (gamma dot) are conventionally used to designate the shear stress and shear rate, respectively.

This relation is known as *Newton's law of viscosity*. When the viscosity is independent of the shear rate, a fluid is called Newtonian. Small-molecule fluids like air, water, mercury, and glycerin are Newtonian. Molten polymers, polymer solutions, and other complex fluids have viscosities dependent on shear rate and are called non-Newtonian.

The viscosity in SI is reported in units of Pa s (Pascal second). Before the introduction of SI, poise was the most frequently used unit (1 Pa s = 10 poise). Here are some other useful conversion factors:

Property	To convert from:	To:	Multiply by:
Density	g/cc	lb/ft^3	62.43
Pressure	lb/in^2	MPa	0.00689
Output	lb/h	kg/h	0.45359
Output	lb/h	g/s	0.125997
Length	inch	mm	25.4
Specific output	lb/h/inch-c	kg/h/mm-c	0.017858
Specific output	lb/h/inch-c	kg/h/mm-d	0.055245
Viscosity	poise	Pa s	0.1
Viscosity	poise	lb s/in^2	68947
Stress	lb/in^2	kPa	6.8947
Stress	lb/in^2	dynes/cm^2	68947
MVTR	g*mil/100 in^2-d	g*mm/m^2-d	0.394
Permeability	cc*mil/100 in^2-d/atm	cc*mm/mm^2/s/atm	4.56E − 06
Permeability	cc*mil/in^2-d/atm	cc*mil-m^2-d/atm	15.5
Weight	lb	kg	0.4536
Peel or seal strength	lb/in	g/in	454
Peel or seal strength	g/in	N/15 mm	0.0059

MVR, moisture vapor transmission rate.

$$1 \text{ Pa s} = 1.45 \times 10^{-4} \text{ lb}_f\text{s/in}^2$$
$$= 0.67197 \text{ lb}_m\text{/s ft} = 2.0886 \times 10^{-2} \text{ lb}_f\text{s/ft}^2$$

The viscosity of water is 10^{-3} Pa s, while the viscosity of most polymer melts during their processing may vary from 10^2 to 10^5 Pa s. The shear stress is measured in units of Pa = (N/m^2) or psi (pounds (lb$_f$) per square inch) and the shear rate in reciprocal seconds (s^{-1}).

One remarkable property of polymeric liquids is their *shear-thinning* behavior (also known as pseudoplastic behavior). If we increase the shear rate (ie, extrude faster through a die), the viscosity decreases (Fig. 6.2). This reduction of viscosity is due to molecular alignments and disentanglements of the long polymer chains. It has been said: "polymers love shear." The higher the shear rate, the easier it is for polymers to flow through dies and process equipment.

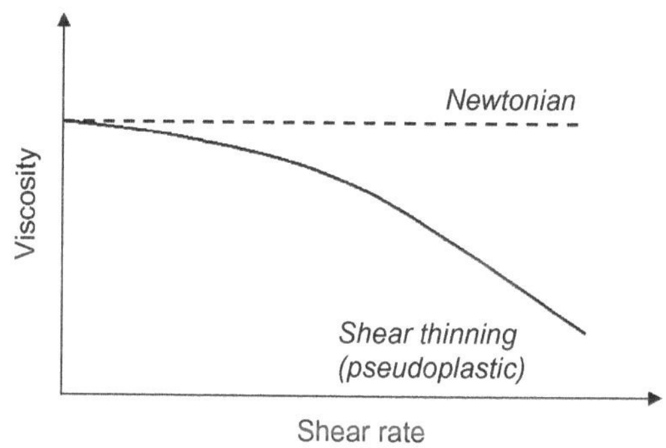

Figure 6.2 Newtonian and shear-thinning viscosity behavior.

The most frequently used model to express the shear-thinning behavior of polymers is the power law:

$$\eta = m\dot{\gamma}^{n-1}$$

This expression is a straight line when plotted on double logarithmic coordinates as shown in Fig. 6.3. The value of the consistency index m can be obtained from the intercept at $\dot{\gamma} = 1$. The exponent $n - 1$ is the slope, because

$$\log \eta = \log m + (n - 1)\log \dot{\gamma}$$

For $n = 1$, the power-law model reduces to Newton's law (constant viscosity). As n decreases, the polymer becomes more shear thinning.

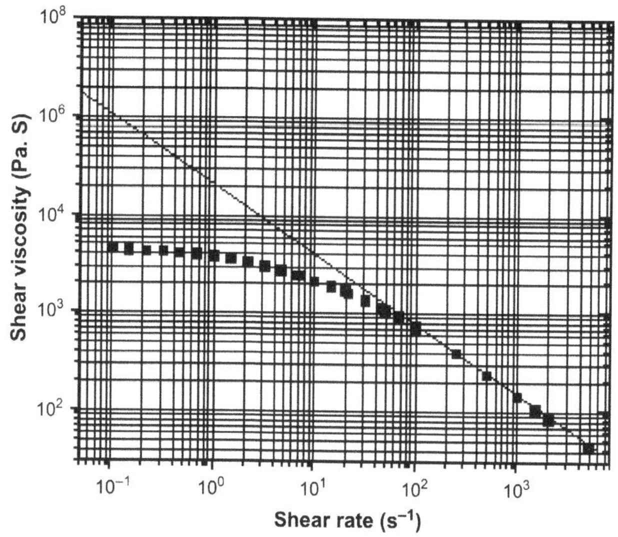

Figure 6.3 Double logarithmic plot of shear viscosity versus shear rate. The straight line is the power-law fit which fails at low shear rates. For this graph the coefficients for the straight line are: $m \approx 20{,}000$ Pa s^n; $n \approx 0.3$.

The power-law exponent varies between 0.8 (for polycarbonate (PC)) and 0.2 (for rubber compounds). For various polyethylene (PE) grades, the range is $0.3 < n < 0.6$ and depends on molecular weight and chain branching. The consistency m for molten polymers varies between 1000 Pa s^n (for some polyethylene terephthalate (PET) resins) and 100,000 Pa s^n for highly viscous rigid polyvinyl chloride (PVC). The value depends on chain mobility and molecular weight and varies exponentially with temperature.

During single-screw extrusion, shear rates may reach 200 s^{-1} in the screw channel near the barrel wall and much higher between the flight tips and the barrel. At the die lip or exit, the shear rate can be as high as 1000 s^{-1}. During cavity filling in injection molding, shear rates can reach 10,000 s^{-1}. Low shear rate on a die wall implies slow polymer melt movement over the metal surface. Some die designers try to design dies for cast film or blown film operations not having wall shear rates less than, say 10 s^{-1}, to prevent the molten material from hanging up. When the wall shear stress exceeds 0.14 MPa, sharkskin (ie, surface mattness) occurs in capillary viscometer measurements using various HDPE grades. At very high shear rates, a flow instability known as melt fracture occurs [2,3].

Melt index (MI), melt flow index (MFI), or melt flow rate (MFR) (for polypropylene) refers to the grams per 10 min pushed out of a die of prescribed dimensions according to an ASTM Standard [4] under the action of a specified load (Fig. 6.4). For PE (ASTM D-1238), the load is 2.16 kg, and the die dimensions are $D = 2.095$ mm and $L = 8$ mm. The experiment is carried out at 190°C. For PP, the same load and die dimensions are used, but the experiment is carried out at 230°C.

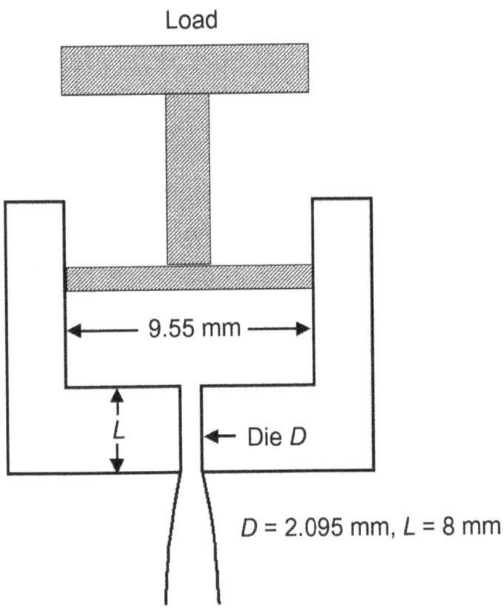

Figure 6.4 Schematic of a melt indexer.

When measuring the melt index with a 2.16 kg load, the wall shear stress can be calculated to be $\tau_w = 1.94 \times 10^4$ Pa (2.814 psi) and the wall shear rate approximately $\dot{\gamma} = (1838/\rho) \times \text{MI}$ where ρ is the melt density in kg/m^3. Assuming $\rho = 766$ kg/m^3 for a typical PE melt, we get $\dot{\gamma} = 2.4 \times \text{MI}$. Low melt index means a high molecular weight, highly viscous polymer. A high melt index means low molecular weight, low-viscosity polymer. When the melt index is less than 1, the material is said to have a fractional melt index. Such materials are used for film extrusion. For some film grades, MI can be less than 0.1. Most extrusion PE grades seldom exceed MI = 12, however, for injection molding, MI is usually in the range of 5–50.

An approximate calculation of both m and n can be carried out by using two melt index values (MI and HLMI). MI refers to standard 2.16 kg weight and HLMI to high load melt index (frequently 21.6 or 10 kg). By manipulating the appropriate equations for pressure drop, shear stress, and flow rate [1], we have

$$\text{Power-law exponent } n = \frac{\log(\text{HL}) - \log(\text{LL})}{\log(\text{HLMI}) - \log(\text{MI})}$$

$$\text{Consistency } m = \frac{8982 \times (\text{LL})}{\left[\frac{1838}{\rho} \times \text{MI}\right]^n}$$

where:

LL	=	low load (usually 2.16 kg)
HL	=	high load (usually 21.6 or 10 kg)
ρ (kg/m^3)	=	the melt density at the test temperature
n	=	dimensionless exponent
m	=	parameter with dimensions of Pa s^n

Intrinsic viscosity, frequently designated as IV, is a relative number that measures the average molecular weight through the Mark–Houwink equation:

$$[\eta] = \text{KM}_v^a$$

where:

M_v	=	the (viscosity) average molecular weight
K	=	constant depending on the polymer solvent system
a	=	constant depending on the polymer solvent system

For example, for PET, IV is determined by dissolving PET at less than 1% in a solvent and measuring the time required for a certain volume of the solution to flow through a capillary. By extrapolating to 0% concentration, the intrinsic viscosity is obtained. The intrinsic viscosity has the dimensions of inverse concentration. The American units are 100 cm^3/g, and the European units are cm^3/g, usually expressed as dL/g or mL/g. The IV for bottle grade PET resins is usually between 0.70 and 0.85 dL/g. This number is of relative importance. High IV means higher average molecular weight and higher melt viscosity.

Polyvinyl chloride polymers are often graded according to their K-value, which is a measure of their molecular weight. It is obtained from intrinsic viscosity measurements in cyclohexanone solution. K-values vary between 35 and 80. Low K-values imply low molecular weight (which is easy to process, but has inferior properties), and high K-values imply high molecular weight which is difficult to process and has outstanding properties.

6.3 Mathematical Relations

For unidirectional flows, the shear rate is simply the derivative of velocity (V_x) with respect to distance y, perpendicular to the flow direction

$$\dot{\gamma} = \frac{dV_x}{dy}$$

For the drag flow shown in Fig. 6.1, the velocity profile is linear (due to the no-slip assumption at top and bottom plates)

$$V_x = \frac{U}{h} y$$

and the shear rate

$$\dot{\gamma} = \frac{dV_x}{dy} = \frac{U}{h}$$

For pressure-driven axial flow (z direction) of a Newtonian fluid in a tube of radius R, the velocity profile is parabolic

$$V_z = V_{\max}\left(1 - \frac{r^2}{R^2}\right)$$

and the absolute value of shear rate

$$\dot{\gamma} = \left|\frac{dV_z}{dr}\right| = \frac{2r}{R^2} V_{\max}$$

varies between zero at the axis of symmetry ($r = 0$) and a maximum value at the wall $r = R$

$$\dot{\gamma}_w = \frac{2}{R} V_{\max}$$

The relation between maximum and average velocity in tube flow is

$$V_{\max} = 2V_{\text{avg}}$$

therefore

$$\dot{\gamma}_w = \frac{4}{R} V_{\text{avg}}$$

and since the volume flow rate Q is related to the average velocity by

$$Q = V_{\text{avg}} \pi R^2$$

we may write

$$\dot{\gamma}_w = \frac{4Q}{\pi R^3}$$

Although this relation is strictly valid for Newtonian fluids, it is also used for non-Newtonian fluids, and it is referred to as the *apparent shear rate*.

A correction is necessary (Rabinowitsch correction) for shear-thinning fluids. For the power-law model, the true (Rabinowitsch corrected) shear rate becomes

$$\dot{\gamma}_w = \frac{3n+1}{4n} \frac{4Q}{\pi R^3}$$

This means that for a material with power-law index $n = 0.4$ (very common), the relation between apparent and true shear rate is

$$\dot{\gamma}_{\text{true}} = 1.375 \times \dot{\gamma}_{\text{apparent}}$$

Capillary dies with 1–2 mm diameter are usually used to measure viscosity. The shear stress is determined from the pressure drop by the equation

$$\tau = \frac{\Delta P}{2L} r$$

which varies linearly from zero on the capillary axis ($r = 0$) to a maximum value at the wall ($r = R$)

$$\tau_w = \frac{\Delta P}{2L} R$$

The ratio of the measured wall shear stress and the corresponding wall shear rate gives the viscosity

$$\eta = \frac{\tau_w}{\dot{\gamma}_w}$$

Fig. 6.5 shows schematically the variation of velocity, shear rate, and shear stress for polymer melt flow in a tube. It should be noted that the shear stress is linear while the velocity profile is quasi-parabolic.

$$V_z = V_{\max}\left[1 - \left(\frac{r}{R}\right)^{\frac{n+1}{n}}\right]$$

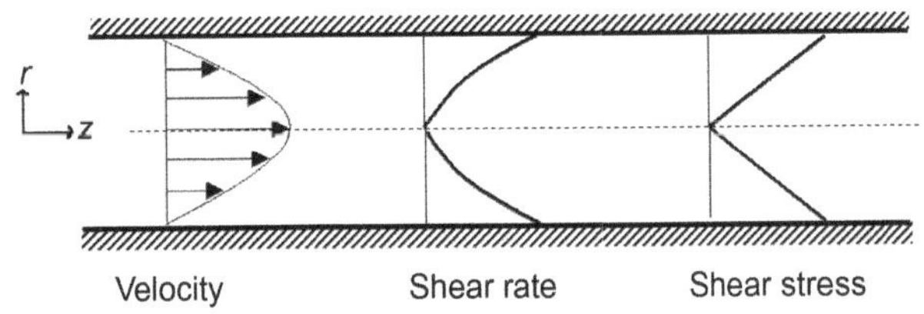

Figure 6.5 Velocity, shear rate, and shear stress profiles for pressure-driven flow through a tube of radius *R*.

The pressure drop ΔP is measured in the reservoir (see Fig. 6.6), and at the entrance to the capillary there is an excess pressure drop ΔP_e. The easiest way to determine the excess pressure drop ΔP_e is to carry out the experiment using a twin-bore viscometer having two capillaries. One of them has very short length ($L = 0$), and the pressure drop recorded is essentially the excess pressure due to the entrance (orifice). This correction is necessary when capillaries are relatively short ($L/R < 50$) and is known as the Bagley correction [2,3]. The Bagley correction is usually expressed as

$$n_B = \frac{\Delta P_e}{2\tau_w}$$

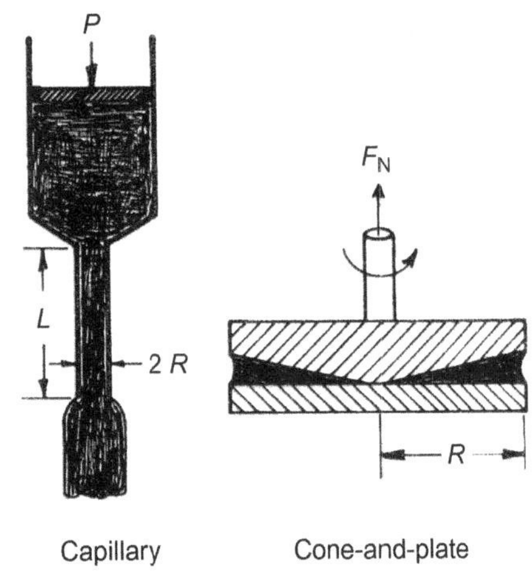

Figure 6.6 Schematic representation of a capillary and a cone-and-plate viscometer.

The Bagley correction (n_B) may reach perhaps 20 when polymeric materials are extruded near the critical stress for sharkskin. For a Newtonian fluid, the value for n_B is 0.587.

The Bagley corrected shear stress becomes

$$\tau_w = \frac{\Delta P_{cap} + \Delta P_c}{2\left(\frac{L}{R} + n_B\right)}$$

To apply the Bagley correction, measurements with at least two capillaries are needed.

The Bagley entrance correction can be relatively large when one compares it to the pressure drop in the capillary

$$\Delta P_{cap} = 2\frac{L}{R}\tau_w$$

For a standard capillary of $L/R = 32$, we have $\Delta P_{cap} = 64\ \tau_w$. If we assume some sort of "average" value $n_B = 10$, then $\Delta P_e = 20\ \tau_w$ or approximately 30% of the pressure drop in the capillary. Obviously, for shorter dies, the entrance pressure drop accounts for a large portion of the total pressure drop recorded from the reservoir to the exit of the capillary.

Without the Rabinowitsch and Bagley corrections, there can be significant errors in viscosity data obtained from capillary instruments.

The power-law model discussed in the previous section is good for fitting the viscosity data at the high shear region (usually from 10 to 2000 s^{-1}). At the low shear region, there is a Newtonian (horizontal) plateau, as shown by the data points in Fig. 6.3.

Two other models are frequently used for better fitting of data over the entire shear rate range:

Carreau-Yasuda

$$\eta = \eta_0(1 + (\lambda\dot{\gamma})^a)^{\frac{n-1}{n}}$$

where η_0 is the viscosity at zero shear, and λ, a, and n are fitted parameters.

Cross model

$$\eta = \frac{\eta_0}{1 + (\lambda\dot{\gamma})^{1-n}}$$

where η_0 is the zero shear viscosity, and λ and n are fitted parameters. Note that, in this model, when $\lambda = \frac{1}{\dot{\gamma}}$, $\eta = \frac{\eta_0}{2}$.

With rotational viscometers (cone-and-plate or parallel plate), the shear stress is determined from the applied torque and the shear rate from the rotational speed and the gap where the fluid is sheared.

For the cone-and-plate instrument of Fig. 6.6, the shear rate is given by

$$\dot{\gamma} = \frac{\omega}{\alpha}$$

and the shear stress by

$$\tau = \frac{3M}{2\pi R^3}$$

where:

ω	=	rate of rotation
α	=	cone angle (usually less than 5°)
M	=	torque
σ	=	radius

Capillary viscometers are usually used for the shear rate range from 1 to 3000 s^{-1}. Rotational viscometers are usually used for the range 10^{-2} to 5 s^{-1}. At higher rotational speeds, secondary flows and instabilities may occur which invalidate the simple shear assumption. Simple shear means the slice-by-slice relative motion in a fluid as if it were composed of imaginary fluid slices, like a deck of cards. For more information about viscosity measurements, the reader is referred to Macosko [2].

The viscosity of polymer melts varies with temperature in an exponential manner

$$\eta = \eta_{ref} \exp(-b \Delta T)$$

The temperature sensitivity coefficient b is usually between 0.01 and 0.1°C. For common polyolefin grades, we may assume that $b = 0.015$. This means that for a temperature increase $\Delta T = 10$°C (18°F), the viscosity decreases by 14%. For HDPE (linear polymer) the value of b is 0.01, while for LDPE $b = 0.03$.

The effects of factors such as shear rate, molecular weight distribution, pressure, filler, temperature, and additives on viscosity are summarized in Fig. 6.7 following Cogswell [3]. Linear narrow molecular weight distribution polymers (metallocene catalyzed) are more viscous than their broad distribution counterparts. Fillers may increase viscosity (greatly). Pressure increases viscosity (negligible under usual extrusion conditions but important in injection molding). Various additives are available and are designed to decrease viscosity. The zero shear viscosity increases dramatically with the weight average molecular weight:

$$\eta_0 = \text{const}\, M_w^{3.4}$$

For some metallocene-catalyzed PEs with long-chain branching, the exponent might be much higher (perhaps 6.0).

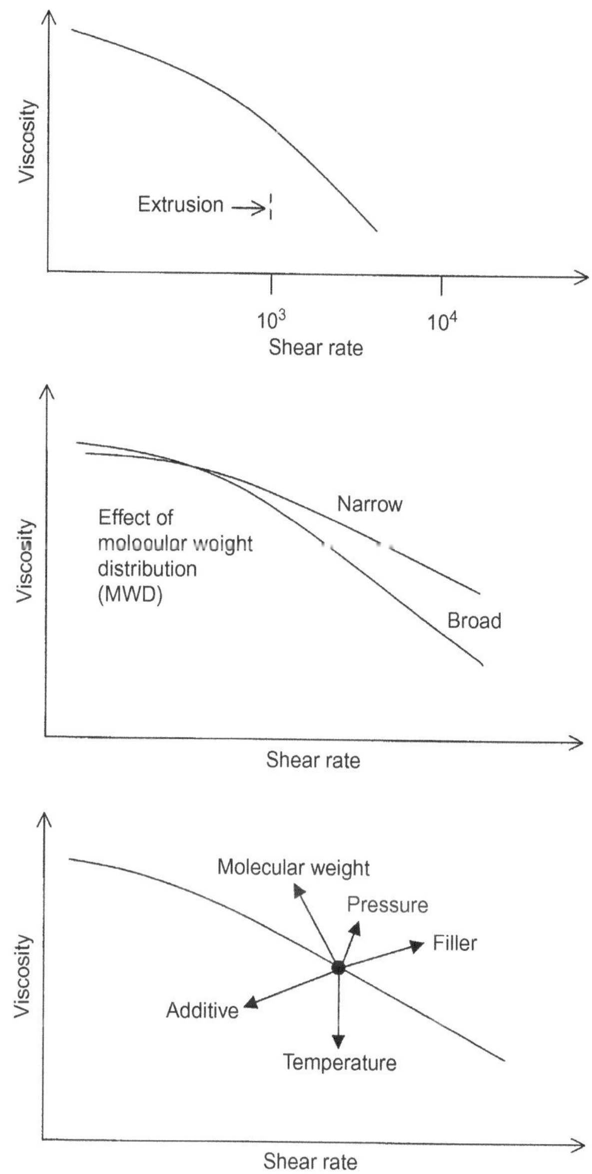

Figure 6.7 The influence of various parameters on polymer viscosity.

In the above discussion of viscosity measurements, the assumption is made that the no-slip condition on the die wall is valid. This is, however, not always the case. In fact, at shear stress levels of about 0.1 MPa for PE, slip occurs. Wall slip is related to the sharkskin phenomenon [5]. Wall slip is measured by the Mooney method [5,6] in which the apparent shear rate ($4Q/\pi R^3$) is plotted against $1/R$ for several capillaries having different radii. In the absence of slip, the plot is horizontal. The slope of the line is equal to $4 \times$ (slip velocity), as explained by Dealy and Wissbrun [6].

6.4 Extensional Viscosity and Melt Strength

Extensional (or elongational) viscosity is the resistance of a fluid to extension [2]. While stretching a low-viscosity fluid-like water is difficult to imagine, polymer melts exhibit measurable resistance. In fact, about 100 years ago, Trouton measured the stretching and shearing resistance of stiff liquids, including pitch, and found that the extensional to shear viscosity ratio is equal to 3.

$$\frac{\eta_e}{\eta} = 3$$

This relation, known as the Trouton ratio, is valid for all Newtonian fluids and has a rigorous theoretical basis that confirms Trouton's experiments.

Measuring elongational viscosity is considerably more difficult than measuring shear viscosity. One device used involves capillary extrusion and subsequent stretching with a pair of rollers. The maximum force required to break the extruded strand is referred to as *melt strength*. In practice, the terms extensional viscosity and melt strength are sometimes confused. Fig. 6.8 compares extensional viscosity as a function of stretch rate ($\dot{\varepsilon}$) and compares it to the shear viscosity as a function of shear rate ($\dot{\gamma}$). Melt strength is an engineering measure of resistance to extension. Several extrusion processes involve extension, such as film blowing, melt spinning, and sheet or film drawing. The stretch rates in film blowing can exceed $10\ s^{-1}$, while in entry flows from a large reservoir into a smaller diameter capillary, the maximum stretch rate is likely to be one order of magnitude lower than the maximum wall shear rate (eg, in capillary viscometer, when $\dot{\varepsilon}_{max}$ about $100\ s^{-1}$, $\dot{\gamma}_{max}$ is about $1000\ s^{-1}$). Frequently, the extensional viscosity is plotted as a function of stretching time (increasing) without reaching a steady value (strain hardening).

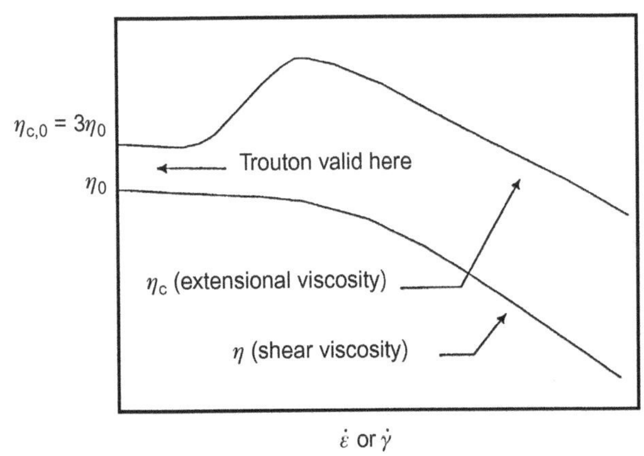

Figure 6.8 Extensional and shear viscosity as a function of stretch and shear rate, respectively.

The excess pressure drop encountered in flow from a large reservoir to a smaller diameter capillary is due to elongational viscosity. In fact, Cogswell [3] has developed a method for measurement of elongational viscosity η_e from excess pressure drop ΔP_e (ie, the Bagley correction):

$$\eta_e = \frac{9(n+1)^2(\Delta P_e)^2}{32\eta\dot{\gamma}^2}$$

at

$$\dot{\varepsilon} = \frac{4\eta\dot{\gamma}^2}{3(n+1)\Delta P_e}$$

Shear and extensional viscosity measurements reveal that linear low-density polyethylene (LLDPE) is "stiffer" than LDPE (branched) in shear, but "softer" in extension. In extension, the linear LLDPE chains slide by without getting entangled. However, the long branches of the LDPE chains result in significantly larger resistance in extension. In the film blowing process, LDPE bubbles exhibit more stability because of their high extensional viscosity. Typical LDPE and LLDPE behavior in shear and extension is shown in Fig. 6.9. LDPE is often blended with LLDPE to improve the melt strength and consequently bubble stability in film blowing. Most PP grades are known to exhibit very low melt strength. However, recent advances in polymer chemistry have led to the production of some high melt strength PP grades (with long-chain branching).

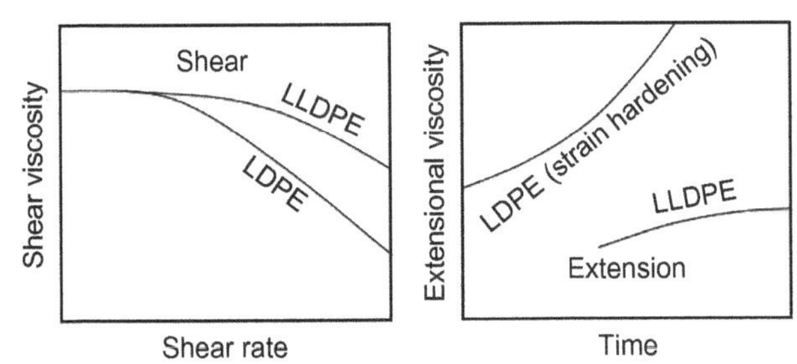

Figure 6.9 Schematic representation of LDPE and LLDPE behavior in shear and extension.

6.5 Normal Stress Differences and Extrudate Swell

Stress is defined as force divided by the area on which it acts. It has units of lb_f/in^2 (psi) in the British system or N/m^2 (Pascal, Pa) in SI. When a force is acting tangentially on a surface, the corresponding stress is referred to as *shear stress*. When a force is perpendicular (normal) to a surface, it is termed *normal stress*. Pressure is a normal stress. When a fluid flows through a conduit, it is acted upon by the normal (pressure) forces and it exerts both normal and shear (stress) forces on the conduit walls. For flow through a planar die, the shear stress is zero at the midplane and maximum at the wall, while the corresponding velocity profile is quasi parabolic.

Weissenberg discovered in the 1940s that polymer solutions and melts, when subjected to shearing, tend to develop normal stresses that are unequal in the x (direction of flow), y and z (normal directions) which are added to or subtracted from the local pressure. But, why are these elusive forces generated? They are generated because polymer's long molecular chains exhibit anisotropic or nonuniform properties when they flow. Any further explanation of the physical origin of normal stresses is likely to be controversial. Here is perhaps an oversimplification: Shearing means motion of a fluid in a slice-by-slice manner. If the imaginary slices were made of an extensible elastic material (like slices of rubber), shearing would also result in extension in the flow direction and uneven compression in the other two directions. So, when an (elastic) polymer solution or melt flows along a pressure gradient, it is less compressed in the direction of flow than in the other two normal directions.

The so-called First Normal Stress Difference N_1 is defined as the total normal stress in the direction of the flow (σ_{xx}) minus the perpendicular (σ_{yy}) stress.

$$\begin{aligned} N_1 &= \sigma_{xx} - \sigma_{yy} = (-P + \tau_{xx}) - (-P + \tau_{yy}) \\ &= \tau_{xx} - \tau_{yy} \end{aligned}$$

The Second Normal Stress Difference is

$$\begin{aligned} N_2 &= \sigma_{yy} - \sigma_{zz} = (-P + \tau_{yy}) - (-P + \tau_{zz}) \\ &= \tau_{yy} - \tau_{zz} \end{aligned}$$

We use normal stress differences rather than just normal stresses to remove the value of the pressure. Experiments show that N_1 is positive for usual polymers (ie, extensive, while the compressive pressure forces are negative). N_2 is negative and of the order of 20% of N_1 for most common polymers. N_1 is very sensitive to the high-molecular-weight tail of a polymer. Broad molecular weight distribution polymers exhibit high N_1 values.

The normal stress differences can be very large in high shear rate extrusion through lips of a die. Some authors suggest a variation for the normal stress difference at the wall in the form

$$N_{1w} = A\tau_w^b$$

The stress ratio

$$S_R = \frac{N_{1w}}{2\tau_w}$$

can reach a value of 10 or more at the onset of melt fracture.

The rod-climbing effect observed by Weissenberg (Fig. 6.10a) when a cylinder rotates in a polymeric liquid is due to some sort of "strangulation" force exerted by the extended polymer chains, which results in an upward movement normal to the direction of rotation (normal stress difference). The extrudate swell phenomenon [7] (Fig. 6.10b) is due mainly to the contraction of the exiting polymer that is under extension in the die due to N_1. The uneven compression in the various directions results in a

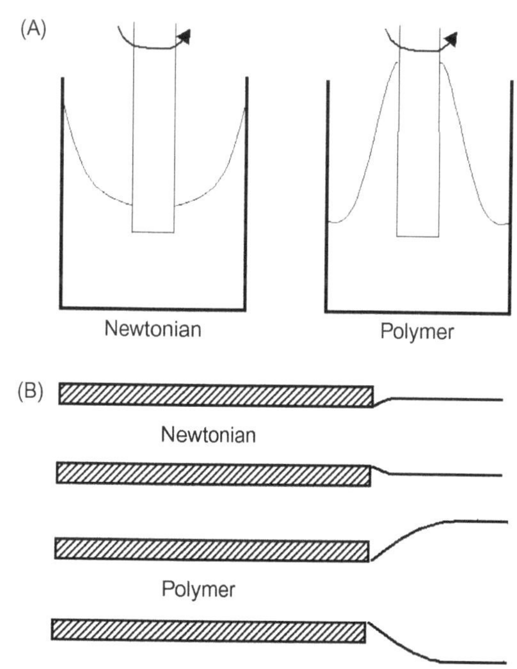

Figure 6.10 (a) Rod-climbing (Weissenberg) effect in polymeric fluids; (b) extrudate swell.

number of unusual flow patterns and instabilities. The secondary flow patterns in square channels observed by Dooley and coworkers [8] are due to the second normal stress difference. Bird et al. [9] in their book state: "A fluid that's macromolecular is really quite weird, in particular the big normal stresses the fluid possesses give rise to effects quite spectacular."

In extrusion through dies, the extrudate diameter (d) is larger than the die diameter (D). Extrudate swell ratios (d/D) reach values of 400% or more, under certain conditions.

This phenomenon (also known as die swell) has been studied by several researchers. While the primary mechanism is release of normal stresses at the exit, other effects are also important. Extrudate swell is largest for zero length dies (ie, orifices). It decreases for the same throughput with increasing die length due to fading memory as the residence time in the die increases. Even Newtonian fluids exhibit some swell upon exiting dies (13% for round extrudates, 19% for planar extrudates). This Newtonian swell is due to streamline rearrangement at the exit. The swell ratio can be influenced by thermal effects due to viscosity differences between the walls and die center. Maximum thermal swell can be obtained when a hot polymer flows through a die with colder walls. Swell ratio of about 5% on top of other mechanisms can be obtained from temperature differences. For linear polymers, the swell ratio increases as the molecular weight distribution broadens.

Several attempts have been made to predict extrudate swell through equations relating the swell ratio d/D (extrudate diameter/die diameter) to the first normal stress difference at the wall N_{1w}. Based on the theory of rubber elasticity, the following equation is obtained [7].

$$N_{1w} = 2\tau_w\left(3\left[\left(\frac{d}{D}\right)^4 + 2\left(\frac{d}{D}\right)^{-2} - 3\right]\right)^{1/2}$$

Based on stress release for a Maxwell fluid (described later in the chapter) exiting from a die, Tanner's equation can be derived [7]

$$N_{1w} = 2\sqrt{2}\tau_w\left[\left(\frac{d}{D} - 0.13\right)^6 - 1\right]^{1/2}$$

Although this equation has a more rigorous derivation and theoretical basis, the rubber elasticity theory is believed to give better predictions. Unambiguous evaluations are virtually impossible to carry out, because there are no reliable methods for measuring N_{1w} at high shear rates and stresses. At low shear rates (up to perhaps 1 s^{-1}), the first normal stress difference can be determined by measuring the separation force that develops in a cone-and-plate instrument (see Fig. 6.6) due to the Weissenberg effect.

6.6 Stress Relaxation and Dynamic Measurements

When flow stops, the stresses become immediately zero for small-molecule Newtonian fluids like water or glycerin. For polymer melts, the stresses decay exponentially after flow stops. Stress relaxation can be measured in a parallel plate or a cone-and-plate rheometer by applying a given shear rate level (rotation speed/gap) and measuring the stress decay after the rotation is brought to an abrupt stop. Such tests, however, are not performed routinely, because of experimental limitations associated with abrupt stopping of strain and stress measurement decay over more than three decades with one transducer.

Dynamic measurements involve the response of a material to an imposed sinusoidal stress or strain on a parallel plate or cone-and-plate instrument. A perfectly elastic material that behaves like a steel spring, by imposition of extension (strain), would develop stresses that would be in-phase with the strain, because

$$\text{Stress}(\tau) = \text{Modulus}(G) \times \text{strain}(\gamma)$$

However, for a Newtonian fluid subjected to a sinusoidal strain, the stress and strain will not be in-phase because of the time derivative (strain rate) involved

$$\begin{aligned}\tau &= \eta\dot{\gamma}\\ \tau &= \eta\frac{d\gamma}{dt} = \eta\frac{d}{dt}(\gamma_0 \sin\omega t) = \eta\omega\gamma_0\cos\omega t\\ &= \eta\omega\gamma_0\sin(\omega t + 90^\circ)\end{aligned}$$

where ω is frequency of oscillation. That is, a Newtonian fluid would exhibit 90° phase difference between stress and strain. Polymeric liquids that are partly viscous and partly elastic (viscoelastic) will be $0 \leq \phi \leq 90^\circ$ out of phase.

We can define

$$G'(\omega) = \frac{\text{in-phase stress}}{\text{maximum strain}} \begin{bmatrix} \text{storage} \\ \text{modulus} \\ \text{(elastic part)} \end{bmatrix}$$

$$G''(\omega) = \frac{\text{out-of-phase stress}}{\text{maximum strain}} \begin{bmatrix} \text{loss} \\ \text{modulus} \\ \text{(viscous part)} \end{bmatrix}$$

where ω ranges usually from 0.01 to 10^3 rad/s. Larger G' implies more elasticity. Further, we can define

$$\eta' = \frac{G}{\omega} \text{ the dynamic viscosity}$$

$$\eta'' = \frac{G}{\omega}$$

and the magnitude of the complex viscosity

$$|\eta^*| = (\eta'^2 + \eta''^2)^{1/2}$$

An empirical relationship called the "Cox–Merz rule" states that the shear rate dependence of the steady-state viscosity η is equal to the frequency dependence of the complex viscosity η^*, that is

$$\eta(\dot{\gamma}) = |\eta^*(\omega)|$$

The usefulness of this rule, which holds for most polymers, is that while steady measurements of shear viscosity are virtually impossible above 5 s^{-1} with rotational instruments, the dynamic measurements can easily be carried out up to 500 rad/s (corresponds to $\dot{\gamma} = 500\ s^{-1}$) or even higher. Thus, the full range of viscosity needed in extrusion can be covered.

Some typical results involving narrow and broad molecular weight distribution samples are shown in Fig. 6.11. The relative behavior of G' versus ω can be used to identify whether a sample is of narrow or broad molecular weight distribution [6]. In fact, from the "crossover point" where $G' = G''$, it is possible to get a surprisingly good estimate of the polydispersity M_w/M_n for PP [43]. For such experiments to be meaningful, the imposed strain amplitude must be low, so that the measured G' and G'' values do not

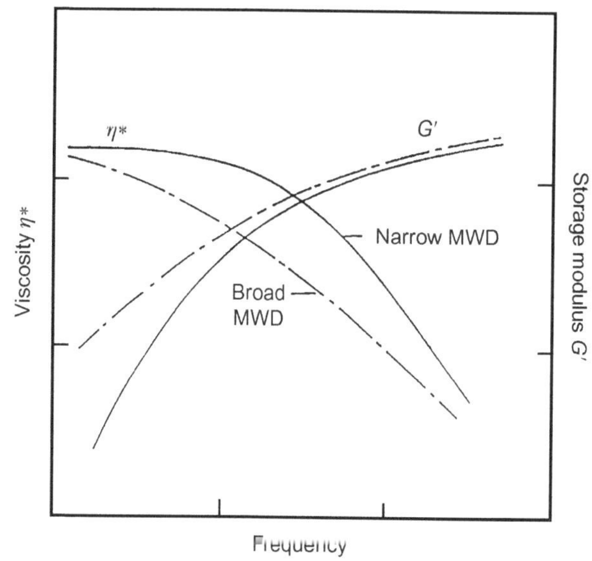

Figure 6.11 Storage modulus G' and dynamic viscosity η^* behavior of broad and narrow molecular weight distribution polymers.

vary with the strain, but they are intrinsic properties of the polymer structure. This is the region of the so-called linear viscoelasticity.

Another interesting result is the relation between storage modulus and first normal stress difference at very small deformations ($\omega \rightarrow 0$, $\dot{\gamma} \rightarrow 0$)

$$2G' = N_1$$

It is possible to measure N_1 using a cone-and-plate rotational rheometer. The Weissenberg effect results in a separating force between the cone and the plate which can be measured to give N_1.

$$N_1 = \frac{2F_N}{\pi R^2}$$

where:

N_1	=	first normal stress difference
F_N	=	normal (separating) force
R	=	cone radius in Fig. 6.6

Instruments capable of measuring N_1 require high precision construction and very sensitive force gauges. However, measurement of G', the storage modulus, can be carried out more easily.

6.7 Constitutive Equations

These are relations between stresses and strains (deformations). In its simplest form, the Newtonian equation is

$$\tau = \eta\dot{\gamma}_{\text{fluid}}$$

where:

η	=	viscosity
$\dot{\gamma}$	=	du/dy = the shear rate

For a shear-thinning material of the power-law type, we have

$$\tau = \eta\,\dot{\gamma} = m\,\dot{\gamma}^{n-1}\cdot\dot{\gamma} = m\dot{\gamma}^{n}$$

where:

m	=	is consistency
n	=	the power-law exponent

However, the above expressions, when inserted into the equation of conservation of momentum, cannot predict viscoelastic effects such as normal stresses, stress relaxation, or extrudate swell. The simplest way to develop viscoelastic constitutive equations is to combine a model for an elastic solid.

$$\tau = G\gamma_{\text{solid}}$$

with that for a Newtonian fluid

$$\tau = \eta\dot{\gamma}_{\text{fluid}}$$

By differentiating the elastic solid equation and adding the two strain rates, we get

$$\frac{\dot{\tau}}{G} + \frac{\tau}{\eta} = \dot{\gamma}$$

or

$$\tau + \lambda\dot{\tau} = \eta\dot{\gamma}$$

where

$$\lambda = \frac{\eta}{G}$$

has dimensions of time (relaxation constant).

This is known as the Maxwell model and shows that the viscoelastic nature of polymers can be described by viscosity and a relaxation time. More advanced models require a spectrum of characteristic time constants to describe the complex molecular processes involved during stress relaxation.

The concept of stress relaxation is very important in material characterization and polymer processing. If in the simple shear flow experiment shown in Fig. 6.1 the force applied was abruptly stopped, the stress would become immediately zero if the fluid was Newtonian, like water. However, in polymeric fluids, the stress would require some time to relax (exponentially) to zero. Polymers with different molecular weight and molecular architecture would require different relaxation times. Those with higher molecular weight and broader molecular weight distribution would require longer relaxation times. In polymer processing, materials with long relaxation times would have larger frozen-in stresses whenever cooling results in rapid solidification.

The relaxation time is characteristic of the elastic nature of polymeric liquids. In rheology, we define the Deborah number as the ratio of a characteristic material time (λ) and a process time (θ).

$$De = \frac{\lambda}{\theta}$$

The characteristic process time can be chosen as the inverse of the shear rate ($\theta = 1/\dot{\gamma}$), because it represents the time for a flow process to be completed, for example, passage through the die lips or filling a mold cavity. For very small Deborah numbers, the material behaves as a purely viscous fluid, and for very large Deborah numbers the material behaves as an elastic solid. In polymer processing, the Deborah numbers are usually neither too small nor too high, but somewhere in between, and such behavior is called viscoelastic.

Choosing a characteristic material relaxation time is difficult. For example, some authors of publications involving blown film use the parameter λ in the Cross model of viscosity presented above, because it has dimensions of time. In fact, some attempts have been made to correlate final film properties to this parameter. This is a very poor choice of relaxation time not supported by either theory or experimental evidence. Actually, in the Cross model, λ is equal to the inverse of the shear rate at a value where the viscosity is half the zero shear viscosity. A spectrum of relaxation times is required to describe the viscoelastic nature of polymers, however, if only one constant is to be used, theory dictates the "longest relaxation time" which is given for steady shear by:

$$\lambda = \frac{N_1}{2\eta\dot{\gamma}^2} \quad \text{for} \quad \dot{\gamma} \to 0$$

which can also be measured from dynamic experiments as

$$\lambda = \frac{G'}{G''\omega} \quad \text{for} \quad \omega \to 0$$

This result suggests that the elastic nature of polymers can be determined by measuring the storage and loss moduli at very low frequencies, preferably less than 10^{-3} (rad/s). Higher λ values mean more elastic polymers.

When describing polymer flow through channels and dies and for the subsequent stretching, orientation, and solidification processes, the viscoelastic models must be expressed in three dimensions and in a proper mathematical frame of reference that moves and deforms with the fluid. The result is a very complicated expression involving dozens of derivatives [10,11].

The most powerful constitutive equation is the so-called K-BKZ integral model that involves more than two dozen experimentally fitted parameters (see, eg, Mitsoulis [12]). Current trends involve the development of models based on macromolecular motions. De Gennes proposed the snake-like motion of polymer chains called reptation [2], illustrated in Fig. 6.12 and deduced from scaling relations that the zero shear viscosity must be $\eta_0 \approx M^{3.0}$, while experiments give $\eta_0 \approx M^{3.4}$ (M is the molecular weight). Based on the reptation concept, Doi and Edwards [13] developed a constitutive equation which leaves much to be desired before it can be used for predicting viscoelastic flow phenomena. Several attempts have been made to fix the Doi–Edwards theory. Perhaps the most advanced concepts have been proposed by Marrucci and coworkers (see, eg, Ref. [13]).

The most talked about viscoelastic model recently is the Pom-Pom polymer model, developed by McLeish and Larson [14]. The motivation for its development was that the K-BKZ equation fails to predict the observed degree of strain hardening in planar extension when certain functions are adjusted to fit the observed degree of strain softening in shear. The failure to describe the rheology of long-chain branched polymers suggests that some new molecular insight is needed into the nonlinear relaxation processes that occur in such melts under flow. The Pom-Pom model uses an H-polymer structure, in which molecules contain just two branch points of chosen functionality and a "backbone" which links the two pom-poms as shown in Fig. 6.13.

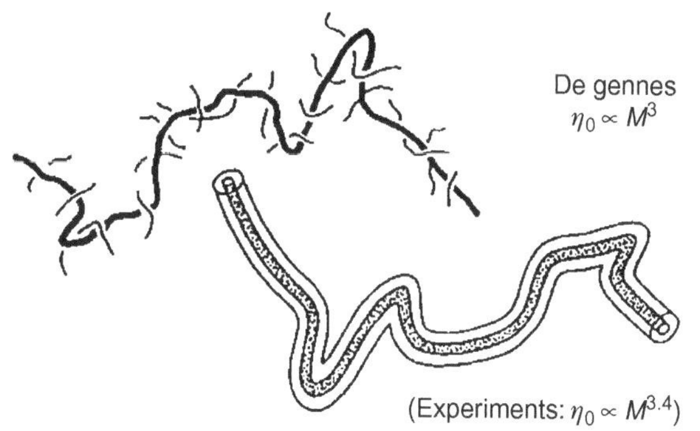

Figure 6.12 Reptation model of polymer chain motion.

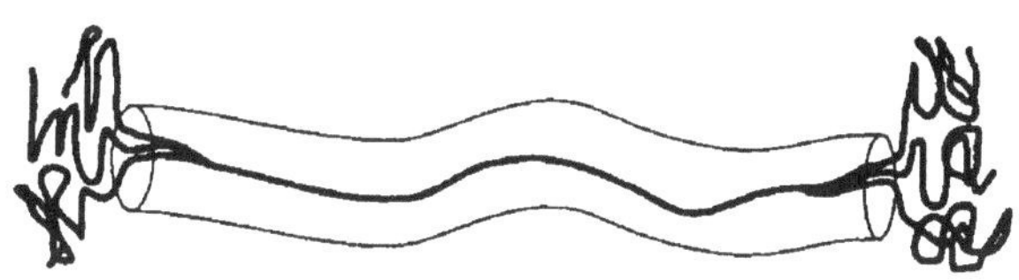

Figure 6.13 The Pom-Pom polymer model.

The Pom-Pom model exhibits rheological behavior remarkably similar to that of branched commercial melts like LDPE. It shows strain hardening in extension and strain softening in shear. It can describe both planar and uniaxial extension. The constitutive equation is integro-differential. For successful application at least 32 parameters must be obtained by fitting experimental rheological data. Of course, fitting 32 or more parameters in a complicated constitutive equation is a mathematical challenge.

Modeling polymer viscoelastic behavior has always been a very controversial subject. While viscoelastic constitutive equations have contributed toward understanding various deformation mechanisms and flow, they unfortunately have not provided us with quantitative predictive power. Very often the predictions depend on the model used for the computations and are not corroborated with experimental observations. Some viscoelastic flow problems can be solved with the appropriate constitutive equations, but this is still an area of academic research with very limited practical applications at the moment.

6.8 Sharkskin, Melt Fracture, and Die Lip Buildup

The term *sharkskin* is used when an extrudate loses surface gloss, also sometimes termed surface mattness. The surface usually exhibits a repetitious wavy or ridged pattern perpendicular to the flow direction. It occurs at a critical stress level of at least 0.14 MPa (21 psi) for most common polymers extruded through capillary dies. With some additives, lubricants, processing aids, or die coatings, the onset of sharkskin can be shifted to higher throughput rates corresponding to higher levels of apparent shear stress.

The prevailing explanation is that sharkskin originates near the die exit and is due to stick-slip phenomena. A critical shear stress near the exit, in conjunction with a critical acceleration, results in skin rupture of the extrudate [15,16]. There was some disagreement over whether slip between the polymer and the die wall causes or helps avoid sharkskin [17]. However, it is now believed that it is slip which helps to postpone sharkskin to higher flow rates. Good adherence is also thought to be potentially beneficial, but stick-slip is always detrimental.

Fluorocarbon polymers, which can cost 5–10× as much as PE, are used in parts per million levels as processing aids with LLDPE. The proposed mechanism is that they deposit on the die surface and allow continuous slip. More recently, boron nitride has been introduced for the same purpose [18]. Other ways to reduce sharkskin and allow higher throughput rates involve heating the die lips to reduce the wall shear stress and polymer viscosity and adding a small exit angle (flaring) at the die exit. Also, special die lip coatings are known to result in sharkskin-free extrusions.

At higher throughput rates, extrudates usually become highly distorted and the pressure in a capillary viscometer shows significant fluctuations. This phenomenon is known as gross melt fracture.

Fig. 6.14 shows LLDPE extrudates for increasing shear rates, illustrating the progression from smooth surface to sharkskin and then melt fracture [19]. It is possible with some polymers (eg, LDPE) to obtain melt fractured extrudates without sharkskin, that is, the surface remains smooth and glossy but overall the extrudate is distorted.

Proposed mechanisms for melt fracture include entry flow vortex instability, elastic instability during flow in the die land for stress ratios ($N_{1w}/2\tau_w$) greater than about 10, stick-slip phenomena, and other interactions between the polymer and the metal die wall. Probably more than one mechanism is responsible [5].

Die lip buildup (also known as die drool) refers to the gradual formation of an initially liquid deposit at the edge of the die exit which solidifies and grows and may partially obstruct the extrudate flow and/or cause defective extrudate surface. Depending on the severity of the problem, continuous extrusion must be interrupted every few hours or days and the solid deposit removed from the die lips. The causative mechanisms are not well known. Observations suggest that die lip buildup is not continuous but intermittent. Tiny droplets of material

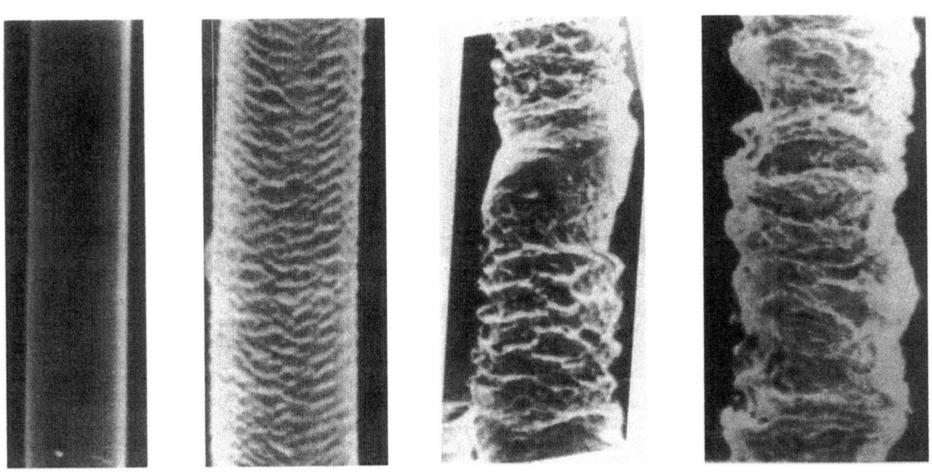

Figure 6.14 LLDPE extrudates obtained from a capillary at apparent shear rates of 37, 112, 750, and 2250 s^{-1}.

come out of the die or perhaps from a rupturing extrudate surface. Some studies suggest that the buildup is rich in lower molecular weight polymer fractions, waxes, and other additives [20].

Remedies for reducing die lip buildup include repairing missing plating and surface imperfections, removing moisture from the feed material, lowering the extrudate temperature, and adding stabilizer to the resin. Fluorocarbon processing aids will sometimes also be helpful, as they are with sharkskin. Die lip exit modification by adding a small die exit angle (flaring) is also known to reduce buildup, for polyethylenes and polycarbonate.

6.9 Rheological Problems in Coextrusion

There are two rheological phenomena that manifest themselves during the flow of immiscible polymer melts through dies: layer nonuniformity and interfacial instability [21,22].

Layer nonuniformity in coextrusion flows is caused mainly by the less viscous polymer going to the high shear region (ie, the wall), thereby producing encapsulation. Fig. 6.15 illustrates this phenomenon for rod and slit dies [23]. Partial encapsulation can occur in common types and sizes of dies, and complete encapsulation is possible for extremely long dies. Differences in polymer wall adhesion and viscoelastic characteristics can also be contributing factors. Weak secondary flows caused by viscoelastic effects (from the second normal stress difference) have been demonstrated to produce layer nonuniformities even when coextruding differently colored polymer streams of the same polymer [24]. This defect can be reduced by choosing materials with the smallest possible differences in viscosity and viscoelasticity (G', G'', extrudate swell) or by adjusting the stream temperatures to bring the polymer viscosities closer to one another.

Layer nonuniformity can also arise in feedblock coextrusion, in which melt streams are merged into a single stream in a feedblock prior to entering the flat die. Uneven flow leakage from the flat die manifold to the downstream die sections can lead to encapsulation of the more viscous polymer by the less viscous, or even the reverse! Feedblock profiling is used to counteract the natural tendency for encapsulation due to viscosity differences [25]. This involves contouring the feedblock flow passages for regions of high or low volumetric throughput (Fig. 6.16). Feedblock profiling combined with eliminating uneven flow leakage from the feeding section of a flat die (or the use of this leakage to counteract the natural tendency for encapsulation) can be used to produce layer-to-layer uniformity in the extrudate. The problem is much more complex

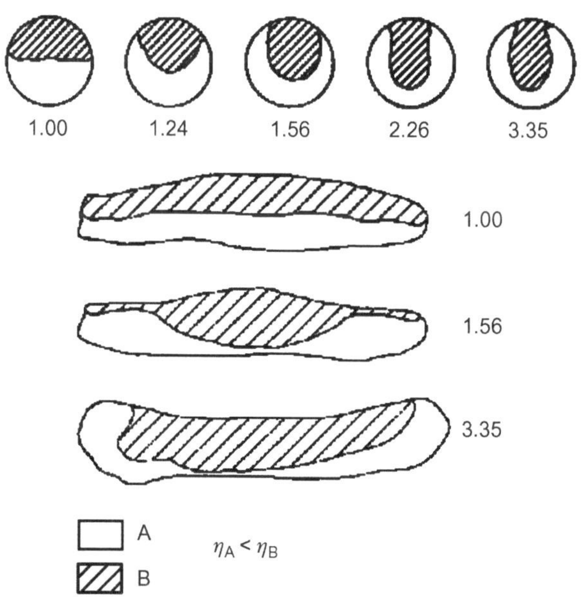

Figure 6.15 Layer-to-layer flow rearrangement as a function of time.

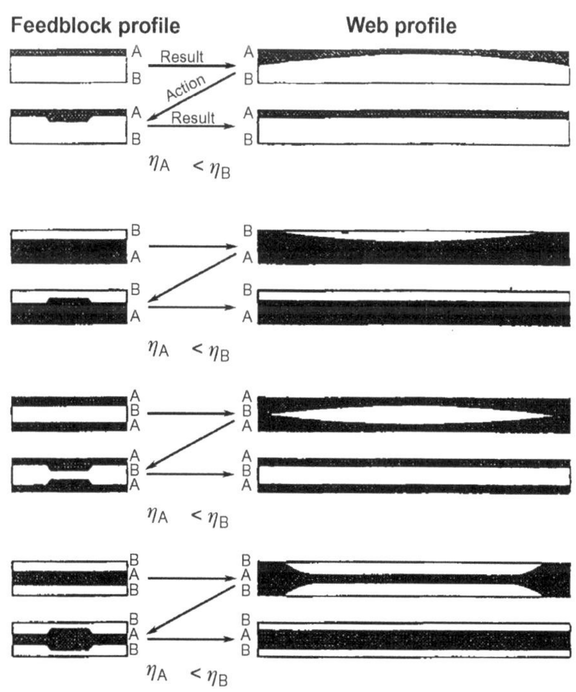

Figure 6.16 Feedblock profiling and the resultant effects.

when coextruding many layers, as profiling for one layer will disrupt the other layers. The influence of a feedblock design change is virtually impossible to predict computationally at present, even with the use of the most powerful 3D finite element flow simulation packages on powerful supercomputers.

Interfacial instability in coextrusion refers to two common defects consisting of highly irregular or sometimes regular waviness which appears in coextruded structures at the polymer/polymer interface. The effect is to reduce significantly the optical quality of coextruded film. It is an internal defect, which distinguishes it from sharkskin, which is a surface defect.

The most frequently encountered interfacial instability is zig-zag (also known as die land) instability, which appears as chevrons pointing in the flow direction. It is initiated in the die land and is characterized by a critical interfacial shear stress, in the range of 40–80 kPa (roughly ¼ to ½ of the critical wall shear stress level for sharkskin). Fig. 6.17 shows the effect of this instability on film clarity [26]. This problem can arise even if adjacent layers are the same material. The mechanism responsible has not been

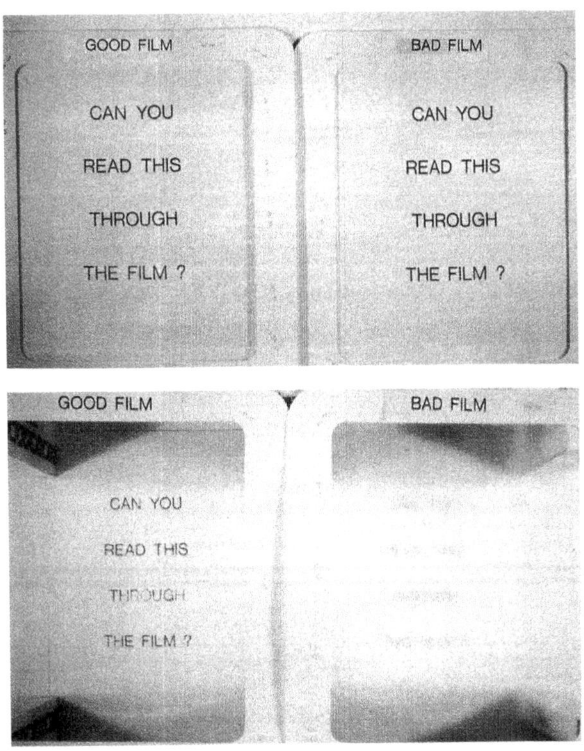

Figure 6.17 The effect of interfacial stability on contact clarity of coextruded films (top) versus see-through clarity (bottom).

conclusively identified. Apparently, there is amplification of certain disturbance wavelengths under high stress conditions [27]. Viscoelasticity is probably a contributing factor, that is, the interfacial normal stress difference is important [22]. Unfortunately, this is impossible to measure and difficult to calculate accurately. The most reliable means of diagnosing zig-zag instability at present is to calculate interfacial shear stress using simulation software.

Zig-zag instability problems are remedied by reducing interfacial shear stresses. The following actions are beneficial:

- decrease the total output rate (this reduces stresses everywhere)
- increase the skin layer thickness (this will shift the interface away from the wall where the shear stress is maximum)
- decrease the skin layer viscosity, for example, by raising its temperature or by using a less viscous polymer (this reduces stresses everywhere)
- increase the die gap (this reduces stresses everywhere)

Matching layer viscosity is a popular remedy that does not always work. In fact, as recommended above, it is often advisable intentionally to mismatch the viscosities by using a lower viscosity resin for the skin layer [28].

The less common interfacial instability is "wave" pattern instability, which appears as a train of parabolas spanning the width of the sheet and oriented in the flow direction. It occurs when a fast moving polymer stream merges with a much slower moving stream in a coextrusion feedblock. When the skin layer is thin relative to the second layer, the wave instability can be more pronounced. Large differences in extensional viscosities between adjacent layers can also make the defect more likely, as can a skin layer with a large extensional viscosity. The instability is aggravated by whatever flow or geometrical asymmetries might be present in the feedblock and die. As well, dies with larger lateral expansion ratios (die lip width divided by manifold entry width) and longer channel lengths (from feed slot vanes to die manifold) are more susceptible [29].

6.10 Rheology of Filled Polymers

Rigid particles added (up to 50% or greater) to polymers either improve the mechanical properties (reinforcements) or reduce the cost (fillers). Calcium carbonate is frequently used as a filler. Multilayer films with 30% by weight calcium carbonate are common. Talc increases stiffness and high temperature creep resistance. Rubber in some thermoplastics increases impact strength. Carbon black, glass fibers, wood flour, and other natural fibers are used for various applications. Good mixing is absolutely necessary to produce good-quality melt. Particle agglomerates must be broken up, (dispersive mixing) and they must be spatially distributed (distributive mixing). Questions regarding mixing will not be dealt with in this chapter. It is assumed that particles are well dispersed and distributed throughout the polymer melt.

Einstein solved the problem of resistance to shearing caused by spheres of neutral density and showed that the viscosity of the suspension η is related to the viscosity of the suspending fluid η_0

$$\eta = \eta_0(1 + 2.5\phi)$$

where:

ϕ	=	volume fraction, defined by the following equation
w	=	weight fraction ratio
ρ_c	=	particle density
ρ_m	=	matrix density

$$\phi = \frac{w/\rho_c}{[w/\rho_c + (1-w)/\rho_m]}$$

Einstein's equation is valid for very dilute suspensions (ϕ up to 0.01).

For higher concentrations the Einstein– Batchelor equation, which has also been theoretically derived, applies [30]

$$\eta = \eta_0(1 + 2.5\phi + 7.6\phi^2)$$

Under certain conditions the coefficient of ϕ^2 is 6.2, but 7.6 is preferable from experience. Theoreticians usually claim that this equation is valid up to $\phi = 0.1$. However, there is some evidence that it can be successfully used with higher fractions and even fibers.

Several semiempirical equations have been proposed for plotting the relative viscosity η/η_0 as a function of ϕ/ϕ_m. The most popular of these models is the Krieger–Dougherty equation [31].

$$\frac{\eta}{\eta_0} = \left(1 - \frac{\phi}{\phi_m}\right)^{\kappa\phi_m}$$

where:

ϕ_m	=	limiting concentration called the "maximum packing fraction" of the filler
κ	=	2.5 for spheres and it is usually referred to as the Einstein coefficient

For spheres packed on a simple cubic lattice, the maximum packing fraction will be

$$\phi_m = (4\pi/3)/8 = 0.524$$

while for random packing $\phi_m \approx 0.64$.

Adding fillers to a polymer melt decreases extrudate swell (d/D) and increases storage modulus G'. The solid polymer stiffness increases (higher tensile modulus) while tensile strength and elongation at failure decrease [32].

6.11 Rheology of Nanocomposites

Polymeric nanocomposites were first developed and commercialized by Toyota in the late 1980s. They dispersed layered silicate nanofillers in a polymer matrix. Nanosized particles have significant influence on end-use properties, such as improved mechanical properties, increased flame retardance, reduced gas and liquid permeability, reduced shrinkage, and altered electrical and optical properties [33].

There are three nanosized reinforcements:

- layered silicates (clays like montmorillonite and kaolinite) [34]
- carbon nanotubes [35]
- graphene [36]

Of particular interest, in packaging films, are polymer/clay nanocomposites due to their ability to form individual platelets when dispersed in a polymer matrix. These nanoplatelets do not allow molecules to diffuse through them, thus the gas molecules follow a tortuous path that effectively increases the film barrier properties. To achieve this, the silicate layers must be exfoliated, which means they must be well dispersed and randomly distributed throughout the polymer matrix.

Rheological trends are similar to other filled polymers, with the following differences:

- Loadings are low (less than 10%).
- The nanoscaled fillers have very large aspect ratios (length/thickness).
- The nanoscaled fillers have extremely large area to volume ratios.
- There are interactions at the molecular level.

For exfoliated silicate layered nanocomposites, the effect of nanoscale filler can best be summarized by the following statement, from Gopakumar et al. [37]: The complex viscosity η^* "increased substantially at low frequencies and the Newtonian plateau disappeared as clay content increased above 5 wt%. Increases of such magnitude arise in conventional composites at much higher filler loadings (above 20–30%) and may be attributed to weak structures that may remain intact at low frequencies."

6.12 Rheometers

The commercially available instruments for measuring the various rheological properties can be classified into the following categories:

- melt indexers
- capillary viscometers
- rotational rheometers
- torque rheometers

Melt indexers are inexpensive, and they are used for quality control. Measurements with the standard load (2.16 kg) and a high load (usually 21.6 or 10 kg) can provide useful information into the polymer's shear-thinning behavior and a rough estimate of power-law parameters as explained above.

Capillary instruments are equipped with one or two pistons for simultaneous measurements through a long and a very short (orifice) die. At least one vendor sells a three-bore instrument. Usually the shear rate is determined from the piston speed. The orifice die is used to measure the entrance pressure drop for calculating the Bagley end correction. These viscometers are usually accompanied by computer software for performing the Rabinowitsch and Bagley corrections. Also, the extensional viscosity according to Cogswell's method (from the entrance pressure drop) is easily calculated from two- or three-bore instruments. Performing the Bagley correction with single-bore instruments is considerably more time consuming. Separate measurements must be carried out with three or preferably four capillaries with different *L/D*. Then, the pressure drop versus *L/D* lines must be extrapolated to obtain the Bagley correction [6]. Capillary viscometers may also be supplemented with a pair of rollers for measuring the melt strength.

Rotational instruments that have either cone-and-plate or parallel disk geometry are operated in either steady or dynamic (oscillatory) modes. The steady rotational speed measurements are limited to lower shear rates (ie, less than 5 s^{-1}) because, at higher shear rates, secondary flows or flow instabilities occur that violate the simple shear flow requirement. In dynamic mode, rotational instruments are used extensively for determination of storage modulus G', loss modulus G'', and complex viscosity η^*, usually over a three-decade frequency range with the same transducer. To extend the range, the so-called time–temperature superposition is used, in which measurements are carried out at different temperatures and the curves are shifted horizontally and vertically [6] and may cover usually an eight-decade frequency range from 10^{-4} to 10^4. The low-frequency range is needed for determining relative elasticity levels, for example, broad molecular weight distribution polymers exhibit higher values of G' at low frequencies. Torque rheometers utilize mixing rotors, and the viscosity is determined from the measured torque and rotational speed [38,39]. Because the rotor blades have a complex geometry, neither the strain rate nor the temperature is uniform within the sample. These instruments are not suitable for accurate measurements, and they are mainly used for quality control purposes.

Extensional viscosity measurements are difficult to carry out. Cogwell's method from entrance pressure drop in capillaries provides only a rough approximation. Several instruments became commercially available over the years, but virtually all of them were plagued by miscellaneous complexities and experimental inaccuracies. A device has been developed [40] that can be attached to conventional rotational rheometers and appears to have captured a significant portion of the market.

A review of various commercially available instruments as well as the names of major suppliers was published [41] in 2004, and another review with more technical details has also appeared [42].

6.13 Concluding Remarks

Polymer resins are frequently sold on the basis of density and MI. However, MI is only just one point on an (apparent) viscosity curve at low shear rate. Plastics extrusion involves shear rates usually up to $1000\ s^{-1}$, in cavity filling in injection molding, the shear rate can reach $10{,}000\ s^{-1}$ and, in some coating operations, even higher shear rates are reached. For equipment and process design, troubleshooting, or optimization, it is important that the polymer's shear-thinning behavior is known. For a rough estimate of shear thinning, different loads in a melt indexer can be used. More accurate measurements require capillary and/or rotational rheometers. In capillary viscometry, the Rabinowitsch correction (for shear rate) and the Bagley correction (for shear stress) are usually necessary. At higher shear stress, around 0.1 MPa, wall slip occurs and slip velocity can be measured with the Mooney method.

Rotational rheometers (cone-and-plate or plate-plate) can also be used from very low shear rates of $10^{-2}\ s^{-1}$ to perhaps up to $5\ s^{-1}$. At higher values, the onset of secondary flows gives erroneous results. Rotational rheometers can be used in dynamic mode to yield storage (G') and loss moduli (G''). The complex viscosity versus frequency curve obtained from dynamic measurements coincides with the shear viscosity versus shear rate curve up to perhaps $1000\ s^{-1}$.

Other measurements might also be necessary to provide insight into a process or for differentiating polymers. These include the extensional viscosity and the first normal stress difference. Both these properties are difficult to measure and require special equipment. For extensional viscosity, the melt strength is an alternative and, for first normal stress difference, the storage modulus G' at low frequencies can be used.

Generally, rheological instruments are expensive and require knowledgeable people to run them. The engineer must understand his/her problem very well before deciding which rheological measurements are necessary for improving the understanding of the process or material behavior. Capillary viscosity measurements are perhaps more useful in understanding the processes, while dynamic measurements give more insight into the structure of polymer resins.

References

[1] Vlachopoulos J, Wagner JR, editors. The SPE guide on extrusion technology and troubleshooting. Brookfield, CT: Society of Plastics Engineers; 2001.

[2] Macosko CW. Rheology: principles, measurements and applications. New York: VCH Publishers; 1994.

[3] Cogswell FN. Polymer melt rheology. Cambridge: Woodhead Publishing; 1996.

[4] Chenoy AV, Saini DR. Thermoplastic melt rheology and processing. New York: Marcel Dekker; 1996.

[5] Hatzikiriakos SG, Migler K. Polymer processing instabilities: control and understanding. New York: Marcel Dekker; 2004.

[6] Dealy JM, Wissbrun KF. Melt rheology and its role in plastics processing. London: Chapman and Hall; 1996.

[7] Vlachopoulos J. Rev Def Beh Mat 1981;3:219.

[8] Debbaut B, Avalosse T, Dooley J, Hughes KJ. Non-Newt Fluid Mech 1997;69:255.

[9] Bird RB, Armstrong RC, Hassager O. Dynamics of polymeric liquids, vol. I. New York: Wiley; 1987.

[10] Tanner RI. Engineering rheology. Oxford: Oxford Engineering Science; 2000.

[11] Baird DG, Collias DJ. Polymer processing principles and design. New York: Wiley; 1998.

[12] Mitsoulis EJ. Non-Newt Fluid Mech 2001;97:13.

[13] Marrucci G, Ianniruberto G. J Rheol 2003;47:267.

[14] McLeish TCB, Larson RGJ. Rheology 1998;42:81.

[15] Rutgers R, Mackley MJ. Rheology 2000;44:1319.

[16] Denn MM. Ann Rev Fluid Mech 2001;33:265.

[17] Ramamurthy AV. In: Proceedings of Xth Intl Cong, Rheo, Sydney; 1988.

[18] Achilleos EC, Georgiou G, Hatzikiriakos SGJ. Vinyl Addit Technol 2002;8:7.

[19] Moynihan RH. The flow at polymer and metal interfaces [Ph.D. thesis]. Dept of Chem Eng, Virginia Tech; 1990.
[20] Gander JD, Giacomin J. Polym Eng Sci 1997; 37:1113.
[21] Han CD. Multiphase flow in polymer processing. New York: Academic Press; 1981.
[22] Meier M. Fliessprobleme in Mehrschichtstroemungen [Doctoral thesis]. Germany: IKV, TH Aachen; 1990.
[23] Minagawa N, White JL. Polym Eng Sci 1975; 15:825.
[24] Dooley J. Viscoelastic flow effects in multilayer polymer coextrusion [Doctoral thesis]. Netherlands: U. Eindhoven; 2002.
[25] Cloeren P. In: Proceedings of SPE Extrusion Division RETEC, Toronto, Canada; 1993.
[26] Shroff R, Mavridis H. Plast Technol February 1991;54.
[27] Perdikoulias J, Tzoganakis C. Plast Eng 1996; 52(4):41.
[28] Butler TJ, editor. Film extrusion manual. Atlanta, GA: TAPPI Press; 2005.
[29] Ramanathan R, Shanker R, Rehg T, Jons S, Headley DL, Schrenk WJ. SPE ANTEC Tech. Papers 1996;42:224.
[30] Housiadas KD, Tanner RI. Rheol Acta 2014;53:831.
[31] Gupta RK. Polymer and composite rheology. New York: Marcel-Dekker; 2000.
[32] Leblanc JL. Filled polymers. New York: CRC Press; 2010.
[33] Bhattacharya SN, Gupta RK, Kamal MR. Polymeric nanocomposites. Münich: Hanser; 2008.
[34] Paul DR, Robeson ML. Polymer 2008;49:3187.
[35] Ma P-C, Siddiqui NA, Marom G, Kim J-K. Compos Part A 2010;41:1345.
[36] Kim H, Abdala AA, Macosko CW. Macromolecules 2010;43:6515.
[37] Gopakumar TG, Lee JA, Kontopoulou M, Parent JS. Polymer 2002;43:5483.
[38] Bousmina M, Ait-Kadi A, Faisant JB. J Rheol 1999;43:415.
[39] Santi CR, Hage Jr E, Correa CA, Vlachopoulos J. Appl Rheol 2009;19:13148.
[40] Sentmanat M, Wang BN, McKinley GH. J Rheol 2005;49:585.
[41] Manolis Sherman L. Plast Technol May 2004.
[42] Hou YY, Kassim HO. Rev Sci Instrum 2005; 76:101101.
[43] Shang SW. Adv Polym Tech 1993;12:389.

PART III

Technologies

7 Coextrusion Equipment for Multilayer Flat Films and Sheets

Eldridge M. Mount, III
EMMOUNT Technologies, Canandaigua, NY, United States

7.1 Designing and Manufacturing Multilayer Flexible Plastic Films with Multilayer Coextrusion Die Technology

7.1.1 Overview

Several generations of coextrusion film dies exist for both tubular and flat dies, as well as several techniques for combining molten polymers for the production of multilayer polymer films. For this discussion, we will focus solely on the production of flat sheets and films. The purpose of a film or sheet die is to produce a polymer film with uniform thickness dimensions across its width which is suitable for any subsequent film process. For a multilayer system, the production of uniform layer dimensions is added to the overall requirement of uniform total thickness. The number of polymers which can be combined is limited primarily by the ability to locate extruders to feed the molten polymers to the coextrusion system and will depend in part on the method used to perform the melt combination.

The most general configuration for coextrusion systems consists of the individual extrusion and melt systems used to prepare the molten polymers (extruders, filters, and melt pumps), the melt pipes used to deliver the polymers from the extruders, the adapter to gather and route the melts from the extruders, a coextrusion feedblock to shape and combine the melts and a die to spread, thin, and distribute uniformly the melt to form the coextruded film.

In producing flat multilayer films, there are two principal approaches: feedblock coextrusion and multicavity die coextrusion. In addition, it is possible and, oftentimes advantageous, to combine feedblock and multicavity dies together. In some instances, the adapter and feedblock can be combined together into a single structure and, if this is done, it becomes especially important to determine the adapter/feedblock size to permit future coextrusion system changes without disturbing the melt pipe and extrusion systems.

Because of this ability to combine sections and technologies for multilayer films and sheets together, there is a fair amount of strategy which must be developed when establishing a coextrusion system, especially in relationship to the number and thicknesses of the layers to be combined. Care should be taken to define adequately the products necessary or desired to be manufactured, so that initial equipment choices do not inadvertently limit equipment changes for future products or prevent cost-effective modifications to the equipment layout. It is of primary importance to size and locate properly the adapter used to collect and route the melt streams to the feedblock and/or die to permit the use of different feedblock and die combinations without the need to replace existing melt pipes or to disturb the existing extruder layouts. Changes which require new melt pipes and extruder relocations are costly and time consuming, and potential costs or lost production may prevent the changes from being made due to costs. To a large extent, the adapter can be viewed as a simple extension of the melt pipes, but adapters can also serve the function of routing various melt streams to various inlet locations on feedblocks or dies. This routing feature has been added to various adapter designs using interchangeable flow plates or selector plugs and adds a great deal of flexibility to coextrusion systems, especially when polymer-specific extrusion systems are required. This feature permits changing a polymer from one surface layer to another surface or interior layer without the need to change

Multilayer Flexible Packaging. DOI: http://dx.doi.org/10.1016/B978-0-323-37100-1.00007-7

extruders. This can be important, for instance, if the screw design or melt transfer pipes are optimized for the particular material being extruded and cannot be easily extruded with another screw or melt pipe design. The alternative to routing in the adapter would be to change the screw in the extrusion system, a more difficult and time-consuming operation than changing an adapter system or adapter plate.

7.1.2 Introduction

Multilayer films have evolved over the last several decades as new polymers and processing technologies have become available. Originally, multilayer packaging structures were produced from single-layer film products which were glued together by several lamination processes or were coated with additional polymer layers. Often times, the polymer films, such as cellophane, which were available, were not melt extrudable, and it was not possible to produce multilayer products directly as it is today by coextrusion. At times, the multilayer structures were produced containing nonpolymer materials such as aluminum foil to supply the desired packaging properties such as light and gas barriers not available in polymer films.

Coextrusion of multilayer films became possible as new polymers were developed which expanded the available polymer properties and, as new copolymerization technologies were commercialized, the compatible polymers were increased. At the same time, new film manufacturing technologies were developed which further enhanced the polymer properties and permitted the production of thinner film combinations. As polymer layers became thinner, it became impractical to combine the layers together after they were produced as films and, in some cases, the required polymer thickness necessary for the layer was far less than would be practical to produce and handle as a single layer for subsequent lamination. In these cases, lamination resulted in more expensive combinations than were necessary. All of these factors drove early film developers to look for better ways to combine various polymers together without the need to produce individual free-standing film layers for later combination, and coextrusion was born.

Polymer processing of thermoplastic materials consists of transforming solid polymers in easy-to-transport pellet form into useful shapes and combinations and is most often performed by an extrusion-based process where the solid polymer is melted, mixed if necessary to form a homogeneous melt and then pumped to a die where it is shaped and, on exiting the die, is solidified by cooling. It is the shaping of the polymer by the die which differentiates the many polymer processes from each other and which is critically important in producing multilayer films. It should be noted at this point that once the polymer leaves the extruder and enters the coextrusion flow system the flow process is said to be pressure flow or Poiseuille flow.

In this chapter, we will focus on the production of thin multilayer films. In some instances, the films will be later oriented and therefore the film passes through a relatively thick sheet stage as it exits the die after the coextrusion process. This cast sheet, which is formed as a precursor to the multilayer film orientation step, is, in many regards, similar to the multilayer sheets formed for subsequent thermoforming. Consequently, much of what is discussed here is applicable to the coextrusion of multilayer sheet products as well.

The molten polymers produced in extrusion are, in general, high viscosity fluids which exhibit low Reynolds numbers, or a general description might be that of creeping, laminar flow in the various materials and flow geometries, and therefore it would appear that the combinations of various polymer layers would be straightforward. However, the polymers are generally viscoelastic which will add some complexity to the flow of the polymer layer combinations and the choice of the layer combination methods. But, in general, the combination of polymer melts into layered flows is fairly straightforward and has been developed successfully over the years into a highly evolved set of techniques and equipment categories. In general, many polymer melts may be successfully combined before a single cavity film die followed by melt spreading in the die by what is known as feedblock coextrusion technology. Alternatively, the melts may be extruded in multicavity dies where the individual molten polymers are first spread and distributed before their combination. Both feedblocks and multicavity dies are common today and are also used in combination together; both technologies have their advantages and disadvantages in terms of cost and performance as well as process flexibility. Consequently, care should be taken to define the most appropriate technology or combination of technologies when establishing or modifying a coextrusion system so that future modifications in die and feedblock technology, product designs, and infrastructure of the film lines are not compromised or excessively expensive. Today, films with two to several hundred individual layers are common.

7.1.3 General Coextrusion Equipment Overview

A coextrusion system is a combination of parts which takes molten polymer from two or more extruders, routes and shapes the melt, and delivers it to the combination point where the melts are stacked and then delivered to a die for spreading to the final width (Fig. 7.1). In its most general configuration, the coextrusion system will consist of an adapter, a feedblock, and a single or multicavity die. Each of these components and their functions will be described in the following sections.

7.1.3.1 The Adapter

The adapter is designed to collect and route the various melt streams to the feedblock and/or die used to produce the multilayer film and sheet. Consequently, the adaptor's selection is a key strategic decision in selecting a coextrusion system and of critical importance as it can be the key to future changes in coextruded product design because it can control the magnitude of the equipment change necessary. Adaptors can be stand-alone parts or can be made integral with feedblocks or dies. The adaptor is the bridge between the extrusion and the coextrusion system. The adaptor can be viewed as an extension of the melt system or as the coupling of the melt system to the coextrusion system.

The adaptor should be sized such that it permits the future addition of layers to an existing coextrusion system without the need to replace existing melt system components (melt pipes, filters, melt pumps, etc.) (eg, in the production of five-layer films and sheets by the use of a three-layer feedblock attached to a three-cavity die). In this case, the adapter should be physically sized such that it permits the future addition of two additional feedblocks to the system without the need to change the

Figure 7.1 General configuration of the coextrusion system consisting of an adapter, a feedblock, and a single or multicavity die.

adaptor dimensions. If this is not done during the initial coextrusion system design, the larger physical dimensions, typical of a feedblock system for multiple die cavities, compared to a single cavity feedblock system, will generally require that new melt pipes be obtained to replace the existing melt pipes. An alternative to new melt pipes would be to relocate the extrusion or downstream equipment to accommodate the increased adaptor section dimensions. This can significantly increase the cost and installation time for an upgraded coextrusion system and, in some cases, makes the change economically unjustifiable.

Adaptors can have fixed flow channels or can be made such that they allow multiple flow paths of the melts between the adaptor inlet and the adaptor exit to the feedblock and/or die. This can be accomplished with segmented adaptors with interchangeable body segments to reroute flow or with removable plugs. The decision between fixed and interchangeable adaptors is a balance between product and process flexibility and melt stream integrity.

In fixed geometry adaptors (Fig. 7.2), there is little to no chance of melt stream cross-contamination. However, changing polymer layer positions requires either a change in adaptor or changes in resin feed streams (and perhaps screw designs) to the extruders. Both require some downtime to make the change and to reestablish the manufacture of acceptable product.

With interchangeable flow path adapters (Fig. 7.3), there is, perhaps, some possibility of cross-contamination between melt streams, but this is generally not a problem. However, the increase in ease of layer position changes can be dramatically increased depending on the adapter design.

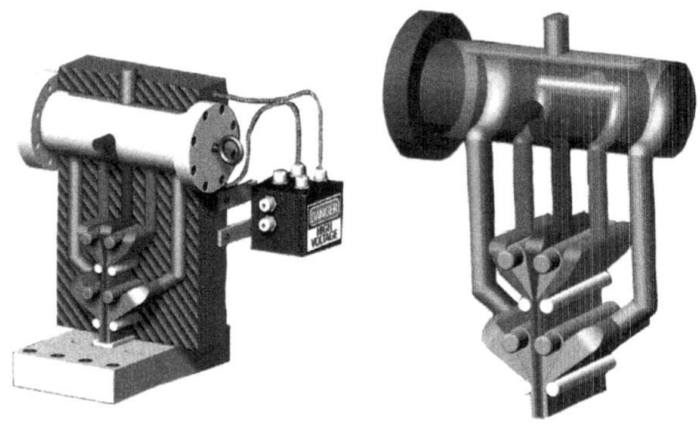

Figure 7.3 Adapter with removable flow selector showing position in adapter and the polymer flow paths.

For segmented adapters (Fig. 7.4), the change will require removal and rebuilding of the adapter, whereas adapters with interchangeable inserts will

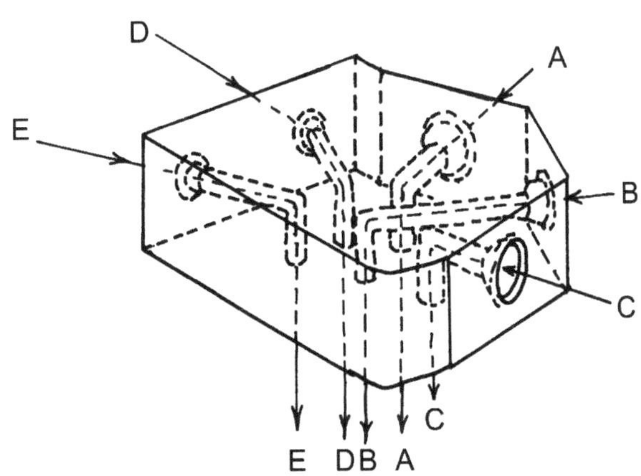

Figure 7.2 Fixed geometry adapter showing inflow from extruders A, B, C, D, and E, and the outflow of each extruder to the feed block. The adapter flow locations cannot be changed without replacing the adapter.

Figure 7.4 Segmented adapter with removable programming plates for rerouting of melt flow from the adapter to the feedblock.

only require removal, some cleaning, and reinsertion of the new flow insert and can be accomplished with minimal losses in production.

Therefore, as with adapter sizing, adapter flow path control is a strategic decision which is dependent on the expected number and frequency of product design changes, the sensitivity of layer polymers to extrusion (screw) technology, the infrastructure of the plant for resin rerouting, and the ability of the die cavities to maintain acceptable layer dimensions across the film or sheet as polymers are changed.

7.1.3.2 The Feedblock

The purpose of the feedblock is to shape and combine the multiple polymers entering from the adapter into a well-defined stack of polymers for delivery to the flat die. Feedblock technology exists in several basic geometries and combination philosophies ranging from the segmented flow of the Dow feedblocks (Fig. 7.5), the stepwise addition of the Welex modular designs (Fig. 7.6) with interchangeable cassette inserts, and the Cloeren style feedblock with adjustable vane designs (Fig. 7.7).

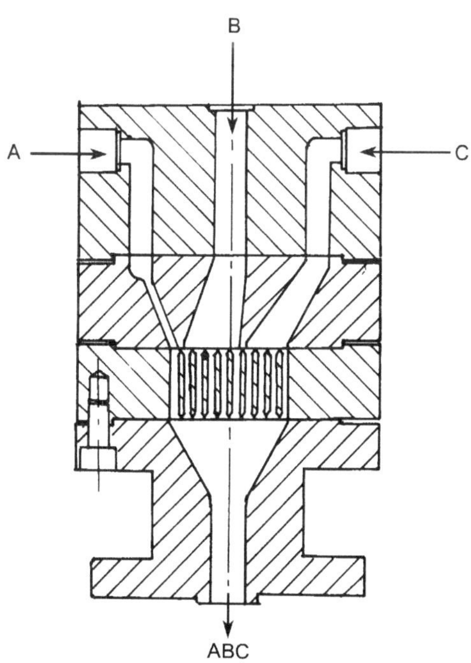

Figure 7.5 Three-layer Dow feedblock showing segmented flow plates.

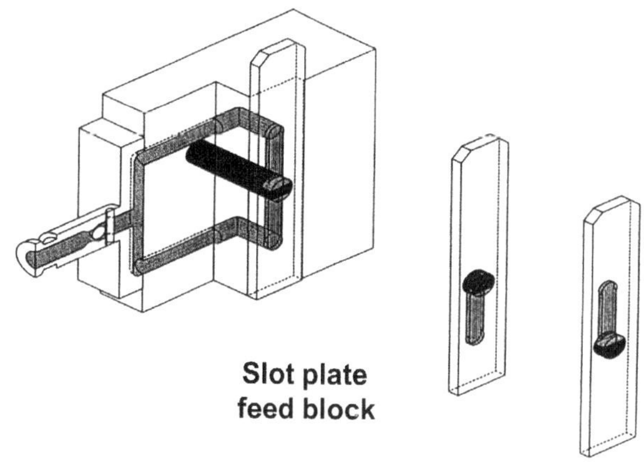

Figure 7.6 WELEX modular adapter showing various interchangeable cassette inserts. Adapters may be combined sequentially to permit multiple layer addition.

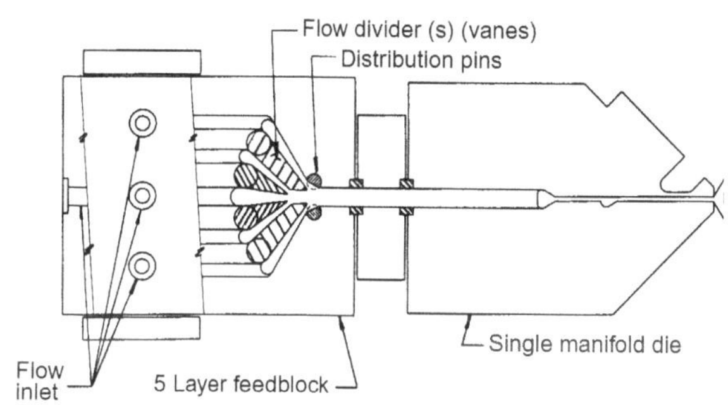

Figure 7.7 Cloeren five-layer adjustable vane feedblock.

Overall, these feedblock designs can be broken down into two design types: fixed geometry feed blocks, Dow and Welex, and the variable geometry feedblock, Cloeren. All three feedblocks can be modified by the use of interchangeable parts, but only the Cloeren vane feedblock may be adjusted during operation.

Feedblock technology has evolved over the last several decades from simple beginnings to the current art where feedblocks are very flexible and effective. The understanding of the design goals for feedblocks and a better understanding of how to accommodate interface migration and layer encapsulation, which is common in simple feedblocks, has evolved. Although not a lot of analytical work has been published, recent experimental efforts are highlighting the impact of flow channel geometry and melt elasticity [1] and elongational viscosity [2] as key factors in maintaining layer integrity.

7.1.3.3 The Die

As with feedblocks, the dies have evolved from simple forms to the modern coextrusion dies. In general, there are two principal approaches to coextrusion with dies: the single-manifold die, combined with a feedblock, and the multimanifold coextrusion die. The multimanifold die may also be combined with a feedblock. Originally, the use of multimanifold dies was limited more by manufacturing methods and the difficulty in producing multiple die body sections, which greatly increased their costs relative to single-manifold dies. However, with the revolution of multiaxis computer numerical controlled (CNC) machining capabilities, the machining difficulties have been greatly diminished and their availability has increased. The improved CNC capabilities also permit the machining of the die manifolds to the optimum flow design with fewer limitations imposed on manifold and preland geometry by the die manufacturer. This has led to the ready availability of multilayer dies with as many as five or more individual manifolds. The number of individual flow cavities in a die is somewhat limited by the die width as the cost and mechanical complexity of the die increases with each additional cavity.

7.1.4 Theoretical Understanding of Die and Feedblock Design Parameters

7.1.4.1 Rheological Background

There are several concepts which must be comprehended before the attempt is made to describe key die and feedblock design parameters. These concepts relate to the material science of polymer melts and the science of rheology. Although a complete description of polymer melts and rheology is beyond the scope of this chapter, it should be possible to state and highlight the key concepts, and the interested reader can then turn to the literature or textbooks on the subjects and find the derivations and background assumptions. Our first observation is that, within the scope of this chapter, we are dealing with fluids and that the fluids are forced to flow through the coextrusion equipment by pressure alone. Because the flow is a pressure-driven flow, we will be interested in the relationship between the volumetric output (Q) and the pressure or pressure drop (ΔP), required to sustain it. It will be from the solutions to the conservation equations (mass, energy, and momentum) from which we obtain the various flow models used to design and trouble shoot coextrusion feedblocks and dies.

The development of the flow models also requires the use of a constitutive equation for the material and an equation of state if the material is compressible. The constitutive equation relates the material flow behavior in a stress field and defines the relationship between the stress components and the rate of strain (deformation) in the fluid. This constitutive relationship must be determined experimentally for each material and is generally depicted as the flow curve which plots the stress versus the rate of deformation or, more commonly, the shear rate. The shear rate is derived from the first invariant of the rate of deformation tensor which will be dependent on the geometry of the flow. For complex flows, the material behavior is difficult to measure and, therefore, it is a common practice to use rheologically "simple" flows to characterize the materials. Flows which meet this rheologically "simple" requirement uniquely relate the stress (τ_{xy}) components in the fluid to the shear rate by a single material constant, the shear viscosity (η). The viscosity is defined as the ratio of the shear stress to the shear rate.

$$\text{Shear viscosity} = \frac{\text{Shear stress}}{\text{Shear rate}} \qquad (7.1)$$

$$\eta = \tau_{xy}\dot{\gamma}$$

Representative simple flows are the pressure flow in a tube (Poiseuille flow), the drag flow in a narrow gap between rotating cylinders (Couette flow), and the flow in a narrow gap between a flat plate and a low angle cone. These three flows form the basis for instruments used to measure material viscosity as a function of temperature and shear rate, and each instrument has an effective range of shear rates over which it can operate. Shear rates of 0.0001–0.1 s^{-1} for a cone and plate, 0.001–10 s^{-1} perhaps for a Couette viscometer, and 0.1–10 000 s^{-1} for a capillary rheometer.

As explained above, it is necessary to obtain the fluid flow behavior in order to perform the die design calculations as well as to derive the theoretical flow models used in the die design. Fig. 7.8 displays the flow curve, log shear stress versus log shear rate, for various types of flow properties found for real materials.

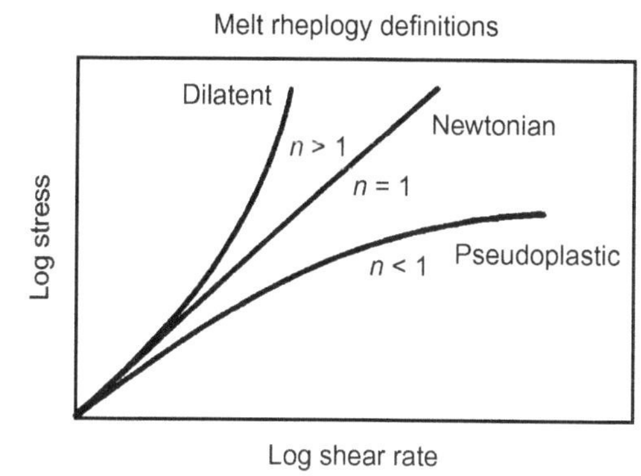

Figure 7.8 Flow curves of ideal materials.

In Fig. 7.8, we see the flow curves for a shear thickening or dilatent fluid ($n > 1$), a Newtonian fluid ($n = 1$), and a shear thinning ($n < 1$) or pseudoplastic material. Also noted in Fig. 7.8 is n, the power-law index which is the slope of the log (shear stress) versus log (shear rate) curve.

An example of a dilatant material would be wet sand or fluid–particle mixtures, such as ketchup, and the shear stress increases in a shear field due to the building or forming of an internal structure in the fluid which retards flow. Newtonian fluids, such as water, show a linear relationship between increasing shear stress and shear rate, whereas pseudoplastic materials, representative of many polymer melts, show a shear thinning behavior where the shear stress decreases with increasing shear rate due to breakdown of the entangled chain molecular structure generally assumed for polymer melts. Real polymer melts show a Newtonian viscosity at very low or zero shear rates with a transition to a pseudoplastic behavior as the shear rate is increased (Fig. 7.9).

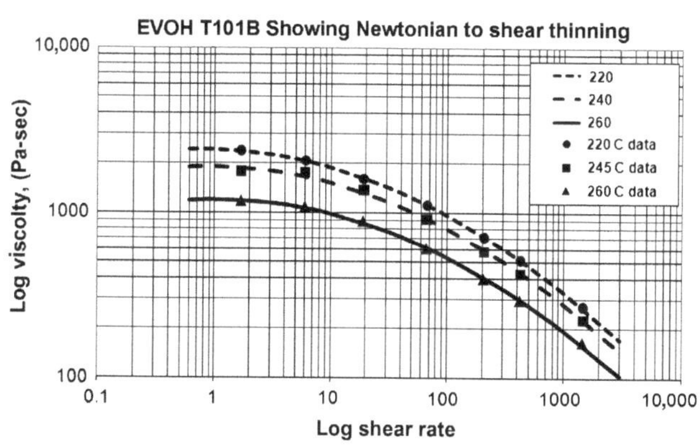

Figure 7.9 Newtonian to pseudoplastic flow for EVOH polymer.

In general, the Newtonian or zero shear viscosity, the shear rate where the transition to shear thinning behavior occurs, is different for different polymers and can occur at shear rates as low as $0.01\ s^{-1}$ to as high as several hundred per second as shown for various materials in Fig. 7.10. As discussed in the following, this variation in apparent shear thinning between shear rates of 10 and $1000\ s^{-1}$ will have an impact on successfully maintaining uniform layer distribution and the interchangeability of polymers in existing dies and feedblocks.

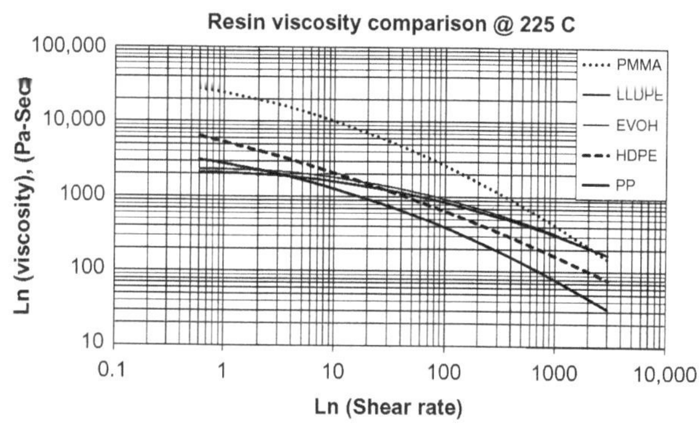

Figure 7.10 Comparison of flow behavior of several resin types.

In polymer melt processing, it is a common engineering practice to use the flow data collected from a capillary rheometer, because the shear rate range of the instruments, approximately $0.1–10{,}000\ s^{-1}$, covers the practical range of shear rates observed in most equipment (Fig. 7.11).

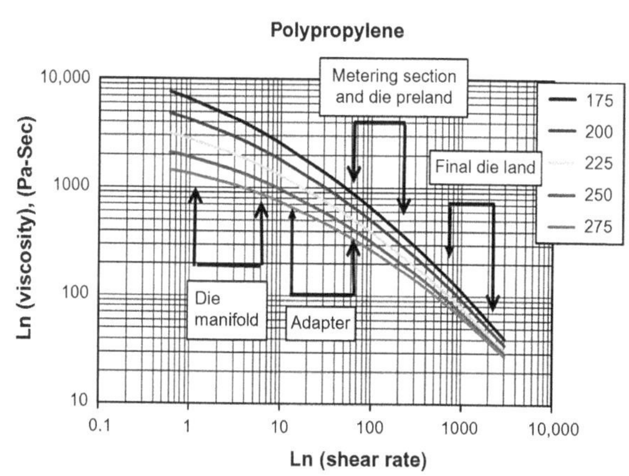

Figure 7.11 Shear rate ranges in various locations in multilayer extrusion equipment.

The capillary rheometer also permits the determination of the fluid viscosity without the need to assume its constitutive behavior by the use of the Rabinowitsch method or correction. Because of this, the Rabinowitsch correction should always be applied to capillary data, eliminating the need to use the apparent viscosity (assumed Newtonian fluid model). However, to characterize completely a polymer melt, it is necessary to obtain cone-and-plate as well as capillary data. Capillary data are generally available and the equipment is relatively easy to operate, but cone-and-plate experiments at low shear rates are extremely time consuming and not readily available. A useful approximation of the zero shear viscosity can be obtained using the method described by Gillespie [3] and would permit an engineering approximation of the zero shear viscosity from capillary data if it were not possible to measure the low shear rate data directly.

It has been a common practice in the development of polymer flow modeling to assume either a Newtonian or power law constitutive equation for the model development. Fig. 7.12 plots the viscosity versus shear rate for a Newtonian, a power law, and a typical polymer melt.

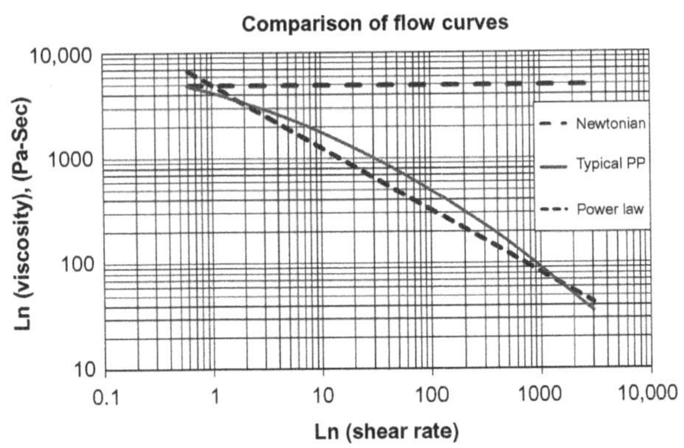

Figure 7.12 The viscosity versus shear rate for a Newtonian, a power law, and a typical polypropylene (PP) melt.

It is clear from Fig. 7.12 that neither the Newtonian or power law fluid accurately represents the real behavior of the polymer melt over the measured shear rate range but, in general, the power law is a much better general representation of the polymer melt at high shear rates. The flow curve can also be piecewise approximated by a series of power law sections and point wise by the first derivative (slope) of the flow curve.

Another aspect of the rheology of dies and feedblocks which will be important in understanding the design and performance of coextrusion equipment is the solutions obtained for the flow in channels with various cross sections. Dies and feedblocks comprised primarily circular, rectangular, and converging flow channels. In addition, die manifolds may have teardrop cross-section flow channels. Therefore, it will be important to understand the relationships between the polymer flow rate, the pressure drops, and the die geometry. This is done by solving the flow problem for the fluid constitutive equation in the geometry describing the die or feedblock. This can be a very complicated calculation and can be done with advanced computer algorithms, a description of which is beyond the scope of this chapter. Instead, we will focus on the analytical solutions obtained from simpler constitutive equations, to gain an improved understanding of the factors relating to the geometry and the fluid characteristics and how they combine to determine the die design. This understanding will aid us in determining when it is possible to interchange two different polymers in a die and what outcome we might expect in terms of system pressure drop as well as layer distribution uniformity.

The primary goal of the feedblock and die designs is to stack and spread the melt layers and have a uniform product exit the die. This is done by designing for a uniform flow rate across the final exit plane of the die (Fig. 7.13). Because the flow is pressure driven, the flow rate is controlled by the pressure drop along all the flow paths from the die entrance to the die exit.

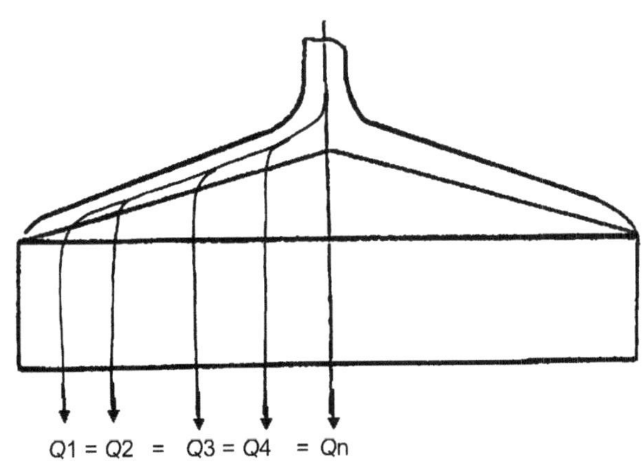

Figure 7.13 Desired uniform volumetric flow rate (Qi) across width of die.

Therefore, understanding the model results for circular and rectangular flow channels will be important. Fig. 7.14 shows the flow problem for flow in a circular pipe, and Fig. 7.15 shows the flow problem for flow in a rectangular slit.

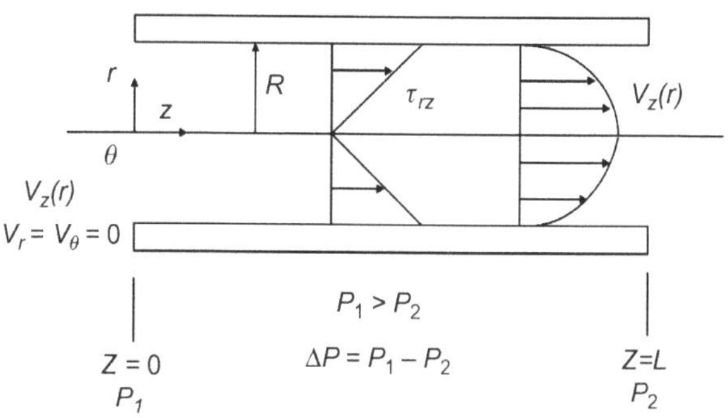

Figure 7.14 Flow problem for circular pipe.

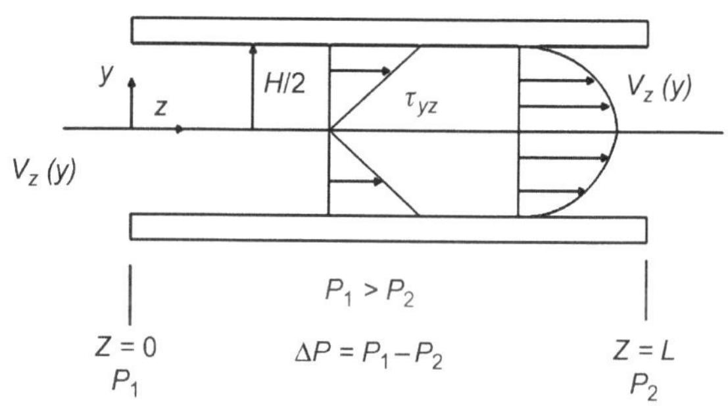

Figure 7.15 Flow problem for parallel plates.

Writing the differential equation for the flow in the pipe, we find:

$$\frac{dP}{dz} = \frac{1}{r}\frac{d[r\tau_{rz}]}{dz} \tag{7.2}$$

Integrating from $r = 0$ to $r = R$ yields,

$$\tau_{rz} = \frac{r\,dP}{2\,dz} \tag{7.3}$$

Equation (7.3) shows that the shear stress varies linearly with position across the tube diameter from a value of zero at the center line to a maximum at the wall where $r = R$. It should be noted that this result is independent of the fluid type. The material constitutive behavior will be important in determining the velocity profile in the tube. The velocity profile (Eq. [7.4]) is determined by substituting the shear stress definition into Eq. [7.3], rearranging and solving for the velocity profile and, at this point, the fluid type, Newtonian, power law, or other fluid type, will be important in determining the velocity profile shape.

$$\begin{aligned} \tau_{rz} &= \eta\dot{\gamma} \\ \eta\frac{dV_z}{dr} &= \frac{r\,dP}{2\,dz} \\ V_z(r) &= \int_0^R \left(\frac{r\,dP/dz}{2\,\eta(T,\dot{\gamma})}\right)dr \end{aligned} \tag{7.4}$$

In order to relate the output, Q, of the tube to the pressure gradient driving the flow, it is necessary to integrate the velocity profile at the tube exit across the radius of the tube,

$$Q = \int_0^R V_z(r)dr \tag{7.5}$$

Substituting in the definitions for a Newtonian and a power law fluid (Eq. [7.6]) into Eq. [7.4] and solving equation (7.5) gives Eq. [7.7] and Eq. [7.8], respectively;

$$\begin{aligned} &\eta = \mu \quad \text{Newtonian fluid} \\ &\eta = \eta_0\left[\frac{\dot{\gamma}}{\dot{\gamma}_0}\right]^{n-1} \quad \text{Power law fluid} \end{aligned} \tag{7.6}$$

For the case of the tube the results for a Newtonian fluid are:

$$\begin{aligned} \nu_z &= \nu_0\left[1 - \left(\frac{r}{R}\right)^2\right] \\ Q &= -\frac{\pi R^4 \Delta P}{8\,\mu L} \end{aligned} \tag{7.7}$$

And for a power law fluid:

$$\begin{aligned} V_z &= V_0\left[1 - \left(\frac{r}{R}\right)^{(n-1)/n}\right] \\ Q &= -\left(\frac{n\pi R^3\dot{\gamma}_0}{3n+1}\right)\left[\frac{R\Delta P}{2\eta_0\dot{\gamma}_0 L}\right]^{1/n} \end{aligned} \tag{7.8}$$

And when $n = 1$ the power law results yields the Newtonian results.

Fig. 7.16 displays the velocity profiles in a tube for a Newtonian fluid and for various values of the power law index, n. For flow of the Newtonian fluid, the velocity profile is parabolic with a maximum melt velocity at the center line of the tube and a zero velocity at the wall. As the slope of

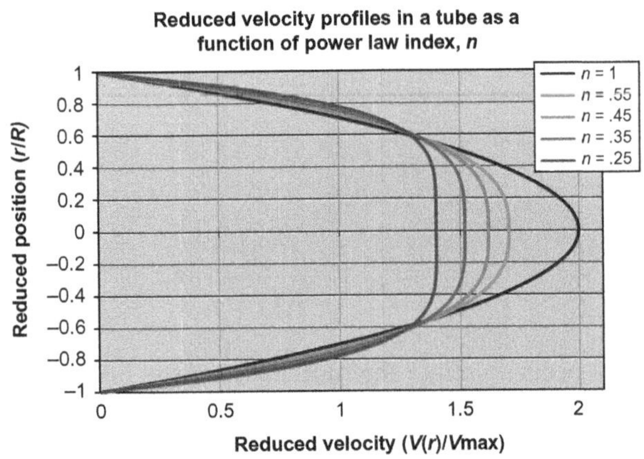

Figure 7.16 Plot of reduced velocity profiles for flow in a tube as a function of the power law index, *n* showing the blunt flow (plug flow) front increasing as *n* decreases.

the velocity profile represents the shear rate in the fluid, we find that the shear rate at the tube center line is zero and the shear rate at the wall is a maximum. For the power law fluid, the velocity profile is a blunt or plug-like flow with the plug-like nature of the velocity profile increasing with decreasing power law index.

Recalling the flow curve for a real polymer in Fig. 7.10, we understand from the shear rate dependence of the viscosity that the melt viscosity will not be uniform across the tube and will be at a maximum viscosity at the center and a minimum viscosity at the wall, even for a single polymer at a uniform temperature. This will be an important point to recall when understanding the need to maintain a decreasing viscosity across the flow of a coextruded structure, and the driving force for flow rearrangement and encapsulation flow in feedblocks and die manifolds are to be discussed later.

For flow in a thin wide slit the differential equation describing the flow in Fig. 7.15 is:

$$\frac{dP}{dz} = \frac{d\tau_{zy}}{dy} \tag{7.9}$$

which on integration from − H/2 to H/2 yields

$$\tau_{zy} = \frac{dP}{dz}y \tag{7.10}$$

Again, the solution gives a linear dependence of stress with position in the slit with a zero shear stress at the center line and a maximum stress at the wall of the slit. Performing the same analysis for the slit as was done for the tube to determine the velocity profile in the slit and output from the slit for a Newtonian fluid yields:

$$\nu_z = \nu_0\left[1 - \left|\frac{2y}{H}\right|^2\right] \qquad Q = \frac{WH^3\Delta P}{12\,\mu L} \tag{7.11}$$

And for a power law fluid:

$$\nu_z = \nu_0\left[1 - \left|\frac{2y}{H}\right|^{\frac{n+1}{n}}\right] \qquad Q = \frac{WH^2}{2\left(\frac{1+2n}{n}\right)}\left(\frac{H\Delta P}{2\eta L}\right)^{\frac{1}{n}} \tag{7.12}$$

Hence, when $n = 1$ is substituted into the power law results the Newtonian results are recovered.

Fig. 7.17 plots the velocity profiles for flow in a thin wide slit and, as was found for the tube, the velocity profiles are parabolic for the Newtonian fluid and plug-like for the power law fluid. Also the shear rate will be zero at the slit center line and a maximum at the slit walls with a maximum viscosity at the center and a minimum viscosity at the slit walls.

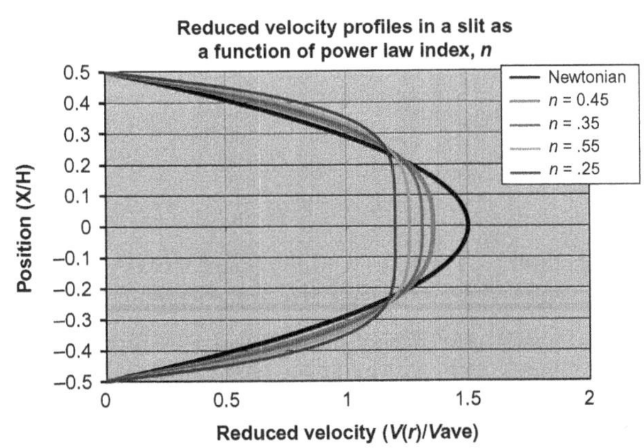

Figure 7.17 Plot of reduced velocity profiles for flow in a slit as a function of the power law index, *n* showing the blunt flow (plug flow) front increasing as *n* decreases.

7.1.5 Feedblock Designs

In general, the feedblock is designed to arrange the melt layers before entry into a die where the layer stack is spread to a much wider and thinner geometry than that in the feedblock. As we will see in the discussion on dies, the die manifold geometry into which the feedblock delivers the melt stack will have a profound impact on the final coextrusion quality.

There are several feedblock designs generally available for use in coextrusion systems today. In general, the melt is combined in rather large cross sections characterized by low shear rates. In general, the melt stack delivered to a die channel for subsequent spreading can have a round or rectangular cross section. Rectangular flows will have an aspect ratio of width to thickness from 2×1 to 4×1 and the corners of the rectangle will be rounded to remove stagnation points and in the limit are oval in cross section (Fig. 7.18) and are often times described as a racetrack entrance.

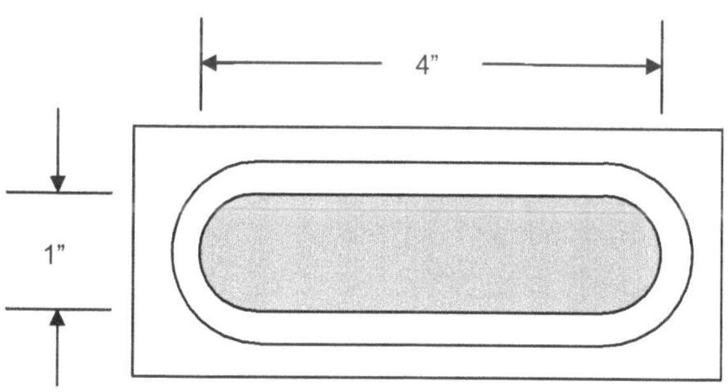

Figure 7.18 Racetrack exit from feed block to match die entrance with rounded corners to prevent degradation.

The first and still common feedblock design is the "Dow" feedblock as shown in Fig. 7.5. The primary parts to this feedblock are:

- the layer arrangement plate where the melt from each extruder is brought to the combination point,
- the flow dividing section where each stream is divided into an equal volume, and
- the adapter section where the combined layers are combined and transported to the die manifold entrance.

There are many forms of the "Dow" feedblock produced and used by many companies because they are simple in conception and manufacturing. In some instances, where rheological problems are encountered, the vanes are sometimes extended into the die adapter section. This is a difficult and expensive approach to implement but is at times effective.

In the "Dow" feedblock, the primary design principle is that the velocity of the various layers in combination should be the same when they are combined. This is accomplished by inserting flow-dividing plates into the flow to divide the flow into perhaps 10 equal volumes. Because the geometry is uniform in cross section of the plates, the velocity at the exit is uniform. A feature of this, and all feedblocks, is that if the layer uniformity exiting the die manifold is poor, then the plates can be replaced with vanes that are machined to various shapes such that more melt is placed in the regions which are found to be thin and less melt in the regions found to be thick so that after spreading there is more material remaining at the thin point and less in the thick point, and the average distribution is thereby improved. This is due to flow rearrangements called "encapsulation" which will be explained below.

All feedblocks permit this melt profiling to improve layer distribution. In general, this profiling is used to overcome poor viscosity matching of the various layers. However, it might also have been possible to improve the layer uniformity with an existing feedblock by modification of the polymer viscosity level. However, profiling is a powerful means of overcoming some flow problems, especially when polymer replacement is not an option. If it is not possible to overcome the poor layer distribution by profiling or polymer selection, then the feedblock/die combination should be replaced by a multimanifold die.

Another widely used feedblock designs are the WELEX designs where polymers are layered using a circular cross section. This permits the easy adaptation to existing dies with circular entrances. Fig. 7.6 shows this feedblock concept. This feedblock is simple in conception and is easy to machine, and they can be stacked in sequence to permit the addition of as many additional layers as space permits. Shaping of the melt layer is accomplished with removable inserts at each combination point. Fig. 7.19 shows another WELEX feedblock

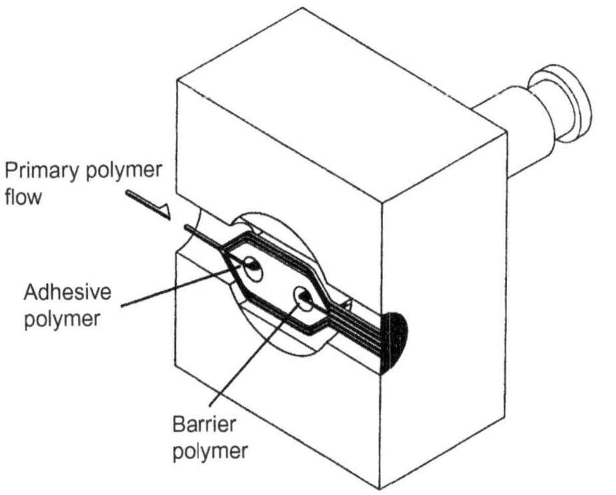

Figure 7.19 WELEX center layer block for introducing an adhesive/barrier/adhesive structure in the center of a flow stream, creating five layers from three flows. Principal flow from the left to the right and adhesive exits counter the primary flow to encapsulate the barrier flow.

design where the flows are divided and stacked around a central torpedo. In this case, one stream is extruded against the flow and is then divided as it is swept downstream by the existing polymer flow. Then a central layer is injected at the downstream exit of the torpedo producing a symmetrical ABA combination into the center of the flow. This could be used to add a central incompatible polymer surrounded by two adhesive polymer layers in to an existing single or multilayer extrusion system.

An adjustable feedblock was invented by the Cloeren Company (Fig. 7.7) which permits the adjustment of the feedblock geometry while the coextrusion system is in operation. The principle behind this design is that the geometry of the combination point can be changed at a constant output rate to change the average melt velocity and shear rate in the melt stream at the point of combination. This permits the modification of the melt viscosity in each layer at the combination point and can be used to eliminate coextrusion defects due to viscosity matching without the need to stop and rebuild the feedblock assembly. This feedblock technology has evolved over the years and been continuously improved with the addition of profiling pins as well as models where fixed but removable insets can be used in place of the adjusting vanes. The Cloeren feedblock systems have been very widely accepted in the coextrusion industry and are extremely versatile.

Extrusion Dies, Inc., also supplies an adjustable feed as well as fixed feedblock designs which are modular in design.

7.1.6 Die Designs

Die designs for film and sheet have evolved over the years to improve the distribution of polymer melt across the width of the film or sheet being extruded. Initially, the dies were developed for single layer films and, initially, these designs were used for coextruded products. Since the beginning of coextrusion, further developments have resulted in die modifications specific to coextruded stacks delivered to the die from coextrusion feedblocks. Recent experimental studies have demonstrated the importance of these die changes to coextrusion flows and the maintenance of undisturbed coextruded layer structures [1].

Initial film dies were "crosshead" slit dies which were fed from one end into a circular manifold and discharged melt through a narrow slit (land) (Fig. 7.20); these are sometimes called infinite cavity designs.

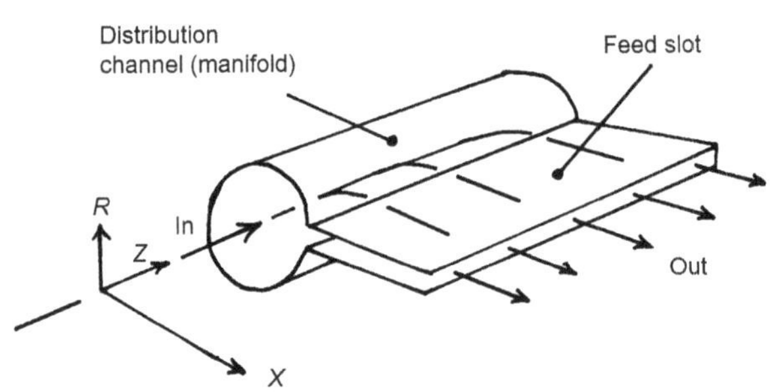

Figure 7.20 Schematic diagram of end-feed die showing flow into and through the manifold and die land.

End-fed dies later evolved into center-fed or "T-slot" dies where the melt typically entered at the center of a circular manifold, spread in two directions and the melt discharged from a slit. It is to be noted that the T-slot die is symmetrical about the entrance port, and the analysis of the manifold can be done as two end-fed dies and, therefore, it is only necessary to evaluate the flow performance of one half of a center-feed die. A die is, in essence, a pipe (or manifold) opened at one end and closed at the other with a slit cut along its length through which the melt leaks. As the melt flows along the manifold, the internal manifold pressure changes due to viscous loss and changing fluid momentum. The viscous resistance will tend to cause the manifold pressure to drop while the momentum changes will tend to increase the manifold pressure and the driving force for flow will vary across the length of the manifold [4]. For a simple crosshead die with uniform manifold diameter and slot width and gap height, this will result in variable flow across the manifold width and a variable product thickness which is not desirable. What is desired is a film of uniform thickness across the die width.

The initial analysis of the crosshead die was performed by Carley [5] for isothermal, Newtonian, and power law fluids is defined as:

$$\dot{\gamma} = \frac{dv}{dx} = f\left(\frac{\tau}{\tau_0}\right)^m \qquad m = \frac{1}{n} = \frac{d\ln\dot{\gamma}}{d\ln\tau} \tag{7.13}$$

Where m, the flow index, varies from one to six for most polymer melts and is more commonly expressed today as $1/n$ where n is the power law index. Inertial effects were ignored in the analysis.

It is instructive to review the analysis for it highlights the basis for a good die design procedure and the results demonstrate the essential findings of flat die analysis. The analysis was focused on determining the uniformity of the flow from the die lips across the width of the die and defined a uniformity index (UI), which is the ratio of extrusion rate at the end of the manifold to the flow rate at the manifold entrance. The results were expressed in terms of the flow index (m), the width of the die manifold (W), and a die constant alpha (α) with dimensions of inch^{-1} for all values of m and which contains the die geometry dimensions of the manifold radius (R), the die gap height, (H), and the die land length, L.

$$UI = \left\{1 - \frac{[(1+0.05m)m\alpha W]^{(m+1/m)}}{(1+0.05m)(m+1)}\right\}^m \qquad \alpha^{m+1} = \frac{(m+3)H^{m+2}}{2\pi m(m+2)L^m R^{m+3}} \tag{7.14}$$

The expression for UI is a function of the dimensionless group αW and is only dependent on the dimensions of the die and m, the flow index (or the power law index as $m = 1/n$) and not the fluid viscosity or the manifold pressure.

This result was compared to the exact solution obtained for the Newtonian UI (Eq. [7.15]) and it was shown that for $m = 1$, UI resembles the first two terms of the expansion of the Newtonian result giving increased confidence in the result.

$$\mathrm{UI} = 1 - \varepsilon = \sec h(\alpha W) \quad \text{for} \quad \mathrm{m} = 1 \quad \mathrm{UI} = 1 - \frac{(1.05\alpha \mathrm{W})^2}{2.10} \tag{7.15}$$

The result of this flow uniformity analysis has all of the essential features of subsequent die design analyses. The Newtonian and power law results show that flow uniformity is independent of the fluid viscosity and the manifold pressure, and dependent on the die geometry and the flow index.

Polymer viscosity is important in the analysis only through the flow index (m) and we can expect that, for a given die design, the extrusion of different polymers with the same value of m (where $m = 1/n$) will yield similar flow distributions, whereas the substitution of a polymer with a different power law index would be expected to give a different flow distribution and, therefore, thickness profile, across the die exit.

A key learning from this analysis is that, for any given film width, a center-feed die will give better overall uniformity due to the decrease in manifold width as the results show that flow uniformity decreases with increasing manifold width, W. The results also demonstrate a decrease in uniformity with increasing die gap, H. Improvements to die flow uniformity may be achieved by increasing the manifold radius, R, or the die land length, L. These changes to R and L improve flow uniformity by minimizing the pressure drop differences across the width of the manifold and die land. However, a balance must be met as the increase in uniformity is offset by an increase in residence time in the die as the manifold radius increases and an increase in extrusion pressure, at constant output (or a decrease in output, at constant extrusion pressure), as the die land length L increases.

McKelvey and Ito [6] next proposed a die with a straight-tapered circular manifold and a three-gap die land where the three land lengths were a function of position across the die width (Fig. 7.21).

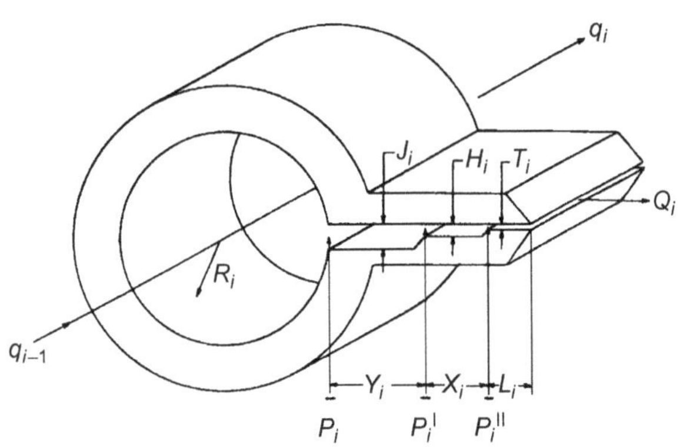

Figure 7.21 Schematic diagram of die with a straight tapered circular manifold and a three-gap die land where the three land lengths were a function of position across the die width.

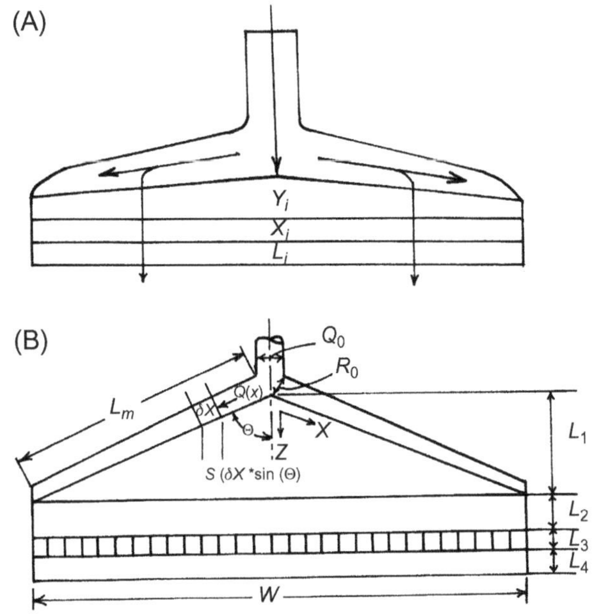

Figure 7.22 (A) Coat hanger die with three variable depth lands (Y_i, X_i, L_i) as in Ref. [6]; (B) coat hanger die of Chung and Lohkamp.

Again, the results of the analysis of this die's performance, the flow uniformity across the die width, showed a dependence only on the die geometry and the flow (power law) index m, as was found by Carley [5]. The analysis showed that the flow variation across the width as a function of m, the flow index for fixed die geometry where the die land gaps, varied linearly across the die width. In principle, based on this analysis, it should be possible to find a die land gap profile combined with a reducing diameter manifold, such that the flow deviations across the die width are reduced to zero.

In the design of flat film and sheet dies, what is critical to controlling uniform flow from the die across its entire width is a uniform pressure drop for all flow paths from the manifold entrance to the die lip exit (Fig. 7.13). This is the primary goal of all die design procedures and when combined with minimizing the polymer residence time by reducing the manifold volume across the die width and varying the die land length, which results in the "coat hanger" design (Fig. 7.22A). In the coat hanger die, the manifold is tapered as well as curved to permit the final land length of the die to vary in length to control the pressure drop. The coat hanger design is a natural outcome from the analysis of McKelvey and Ito [6].

When the coat hanger manifold concepts were first developed, it was difficult and expensive to machine such dies and, therefore, a series of design compromises were developed to aid in the manufacturing and operating characteristics of the flat film die. Die manifolds were generally circular in cross section with a streamlined flow from the manifold into the coat hanger gap (giving a tear drop shape manifold), and die land gaps generally decrease as the die exit is approached to accelerate the melt as it moves through the die to aid in preventing plate out from the melt onto die surfaces.

A compromise coat hanger design is the "fish tail" manifold configuration, where the manifold diameter and the die land height is reduced linearly across the die width. These dies have better flow uniformity than T-slot dies but, generally, do not give a uniform melt distribution across the die width [7] as does the true coat hanger.

Chung and Lohkamp [8] present a die and analysis aimed at improving the design of fixed angle coat hanger dies with circular cross-section manifolds (Fig. 7.22A) to give uniform flow rates across the width of the die. The analysis yields a variable manifold radius $R(\xi)$ where the manifold radius is found to be a function of reduced manifold position (ξ) and the power law index (n).

$$\frac{R(\xi)}{R_0} = (1+\xi)^{n/(3n+1)}$$
$$\xi = \frac{x}{L_m} \tag{7.16}$$
$$L_m = \frac{W}{2\sin\Theta}$$

Here, again, we find that the power law index n is the most significant factor in determining flow uniformity from a film and sheet die.

Coat hanger die manifold designs, which are independent of operating point and material and which give uniform flow rate and uniform stress on the melt, [9] as well as coat hanger dies with uniform flow rate and residence times [10] may be designed. But these designs typically give a die with long coat hanger lengths and, as the width increases such that the dies becomes physically too long, are difficult to seal to prevent leakage.

The limitation in building dies with circular cross-section manifolds are twofold as discussed by Chung and Lohkamp [8] and Dooley [1]. First, in the transition from the manifold to the exit slit, the flow is not streamlined due to the abrupt transition from the manifold to the thin land area (Fig. 7.21). Therefore, it was typical of die manufacturers to create a streamlined transition (Fig. 7.23).

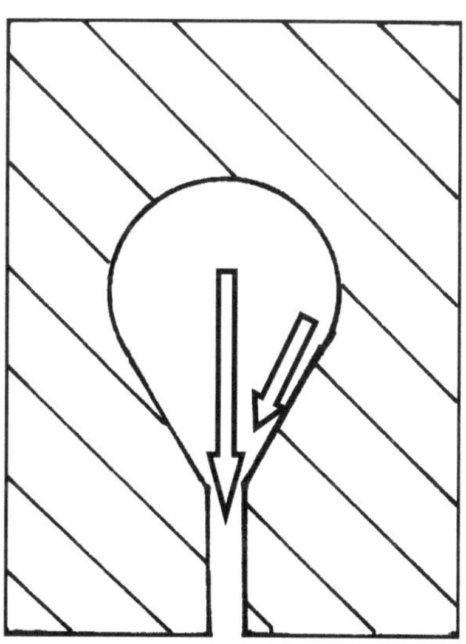

Figure 7.23 Streamlined manifold design eliminating sharp transition from manifold to land as in Fig. 7.21.

Also, as described by Chung and Lohkamp [8], it is difficult and expensive to fabricate coat hanger dies with circular manifolds with varying radii from fixed radii cutting tools. Consequently, the back plane of the manifold was often cut with a flat back using rectangular cutters and the corners cut with a radius cutter to produce the familiar teardrop manifold often found for die manifolds. As will be explained later, the shape of the manifold has a profound effect on the flow of coextruded structures entering the die manifold from a feedblock. As disclosed by Cloeren [11] and later confirmed experimentally by Dooley [1], it has been shown experimentally that the rectangular feedblock is optimum for coextruded structures entering a die manifold from a multilayer feedblock and that the teardrop manifold creates layer nonuniformities.

These findings led to the invention of the "inverted preland" die [12] for coextruded structures (Fig. 7.24) which combines the rectangular manifold with a constant length land of various gap heights, as found in McKelvey and Ito [6], with the exception that the initial land of the die is of a narrower gap than the second land, and the shape of the transition between the first and the second damming zones of the preland are rheology dependent.

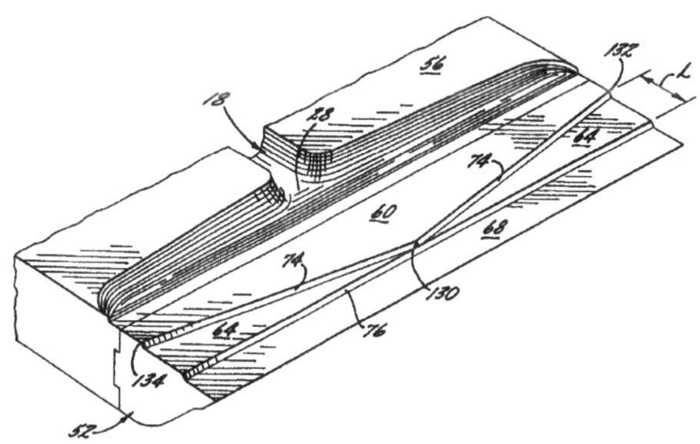

Figure 7.24 Cloeren Inc., inverted preland die design (figure 3 of US Patent 5,256,052) showing initial damming land (60) followed by deeper lands 64 and 68. The transition from 60 to 64 is a function of position across the die width and is dependent on the melt rheology.

From our earlier review of die design calculation models, it should be understood that the shape of the inverted preland transition will be dependent only on the power law index (n) of the polymer being extruded as was found by McKelvey and Ito [6].

Aside from the design for uniform flow across the exit, it has been established that there are several other important considerations for a die design which are polymer dependent, such as uniform residence time or a maximum residence time for temperature sensitive polymers, maintaining die cleanliness with a minimum melt velocity to prevent component or degraded polymer plate out on the die surfaces and lip, as well as to minimize internal die pressures to prevent die body distortion and to maintain stress levels below critical stress for melt disturbance. All these considerations have come a series of die designs which are represented in the literature as the T-slot, the fishtail, various coat hanger, and inverted preland dies.

All die designs require compromises in the fabrication due to limitations in machining capabilities. Early in the evolution of dies, these compromises were more significant in magnitude due to limitations from the manual setup of machining. Such design and fabrication compromises resulted in dies which were not optimum for the polymers available and had to be made mechanically adjustable to overcome the design and fabrication limitations. Such a mechanical compromise was the incorporation of an intermediate adjustable flow restriction, called a choker bar, downstream of the manifold but before the final die land and the flex lip. The idea of a choker bar is to have an internal die gap which is adjustable across the die width so that any deficiencies in the manifold design or machining, or changes in the polymer power law index, poor polymer temperature uniformity, significant output changes, etc. might be accommodated by adjustment and a flat film profile obtained. Choker bars have been used for many years and can be difficult to adjust and create a potential source of polymer degradation in sensitive polymers or critical products. However, with multicavity coextrusion dies, it becomes very difficult to incorporate choker bars easily and inexpensively into the die construction as the number of die manifolds increases. Consequently, better die design procedures should be used to eliminate the need for a choker bar. It is my personal opinion that their use should be avoided whenever possible, and the newer die designs and the improvement in fabrication capabilities of CNC machining permit the elimination of choker bars.

While the coat hanger die is the only manifold and land combination which, theoretically, may be designed to give uniform flow across the die width such that the manifold design is independent of the operating point [7], previous observations [13] of the performance of layer spreading with teardrop manifold dies in combination with feedblocks, and recent experimental work in coextruded structures related to the secondary flows in these manifolds [14] indicate that they should be avoided when used in combination with the stacked multilayer melt streams exiting a feedblock.

7.1.7 Coextrusion Systems for Flat Films and Sheet

As described above, the choice of die design and its impact on spreading of melt stacks should be taken into consideration when feedblocks are combined with a die. However, the coextrusion of flat films and sheets can be successfully accomplished with multilayer feedblocks and single cavity dies, multicavity dies and combinations of feedblocks, and multicavity dies. Which system is the best choice for any given application is dependent upon:

- the specifics of the application,
- the polymers to be combined, and
- the uniformity or quality requirements of the coextrusion which is to be produced.

Combinations of multicavity dies and feedblocks give the most flexible systems when multiple products are to be produced with a wide range of polymer types.

Originally, feedblocks with single cavity dies were the most cost-effective form of coextrusion system as the use of multicavity dies was prohibitively expensive due to the difficulties associated with the machining of the relatively complex combination of body parts required for the multicavity dies. Today, with CNC machining capabilities, the machining of five (or more) cavity dies has been greatly improved and five cavity dies are readily available. Therefore, the choice of coextrusion system has been broadened and the coextrusion system choice becomes strategic in terms of current and future products and the ultimate flexibility required for the system. If a single coextruded product is to be run without modification, then the simplest system capable of supplying the desired product quality can be chosen relatively straightforwardly. The following examples will illustrate this.

For combinations of polymers with similar flow curves and viscosity levels, the use of a feedblock with a single cavity die can be an effective choice for the coextrusion. However, based on the work of Dooley et al. [1,14], the use of teardrop manifolds should be avoided to minimize layer rearrangement and spreading problems due to secondary flows in the manifold. The amount of layer rearrangement seen in the manifold is related to the inherent elasticity of the material, with polystyrene being greater than polyethylene and polycarbonate showing little rearrangement in the study of Dooley et al. [14].

If, however, the polymers in each layer have widely different viscosity levels and flow curves, then the layer spreading in the single cavity due to viscous effects alone will lead to poor layer uniformity and, perhaps, to other flow defects. In this case, the individual layers should be separated and individually spread to the final width and layer thickness in a single die cavity designed specifically for that polymer, before combining the layers.

7.1.8 Layer Instabilities, Causes, and Prevention

Coextrusion layer instability is often times called a "melt disturbance" and there can be several sources, some of which are not directly related to the flow of the layers structure in the die, but rather to other processing problems. Therefore, it is important to eliminate all sources of poor melt quality which can be causing defects in the sheet or film, such as unmolten polymer, gels in the resin, large temperature gradients in the melt stream, before assuming that a rheological problem exists.

When you are trouble-shooting a "melt disturbance," the first approach should be to determine "what has broken" in the process, especially if the problem has recently arisen in an existing product or process. You should first determine that all temperature controllers are functioning properly, and that there are no over temperature or burned-out heaters causing large temperature gradients in the melt entering the die. Next, you should determine that the screw is still performing correctly and that there is no unmolten polymer in the melt stream, and hence the melt temperature and pressure entering the die or feedblock is uniform. Then, you should determine if there has been a change in the resin, either a resin grade change or an out of specification resin lot. Once this is done, it is possible to begin trouble-shooting the rheological problems causing the melt disturbance.

There are several coextrusion problems associated with the flow of layered polymer melts which are generally related to the relative viscosity of the various layers and their relationship to each other. This is easily illustrated by considering the flow of a single polymer material with a homogeneous makeup and temperature in a gap or pipe. Recalling the discussion above for flow in a gap or pipe, the velocity profile of the flow may be calculated and is displayed in Figs. 7.16 and 7.17. Because polymer melts are generally pseudoplastic or "shear thinning," we can examine the viscosity of this single fluid across the flow profile. Recalling that the slope of the velocity profile is the shear rate, we find that the shear rate at the center of the flow is zero (0) and that the shear rate at the wall is a maximum. This will mean that, due to the shear rate dependence of the polymer, the viscosity of a single fluid at uniform temperature is not uniform across the flow channel and the flow will show a viscosity profile ($\eta(y)$) across the flow channel. In particular, we can see that the viscosity at the center line will be at the maximum or zero shear viscosity, η, and the viscosity at the wall will be a minimum due to the increase in shear rate at the wall. This decreased viscosity at the wall is a self-lubricating effect in the flow as it will lower the pressure drop in the system and is therefore the lowest energy state for the flow.

Therefore, when we stack polymers in a layered coextrusion flow, we should ensure that the viscosity of the individual layers decreases as the flow channel wall is approached. If this is not done, the layers will attempt to rearrange themselves to place the lowest viscosity layer on the outside of the flow against the wall, at the region of the highest shear rate, to lower the shear stress (pressure drop) in the system. This flow problem, where the viscosities are not properly stacked, leads to layer rearrangement or encapsulation.

Encapsulation has been reported and demonstrated experimentally in a tube and slot by several authors [15–17] by stacking high- and low-viscosity melts together. It has been found that the lower viscosity material will encapsulate the higher viscosity material given sufficient time. Han [17] has demonstrated a flow inversion in a tube by stacking a high- and low-viscosity melt together with the high viscosity melt against the tube wall and a low-viscosity melt in the core of the flow. He demonstrated that, if the flow channel is long enough, the lowest viscosity material will migrate to the wall of the flow channel and completely surround (encapsulate) the higher viscosity components of the flow. This encapsulating effect is often seen in feedblock coextrusion and in die manifolds and die lands when the layer viscosities are not well matched or improperly positioned. This is the effect which feedblock profiling, described earlier, is attempting to accommodate.

Layer rearrangement in coextrusion flows in die manifolds and transfer conduits (piping) can also be due to viscoelastic effects as has been shown by Dooley et al. [18,19]. In these instances, circular transfer piping causes the least layer rearrangement due to balanced stresses resulting from melt elasticity forces. In rectangular melt transfer systems, sometimes used in multilayer combining adapters, many layers are obtained by dividing and stacking simpler structures [20] (Fig. 7.25).

The layer rearrangements are driven by the secondary elastic flows and are quite spectacular giving multilobed flower-like layer structures from circular feed streams in the rectangular flow channels (Fig. 7.26).

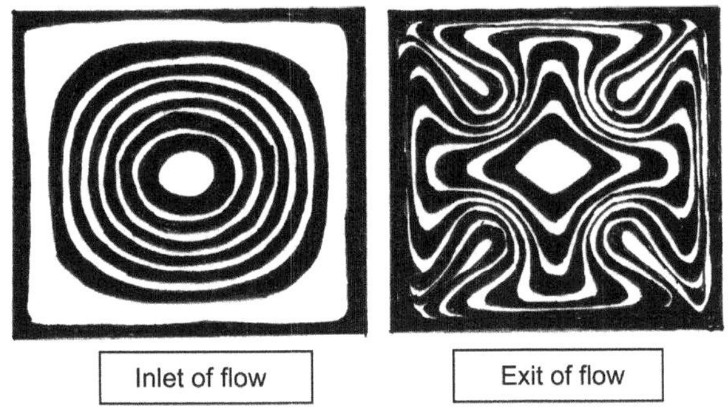

Figure 7.26 Secondary flow impact on layer rearrangement in square melt pipe flow.

Of course, the extent of layer rearrangement will depend upon the length of the transfer conduit. This melt elasticity–driven layer rearrangement is also a significant problem in teardrop-shaped die manifolds [18]. Therefore, when producing layered sheets or films in single cavity dies being fed by multilayer feedblocks, it is desirable to have rectangular manifolds to minimize layer rearrangement due to the more balanced shear stresses at the polymer/metal surfaces of the die.

Layer instabilities or "melt disturbance" which result in a deformation of the interface and perhaps the sheet or film dimensions, manifest themselves

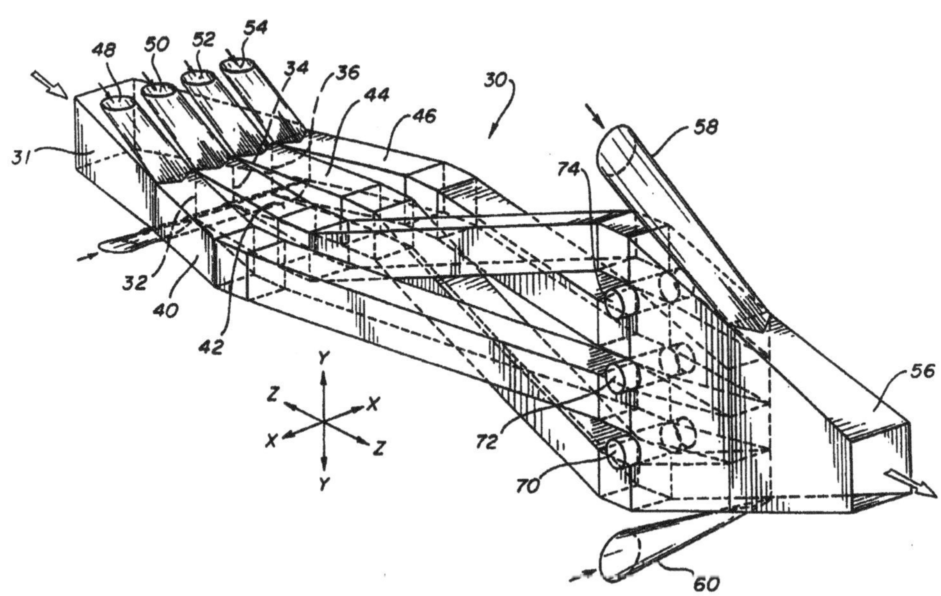

Figure 7.25 Figure 2 of US Patent 2,269,995 showing square cross-section flow channels used for flow dividing and stacking to produce multilayer flows.

Figure 7.27 Melt disturbance of a three-layer sheet showing distortion of the sheet and the chevron appearance of the flow.

in many forms. At its onset, the sheet may still maintain good dimensional characteristics but the optical properties begin to deteriorate such that the clarity or "see through" properties, of the film are impacted [21]. This instability exists primarily at the interface of the flowing polymers and is termed an "interfacial instability" which is used interchangeably with melt disturbance. This incipient interfacial instability manifests itself as a wavy or sinusoidal interface (Fig. 7.27) which results in light scattering of the image passing through the film giving a blurred image.

An example of the optical defect formed is shown in Fig. 7.28.

Figure 7.28 Looking through a film containing melt disturbance and held approximately 8 inches in from of the lettering, showing the image distortion caused by the variable film thickness from the melt disturbance.

As the level of the instability increases, the interface can begin to form a pronounced wave which, on cross section, appears to be breaking over itself (Fig. 7.29).

As the level of the instability increases, the interfacial instability can propagate to the surface deforming the sheet. This larger instability is known by several names such as "zig-zag," "arrow heads," "fish scales," or "chevron" depending on their severity. Looking through the film, this instability may appear as a wide parabolic wave front extending across the full width of the film (Fig. 7.30) and result in a pronounced thickness variation in the sheet.

Alternatively, the wave may be broken into several narrower bands of waves or into high-frequency smaller waves which look like fish scales or arrow heads, etc.

This defect has been isolated to the final die land and appears to be due to exceeding a critical shear stress for the coextruded polymer pair in the final die land, as was demonstrated experimentally [22] and is shown schematically in Fig. 7.31.

This defect is addressed by lowering the shear stress in the final land by several approaches including, lowering extrusion rate, increasing melt or die temperatures, opening the die gap, and lowering the polymer viscosity with resin replacements.

Coextrusion instabilities and defects related to viscosity mismatching of layers is the most common form of coextrusion defect which I encounter. In general, the only acceptable level of interfacial flow defect is "none" as a light and sporadic

Figure 7.29 Cross section of wave-like melt disturbance of core layer entering the skin layer.

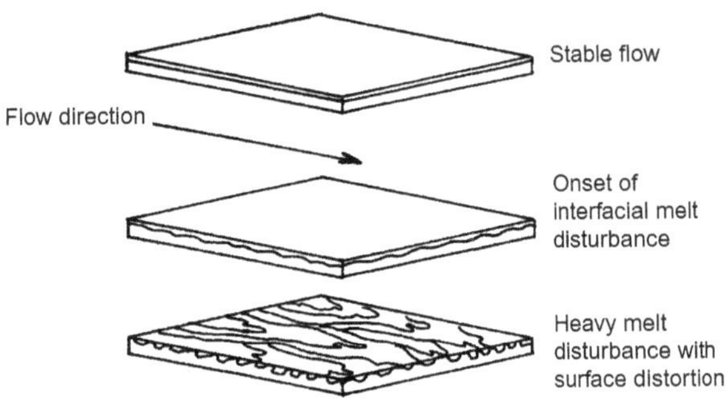

Figure 7.30 Schematic diagram of two-layer flow showing the transition from stable to heavy melt disturbance with film distortion.

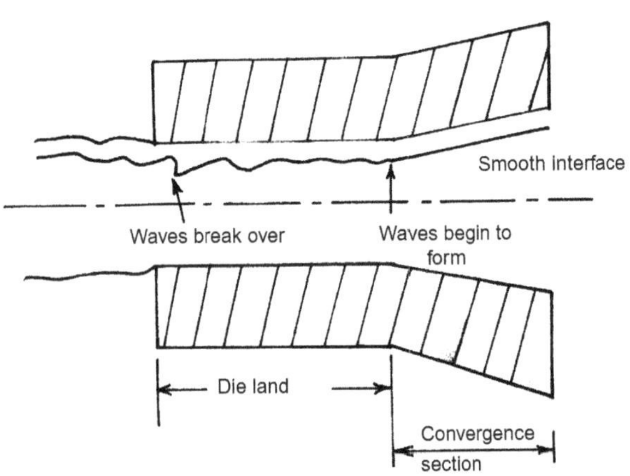

Figure 7.31 Onset of "zig-zag" type melt disturbance to stress level in the final land.

appearance of the defect will often times render the product unusable for both technical or esthetic reasons depending on the application. For instance, an incipient interfacial defect appearing in a lane or sporadically across a film may not be readily visible in transmitted light in the clear film but, if the film is printed with a dark ink or metallized, the contrast between the film regions containing the defect and defect-free regions will enhance contrast between the defect and surrounding film which can make it very prominent.

In many commercial film and sheet products, coextrusion is used to modify the surface regions of a product by making use of thinner outer layers of special or modified polymers. Because the outer layers tend to be thin to minimize the use of the surface polymers, the interfaces of these thin layers are generally located near the flow surfaces where the shear rates are highest and changing rapidly as shown in Figs. 7.16 and 7.17 and, in general, the interfacial instability will appear in a product as outputs are increased. This is because at higher output rates the die surface and layer interfacial shear stresses increase. Also, because the average melt temperature of each layer will change as the extruder screw speeds are increased as the overall output is increased. Usually, the increased extrusion rate will impact the thicker interior layers more significantly than the outer layers as the extrusion rate is increased, resulting in an increase in melt temperature and a drop in melt viscosity of the thick interior layers, whereas the outer layers may change only a small amount.

With higher production rates, the thicker interior layers' viscosity tends to drop faster than the thinner outer layers for several reasons. First, the higher screw speeds and generally larger diameter extruders of the interior layers will heat the polymer melt more than with a smaller diameter extruder due to decreasing energy efficiencies of higher speed and larger diameter extruders [23]. Screw design and its impact on average melt temperature as a function of output level will therefore be an important consideration when increasing the extrusion rate of interior layers. Also, the interior layers may include reclaimed polymer, which is generally lower in viscosity due to the additional extrusion history used to make it. Therefore, the combined effects of higher melt temperature and lower component viscosity can result in a situation where the viscosity of the core layer becomes lower than that of the outer layers as overall output is increased.

Therefore, as productivity is increased, it becomes important to monitor the changes in rheological properties of the various layers' materials at the increased rates. If an interfacial instability appears, it must be addressed by process (extrusion temperatures, die temperatures, reclaim levels, line speeds, etc), product design (layer thicknesses), material (reclaim quality, average viscosity changes), or equipment changes (screw design, die, or feedblock) which result in a proper balance of layer viscosity and interfacial stresses.

7.1.9 Practical Examples

7.1.9.1 Example 1

It is desired to produce coextruded films containing from three to five polymers. The film may contain layers of Nylon and/or ethylene vinyl alcohol polymer (EVOH) barrier layers, either separately or in combination, adhesive resins, and various polyolefins. The films are to be 2000 mm (78 in) wide and range in thickness from 0.051 to 0.178 mm (2–7 mils). Structures are to include split barrier layers and both buried and surface nylon layers and symmetric and asymmetric structures such as:

- Nylon/tie/nylon/tie/LLDPE/LLDPE/m-LLDPE
- Nylon/tie/nylon/tie/PP/PP/EP copolymer
- Nylon/EVOH/nylon/tie/LLDPE/LLDPE/ m-LLDPE
- LLDPE/tie/nylon/EVOH/nylon/tie/m-LLDPE
- PP/tie/nylon/EVOH/nylon/tie/EP copolymer
- LLDPE/tie/nylon/EVOH/nylon/tie/LLDPE
- PP/tie/nylon/EVOH/nylon/tie/PP
- LLDPE/tie/EVOH/tie/EVOH/tie/LLDPE.

There are several possible coextrusion equipment combinations which can be contemplated to produce these various film combinations and, although not all will be practical, several potential combinations are:

- a seven-cavity coextrusion die,
- a five-cavity coextrusion die combined with a three-layer feedblock,
- a three-cavity coextrusion die with two, two-layer and one three-layer feedblock, and
- a seven-layer feedblock and a single cavity die.

Choosing between the various combinations, the rheological compatibility, die mechanical complexity, and the total equipment cost must be considered when selecting the coextrusion equipment. If we assume as a rough rule of thumb die cavities cost approximately \$1000/25.4 mm (\$1000/inch) and feedblocks approximately \$20,000/layer, then the approximate capital cost of each of the four ways to product the 2000 mm (78 inch) wide seven-layer films can be estimated and are approximately:

- 7 × \$78,000 = \$546,000 for a seven-cavity coextrusion die,
- 5 × \$78,000 + 3 × \$20,000 = \$450,000 for a five-cavity coextrusion die combined with a three layer feedblock,
- 3 × \$78,000 + 7 × \$20,000 = \$374,000 for a three-cavity coextrusion die with two, two-layer and one three-layer feedblock, and
- 1 × \$78,000 + 7 × \$20,000 = \$218,000 for a seven-layer feedblock and a single cavity die.

Therefore, the most cost-effective way to produce this film would be with the fourth option, a seven-layer feedblock and a single cavity die, and this is the option most often found in practice.

However, the third option, a three-cavity coextrusion die with two two-layer and one three-layer feedblock would likely be the most versatile combination because of the ability to combine various skin and tie resins with various core barrier combinations permitting independent control of layer distributions of the various combinations before the final combination to form the seven-layer combination. Three- to five-cavity dies are readily produced, and the 2000 mm width is not a machining or mechanical constraint.

While there are many five-cavity and some seven-cavity dies being produced at this time, they are mechanically complex, large in physical dimensions and increased weight, and the challenge of insuring the large area sealing surfaces between the multiple manifolds for a 2000-mm wide die would make manufacturing and the routine cleaning and reassembly of the die very difficult. In addition, it is generally more time consuming to heat the multi-cavity dies at start-up. Also, as the number of cavities increases, there is a significant increase in the internal surface areas wet by polymer, which creates increased opportunities for polymer degradation, contamination, and film defects. These reasons, as well as the lower cost, are why the seven-layer feedblock in combination with a single cavity die is the favored approach in applications of this type.

7.1.9.2 Example 2

It is desired to produce a three to five polymer, five-layer oriented polypropylene film containing a continuous or cavitated PP core, clear or pigmented layers, and various surface layer combinations, where possible surface layers are PP, ethylene propylene (EP) copolymer, ethylene propylene butene (EPB) terpolymer, HDPE, LLDPE, EVOH, Nylon:

- EP copolymer/PP/PP/PP/EPB terpolymer
- EVOH/tie/PP/PP/EPB terpolymer [24,25]
- EVOH/tie/cavitated PP/pigment PP/EPB terpolymer
- HDPE/PP/PP/PP/EPB terpolymer
- EP copolymer/PP/EP copolymer/LLDPE/ EP copolymer
- Nylon/tie/PP/PP/EPB terpolymer
- PP/PP/PP/PP/PP.

There are several possible coextrusion equipment combinations which can be contemplated to produce these various film combinations:

- a five-cavity coextrusion die,
- a three-cavity coextrusion die combined with a three-layer feedblock,
- a three-cavity coextrusion die combined with two, two-cavity feedblocks, and
- a five-layer feedblock and a single cavity die.

While the cost considerations are still important, these dies will generally be 40 inches or narrower in width (depending on the width of the orienting line) and, in the context of a PP-orienting film line costing from $20,000,000 to $40,000,000, will be a relatively small cost. Therefore, in this particular instance, improved layer uniformity, resin change flexibility and layer rheological considerations generally outweigh coextrusion feedblock and die cost considerations.

Currently, the first and second options are widely used to produce these films. The last option would be a potential candidate if the desired film layer combinations were significantly limited to combinations of PP, EP copolymer, EPB terpolymer, and HDPE only. Depending on its layer location, the addition of an LLDPE layer might be possible with minimal modifications to the feedblock or to the outer die cavity if it was to be a surface layer.

In my opinion, the second and third options are the most flexible approaches for this example, because they permit the widest possible combination of materials with minimum die modifications. In the second case, the three-layer feedblock could be located on the central die cavity or one of the outer die cavities, whereas, in the third case, the two-layer feedblocks could be located on both of the outer flow cavities or on an outer die cavity and the central die cavity.

When deciding on which feedblock/die cavity combination to choose, it is possible that by properly sizing the feedblock, several feedblocks could be purchased and all of the potential feedblock/die cavity combinations could be possible. This illustrates that a good deal of product design strategy needs to be taken into consideration when choosing the coextrusion equipment combination. To ensure the maximum flexibility in future product manufacturing, the feedblock/adapter combination should be sized such that the various potential feedblock/die cavity combinations can be substituted for each other without the need to modify or change extruder melt pipe connections to the adapter/feedblock combination. If this is not done initially, then changes between one feedblock/die cavity combinations and another may be economically difficult due to a need to replace all existing melt pipe connections between the extruders and the adapter/feedblock.

In the first three options, the full range of potential resins to be employed will likely require the use of more than one die if layer uniformity is to be optimized for all resins. This is because the die cavity designs for optimum layer uniformity are dependent on the shear thinning (power law index n) behavior of the various polymers as previously discussed. Fig. 7.10 displays the flow characteristics of various polymers contemplated for this example. What is easily seen on Fig. 7.10 is that the PP, EP copolymer,

EPB terpolymer, and HDPE resins show very similar shear thinning behavior (n values), whereas the EVOH and LLDPE show a significant difference (reduction or higher n values) in shear thinning. This means that if the EVOH or LLDPE resins are extruded through a die cavity designed for a PP-based polymer (or a PP-based polymer extruded through a die cavity designed for an EVOH- or LLDPE-based polymer), the layer distribution will not be optimum. Therefore, it may be necessary to have more than one die with various die cavities designed for a range of polymer products. Alternatively, depending on the polymer layer combinations, modifications to the feedblock combination geometry might be possible to profile the shape of the melt stack exiting the feedblock to permit improved layer distributions with existing die cavities. Which approach is the best, acceptable, or necessary, will be dependent on the permissible layer thickness variations permitted by the product end use application.

7.1.9.3 Example 3: Production of a Three- to Seven-Layer Sheet for Thermoforming

It is desired to produce coextruded thermoforming sheet containing from three to five polymers. The sheet may contain layers of Nylon, EVOH, polyvinylidene chloride (PVDC) or other barrier layers separately or in combination, adhesive resins, and various polyolefins. The sheets are to be 1143 mm (45 inches) wide and range in thickness from 12 to 45 mils. Structures are to include monolithic and split barrier layers and symmetric and asymmetric structures such as:

- LLDPE/tie/nylon/tie/nylon/tie/LLDPE
- PP/tie/nylon/tie/nylon/tie/EP copolymer
- LLDPE/tie/nylon/EVOH/nylon/tie/LLDPE
- LLDPE/tie/nylon/EVOH/nylon/tie/m-LLDPE
- PP/tie/nylon/EVOH/nylon/ethylene vinyl acetate (EVA)/LDPE
- LLDPE/tie/EVOH/tie/EVOH/tie/LLDPE
- HDPE/tie/EVOH/tie/EVOH/tie/HDPE
- PS/scrap/tie/PVDC/Tie/scrap/PS
- PP/scrap/tie/EVOH/tie/scrap/PP
- Filled PP/Scrap/filled PP.

As with the film example, there are several possible coextrusion equipment combinations which can be contemplated to produce these various film combinations. However, because the layers are relatively thick, layer uniformity in layers other than the barrier layers are less critical. This permits the assumption that the coextrusion system will contain a combination of feedblocks and multicavity or single cavity dies:

- a three-cavity coextrusion die with two two-layer and one three-layer feedblock,
- a three-cavity coextrusion die, with a five-layer feedblock,
- a seven-layer feedblock and a single cavity die,
- a five-cavity coextrusion die combined with a three-layer feedblock, and
- a two-cavity die with a five-layer and two-layer feedblock.

Although no single system will be able to produce all possible layer combinations and thickness ranges, a wide range of products is possible with the three-cavity die and various multilayer feedblock combinations.

7.2 Summary

Multilayer coextrusion systems for flat film and sheet products are well developed and readily available in many configurations and from many suppliers. Care should be exercised in determining the physical configuration of the installation to minimize costs and physical constraints for future product configurations. Also, a great deal of thought should go into the specification for the system, and accurate information should be supplied to the manufacturer (polymer type, viscosity curves, melt temperatures, etc.) to insure the best outcome.

When choosing a coextrusion supplier, it is important to interview and judge them based on the science and technology used for the design and manufacture of the system as well as the range of coextrusion technology which they employ. For instance, if a manufacturer is not interested in the polymer rheology, it is hard to imagine that they are truly designing the coextrusion system but rather just building it. If the supplier offers only limited possibilities, such as multicavity dies and no feedblocks, then they are likely not able to design coextrusion feedblocks successfully or cannot design the feedblock/die cavity combination.

Coextrusion is a powerful method to produce multilayer films and sheets for a wide range of applications. There are a wide range of approaches which can be used to produce these products, and all are usually capable of producing a wide range of products within limitations. A thorough understanding of the coextrusion science and technology is required to achieve the best results.

References

[1] Dooley J. Viscoelastic flow effects in multilayer polymer coextrusion. In: (2002) PhD Thesis, Technische Universiteit Eindhoven.

[2] Zatloukal M, Vlcek J, Tzoganakis C, Saha P. The effect of layer stretching on the onset of "wave" interfacial instabilities in coextrusion flows. SPE ANTEC Technical Papers, paper 652, 2001.

[3] Gillespie T. The flow of a molten polymer. Transact Soc Rheol 1965;9(2):35–47.

[4] Acrivos A, Babcock BD, Pigford RL. Flow distributions in manifolds. Chem Eng Sci 1959;10:112–24.

[5] Carley JF. Flow of melts in "crosshead"-slit dies; criteria for die design. J Appl Phys 1954;25:1118–23.

[6] McKelvey JM, Ito K. Uniformity of flow from sheeting dies. Polym Eng Sci 1971;11 (3):258–63.

[7] Michael W. Extrusion dies: design and engineering. New York (NY): Hanser Publishers; 1984.

[8] Chung CI, Lohkamp DT. Designing coat-hanger dies by power-law approximation. Modern Plastics March, 1976; 52–55.

[9] Michael W. Extrusion dies: design and engineering. New York (NY): Hanser Publishers; 1984.

[10] Matsubara T. Geometry design of a coat-hanger die with uniform flow rate and residence tine across the die width. Polym Eng Sci 1979;19(3):169–72.

[11] Cloeren P. An overview of the latest developments in flat die coextrusion. Proceedings of Advances in Extrusion Technology. RETEC Extrusion Division and the Ontario Section of the Society of Plastics Engineers (1993) October 18–19.

[12] Cloeren P. Extrusion process and apparatus with modified preland. US Patent 5,256,053, issued October 26, 1993.

[13] Personal communication of P. Cloeren to author.

[14] Dooley J, Hyun KS, Hughes K. An experimental study on the effect of polymer viscoelasticity on layer rearrangement in coextruded structures. Poly Eng Sci 1998;38(7):1060–71.

[15] Southern JH, Ballman RL. Stratified bicomponent flow of polymer melts in a tube. Appl Polym Sci 1973;20:175–89.

[16] White JL, Lee B-L. Theory of interface distortion in stratified two-phase flow. Transact Soc Rheol 1975;19:457–79.

[17] Han CD. Multiphase flow in polymer processing. New York (NY): Academic Press; 1981. p. 387–418.

[18] Dooley J, Hughes K. Measurement of layer deformation in coextrusion using unique feedblock technology. SPE ANTEC Proc 2000;1:36–40.

[19] Dooley J, Jenkins S, Naumovitz J. An experimental study of the flow of an encapsulated polymer melt through a unique blown film die. SPE ANTEC Proc 2001;386–90.

[20] Ramanathan R, Schrenk WJ, Wheatley JA. Coextrusion of multilayer articles using protective boundary layers and apparatus therefore. US Patent 5,269,995 14 (1993).

[21] Mavridis H, Shroff RN. Multilayer extrusion: experiments and computer simulation. Polym Eng Sci 1994;34(7):559–69.

[22] Schrenk WJ, Bradley NL, Alfrey Jr. T. Interfacial flow instability in multilayer coextrusion. Polym Eng Sci 1978;18:620–3.

[23] Mount III EM, Watson JG, Chung CI. A scientific approach to screw design. Polym Eng Sci 1982;22:729.

[24] Foster RH. US Patent 4,561,920, December 31 (1985).

[25] Migliorini RA. US Patent 5,194,318, (only if film is metallized) March 16 (1993).

8 Multilayer Blown (Tubular) Film Dies

John Perdikoulias
Compuplast International Inc., Zlín, Czech Republic

8.1 Introduction

Multilayer dies are used for coextrusion which started to become more popular among many processors in the late 70s and early 80s when OEMs started to include these items in their product list. Before this, advanced processors developed their own coextrusion equipment, and the technology was very closely guarded (and still is today). To understand how multilayer dies developed it helps to divide tubular coextrusion into two categories: barrier and nonbarrier.

Nonbarrier coextrusion would contain coextrusion products made with different polyolefins combined for aesthetics, strength, cost, etc. Barrier products are developed primarily for barrier functions such as food or chemical packaging.

In the nonbarrier category, almost all OEMs have provided successful two- and three-layer coextrusion dies since the early 70s. However, barrier materials have more limited processing conditions and therefore present more difficult processing conditions for barrier coextrusion. Barrier resins such as PA, EVOH, and PVDC have much smaller "processing windows" than the polyolefins that most equipment manufacturers were used to. Barrier resins were also more prone to degradation making them less forgiving to poor die design as well as poor operating practice. As a result, successful coextruded barrier films require good die design, good screw design, and good operating practices. It should also be mentioned that the continued and increased successful coextrusion parallels improvements in resin development. Over the years, resin companies continue to improve the processing behavior and thermal stability of barrier polymers.

There has also been a trend to increase the layers in a structure in order to both improve the film properties and reduce cost. For example, an excellent five-layer barrier structure could consist of:

PA/tie/EVOH/tie/Ionomer.

However, a corresponding 10-layer structure could be:

PA/Tie/LLDPE/Tie/PA/EVOH/PA/Tie/EVA/Ionomer.

It can be demonstrated that the 10-layer structure will provide improved barrier performance at equal amounts of barrier material as well as offer the possibility of reducing or replacing more expensive polymer with lower cost resins that provide adequate performance.

8.2 Conventional Coextrusion Dies

Probably, the most influential company to initially supply coextrusion equipment was Barmag, which is said to have provided more than 150 coextrusion dies by the mid to late 80s. These dies were based on conventional cylindrical designs with rotating "block" systems. The "block" is a term used for the device that simply redirects the polymer from the horizontal direction (when it leaves the extruder) to the vertical direction (to enter the die). In a single-layer die, it can be a relatively simply 90-degree bend. However, for a multilayer die, the block can be much more complicated as it may be required to uniformly divide the flow from each extruder and direct it evenly to the spiral distribution system for each layer. The block complexity can further increase when the die system is required to rotate or oscillate as a means of randomizing thickness variation.

Fig. 8.1 is a cross section of a conventional style three-layer coextrusion die showing both a stationary (left) and a rotating/oscillating (right) block system produced by Brampton Engineering in the 80s.

Multilayer Flexible Packaging. DOI: http://dx.doi.org/10.1016/B978-0-323-37100-1.00008-9

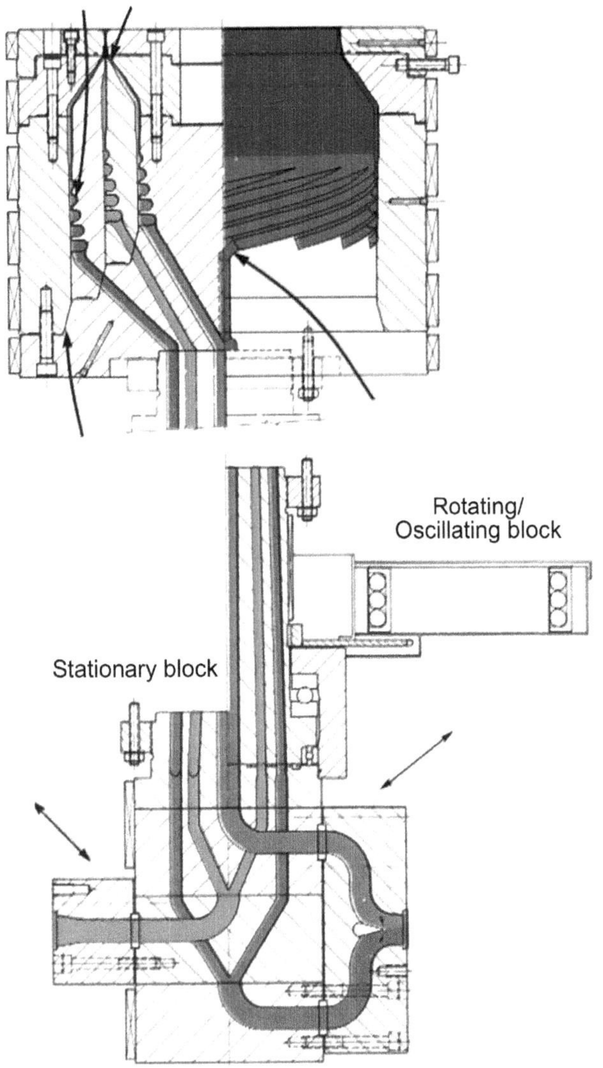

Figure 8.1 Conventional spiral mandrel type coextrusion die with rotating block.

The rotating die systems (that were required for gauge randomization) posed a maintenance problem. Apart from the periodic maintenance (often resulting from failure) of the sealing system the rotating blocks generally provided additional areas for polymer "hang-up" and degradation. Combined with the additional flow length in the rotating block, this resulted in poor purge characteristics and long changeover times. In fact many barrier film processors had a spare die that allowed them to continue production while another die was being cleaned or serviced. This also limited processors who developed coextruded products to those who had the financial strength to cover the intense, development costs, production efforts, and maintenance costs.

Despite the above-mentioned difficulties, the advantages offered by coextrusion encouraged R&D efforts by processors, material suppliers, and equipment manufactures alike. The late 80s and early 90s were witness to many new materials, die designs, and coextrusion products. One of the most important developments was the widespread introduction (due to patent expiration) of the oscillating haul off. While this device is downstream from the die system, it removed the gauge randomization requirement from the die leading to great improvements in die passage streamlining. Without the need to rotate the die, manufactures made improvements by streamlining the flow passages and lowering residence times. However, commercially available coextrusion dies were still essentially comprised three or five single-layer dies (concentric cylinders) that were nested together as represented by Fig. 8.2.

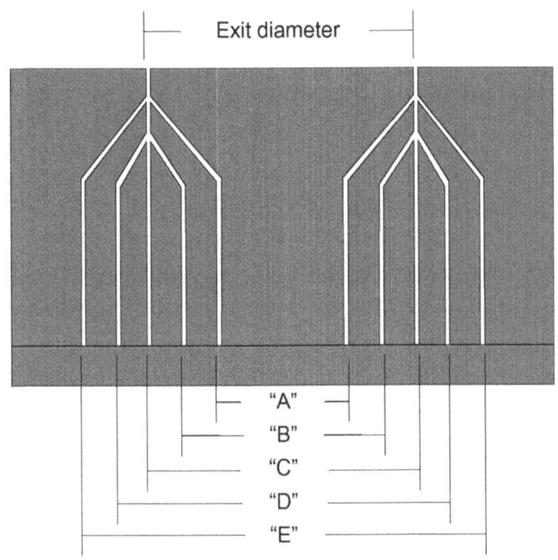

Figure 8.2 Conventional coextrusion die consisting of five nested cylinders.

Each one of the cylinders could be considered as a conventional single-layer spiral mandrel distribution system as shown in Fig. 8.3.

The polymer melt would be distributed by the controlled material leakage out of the helical grooves (spirals). The subsequent layering effect also had benefits with respect to more uniform appearance and physical properties. While these dies have been successfully used for many years the basic concentric cylinder design resulted in dramatically increased surface area as the number of layers increased [1]. Since the

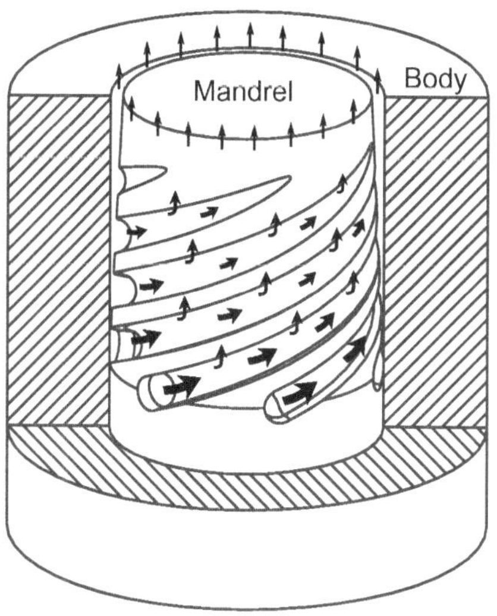

Figure 8.3 Conventional cylindrical spiral mandrel distribution system.

total output of the die is typically limited by cooling, which is related to the surface area or bubble diameter, having more layers simply meant that there was less flow per layer. The resulting larger surface to volume ratio has a higher degradation potential.

8.3 Stacked Dies

In 1989, Brampton Engineering introduced a relatively new concept in coextrusion die design for conventional tubular (blown) film, which used a stacked layer distribution system. Unlike the conventional systems, the stacked dies arrange the layers vertically as shown in Fig. 8.4.

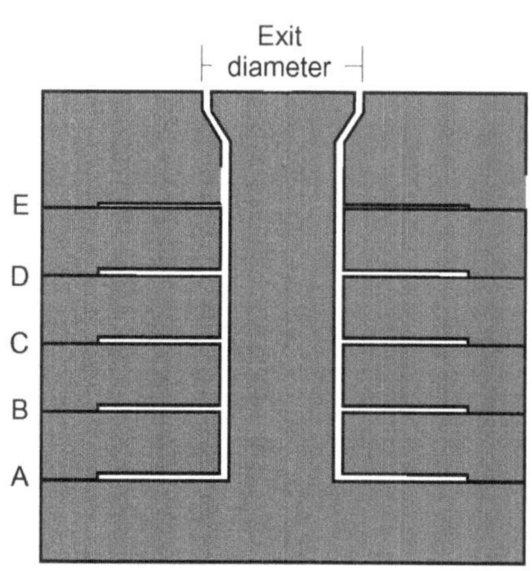

Figure 8.4 Stacked type of coextrusion die.

It has been reported that some processors had developed similar systems for their own use, but it is believed that this was the first time that such technology was available commercial by an OEM for conventional blown film.

It should be pointed out that "stacking" the layers is not completely new concept. In fact, the "stacked" configuration has been used with blow molding dies since the late 50s and 60s [2,3]. However, these designs had some disadvantages, which made them unattractive for blown film applications. In general, these designs used a simple side-feed distribution system as shown in Fig. 8.5.

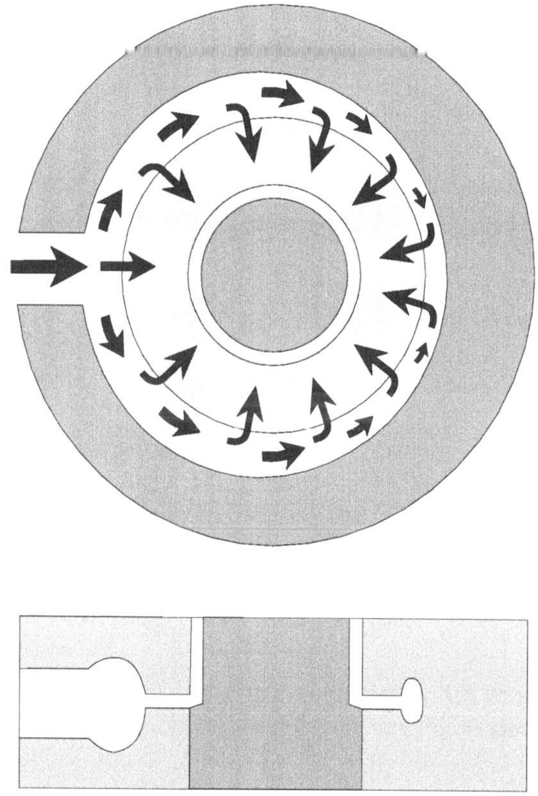

Figure 8.5 A typical side-feed distribution system.

The polymer is essentially split into two flow streams, which are distributed around the die circumference and meet opposite to the entrance point. This would resemble a flat die distribution system that has been wrapped around so that the edges would connect. This distribution system, however, suffers from some drawbacks. First, where the streams merge, there is typically a stagnant flow region where the polymer could degrade. Second, the flow streams do not overlap as in a conventional spiral mandrel die. This means that

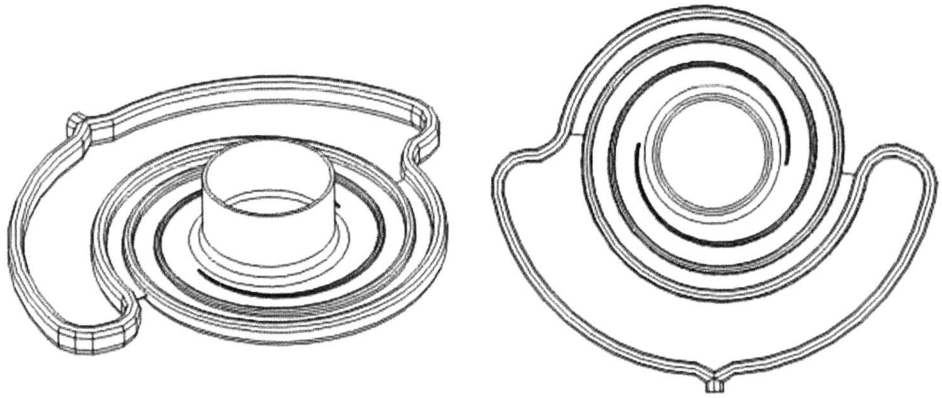

Figure 8.6 A flat spiral distribution system with two spirals.

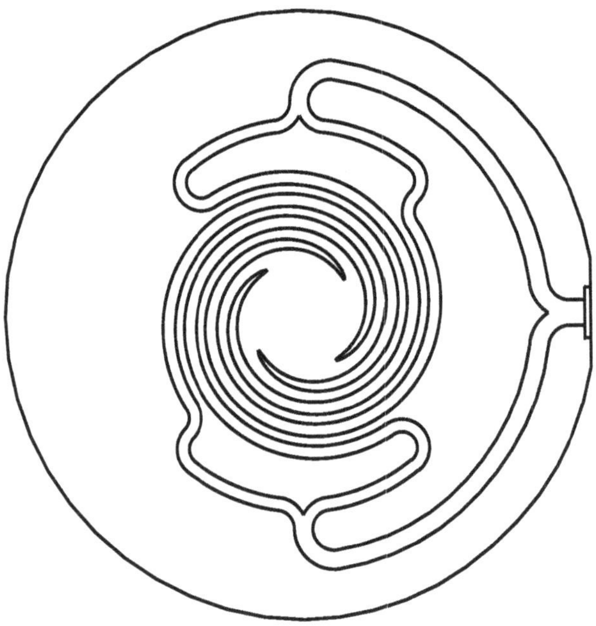

Figure 8.7 A flat spiral distribution system with four spirals.

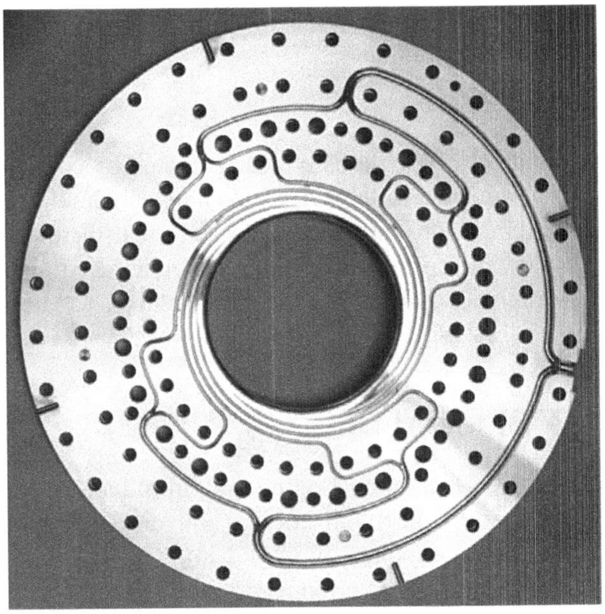

Figure 8.8 A photograph of a Brampton Engineering "SCD" die distribution plate with eight spirals.

the weld line orientation would be perpendicular to the die axis and was more likely to result in a weak point or a visual defect in the final product. In addition, the overlapping or layering provides some additional "mixing" of the polymer melt. Having no overlapping makes the die more sensitive to any inhomogeneities in the melt stream from the extruder. For these reasons, it was desirable to incorporate a spiral-type distribution system into the new configuration. Fig. 8.6 shows a flat spiral distribution system used in a small stacked die.

Fig. 8.7 is a drawing of a flat spiral distribution system with four spirals, each of which travels 360 degrees.

Fig. 8.8 is a photograph of a flat spiral distribution system with eight spirals.

In any spiral-type die, it is highly desirable that the polymer entering the die be divided equally among the spirals. This is done with a binary distribution system before the spirals as shown in Figs. 8.7 and 8.8. In the binary distribution system, the polymer entering the die is first divided into two identical flow streams. Then each of these streams is also divided into two streams resulting in four streams of equal flow as shown in Fig. 8.7. This can be repeated again to get 8 or 2^3 where "3" is in the number of divisions or splits; hence the binary system. The flat spiral distribution system then distributes the flow in a similar manner, as would a conventional cylindrical spiral distribution system. The goal is to obtain a circumferentially uniform radial flow stream at the end of the spiral section. Thankfully, there exist advanced computer-aided engineering tools that can help designers optimize the flow channels and ensure uniform distribution with acceptable residence time and pressure drop.

This radial stream can then be redirected to flow in the die's axial direction and ultimately join with other layers before exiting the die. More details on this system can be found in a dissertation by Perdikoulias [4].

The primary advantage of the stacked die system was that it offered the potential to minimize each layer's wetted surface area thereby minimizing the residence time and improving the system purge characteristics. Another advantage of the stacked die system was that it resulted in a modular die design, which had a few more potential advantages. First, it could theoretically allow a die to be more easily modified to conform to various structures by exchanging modules. The other potential advantage is that the individual flow modules could be operated at different temperatures so as not to expose thermally unstable materials to excessive temperatures for a prolonged period. These advantages have been successfully incorporated in tubular (blown) film dies of up to 10 discrete layers as shown in Fig. 8.9.

After the success of dies provided in the early 90s, many manufactures followed suit and provided their own versions of stacked die systems [5,6].

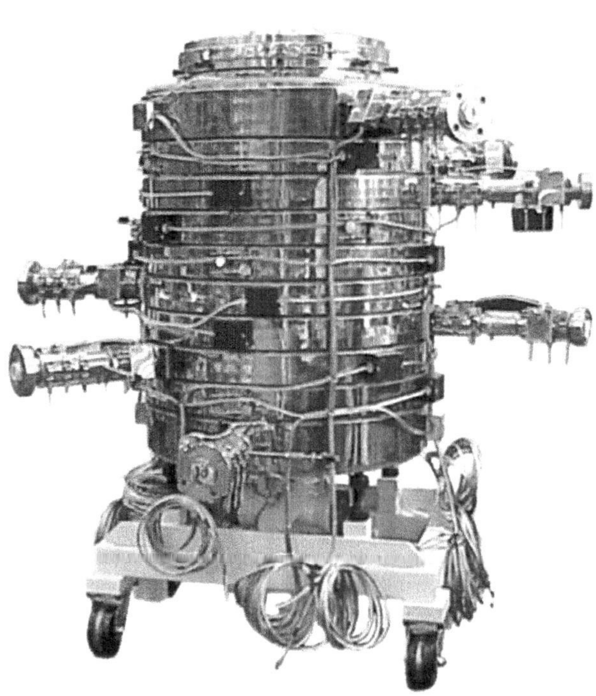

Figure 8.9 Photograph of Brampton Engineering 10-layer stacked die.

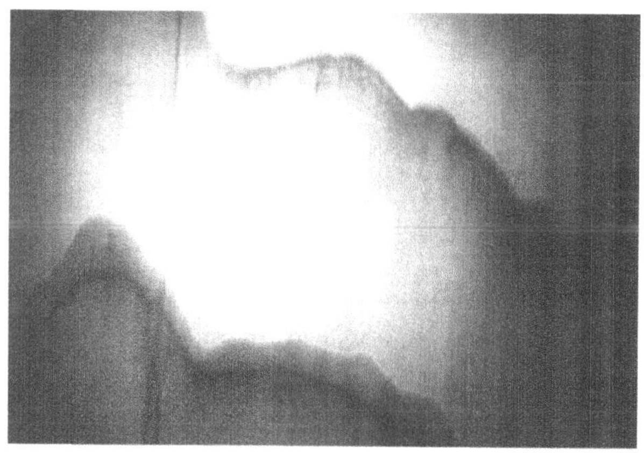

Figure 8.10 Two-layer tubular film with interfacial instability.

8.4 Interfacial Instability

Along with the additional layers used in coextrusion, there can appear some additional and unique problems. The potential degradation problems and residence time distribution have been mentioned earlier. Another common problem is interfacial instability. When materials are coextruded, it is possible that the interface between the two materials does not always remain smooth and uniform. In fact, the phenomenon can be quite dramatic resulting in intermittent layer flow rendering the product useless.

Fig. 8.10 shows a two-layer film sample that exhibits a strong interfacial instability. In one layer, carbon black at a low level has been added to assist visualization. If there was no instability, the film sample would have a uniform gray appearance. Fig. 8.10 clearly shows that this is not the case. In fact, the dark "bands" or "waves" indicate that the carbon black layer has an erratic periodic flow behavior.

Instability phenomena in coextrusion have been the topic of many studies [4,7–17]. However, it has only been relatively recently that the phenomenon was understood well enough for the development of die design criteria that could be used to correct or avoid the problem based on the material rheology and flow ratios [18–20]. There is now commercially available flow simulation software that can be used to predict the possibility of interfacial instability based on die geometry, material properties, and processing conditions [21].

In conclusion, while multilayer film extrusion poses more challenges for material suppliers, equipment manufacturers, and processors, the benefits of improved performance and/or lower cost continue to drive developments in materials, equipment, and process "know how" to the point where eight- and nine-layer lines, with multicomponent blenders for each layer, are becoming more and more common.

References

[1] Perdikoulias J. Annular coextrusion die design, Tappi J, 1992, TAPPI Press, p. 212.
[2] Colombo, R. US Patent, 2,820,249, 1958.
[3] Schrenk, W.J. US Patent 3,308,508, 1967.
[4] Perdikoulias J. Analysis and design of annular dies for mono and multilayer annular flow. PhD thesis. Waterloo, ON: University of Waterloo; 1997.
[5] Mavridis, H. Modular coextrusion stacked dies for blown film, SPE ANTEC Proceedings, p. 101–105, 1996.
[6] TAPPI PLACE Film Extrusion Manual, 2nd ed. TAPPI Press, 2005.
[7] Schrenk WJ, Bradley NL, Alfrey T. Interfacial flow instability in multilayer coextrusions. Polym Eng Sci 1978;18:620.
[8] Mavridis H, Shroff RN. Polym Eng Sci 1994;34:559.
[9] Han CD, Shetty R. Polym Eng Sci 1978;18:180.
[10] Ramanathan R, Shanker R, Rehg T, Jons S, Headley DL, Schrenk WJ. "Wave" pattern instability in multilayer coextrusion—An experimental investigation, SPE ANTEC Proceedings, 224; 1996.
[11] Perdikoulias J, Tzoganakis C. Interfacial instability phenomena in blown film coextrusion of polyethylene resins. TAPPI Proceedings; 1995;97.
[12] Dooley J, Hyun KS, Hughes K. An experimental study on the effect of polymer viscoelasticity on layer rearrangement in coextruded structures. Polym Eng Sci 1998;38(7):1060.
[13] Dooley J, Hughes K. Measurement of layer deformation in coextrusion using unique feedblock technology, SPE ANTEC CD-ROM Proceedings; 2000.
[14] Tzoganakis C, Zatloukal M, Perdikoulias J, Saha P. Viscoelastic stress calculation in multi-layer coextrusion dies, SPE ANTEC CD-ROM Proceedings; 2000.
[15] Zatloukal M, Vlcek J, Tzoganakis C, Saha P. The effect of layer stretching on the onset of "Wave" interfacial instabilities in coextrusion flows, SPE ANTEC CD-ROM Proceedings; 2001.
[16] Zatloukal M, Tzoganakis C, Vlcek J, Saha P. Numerical simulation of polymer coextrusion flows. Int Polym Process 2001;198–207.
[17] Martyn MT, Gough T, Spares R, Coates PD. Visualisation of melt interface in a coextrusion geometry, SPE ANTEC CD-ROM Proceedings; 2001.
[18] Zatloukal M, Perdikoulias J, Tzoganakis C, Saha P. The influence of die design and extensional rheology on the onset of "Wave" interfacial instabilities in coextrusion flows, TAPPI CD-ROM Proceedings; 2001.
[19] Zatloukal M, Vlcek J, Tzoganakis C, Saha P. Viscoelastic stress calculation in multi-layer coextrusion dies: die design and extensional viscosity effects on the onset of "Wave" interfacial instabilities. Polym Eng Sci 2002;42 (7):5120.
[20] Xiao K. The effect of die geometry on interfacial instability in multilayer streamlined coextrusion die (SCD), TAPPI CD-ROM Proceedings; 2003.
[21] Virtual Extrusion Laboratory V. 6.1, Compuplast International Inc.; 2006.

9 Process Engineering

Eric Hatfield[1] and Vincent Vezza[2]
[1]Clopay, Mason, OH, United States; [2]Independent Consultant, Rochester, NY, United States

9.1 Managerial Summary

What process engineers do is one of the most misunderstood jobs of the engineering profession. Project engineers usually install equipment. Product engineers develop and enhance products for the customer. Maintenance engineers maintain and extend the life and reliability of manufacturing equipment. But what does a process engineer do?

The process engineer is the one who has the know-how to take the raw materials that the product engineer feels the customer requires in his/her product and sets the operating conditions and runs it through the equipment the project engineer has installed. He/she has material knowledge and what process levers his/her equipment gives him/her to obtain the properties the customer desires.

A process engineer sees the manufacturing equipment and materials as one entity in a state of flux. To him/her the process is as if it is alive and he/she is the doctor who can adjust the conditions and inputs to get the desired healthy patient or product.

A process engineer usually (though far from always) reports to operations rather than engineering or R&D. His/her focus is on transforming raw materials to the finished product on the process in question. He/she does this in the most economic manner possible to get the highest quality product possible.

In short, a process engineer's job is to make more of a higher quality product for less money or cost.

9.2 Process Engineering: What Is It and Why Is It Essential?

When your machine breaks down, you call the maintenance engineer. When your customer wants to change a product, you call your product engineer. When you cannot make the product correctly and are running scrap, you call your process engineer.

he process engineer knows the strengths and limitations of the equipment and the product as it is being run on the equipment. The process engineer knows the age and wear on the extruder screws, nips, winders, and other associated equipment, and knows how long they can be pushed and what temperature and output adjustments need to be made as they wear.

Sometimes process engineering is confused with process control. With process control, a set point is maintained. However, a process engineer would select the set point.

Many (but by no means all) process engineers are chemical engineers. Chemical reactions are in a state of flux, not static and this is how a process engineer must look at the process. It is in a state of flux with the raw materials reacting to the equipment and the conditions they are run at to get a film with the desired attributes.

9.3 Some Process Engineer Tools

There are many tools that a process engineer uses. Some are listed below:

- *Engineering principles:* A process engineer applies engineering principles to what is happening to the raw materials as they go through the manufacturing equipment and are transformed into the finished product. He is not a big believer in making things as an "art." Everything has a scientific reason for why it happens. One just needs to figure it out.
- *Statistics:* A process engineer must deal with large quantities of often-conflicting data when troubleshooting a problem or getting a new product to run. Applying statistical techniques allows him/her to see "the forest for the trees" and make scientific sense of the data. For example, using some simple data smoothing (eg, moving average) for an extruder pressure fluctuation can reduce the noise in the data coming from the pressure transducer. Then, being able to determine more clearly the frequency and pressure fluctuations, the process engineer

Multilayer Flexible Packaging. DOI: http://dx.doi.org/10.1016/B978-0-323-37100-1.00009-0

can go looking for other things in the process with similar frequencies and magnitudes.

There are many statistical tools available to the process engineer. Some of these tools are:

- Descriptive statistics
  - Mean
  - Standard deviation
  - Histogram
  - Scatter diagram
- Design of experiments (DOE)
- Statistical process control (SPC)
- Multivariate statistical process control (MSPC)

Most engineers know what the mean and standard deviation are; they will not be covered in this chapter. Some of the following tools will be familiar. However, it is worth covering them.

9.3.1 Histogram

The histogram presents the data visually. The frequency at which measurements or range of measurement occur is recorded. The measurement is usually recorded on the horizontal axis with the frequency recorded on the vertical axis. The general rule of thumb for constructing a histogram is: take the square root of the number of data points; this will give you the number of cells (columns) on the histogram. There are other rules for finding the number of cells. However, they will not be covered here. Many software packages will do this automatically. The histogram in Fig. 9.1 was generated in Minitab using a normal generator with a mean of 10 and standard deviation of 2.0. There are 500 data points.

A histogram reveals valuable information about the process. If a histogram shows a normal distribution, bimodal, and multimodal of the raw process data, this could indicate that the process is in control or there was a process shift or multiprocess streams (different machines). Therefore, a quick look at the process data histogram could reveal some valuable information about the process.

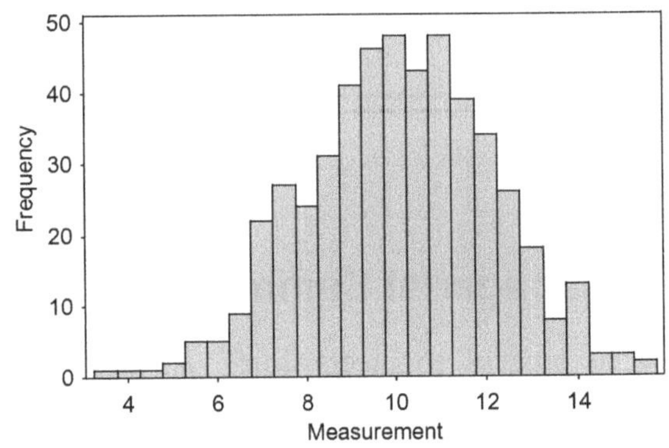

Figure 9.1 Frequency versus measurement histogram for mean = 10 and std. dev. = 2.

9.3.2 Scatter Diagram

The scatter diagram is used to show the relationship between two variables. A scatter diagram plots pairs of observation on a two-dimensional plot. If two variables move together, a scatter diagram can show this fact.

The scatter plots show no or low correlation in Fig. 9.2, positive correlation in Fig. 9.3, and negative correlation in Figure 9.4. However, correlation does not mean cause and effect. The process engineer should do more work to confirm that there is a cause and effect.

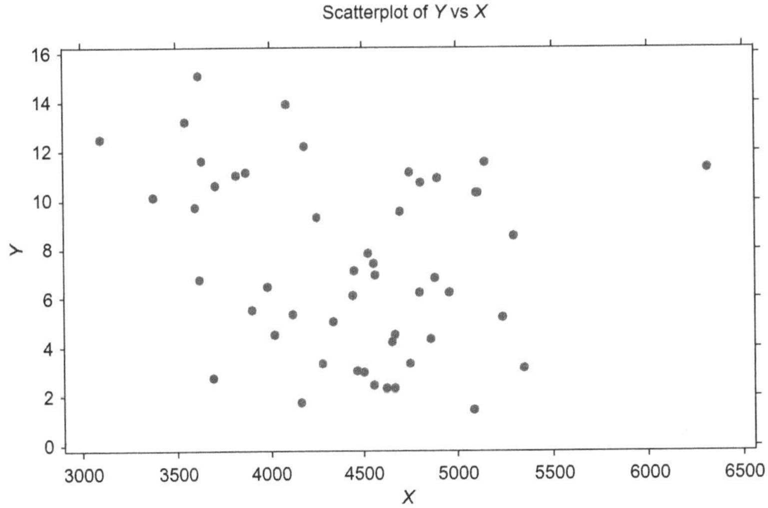

Figure 9.2 *X* versus *Y* plot that shows low or no correlation.

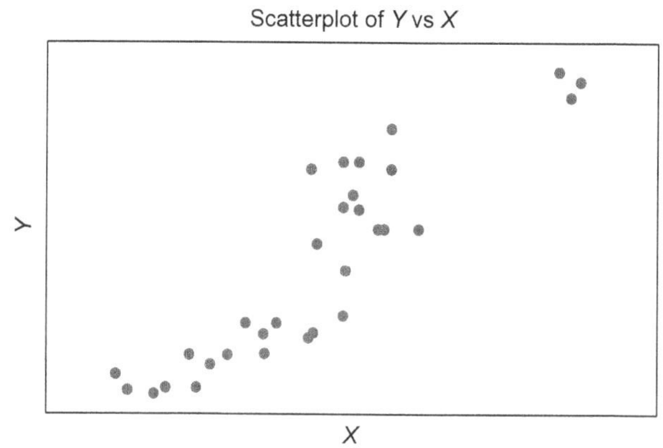

Figure 9.3 *X* versus *Y* plot that shows positive correlation.

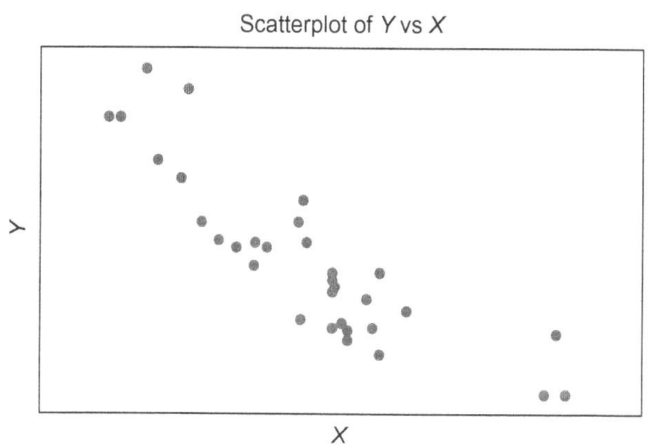

Figure 9.4 *X* versus *Y* plot that shows negative correlation.

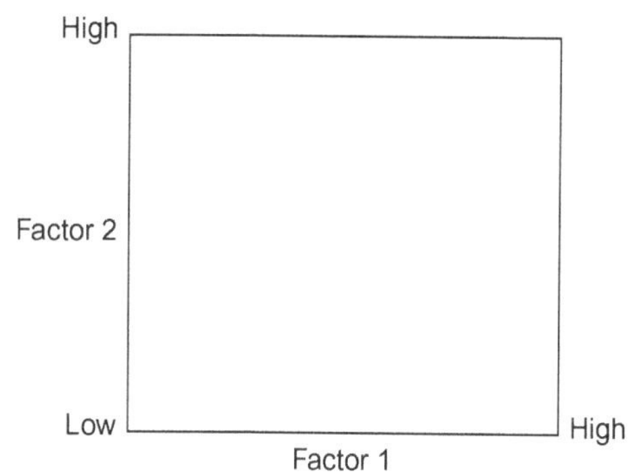

Figure 9.5 Two-level design with two factors 1 and 2.

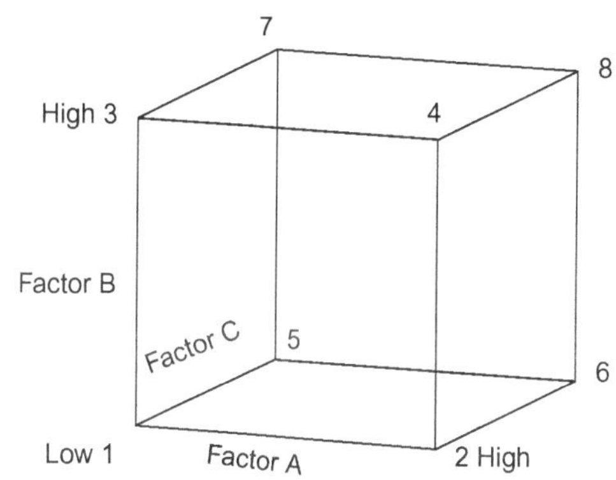

Figure 9.6 Spatial representation of a three-factor two-level design.

9.3.3 Design of Experiments

A statistically designed experiment is a useful tool for helping one to identify the important factors/variables in process. Not only can they help to identify these factors/variables but will help you understand how these factors influence each other when taken as a whole. Most DOEs are two levels (2^n, n being the number of factors). There are other designs like three-level designs that would not be covered. You can include center points to check for curvature between measured points and response. One should not assume that the response is linear. Figure 9.5 is a two-level design with two factors. Figure 9.6 is a spatial representation of a three-factor two-level design. Figure 9.7 adds the center point to the three-factor two-level design. Figure 9.8 adds the star points and is considered a central composite design (CCD).

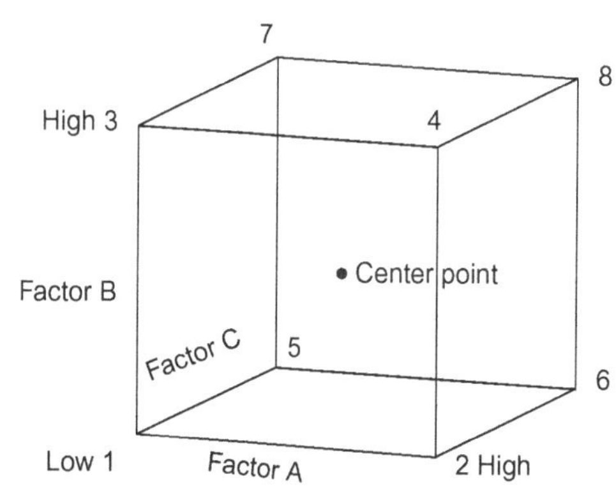

Figure 9.7 Spatial representation of a three-factor two-level design with a center point.

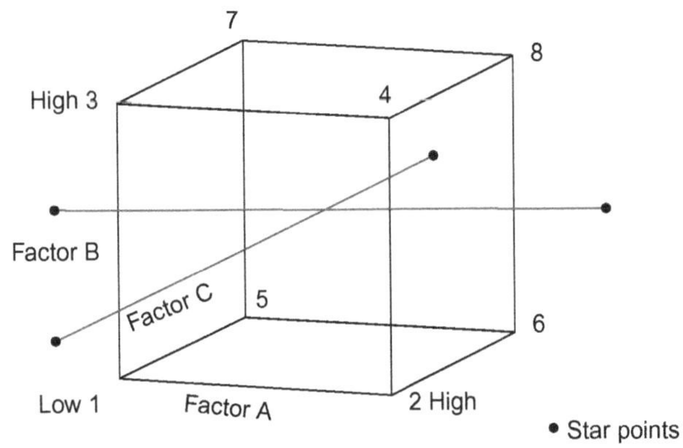

Figure 9.8 Spatial representation of a three-factor two-level design with center point and star points. This design is also known as a central composite design.

9.3.4 Statistical Process Control

Control charts are used to monitor a process and identify whether a common or special cause of variations is happening through the choice control limits.

Two variable control charts are:

- Univariate variable control charts
- Multivariate variable control charts.

9.3.4.1 Univariate Control Charts

Univariate control charts help the process engineer identify the difference between common and special cause variation. However, if there is special cause variation, it is up to the process engineer to determine the cause with the help of these charts and scatter diagrams and design of experiments. These tools are mentioned earlier.

The two univariate control charts are for:

- variable data
- attribute data.

Control charts for variable data are used when variable data are available.

Control charts for attribute data are for counting, or conversion of counts for proportions of percentages or the presence or absence of characteristics. I will mention only one attribute chart because I think it is important to flexible film packaging.

All control charts usually consist of a center line and an upper and lower control limit.

Variable control charts are based on the normal distribution. However, the normal distribution and how it relates to control charts will not be covered here. There are many books on SPC giving the theory behind using the normal distribution for variable control charts.

When using a variable control chart to locate a process is paired with a chart for the variability within a sample, they are referred to as *X*bar/*R* charts or *X*bar/*S* charts depending on the variability chart used. For the *X*bar/*R* chart, the *R* is the sample range and is easy to calculate. If the *S* chart is used, a computer is usually used because the sample standard deviation is calculated and a sample size of 10 or greater is used.

Fig. 9.9 shows an *X*bar/*R* chart. The data was enter into Minitab and then the *X*bar/*R* chart was generated. From the chart, one sees that the sample variability was within limits, whereas the sample location or average was out of the upper limit at sample 16. However, the process steadily decreased after sample 16. This steady decrease indicates an out of control situation. A process intervention might be appropriate in this case.

This is a powerful tool when monitoring a process with only one or two uncorrelated critical process variables.

Another important control chart is based on the Poisson distribution. Control charts based on this distribution is the *c* chart. It is based on the imperfections (defects) on a unit of measure, where the unit of measure is the defects per square foot, defects per square meter, or defects per part. This might be a useful control chart for thin flexible film (Fig. 9.10).

One word of caution, when a sample is outside the control limits, there is an indication that something has changed in the process. However, because a sample is outside the control limits, it does not mean that the sample is out of specification unless the control limits are outside the specification. In all cases, subject matter experts should be consulted for the product disposition.

9.3.4.2 Multivariate Statistical Process Control

In order to do MSPC, a computer and the proper software is necessary.

Multivariate control charts are best used when correlated process variables exist. Many chemical processes are suited for multivariate control charts because most process variables are interrelated or correlated to some extent. If there is no correlation,

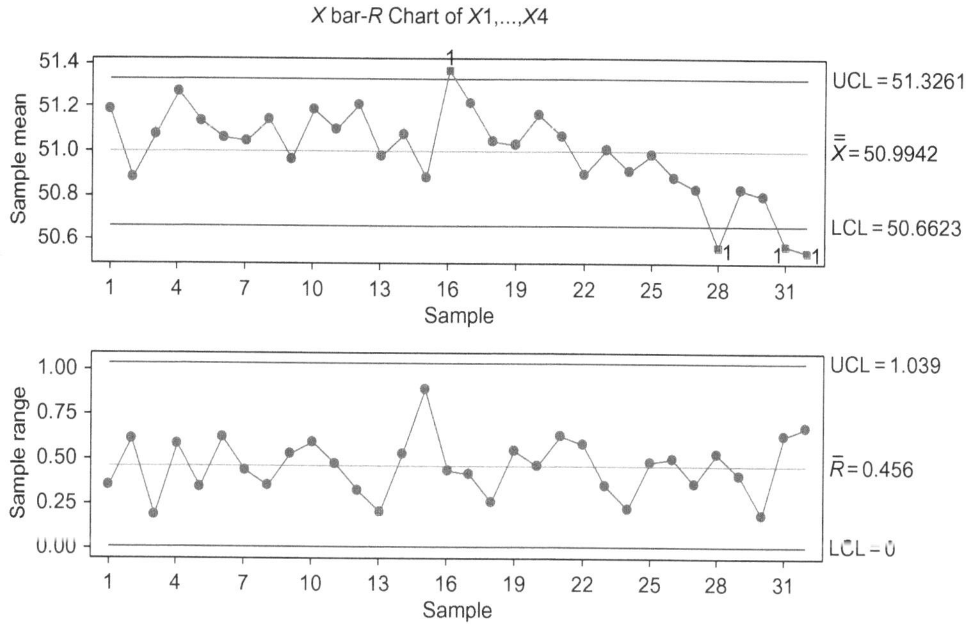

Figure 9.9 *X*bar/*R* chart.

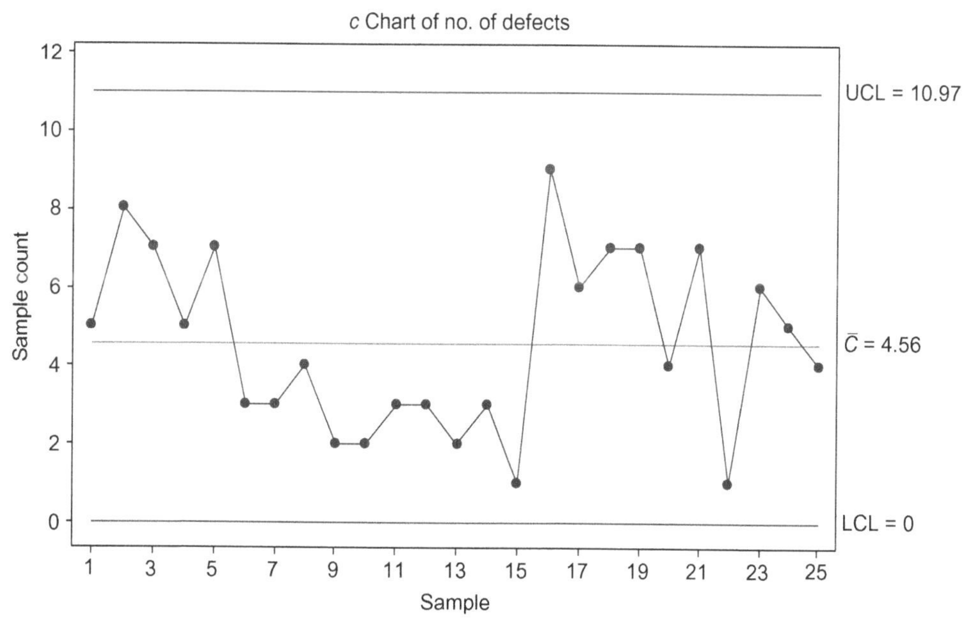

Figure 9.10 *c* Chart for the number of defects.

the univariate control chart should be used. The technique that mentioned here is plotting the Hotelling's T^2 statistic of the raw data (vector) as a way to monitor a multivariable process. When an out of control exists, the engineer must determine which variable or variables are causing the out of control situation. However, this is not always an easy task. Experience using multivariate control charts and process knowledge are important. Fig. 9.11 shows two points outside the control limits points (19 and 23).

- *Operator "interrogation":* Good operators often know what happened, but not always *why*. They are often taught and learn that if *that* happens do *this*. Many times this is the correct thing to do; however, they do not

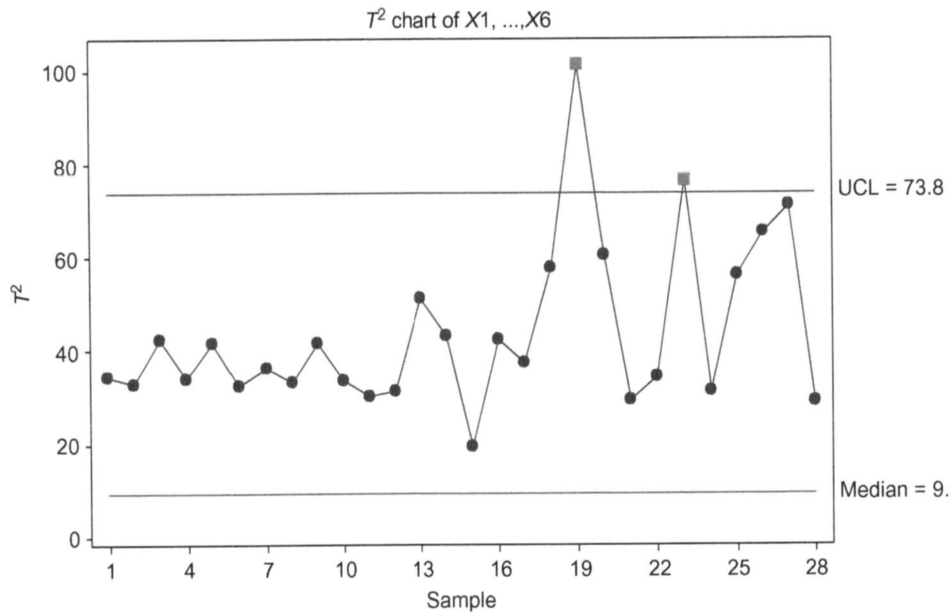

Figure 9.11 T^2 multivariate control chart with six variables.

always know why they do *this* to fix *that*. A process engineer can take this knowledge and use it to learn the scientific reasons why it works and also know when not to make the corrective action when it only appears to be the same problem. Operators and supervisors often see what is happening when a problem occurs. They will frequently draw their own conclusions as to what is making it happen. This can cloud the real cause and slow finding a solution. However, careful listening by the process engineer to what the operator has seen and done, along with the response, can quickly lead to the problem root cause. This requires tact and careful questioning of the operator to get at the truth. Not unlike a detective questioning a witness. For example, suppose the film thickness is varying as much as 30% in the machine direction. The operator feels that this is because one extruder in the multilayer structure is surging. The process engineer checks the extruder in question and it is, indeed, surging, however, the layer is only 5% of the structure, the pressure surge was in the order of 15%. This is nowhere near enough to cause a 25% gauge change. Plus the frequency was repeating in a constant manner and the gauge variation frequency appeared to be random. Upon reviewing the other instruments, it was found that the first pull roll amps were surging a great deal. After tacking the roll speed, it was found to be changing from 20% to 30%. Upon calling maintenance, it was discovered that a drive belt was loose and slipping. After getting this fixed and the line back into production, the process engineer can start investigating the cause of the extruder surging.

- *Data "interrogation":* Very similar to "operator interrogation" as described earlier, a process engineer needs to be able to sort through a menagerie of data to find which things are important.
- *Past experience:* Past experience is invaluable to the process engineer. A problem that one has seen in the past on a particular product or machine can rear its ugly head again. For example, the quality control (QC) laboratory technician tells the process engineer that the coex cereal liner film on one of the lines smells bad. This particular film has butylated hydroxytoluene (BHT) added to it for the customer. Past experience has shown when this happens the BHT had either exceeded its shelf-life or the extruder is running too hot. The process engineer checks the date on the BHT box and finds it is fine. He then looks at the melt temperature on the extruder and finds

it too hot. After slowing the extruder down some and lowering some of the barrel temperatures, the foul odor is no longer in the film. Past experience permitted the process engineer to fix the problem quickly. If changing the extruder's temperature had not fixed the problem, then the process engineer would have had to dig deeper.

- *Knowledge of critical fitness for use (FFU) product demands:* Knowledge of what properties are important to the film customer is important for the process engineer to understand. This knowledge allows him/her to know which problem to address first when confronted with multiple problems. It also tells him/her which properties to improve or make more robust when working continually to improve the film products. A process engineer is always running little experiments to see how the process responds and what happens to the final product. Most can be done while keeping the final product within its specification window. These little experiments often permit the specification to be tightened to provide the film customer with a more uniform and consistent product.
- *An open, inquiring mind:* A process engineer needs to have most importantly an open and inquiring mind. In troubleshooting, he/she needs to listen to what is said by all and to review the data without prejudice and, when things seem to be running well, he/she continues to look for problems and interactions. Always studying the process and the product being produced.

9.4 Examples of How Process Engineers Can Save Time and Money

- *Prioritize data or process changes:* When changes need to be made to the process, it is the process engineer who determines what changes are made and in what order. This is only natural since he/she is the one who best understands the process and its strengths and weaknesses.
- *Provide leadership and hands-on training to operators and supervisors:* As the process engineer goes about his/her daily work, solving problems and running little experiments, he/she also teaches the frontline supervisors and operators. This often just happens naturally since the process engineer is on the floor working next to the operator and having him/her to make changes. Then the operator or supervisor is also there to see what the results are. This is in conjunction with the fact that often the process engineer is asked "why" are we doing this? He then has a perfect opportunity to teach.
- *Explain management goals to the operating personnel:* This is not commonly recognized and is normally not in a process engineer's job description. In many cases, the process engineer is the best and often most efficient means of transmitting management information and strategy to the frontline employees. This is because not only is the process engineer part of management and informed of upper management's plans and desires, he/she is also working side by side with the production operators. In many cases, they spend time together in troubleshooting and experimenting. This provides opportunities to talk and explain why management is doing certain things and what the company's plans and needs are.
- *Make R&D and new products work and aid in their commercialization:* In many cases, a new or enhanced product is first developed on a pilot line. When it comes to the manufacturing line, it is far from ready for day-to-day production. This is partly because the product development or R&D person has different priorities and perspective than those in manufacturing. (Remember the process engineer often reports through operations.) Making the product on the pilot line is just a beginning step in making a real product. The pilot line makes samples that are tested for the perceived attributes. It is then run on the customer's machine to see if it runs all right and the properties are indeed there. When it gets to the plant, it is in a rough form. Little or no thought has been given to operability, line speed, change over time, expected/possible yields, etc. In short, other than a raw material cost calculation versus selling price, little thought has gone into whether this product can be manufactured at a profit.

This is where the process engineer goes to work.

- Often different resin grades have been used. Sometimes this is necessary, but often it is not. A standard high-density polyethylene (HDPE) grade can be used. This allows the plant to have fewer grades in the plant and allows purchasing to get lower prices by offering larger volumes.
- The rate and yield at which a product can be produced must be determined and then pushed. Added to this, in many R&D pilot lines, or even trials on production machines, recycle is not put into the product. When the reclaim is added, the physical properties may change and the film may run differently on the customer's machine.
- These are just a few of the areas that a process engineer must address to obtain the product to a commercial state that can be profitably produced.

- *Helps get new products accepted by plant:* Many people are adverse to change and do not want to be bothered with trials for new products. Often the R&D guy is seen as an "outsider" or "from corporate." It is the process engineer who acts as the go between and gets the new product into the plant's portfolio.
- *Bring new, or help implement technology onto the plant floor:* New technologies are brought to the plant in much the same way as new products are. The process engineer's roll is much the same for new technologies as it is with new products.
- *Help QC and manufacturing address quality problems quickly and fix them:* QC technicians come to the process engineer first to fix a problem or the supervisor does after being informed of a possible quality problem.

9.5 Conclusion

The process engineer is the one that ties all the other engineering disciplines together to make the final product for the customer. He is the one who uses the equipment that is installed by the project engineer, to make the product out of the materials selected by the product engineer, to make the film for the final customer who, in turn, pays for it so the cycle can begin all over.

10 Blown Film, Cast Film, and Lamination Processes

John R. Wagner, Jr.
Crescent Associates, Inc., Rochester, NY, United States

10.1 Blown Film

Blown film is a major processing technique for producing a biaxial melt drawn film. This technique uses air pressure to produce a transverse direction draw and a higher speed haul-off roll speed to provide a machine direction draw. Several billion pounds of polymer, mostly polyethylene, are processed annually by this technique [1]. Other polymers that can be processed by blown film extrusion are polypropylene (PP), polyvinyl chloride (PVC), polystyrene (PS), ethylene vinyl acetate (EVA), ethylene vinyl alcohol (EVOH), polyamide (PA), and polyurethane (PU). Markets that blown film extrusion serves are as follows:

- industrial films and bags
- agricultural and construction films
- barrier films
- stretch films
- PVC cling films
- laminating films
- can liners
- high barrier small tube systems.

Coextrusion to produce multilayer films is a major component of blown film production. Production line output goes from 100 pounds per hour (PPH) to 10,000 PPH. Coextrusion systems can have from two to nine extruders making 2- to 17-layer films. If one has a four-extruder system, it is common to make a symmetrical seven-layer [ABCDCBA] structure from the four different polymers streams.

There are many manufacturers who produce excellent equipment for making multilayer blown films. Fig. 10.1 [2] is a simplified single-layer blown film line layout. Fig. 10.2 is a 3D overview of a seven-layer blown film line. Fig 10.3 shows a nine-layer blown film die and bubble with three of the extruders. Fig. 10.4 shows a blown film tower and collapsed film going to the winder.

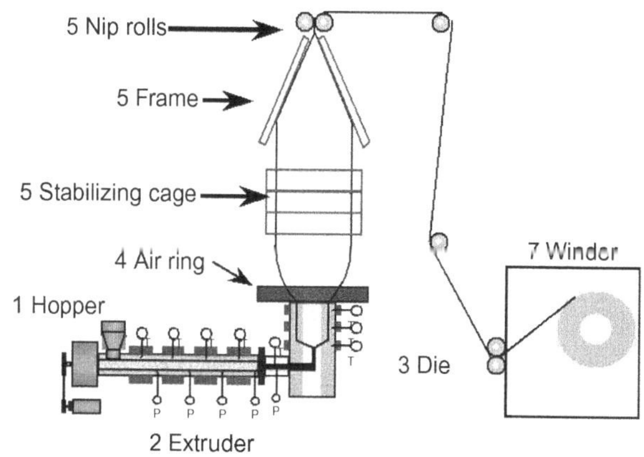

Figure 10.1 Blown film process.

The components of a coextruded blown film line are:

- resin feed system
- extruders
- coextrusion die
- air ring
- internal pressure control for adjusting bubble diameter
- collapsing frame
- take-up or haul-off roll which set the MD draw
- treatment system
- winder.

Multilayer Flexible Packaging. DOI: http://dx.doi.org/10.1016/B978-0-323-37100-1.00010-7

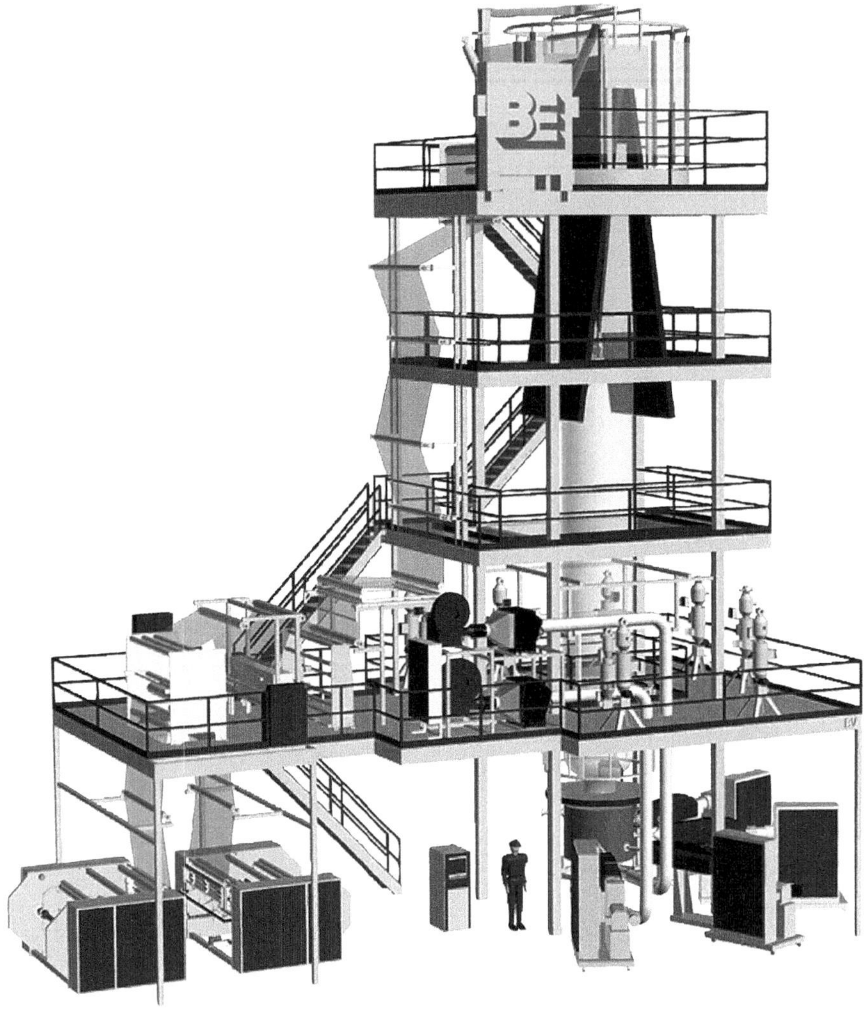

Figure 10.2 Seven-layer blown film 3D layout. *Image provided by BE (Brampton Engineering), www.be-ca.com.*

The design features that are important in producing quality film at a competitive price are:

- an efficient and properly sized resin handling and feed system
- an efficient screw design that gives quality melt with:
  - low and uniform temperature
  - stable pressure
  - at high rate
- an optimized die that provides good layer control and thickness uniformity. The die must also be designed for ease of maintenance and durability
- air rings that provide excellent cooling control and uniformity
- automated web handling systems for improved efficiency and reduced changeover times
- modular design features for product changeovers
- integrated control systems that are intuitive, operator friendly, and keep the process parameters on target.

For more detailed information, the reader can find many excellent works on the blown film process and operation. Guiles has a troubleshooting guide [3]

Figure 10.3 Nine-layer blown film die and bubble. *Image provided by BE (Brampton Engineering), www.be-ca.com.*

and a blown film process description [4]. Butler et al. [5] is an excellent reference. Vlachopoulos and Wagner [6] also have a chapter by Butler on the blown film process and troubleshooting. Calhoun in *SPE Plastics Technicians Toolbox Book 6* [7] has a chapter on blown film extrusion and Cantor in his book [1] includes a CD with a PE blown film simulation that is an excellent training aid.

10.2 Cast Film

Compared to the blown film process, a cast film process quenches the molten extrudate on a chilled steel roller or rollers after it exits the die. Sheet can also be made by this process. However, with a sheet line, the line speeds are slower and there are multiple chill rolls to provide the required heat removal.

Figure 10.4 Blown film tower and collapsed film going to winder. *Image provided by Davis-Standard, http://www.davis-standard.com/.*

The distinction between film and sheet usually is at 0.010″ or 10 mils (250 μm). Film is considered to be less than 0.010″ and sheet greater than 0.010″.

Fig.10.5 [8] shows the cast film process.

Different compatible resins can be combined with feedblock technology and/or in a multilayer die and extruded onto a chill roll where the film is solidified. There is usually a corona treatment station to enhance wettability and adhesion, plus a gauging system for measuring and controlling the gauge profile, a slitting section to remove the edge beads, and then a combination slitter winder to make different width rolls. The trim is usually ground up and recycled back to one of the extruders.

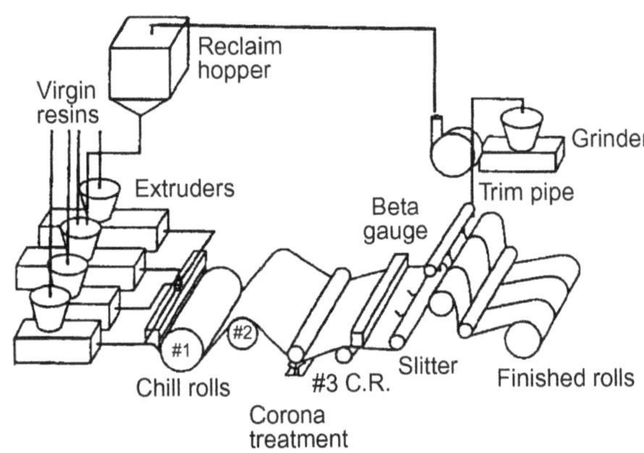

Figure 10.5 Cast film process.

Fig.10.6 shows the die exit and first large chill roll and a second smaller chill roll. Fig. 10.7 shows the melt pipes that feed a multilayer die. In this case, there are four melt pipes for four different resins that can be combined in up to seven layers. Fig. 10.8 is a cast film winder that features in line slitting.

The reader is referred to Refs. [3,5,6,7] that provide more detailed information on the cast film process.

Figure 10.6 Cast roll. *Image provided by Davis-Standard, http://www.davis-standard.com/.*

Figure 10.7 Four melt pipes feeding multilayer die. *Image provided by Davis-Standard, http://www.davis-standard.com/.*

Figure 10.8 Cast film winder. *Image provided by Davis-Standard, http://www.davis-standard.com/.*

10.3 Extrusion Coating and Lamination

The difference between extrusion coating and extrusion lamination is the presence of the second or auxiliary web that sandwiches the melt on the second side. For extrusion coating, only one substrate is used and the extrudate coats the surface and is quenched on a chill roll much like a cast film process. Fig. 10.9 [9] shows the detail for extrusion laminating where the auxiliary substrate is coming from the right side onto the chill roll and the extrudate is squeezed between the two substrates between the pressure roll and the chill roll.

Fig. 10.10 [9] shows a typical laboratory extrusion lamination line. This design has three extruders for making multiple layers in the extrudate to facilitate the coated or laminated product properties.

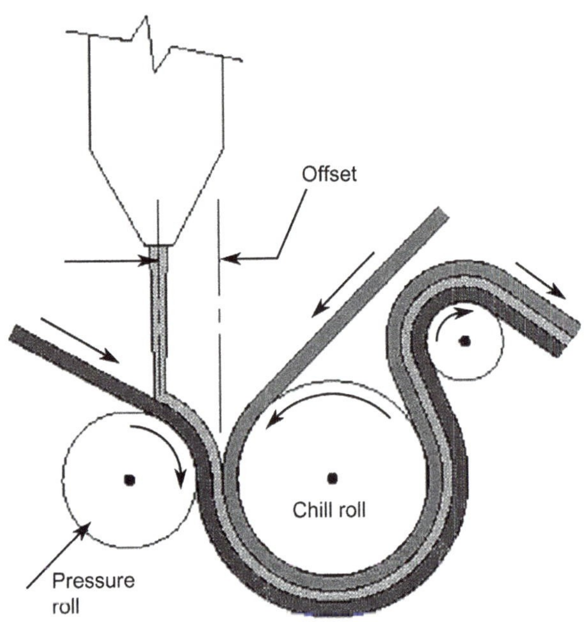

Figure 10.9 Extrusion lamination detail.

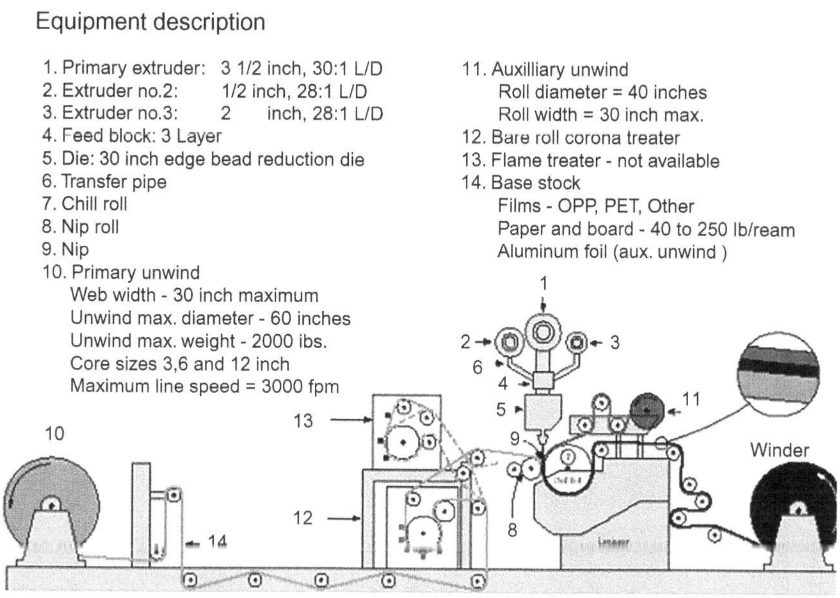

Figure 10.10 High-speed laboratory extrusion lamination line.

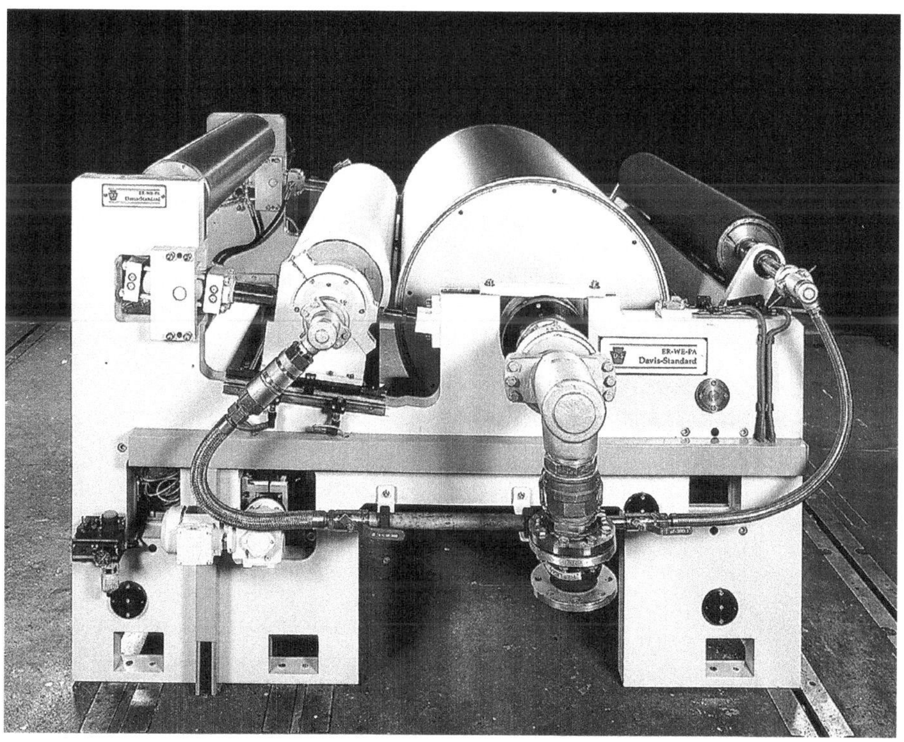

Figure 10.11 Extrusion coating/laminating station. *Image provided by Davis-Standard, http://www.davis-standard.com/.*

Fig. 10.11 shows an extrusion coating/laminating station with a large cooling roll, incoming rubber-backed cooled nip roller, cooled take-off roll, and idler tension measuring roll.

Fig. 10.12 shows a multilayer extrusion coating process in operation. You can see the main and two coextruders, combining block on the deckled single-cavity die, chill roll, and take-off section.

Figure 10.12 Multilayer extrusion coating station in operation. *Image provided by Davis-Standard, http://www.davis-standard.com/.*

Extrusion coating and laminating lines, like many processes, require:

- excellent melt quality
- proper configuration to fit products to be made
- good gauge control to promote high quality and economical resin usage
- automation to help maintain quality and repeatability
- flexible design to accommodate new products
- intuitive state of the art and updatable control systems to provide reliable process control.

Products typically made with extrusion coating and laminations are:

- lidding stock
- candy wrapper
- snack food bags
- medical packaging
- condiment packages
- soup sachets
- toothpaste tubes
- cable wrap.

The reader is referred to Cooper [9] and Giles et al. [10] who provide further insight to this process.

References

[1] Cantor K. Blown film extrusion. Munich: Carl Hanser Verlag; 2006. p. 1.

[2] Butler TI. Blown film extrusion. In: Vlachopoulos J, Wagner Jr. JR, editors. The SPE guide on extrusion technology and troubleshooting. Brookfield, CT: Society of Plastics Engineers; 2001.

[3] Guiles Jr. HF, Wagner Jr. JR, Mount III EM. Extrusion the definitive processing guide and handbook. 2nd ed. Norwich, NY: William Andrew Inc.; 2014. p. 301–6.

[4] Guiles Jr. HF, Wagner Jr. JR, Mount III EM. Extrusion the definitive processing guide and handbook. 2nd ed. Norwich, NY: William Andrew Inc.; 2014. p. 453–63.

[5] Butler TI, Ealer GE, Marks SB, Oliver GDJ. Perdikoulias. Film extrusion manual. 2nd ed. Atlanta, GA: TAPPI; 2005.

[6] Vlachopoulos J, Wagner Jr. JR. The SPE guide on extrusion technology and troubleshooting. Brookfield, CT: Society of Plastics Engineers; 2001.

[7] Calhoun AR, Wagner Jr. JR. Extrusion processes. Plastics technician's toolbox, book 6. Brookfield, CT: Society of Plastics Engineers; 2004.

[8] Ivey J. Cast film and sheet extrusion. In: Vlachopoulos J, Wagner Jr. JR, editors. The SPE guide on extrusion technology and troubleshooting. Brookfield, CT: Society of Plastics Engineers; 2001.

[9] Cooper J. Extrusion coating. In: Vlachopoulos J, Wagner Jr. JR, editors. The SPE guide on extrusion technology and troubleshooting. Brookfield, CT: Society of Plastics Engineers; 2001.

[10] Guiles Jr. HF, Wagner Jr. JR, Mount III EM. Extrusion the definitive processing guide and handbook. 2nd ed. Norwich, NY: William Andrew Inc; 2014. p. 465–8.

11 Machine Direction–Oriented Film Technology

Eric Hatfield
Clopay, Mason, OH, United States

11.1 Managerial Summary

Machine direction orientation (MDO) of films has been studied for decades. In the 1950s and 1960s, the academics studied property improvements when a film is stretched in the machine direction (MD). It has taken many years for the industry to begin to commercialize the technology.

When a film is MD oriented, it is stretched in the MD only. In either simultaneous or sequential biaxial orientation, the film is stretched in both the machine direction and the transverse direction (TD).

An MDO machine is a stack of usually separately driven rolls mounted in side frames for support. The rolls are heated and/or cooled in separate temperature zones.

When a film is correctly MD oriented, many of its physical and barrier properties are greatly improved. For instance, the moisture barrier of high-density polyethylene (HDPE) can be doubled. This can open the opportunity for down gauging or creating a film with much higher barrier at the same gauge.

Other film properties that can be improved are:

- optics: clarity, haze, and gloss
- tensiles: increase in break and tear
- controlled shrink levels
- stiffness: secant modulus can increase from two to three times in both TD and MD directions.

The effect of this technology on multilayer flexible plastic films is to enhance their properties and increase the places where the advantages of multiple properties that you get from a single multilayer sheet can be utilized. Also some MDO products are best made as multilayer films.

11.2 Description of MDO Hardware and Technology

MDO of film has been around for decades, but has been slow to gain prominence. This is due to the process subtleties that do not always appear to be straightforward. Despite this, there have been a number of unique (and profitable) products made using MDO.

Unlike the MDO *process*, the machine is relatively straightforward. The machine is basically a roll stack with separately driven rolls and multiple heat zones. A typical MDO will have 8–12 rolls.

The four sections in the MDO machine and process are as follows:

- *The preheat section:* These rolls heat the incoming film to the desired temperature for orientating the film.
- *The draw section:* In the draw section, the film is oriented, or drawn (stretched), between a set of rolls with, typically, a narrow gap between them. This gap is usually less than 3 mm and sometimes less than 1 mm. The rolls used in the draw section are usually smaller in diameter than the other rolls in the MDO. This is to get the tangent points closer together and thus reduce the distance over which the film is oriented.
- *The annealing section:* The annealing section heat sets or anneals the film to prevent it from shrinking back to its preoriented state.
- *The cooling section:* This section cools the film down from the annealing temperature to ambient and sends the film on downstream to the next process.

Multilayer Flexible Packaging. DOI: http://dx.doi.org/10.1016/B978-0-323-37100-1.00011-9

11.3 How a Machine Direction Orientor Works

There are four main steps or parts in an MDO (Fig. 11.1).

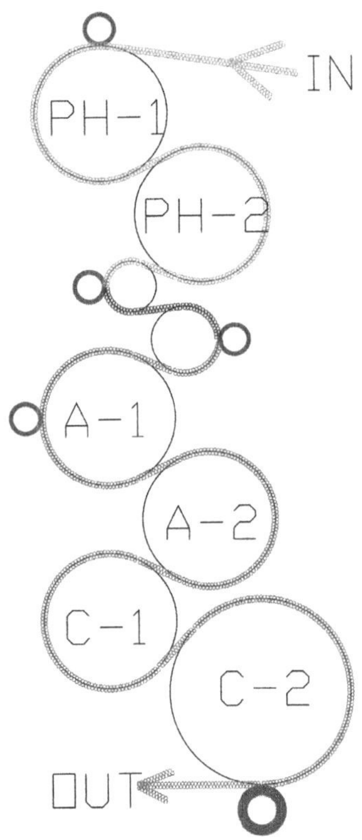

Figure 11.1 MDO sketch.

11.3.1 Preheat

The preheat rolls are labeled PH-1 and PH-2. Their function is to raise the film temperature uniformly to orientation temperature without putting hard wrinkles into the film. The first preheat roll (PH-1) has a nip roll to control the incoming web tension.

11.3.2 Drawing

The draw rolls are the rolls between PH-2 and A-1. After the film is at the desired orienting temperature, it is stretched between the draw rolls. These are smaller diameter than the preheat rolls so that the distance between the tangent points is less. The gap between these rolls is adjustable for the gauge and desired film properties and appearance. The film can be stretched up to 10× or more in this section. There are nip rolls on the slow roll and the fast roll for tension control.

11.3.3 Annealing

The annealing rolls are A-1 and A-2. This section heat sets the film and "locks" the property changes from the orientation into the film. It also controls how much the film will shrink back when later exposed to heat. There is a nip roll on the first annealing roll for tension control.

11.3.4 Cooling

The last section (rolls C-1 and C-2) cools the film to near-ambient temperature.

11.4 The Process and Its Effect on the Film

The film is heated to increase the polymer molecular mobility. This permits the film to be drawn at higher draw ratios. The film properties are a function of the orientation temperature.

There are various theories as to what happens on the molecular level when the film is stretched. The one presented here has been used for HDPE and can be extrapolated to other polymers.

Before film is oriented via MDO, it has regions of ordered crystals (Fig. 11.2). These are tied together with molecules that go from one crystal, through the adjacent amorphous region, into another crystal. Around the crystals are amorphous regions

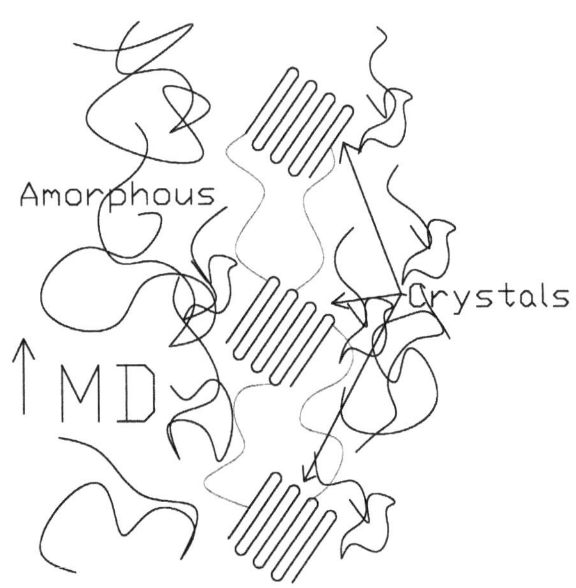

Figure 11.2 Crystal and amorphous arrangement before orientation.

where the molecules are arranged randomly. The crystals are plates where the molecules are tilted at an angle (34.4 degree) [1] to the film surface.

As the film is stretched in the MD, the random amorphous molecules begin to align themselves in the orientation direction (Fig. 11.3). This orientation continues until the molecules tying the crystals together are fully extended (Fig. 11.4).

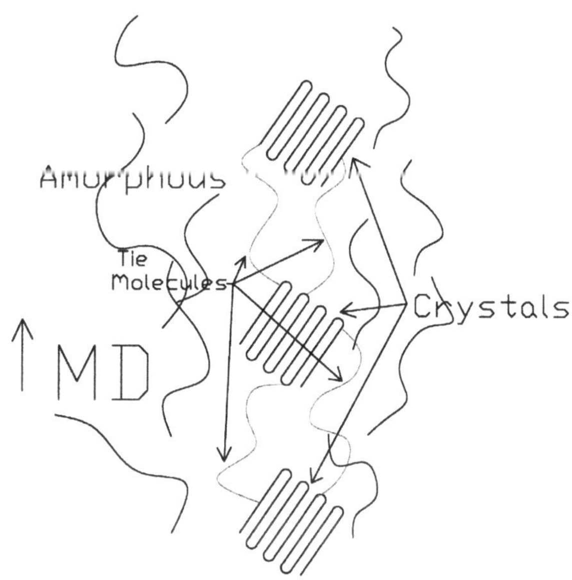

Figure 11.3 Initial tie molecule orientation as the polymer starts to be stretched.

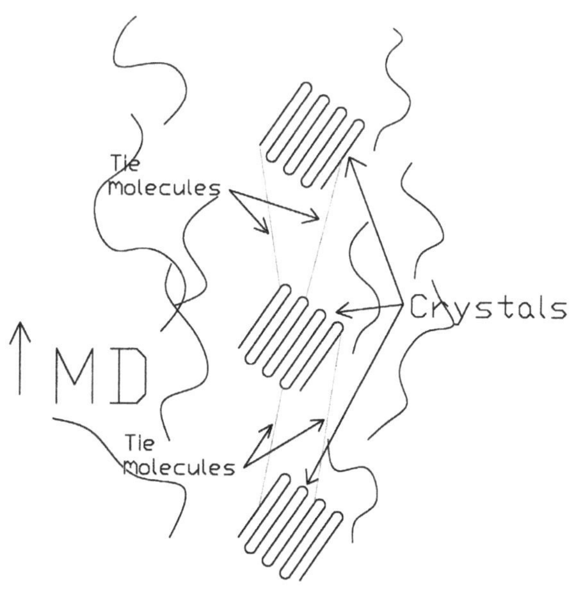

Figure 11.4 Continued tie molecule orientation as the polymer is stretched further.

In the next orientation phase, the angle or tilt of the molecules within the crystals changes such that they are aligned parallel to the film surface (Fig. 11.5).

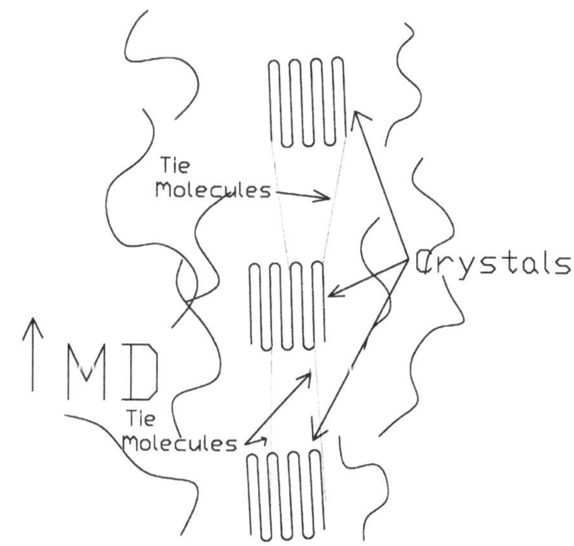

Figure 11.5 Fully stretched tie molecules rotate crystals to bring them into alignment.

In the third orientation phase, the crystals begin to unravel. This continues until the film breaks.

The above orientation phases take place at different draw ratios for different resins. This results in different properties. For a coextrusion, properties can be optimized at a given draw ratio using different resins in the individual layers.

11.5 Properties of MD-Oriented Films

The following properties can be improved when a coextruded film is oriented in the MD:

- optics: clarity, haze, and gloss
- tensiles: increase in break and tear
- controlled shrink levels
- stiffness: secant modulus can increase from two to three times in both TD and MD directions inclusively
- barrier: both oxygen and water vapor
- dead fold
- film “toughness” and ability to withstand puncture and heavy loads.

Detailed data for some of the properties follows.

11.5.1 Improved Optics

In Fig. 11.6, one can see how the haze decreases and gloss increases as the draw ratio is increased to 6.

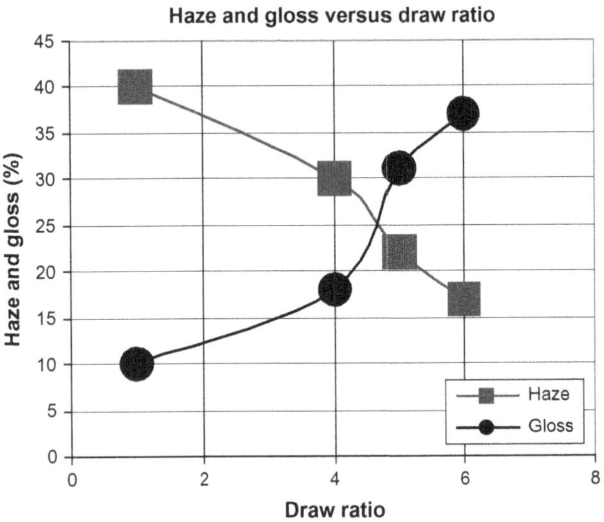

Figure 11.6 Haze and gloss versus MDO draw ratio [2].

This improvement is demonstrated in the see-through clarity which is shown in Fig. 11.7 (unoriented film) and Fig. 11.8 (oriented film).

Figure 11.7 Poor see-through clarity of unoriented film [2].

Figure 11.8 Improved see-through clarity of oriented film [2].

11.5.2 Increased Tensiles

Fig. 11.9 shows the MDO tensile strength at break is greatly increased when linear low-density polyethylene (LLDPE) is oriented in the MD. The TD tensile strength at break is only slightly reduced. Unlike HDPE, the TD elongation to break is still several hundred percent. When an item is packaged in an MDO film, such as the one in Fig. 11.9, the film must elongate in both the TD and MD directions. This permits the much higher MD tensile strength to come into play. This creates a much tougher package.

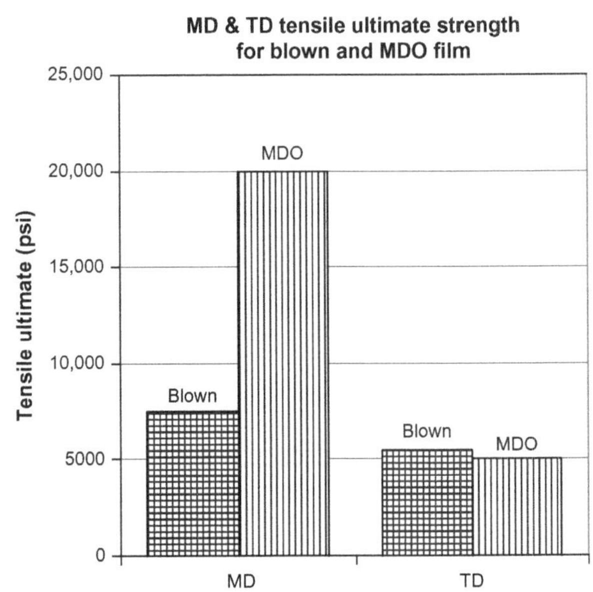

Figure 11.9 MD and TD tensile ultimate strength for blown and MD-oriented LLDPE [3].

11.5.3 Improved Stiffness

The stiffness (as measured by secant modulus) is greatly increased for both LLDPE and HDPE. Fig. 11.10 presents secant modulus versus MD orientation for LLDPE, and Fig. 11.11 shows MD modulus versus draw ratio for HDPE.

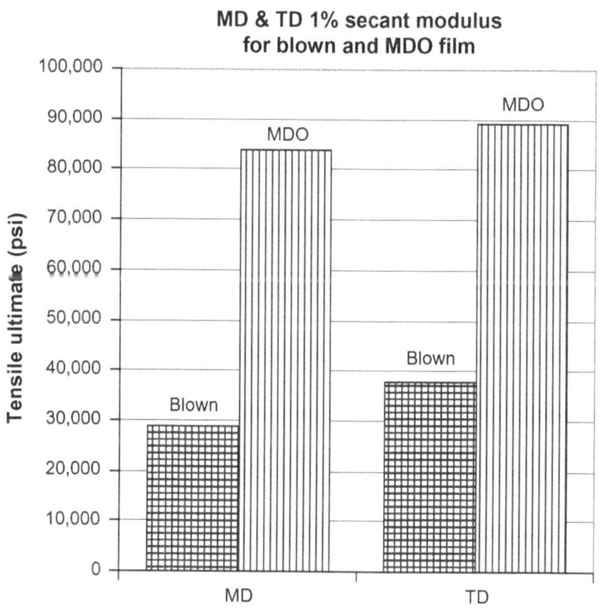

Figure 11.10 MD and TD 1% secant modulus for blown and MD-oriented LLDPE [3].

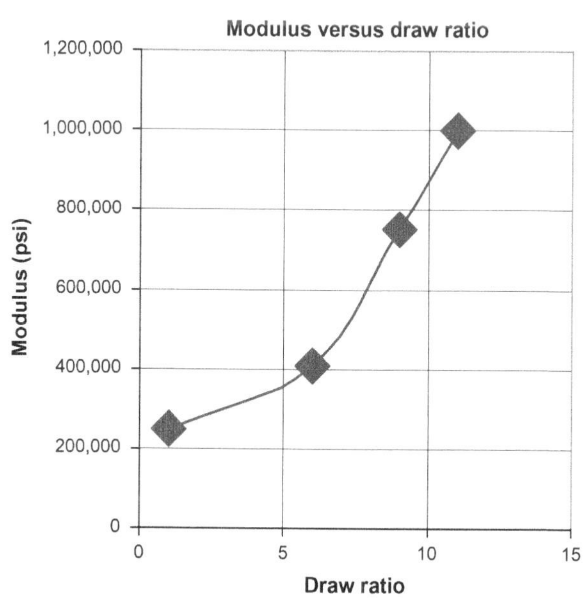

Figure 11.11 MD modulus versus draw ratio for HMW HDPE [4].

11.5.4 Improved Barrier Properties

When HDPE and LLDPE are MD oriented, their oxygen and water vapor barrier properties are increased two times or more. Table 11.1 shows the moisture barrier improvement for a 2-mil LLDPE film after MDO and for a 1.5-mil HDPE film at 6:1 draw ratio. Similar improvements are seen in oxygen transmission rates.

Table 11.1 Moisture Barrier Improvement for a 2-mil LLDPE Film After Machine Direction Orientation and for a 1.5-mil HDPE Film at 6:1 Draw Ratio

	WVTR (gm/100 in^2–24 h/atm)	
	No MDO	MDO
LLDPE (2 mil)	0.95	0.56
HDPE (1.5 mil)	0.27	0.14

11.5.5 Recent MDO Advances

Since 2009 there have been many new MDO installations. With these new machines much process and material knowledge has been gained. This knowledge growth has mostly been evolutionary and will continue with continuous improvement.

One revolutionary change is improved thickness uniformity which greatly improves first pass yields off MDO lines. This improvement addressed one inherent consequence when you stretch film. It necks in. There are many theories as to why this happens, and they are beyond the scope here.

When film necks in it gets thicker. (The material must go somewhere). Unfortunately, the thickness increase is not uniform across the web. The edges get thicker than the center. This thickness difference gets less as you move closer to the web center. This makes it difficult to wind rolls. The old solution was to trim the thick edges off. This caused low first pass yields and increased the MDO film material costs. Thus, film applications were limited due to cost considerations.

The solution was to make the edges thinner going into the MDO. When the film was then stretched the edges came out at the desired thickness. This brought yields back to normal blown film levels, and increased the areas where MDO film could compete.

One way to thin the edges was to heat the bubble area that would become the edges going into the MDO. The extra heat would cause the film to stretch more while still molten and become thinner. Adding a thickness gauge after the MDO provides the possibility for closed loop control.

11.6 Summary

Putting films through the MDO process greatly enhances their properties. Barrier properties, both water and oxygen, can be doubled with the right orientation and annealing conditions. Film stiffness (measured by secant modulus) can be increased by two to three times and, in some cases, four times.

Optical properties (haze, gloss, and clarity) can also be enhanced, as well as tensile properties. Films can also be made with controlled shrink levels and temperature.

The MDO machine is relatively simple in nature. However, the process conditions and materials selected can be far from straightforward.

References

[1] Breeze R. Modeling the effects of solid state orientation on blown high molecular weight high density polyethylene films: a composite theory approach [Thesis]. University of Cincinnati; 2005.

[2] Hatfield E, Tate R, Williams K, Todd B. New MDO medium molecular weight high density polyethylene films. In: ANTEC proceedings of the 60th annual technical conference & exhibition, XLVIII, San Francisco, CA; 2002 May 5–9.

[3] Hatfield E. Unique MD oriented multilayer barrier films for food & industrial applications. Chicago: FLEX-PAK AMERICA; March, 2002.

[4] Breeze R. Effects of machine direction orientation (MDO) on the moisture and oxygen barrier properties of HMW-PE films. In: Annual technical conference. The Association of Paper & Plastics Industry (TAPPI); 2005.

12 Oriented Film Technology

Jürgen Breil
Brückner Maschinenbau GmbH & Co. KG, Siegsdorf, Germany

12.1 Introduction

Today, oriented plastic films are produced industrially in large quantities. As such, the majority are biaxially oriented, that is, films are stretched in both directions, which is understandable when taking the improvement in characteristics into consideration [1]. The characteristics attained with biaxially oriented films ideally meet the demands on modern flexible packaging. As illustrated in Fig. 12.1, flexible packaging has to fulfill the protective function and product design in line with economic and, increasingly, environmental aspects. The required packaging product protection is attained by the excellent barrier properties against gases (water vapor, oxygen, and others) as well as good-quality seals. The product appearance is attained by high-gloss and transparent thin packaging film as well as by excellent printability. The requirements for sought-after economic packaging are met by good material utilization and the fulfillment of the demand for high-speed packaging lines. Environmental aspects, which play an ever increasing role, can be satisfied by utilizing environmentally friendly materials (such as polyolefin), as well as the optimum raw material yield, thus ensuring maximum packaging effect with minimum material cost.

To what extent biaxial orientation improves properties is shown in Fig. 12.2 using biaxially oriented polypropylene (BOPP) as an example. Along with a significant increase in the mechanical properties (Young's modulus and tensile strength), a considerable improvement in the optical (haze and gloss) as

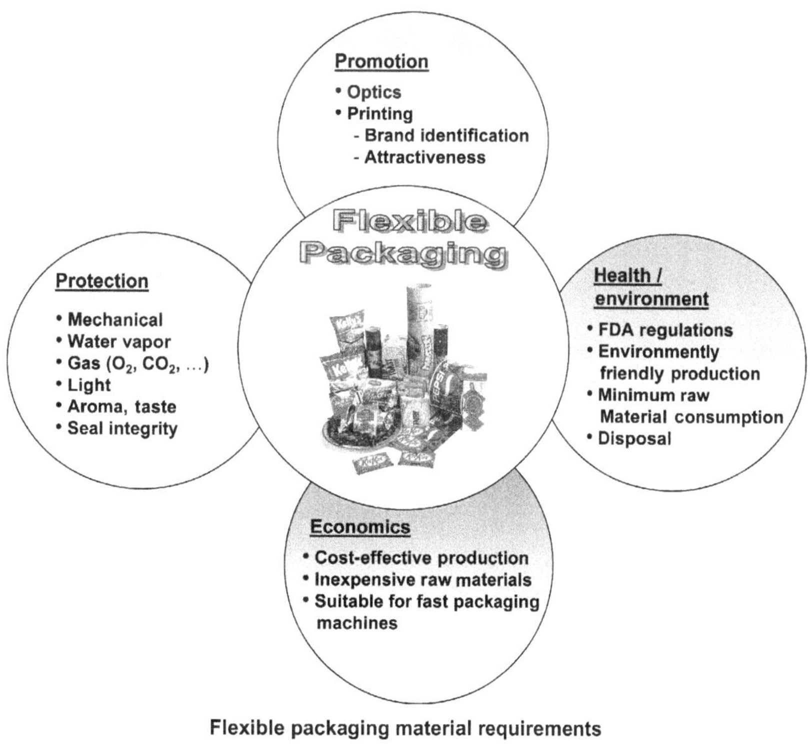

Figure 12.1 Flexible packaging requirements for protection, promotion, health/environment, and economics.

Multilayer Flexible Packaging. DOI: http://dx.doi.org/10.1016/B978-0-323-37100-1.00012-0

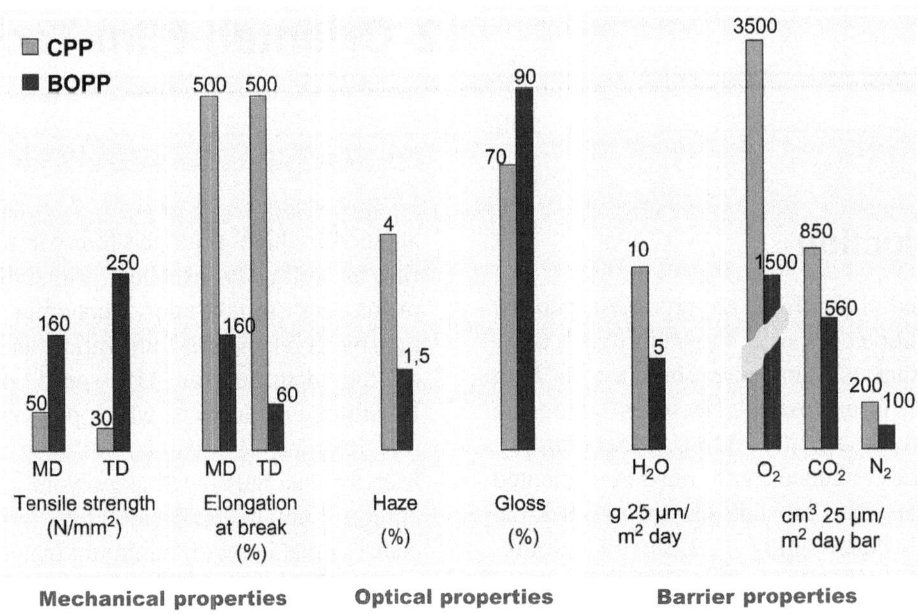

Figure 12.2 Property improvement for biaxially stretched PP (BOPP) versus cast PP (CPP).

well as the barrier properties can be seen. The overall improved barrier properties attained are due to the orientation of the molecule chains which, for a non-oriented polymer, are random, whereas in the stretching process, a clear molecule chain orientation occurs. As such, biaxial orientation of plastic films represents a refinement process which is applicable for almost all plastics. Semicrystalline plastics in particular, such as polypropylene (PP) and polyester, also augment the crystallinity by the stretching process, which considerably improves the mechanical values.

Fig. 12.3 shows the increase in the Young's modulus in machine and transverse directions as a result of the biaxial orientation process for polypropylene and polyester. The exceptional mechanical properties in combination with the barrier and optical properties with comparably low raw material costs have led to the fact that BOPP and biaxially oriented polyethylene terephthalate (BOPET) films account for the largest and most significant share among stretched films.

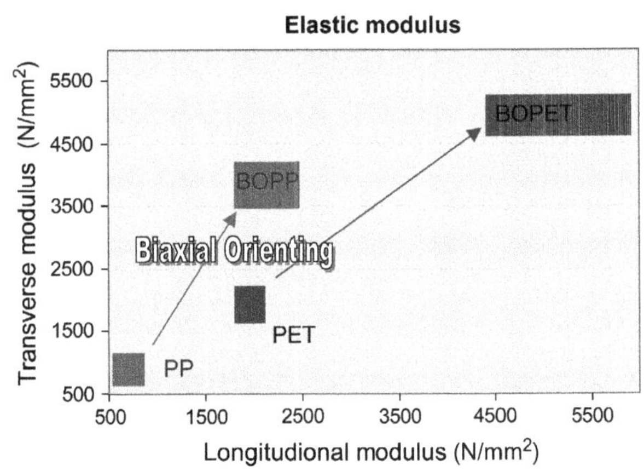

Figure 12.3 Mechanical property enhancement for BOPP and BOPET.

12.2 Orienting Technologies

In orienting technologies, one can, in general, differentiate between the orientation draw direction and the related stretching process. The stretching processes shown in Fig. 12.4 (longitudinal, transverse stretching, sequential-biaxial stretching, and simultaneous-biaxial stretching) do not depict competitive, but rather supplementary features in order to attain specific film characteristics. As such, the required stretching equipment varies depending on the process. Stretching in the machine direction is normally done by means of a machine direction orienter (MDO) via rolls with increasing speed. Typical products are, for example, tear stripes or polypropylene adhesive tapes. For all transverse-oriented films, the stretching process takes place by means of a transverse direction orienter (TDO), where the film is fixed on both ends and, upon passing through an oven at various temperatures, is stretched in a transverse direction.

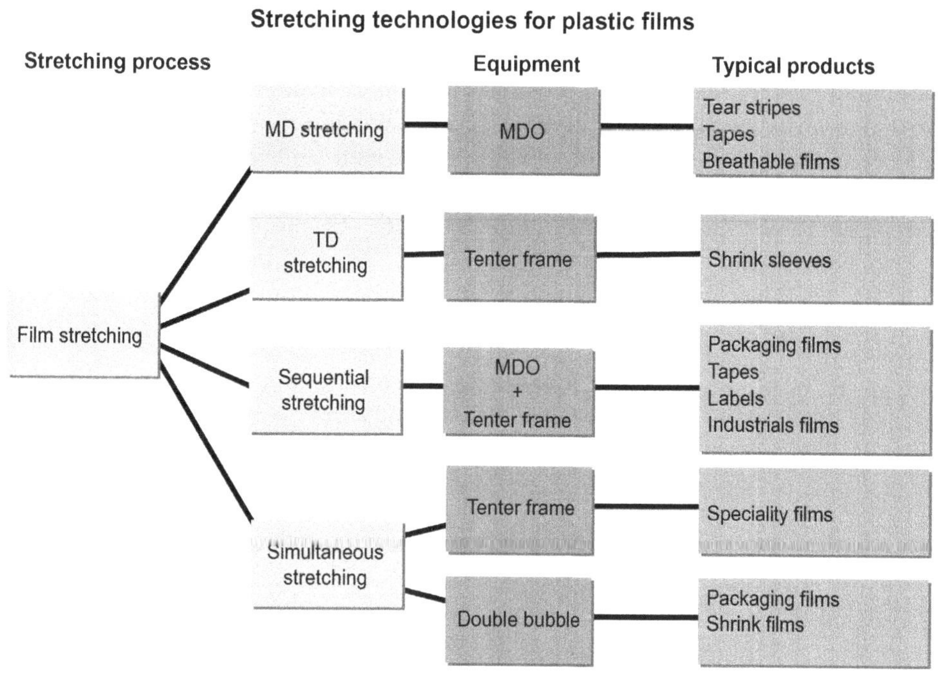

Figure 12.4 Plastic film stretching.

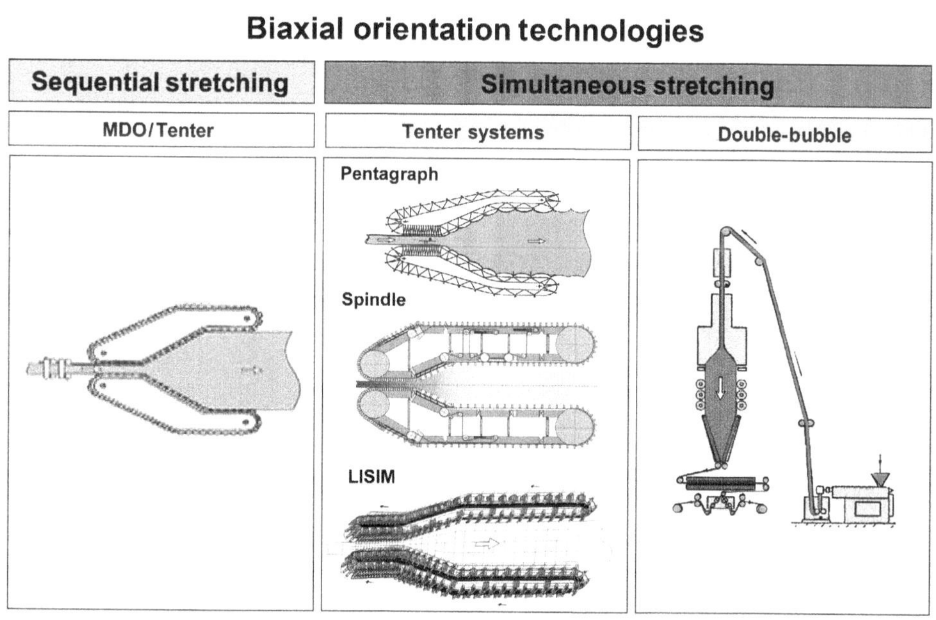

Figure 12.5 Biaxial orientation.

Typical examples of transverse stretched film types are shrink sleeves, where shrinkage merely occurs in the transverse direction. Biaxial orientation, in the machine and transverse directions, can be done either sequentially or simultaneously. In the sequential process, an MDO and a tenter frame are successively utilized [2,3]. This process has the widest prevalence for all stretched film types. All types of packaging films, tapes, labels, and industrial films are manufactured with this process.

Simultaneous orientation, however, is possible with the tenter and the blown process (Fig. 12.5). The blown process is a so-called double bubble process where, initially, a tube is extruded, then rapidly cooled and then heated to the stretching temperature. A synchronous increase of the draw-off speed and bubble expansion by internal pressure results in the required simultaneous orientation process.

Sequential stretching first in the machine direction and then in the transverse direction utilizing an MDO and a TDO is the most prevalent process in use today. Fig. 12.6 is a cut-away view of a three layer–coextruded BOPP line that shows the main extruder, two coextruders, die and casting station, MDO, TDO, gauging station, treatment and full-width winder. Fig. 12.7 shows a typical BOPP process temperature profile. First, the PP resin is melted in the extruders, then quenched on the casting roll, transferred to the MDO where it is reheated and stretched in the machine direction. There is some annealing between the MDO and the TDO. In the TDO, the web is reheated before transverse stretching, annealed, and cooled down before winding. Fig. 12.8 shows a production line

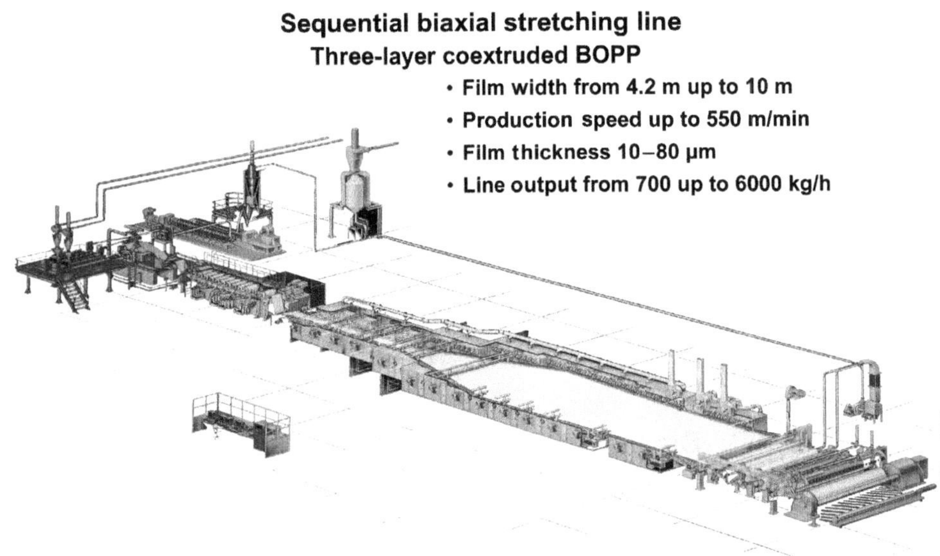

Figure 12.6 A three layer–coextruded BOPP sequential stretching line.

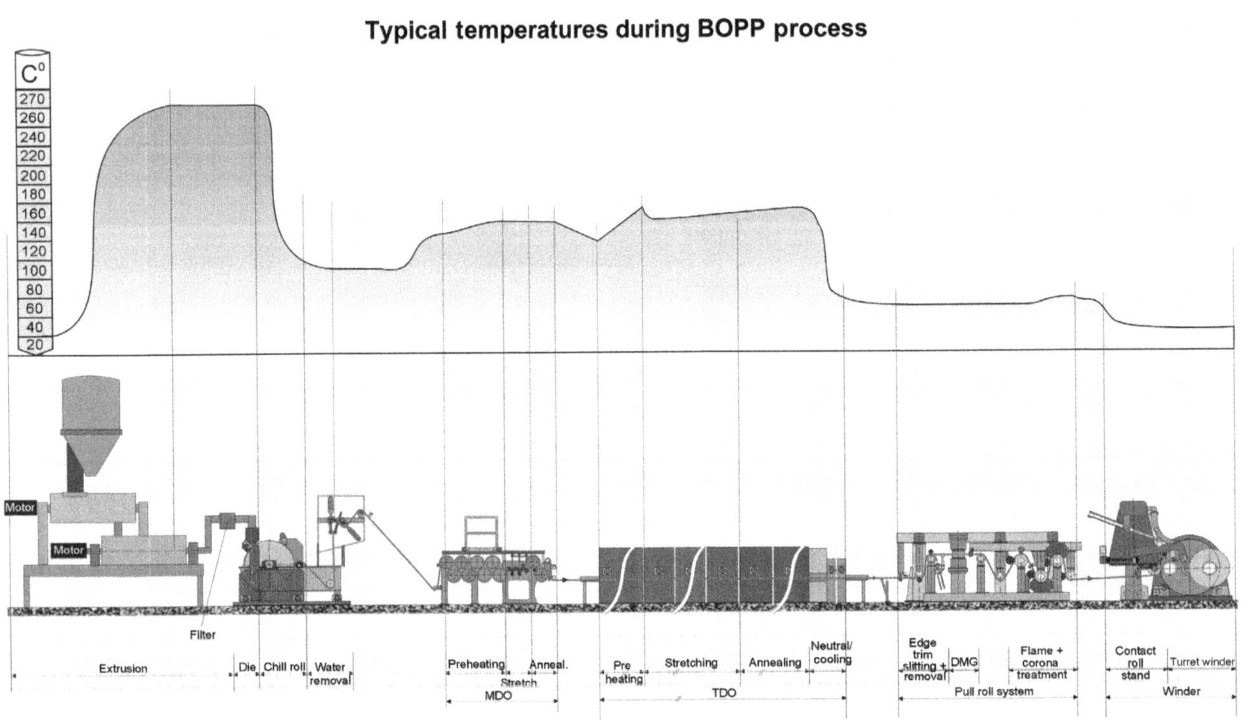

Figure 12.7 Typical temperatures during the BOPP process.

Figure 12.8 Machine direction orienter with entrance.

Figure 12.9 Finished 10-m BOPP mill.

MDO and entrance to the TDO. Fig. 12.9 shows a finished 10-m mill roll after being removed from the winder.

Simultaneous orienting technology represents an alternative to the prevalent sequential orientation. Fig. 12.10 compares for BOPP the useable range for the stretch ratios between the sequential and simultaneous processes.

In sequential orientation, the stretching process occurs in two steps and a relatively small process window in terms of temperatures and stretching ratios is available. During simultaneous stretching, however, the usable stretching ratios are considerably larger. For instance, in machine and transverse directions, it is possible to set identical stretching ratios or even realize a higher stretching ratio in machine direction in order to achieve improved machine direction mechanical properties. A further advantage is the possibility to relax in simultaneous orientation by diminishing the clip spacing, not only in the machine but also in the transverse direction. Furthermore, it can be emphasized that, as a contact-free process, simultaneous orientation avoids the limitations of

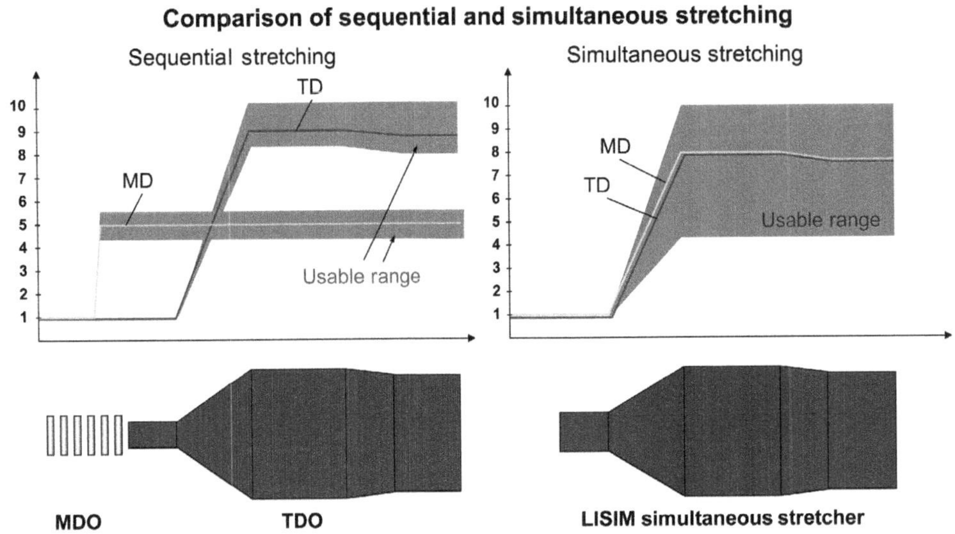

Figure 12.10 Usable stretch ratios for sequential and simultaneous stretching.

stretching via rolls. Such differences lead to several advantages in terms of the product characteristics as shown in Fig. 12.11. In particular, a new developed system called LISIM (linear motor simultaneous stretching system) offers the following advantages [4]:

- high productivity (speed and width)
- high flexibility (stretching ratios and relaxation rates) in MD and TD
- high reliability.

These features are achieved by individually driven clips with linear motor technology. The improved mechanical properties are due to higher stretching ratios. The shrink characteristics are controlled by the unrestricted relaxation and tensilizing stretching possibilities. Barrier properties can be improved upon considerably by using coextruded barrier materials, where the process is particularly advantageous for those materials which cannot be stretched sequentially due to the crystallinity created

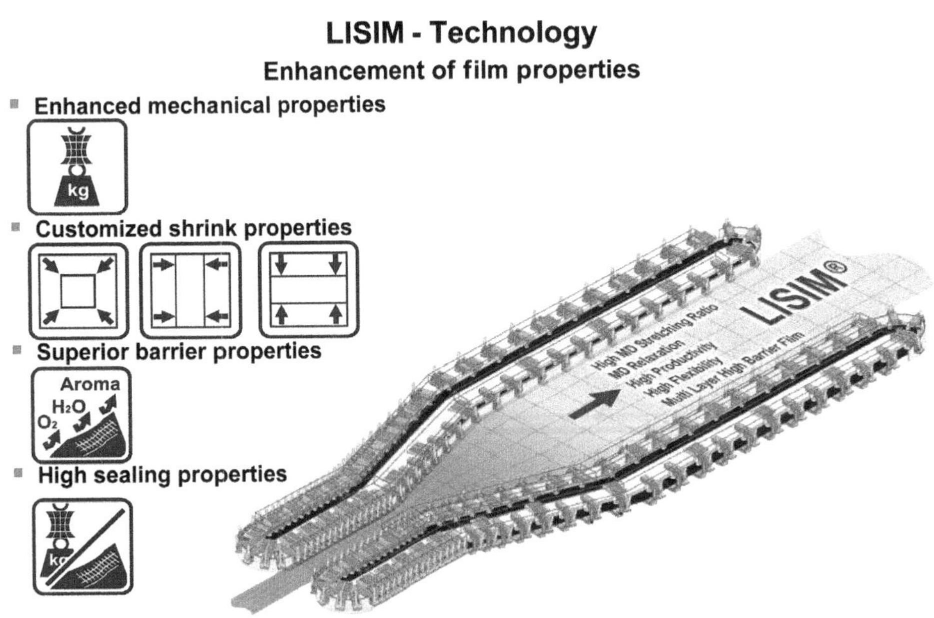

Figure 12.11 Enhanced film property possibilities with LISIM.

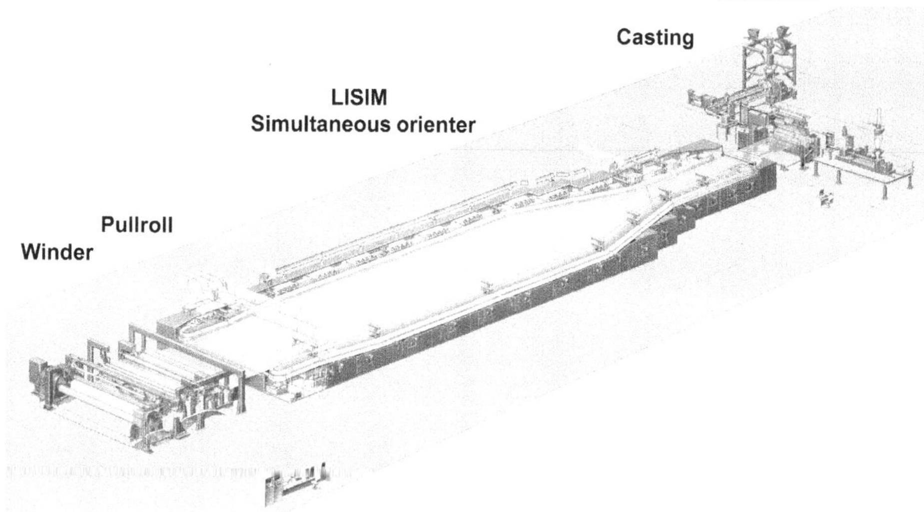

Figure 12.12 Simultaneous stretching line.

by the first MD stretching process. Improved sealing properties are made possible because low seal temperature copolymers can be applied. These low-temperature heat sealing polymers are not processable in a standard MDO as they stick to the rolls during the machine direction orientation. LISIM technology has been scaled up from laboratory scale to production line dimension. Lines equipped with this technology for the production of polypropylene and polyester film have been running successfully and reliably for several years. The overall line layout is shown in Fig. 12.12. Apart from the orienter, the components of such a line are similar to those of a sequential line. In particular, those components at the front end, that is, raw material supply, extrusion and casting unit as well as those at the rear, that is, pull roll and winder are, apart from minor details, identical. Only the orienter components are different. Instead of MDO and TDO, a simultaneous orienter is applied. Typical output figures for sequential and simultaneous stretching equipment, representing today's state of the art are shown in the list that follows. Basically, in terms of output data, it can be said that the efficiency of these high-speed lines is increasing, since the output capacity for certain film thicknesses is merely a matter of speed and working width. Over the past 40 years, ever since this technology was implemented on an industrial scale, constant efforts have been made not only to increase the working width but also the line speed. As such, new technological challenges are constantly arising with the aim to overcome the bottlenecks in the line components. Today, the state of the art for BOPP lines features:

- working width 10 m
- speed 530 m/min
- output capacity 6000 kg/h

The trend for even higher output capacities will continue in the future. Nowadays, line concepts for even higher speeds and output capacities of 7 tons/h and above are being designed.

12.3 Oriented Film Types—Applications

In addition to the orientation technology outlined previously, the film products and their applications will now be explained. In most cases, stretched films for packaging applications are further processed. The most significant converting processes are vacuum coating (metallizing, SiO_x, AlO_x), offline coating (acrylic, polyvinylidene chloride, polyvinyl alcohol, etc.), lamination with other stretched films, polyethylene (PE) sealing layers and printing (front printing and reverse printing). Such downstream processing will not be discussed in this chapter.

Considering the market for oriented films in general, the various raw materials used can be distinguished.

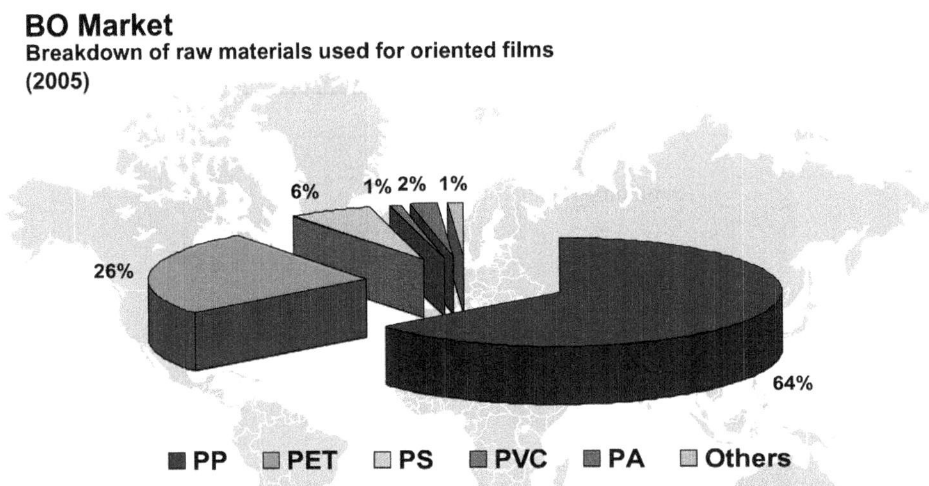

Raw material	PP	PET	PS	PVC	PA	Others	Total
1000 tpa	6089	2181	556	120	185	61	8972

Based on Name Plate Capacity (world) incl. 2004 contracts

Figure 12.13 Biaxial-oriented film.

Fig. 12.13 shows a breakdown of oriented films manufactured worldwide. BOPP film constitutes by far the largest share with over 6 million tons per annum. In view of the favorable relation between raw material prices and film properties, it can be assumed that the steady growth of 7% per annum will continue. Biaxially oriented PET film has the second largest share followed by polystyrene (PS), polyvinylchloride (PVC), and polyamide (PA).

12.3.1 Films Oriented Biaxially

Among oriented films, the biaxial orientation is the most preferred technology as it leads to improved properties in both (MD and TD) directions. This can be recognized by comparing the properties of the most common biaxially oriented film types, which are BOPP, BOPET, and BOPA (Table 12.1).

The various film types mainly differ with regard to the mechanical, thermal, and barrier characteristics and determine the particular application. Similarly, other properties, like thermal resistance or electrical properties, differentiate the film types and predestine them for specific applications.

12.3.1.1 BOPP Films

With a worldwide consumption of over 6 million tons, BOPP films constitute by far the largest share in biaxially oriented film. The applications are very diverse and can basically be split into packaging applications, not only in the food but also in the nonfood sector. Fig. 12.14 shows a few of these applications which play an important role in everyday life.

A further classification of such applications can be made in terms of the thickness range and the number of layers. The thinnest films are required for electrical applications, such as capacitor film, with a thickness of at least 3 μm. The thickest films are available within the synthetic paper sector up to 180 μm. As shown, films with a thickness range between 15 and 35 μm are widely applied for the varied packaging applications (Figs. 12.15 and 12.16).

One differentiates between one layer and multilayer, where the three layer–coextruded film has the largest share. The core layer of PP homopolymer is coextruded with the outer PP copolymer layers. The outer layers have a lower melting point thus ensuring

Table 12.1 Film Properties of Common Biaxially Oriented Films

Mechanical Properties		Unit	20 μm BOPP	12 μm BOPET	15 μm BOPA
Tensile strength	MD	N/mm^2	140	230	250
	TD	N/mm^2	280	260	280
E-Modulus	MD	N/mm^2	2000	4400	3500
	TD	N/mm^2	3500	5200	3800
Elongation	MD	%	220	110	110
	TD	%	70	90	100
Impact strength		kg/cm	5	5	15
Tear propagation		g	3.5	3.5	7.5
Thermal shrinkage		%	5% at 135°C	2% at 190°C	2% at 160°C
Density		g/cm^3	0.91	1.393	1.16
Yield		m^2/kg	55	59	58
OTR		cc/m^2 day	1600	90	40
WVTR		g/m^2 day	6.0	8.5	270–300
Surface tension		Dyn	40	50–55	50–55

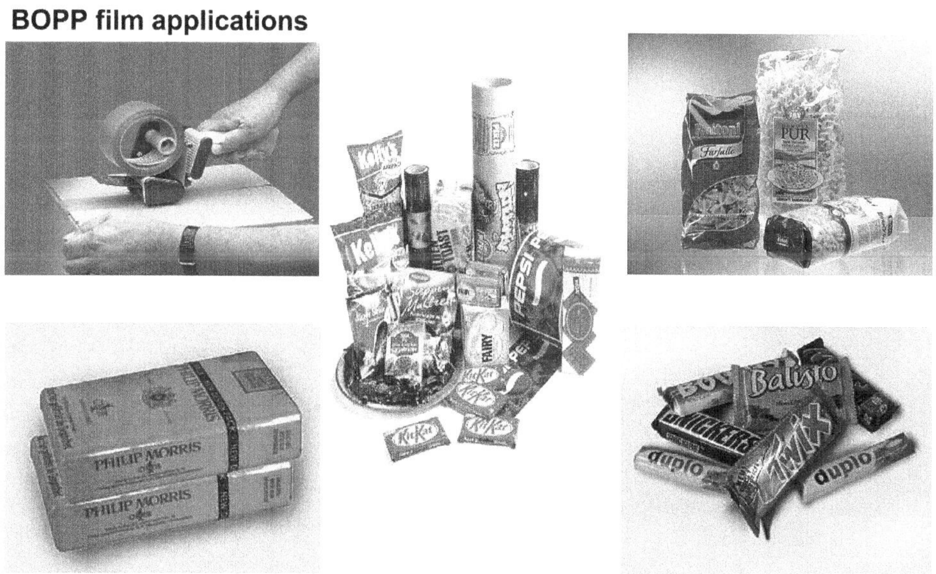

Figure 12.14 BOPP film applications.

that the sealing process necessary for packaging applications can take place at temperatures that do not deform the main layer. In the last few years, there has been a strong trend toward five-layer and, in certain cases, also to seven-layer films [5]. The advantages of five-layer technology are, on the one hand, improved characteristics, such as better optical, gloss, transparent, opaque properties, as well as cost advantages, expensive additives are predominantly added in the thinner intermediate layers. In Chapter "Web Handling and Winding," multilayer-oriented films, the various structures and applications are further described. Biaxially oriented PP films are widespread, not only the transparent applications but also the white opaque film types which are mainly applied for packaging and labeling. Inorganic additives (eg, calcium carbonate) are implemented in the polymer matrix [6]. These particles lead to an initial

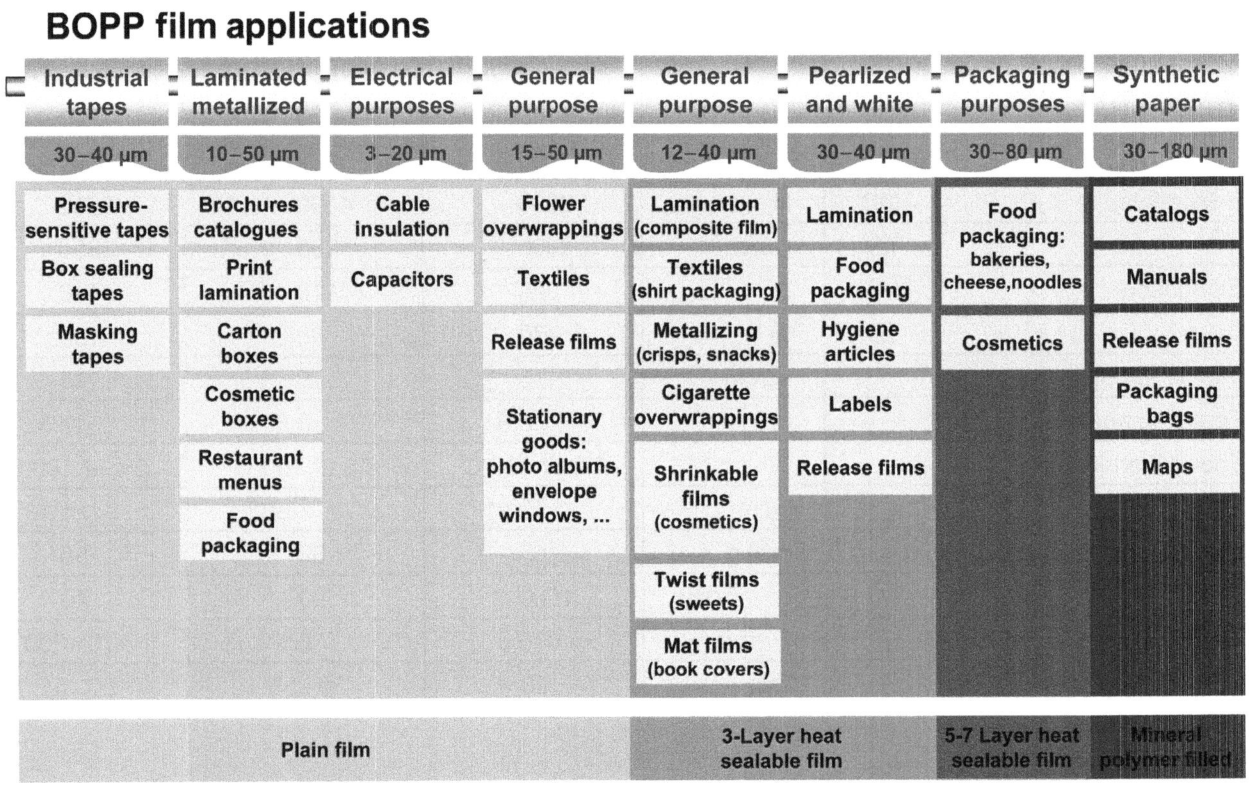

Figure 12.15 BOPP film applications.

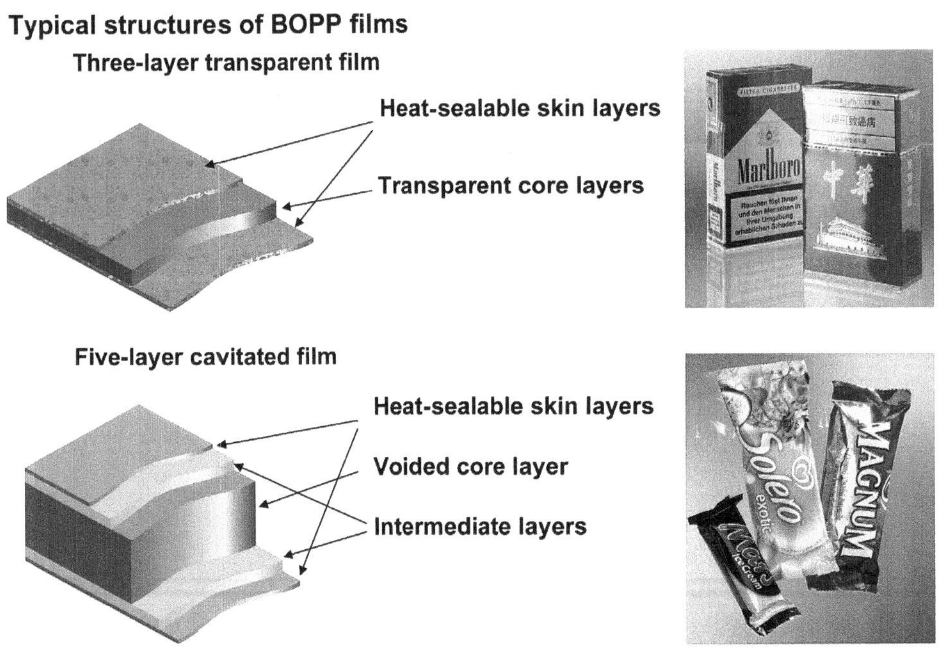

Figure 12.16 Typical BOPP film structures.

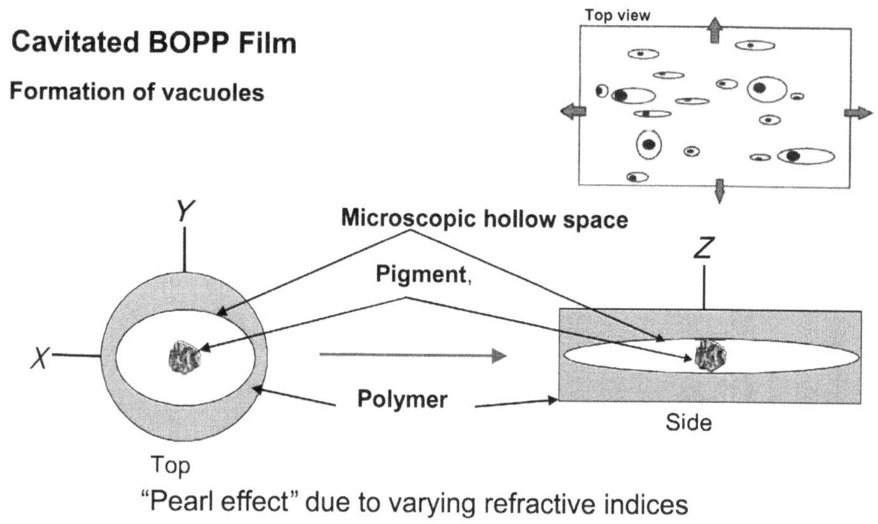

Figure 12.17 Cavitated BOPP film mechanism.

flaking/separation from the polymer matrix during the machine direction orientation, so that, during consequential transverse direction orientation, small cavities occur (Fig. 12.17).

In view of these so-called vacuoles, the light is refracted in varying ways such that the required pearl effect arises. At the same time, the density reduction gives rise to the fact that, with the use of the same raw material, compared to noncavitated films, an enlarged thickness occurs. Both aspects are mainly used for confectionary, chocolate bars, ice-cream, etc. Synthetic paper takes a special role among cavitated BOPP films. The effect of vacuole formation during orientation is also made use of, where a larger density range of 0.6–0.9 g/cm^3 can be produced. Applications for synthetic paper are extremely versatile and cover a large thickness range (Table 12.2). Three- and five-layer films are coextruded where the surface is optimized in order to attain good printability.

Fig. 12.18 shows a 100-μm synthetic paper cross-section showing the calcium carbonate particles, the cavities, and the nondensity-reduced skin layers. Synthetic paper is frequently coated in further processing in order to attain a better absorption and a quicker drying of the printing inks.

Table 12.2 Thickness Range for Synthetic Paper

Thickness (μm)	Recommended Applications
50–180	Pressure-sensitive, cut and stack and wrap-around labels; release liners, posters, inkjet printing base
75–100	Pressure-sensitive, wrap-around, and in-mold labels
75–200	Cut and stack and wrap-around labels, posters, maps, shopping bags, business cards, calendars, banners
75–250	Labels, books, posters, calendars
75–400	Maps, posters, tags, cards, charts, menus, phone cards, calendars, banners
130–700	Carriers, files, folders
250–1000	Cards, tags, book covers, folders, charts, maps

12.3.1.2 BOPET Films

Biaxially oriented polyester films (BOPET), with approximately 2 million tons per year, are the second most common oriented film following BOPP. In the past, BOPP films dominated in packaging applications and BOPET films dominated in technical applications. Biaxially oriented polyester film, with its high rigid properties, was ideal as a carrier film for magnetic tapes, floppy disks, and capacitors. Since this magnetic recording medium has been substituted by the optical data medium, this application has strongly declined over the last few years. At the same time, however, an increase in the prevalence of BOPET films in the packaging industry has taken place, resulting in a worldwide growth of 4–5%. Upon reviewing the breakdown chart for the various applications (Figs. 12.19 and 12.20), one can conclude that with approximately 40%, packaging applications represent the largest share.

Cross-section of BOPP synthetic paper
Three-layer coextruded, Thickness 100 µm

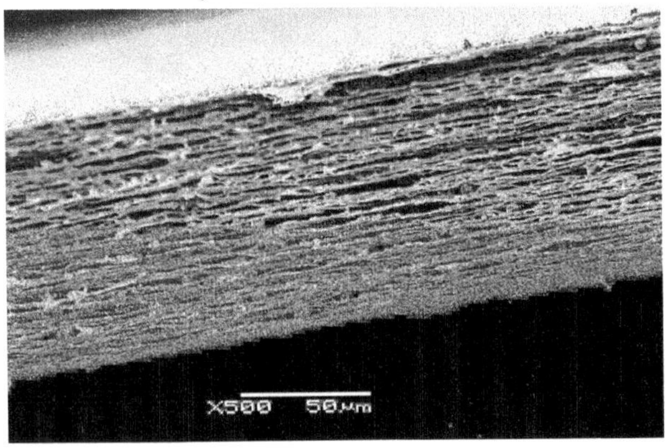

Figure 12.18 Cross section of a 100-µm BOPP synthetic paper showing the calcium carbonate particles, cavities, and uncavitated skin layers.

The basic characteristics:

- high mechanical strength
- good temperature and chemical resistance
- dimensional stability over a broad temperature range
- adjustable friction coefficient
- excellent optical clarity
- good printability

reflect the specific beneficial features for the various applications accordingly (Fig. 12.21).

Variants ensue from the different stretching processes, recipes, coextrusion, and coating processes. In the sequential stretching process, the longitudinal-transverse (MD/TD) process is dominant [7]. However, the transverse-longitudinal (TD/MD) process and the longitudinal-transverse-longitudinal (MD/TD/MD) processes are also applied. For the MD/TD/MD process, higher stiffness values in machine direction can be achieved. The simultaneous stretching process is applied for very thin films, for example, for capacitor films,

BOPET film applications

Figure 12.19 BOPET film applications.

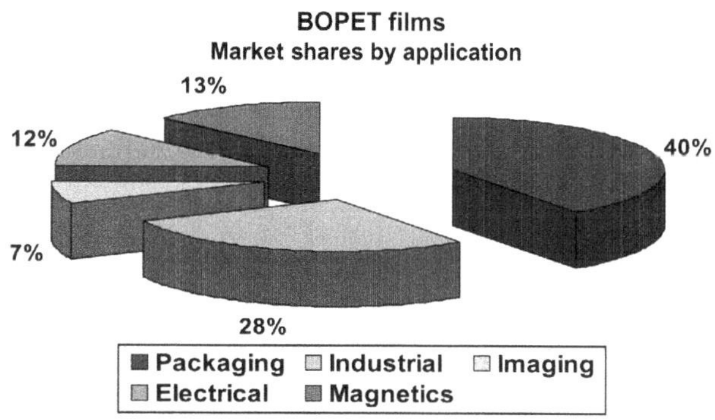

Figure 12.20 BOPET films market share.

and the contact-free stretching technology allows for high-quality optical uses.

In view of the good stiffness values and sliding properties in the packaging sector, the benefits, such as excellent machinability plus good printability and optical appearance, are applied. With coextrusion, sealable or matte surfaces can be attained (Fig. 12.22).

Furthermore, a frequently applied advantage of coextrusion technology is the application of inorganic additives in the thin outer layers, in order to adjust the required friction coefficient without

BOPET film applications

Magnetic tapes 6–12 μm	Magnetic use 6–76 μm	Packaging as laminates 8–25 μm	Metallized films 6–19 μm	Electrical purposes 0.5–350 μm	Graphic arts and X-Ray 20–200 μm	Other applications 10–125 μm
Audio tapes	Audio tapes	Food packaging	Metallic yarns fashion	Cable wrapping	X-ray foto film	Adhesive tapes
Video tapes	Video tapes	Pouches	Packaging	Electro insulation	Micro films	Green houses
Cassettes, computer tapes	Computer tapes	Fatty and oily foods	Hot foil stamping	Slot liners motor insulation	Litho films	Stiffeners
Micro cassette tapes	Floppy discs	Coffee bags (under vacuum)	Solar control mirrors	Capacitors	Graphic arts	Release films
Carbon ribbon		Shrinkable films	Wall covering decoration	Flexible circuits	Drafting films engineering	Label cards
		Medical supplies A-PET/C-PET		Thermal transfer tapes	Overhead projection	
					Foto resist	
					Prepress	
Tensilized film	Balanced film					

Figure 12.21 BOPET film applications.

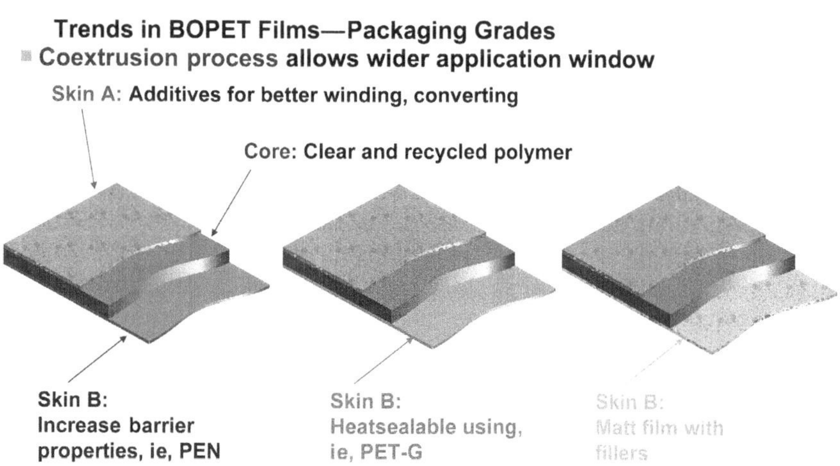

Figure 12.22 Coextrusion trends in BOPET.

having a negative influence on the transparency. In-line coating processes are also widespread which ensure optimum printing ink adhesion. A common downstream processing phase of BOPET film is metallizing, which is mainly used not only to improve the barrier properties but also to attain an attractive visual appearance. For numerous food wrappings, the barrier properties, in terms of oxygen and aroma, are particularly vital criteria to ensure that the required minimum shelf-life is attained. With metallizing, an oxygen permeation value of $<1\ cm^3/m^2$ day bar can be reached (Fig. 12.23).

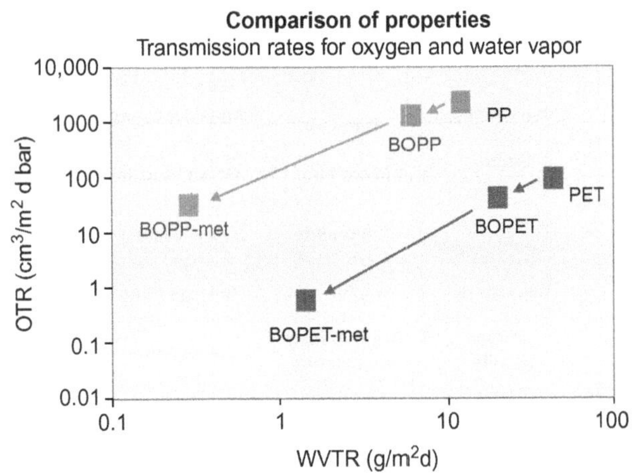

Figure 12.23 Comparison of transmission rates for PP and PET.

Biaxially oriented polyester packaging films are usually laminations, that is, in a further process they are laminated with BOPP, PE film, aluminum foil, or other packaging material. A typical laminate structure is shown in Fig. 12.24 as an example for coffee wrapping. The polyester film is reverse-side printed and laminated with aluminum foil as a barrier layer and polyethylene film as a sealing layer. Good transparency, high gloss, and the print quality is thus reflected in the image appearance.

Apart from the packaging sector, there are numerous other industrial applications for BOPET films. For example, thermotransfer films for bar code and ticket printers to name a few. The high-temperature resistance is an excellent benefit. Biaxially oriented polyester is also widely used for capacitor and electrical insulating film, with thickness ranges from 0.5 to 350 μm.

In recent years, additional other applications for optical films have been gaining significance. In particular, LCD screens and flat screen TVs are undoubtedly ensuing good growth possibilities for high-quality BOPET films in the future.

12.3.1.3 BOPA Films

With a worldwide volume of 250,000 tons per annum, BOPA (polyamide or Nylon) films represent

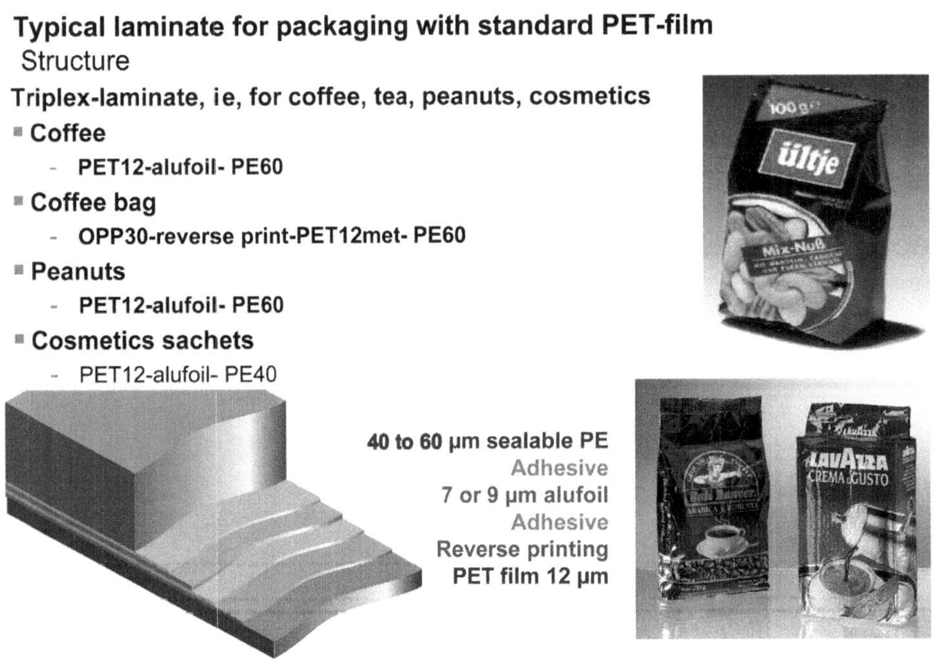

Figure 12.24 Typical PET film.

a small specialty segment, predominantly used in the packaging sector [8]. Particularly in view of the excellent puncture resistance along with good oxygen and aroma barriers, BOPA is primarily processed for flexible wrappings for sausages, cheese, fish, and liquid contents (Fig. 12.25). Thickness is normally in the range of 12–25 μm. Special application, such as gas-filled balloons, is primarily made from metallized thin BOPA film (10–12 μm).

In principle, all above-mentioned stretching processes are suitable for manufacturing BOPA films, that is, not only sequential but also simultaneous and double bubble lines are used. Sequential stretching lines with longitudinal-transverse process and a working width of 4–5 m are widely abundant. The stretching ratio is approximately 3 × 3, process temperatures are shown in Fig. 12.26.

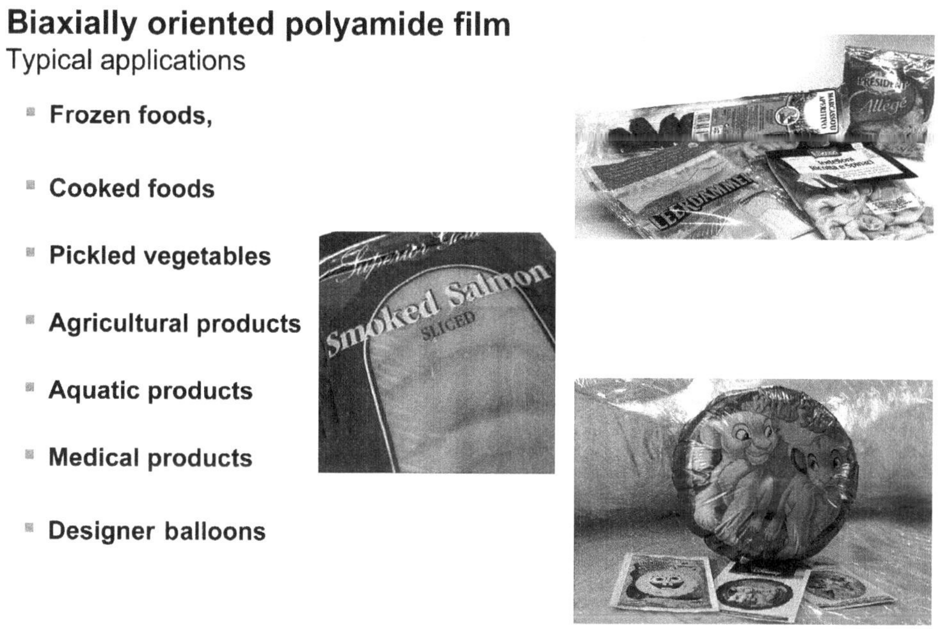

Figure 12.25 Typical applications of BOPA film.

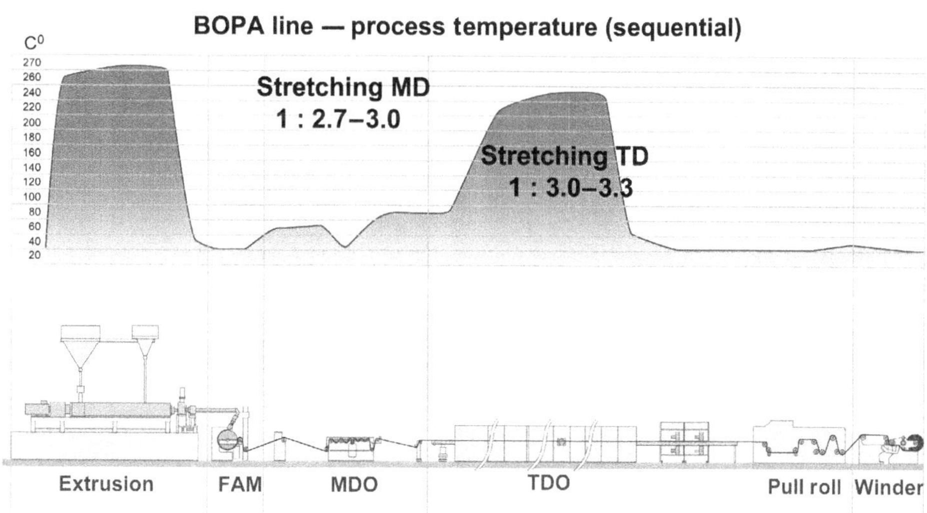

Figure 12.26 Typical process conditions for MD/TD BOPA.

For packaging applications, BOPA films are laminated with other films, mainly PE, in order to ensure the sealability for bag manufacture. Typical laminate structures are shown in Fig. 12.27.

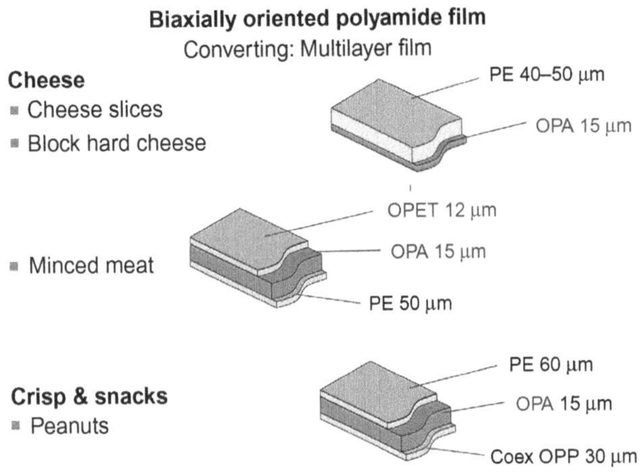

Figure 12.27 Typical BOPA.

12.3.1.4 BOPS Films

The worldwide market demand for BOPS film (biaxially oriented polystyrene) amounts to approximately 600,000 tons per annum and is basically split into two market segments [9]. Thinner 30- to 150-μm films are suited for applications such as envelope windows and separating film for photo albums, and thicker 150- to 800-μm films are mainly cover applications such as deep draw vacuum packaging film (Figs. 12.28 and 12.29). Thinner film types very often require a matte surface and deep draw applications require high transparency and luster. In addition, a good deep draw performance has to be ensured and can be adjusted via the stretching parameters.

Biaxially oriented polystyrene films are produced exclusively with the sequential process (longitudinal-transverse). In order to make the cast sheet, a roll stack is used to ensure that the thick film has optimum surface quality. The temperature is controlled to such an extent, thus ensuring that processing is performed at temperatures higher than the glass transition, as otherwise, polystyrene would be too brittle (Fig. 12.30).

Figure 12.28 BOPS film applications.

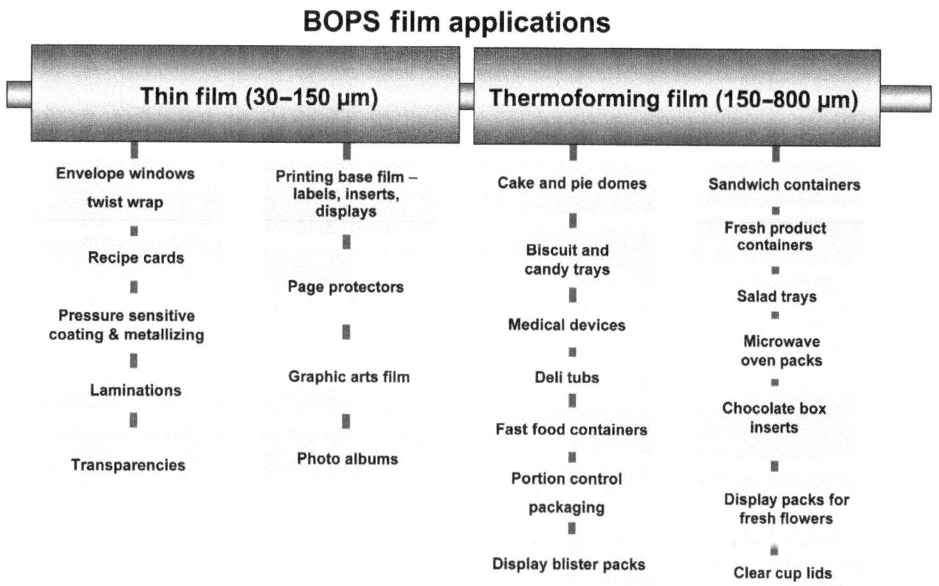

Figure 12.29 BOPS film applications.

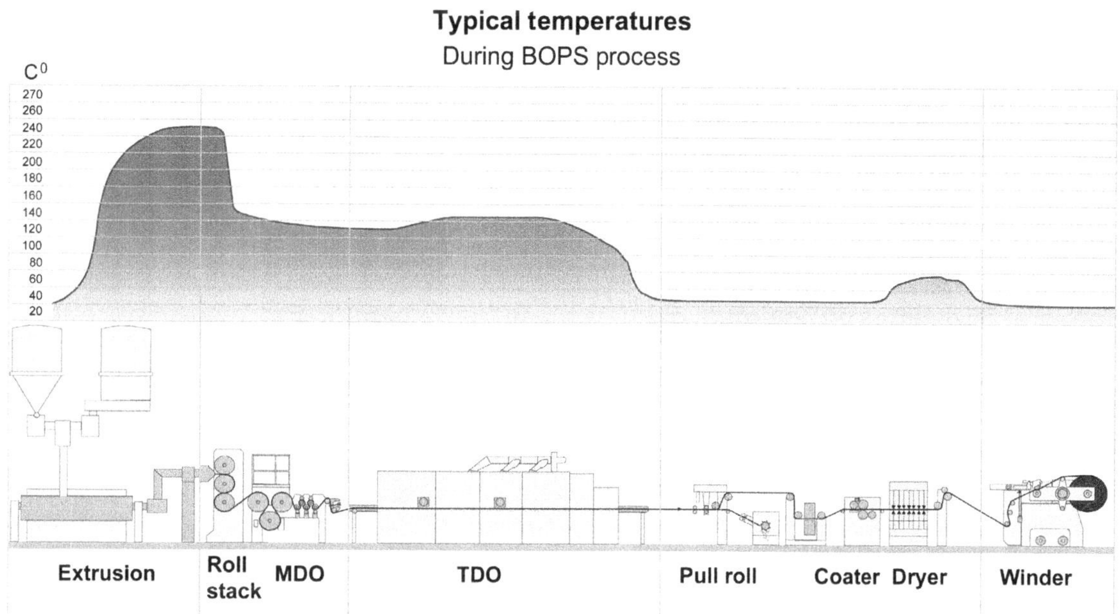

Figure 12.30 Typical BOPS process temperatures.

12.3.1.5 Other BO Films

Besides the oriented films previously mentioned (BOPP, BOPET, BOPA, and BOPS), various other specialty film types need to be mentioned.

Biaxially oriented polyethylene films (BOPE) are solely in use as shrink film applications, where there are many different products varying in layer structure, recipe, and process parameters. In principle, each application has its own tailored shrink values, shrink forces, strengths, and barriers. Barrier properties preferably are attained by coextrusion with ethylene vinyl alcohol (EVOH). For the production of BOPE shrink films, the double bubble process is almost solely used.

Oriented films from renewable resins represent another even more exotic film type on the market at present and are biodegradable. Polylactide (PLA) is the major resin used because it has attractive properties and is already available in large quantities. The raw material is primarily based on corn. Similar to PS film, PLA film can be oriented and yields an attractive property spectrum (Fig. 12.31).

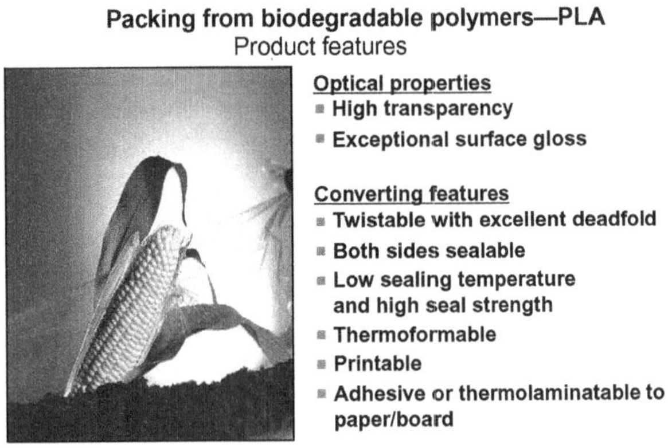

Figure 12.31 Product features for biodegradable polymers (PLA).

In particular, the excellent visual appearance has made it an interesting alternative for packaging. Furthermore, the deadfold characteristics should be noted which are a prerequisite for twist-wrap. Compared with other packaging films, the water vapor barrier, however, is considerably inferior, although this to some extent can be compensated by means of metallizing or SiOx coating. Further uses ensue in view of the permeability for water vapor and thus, such a characteristic is most suited for bread and vegetable packaging (Fig. 12.32).

12.3.2 Film Oriented in Machine Direction

Films oriented solely in machine direction account for a small market share, since this stretching method is only interesting for certain special applications. As such, monoaxially oriented propylene films (MOPP) are used for decoration ribbons, banderoles and tear strips for cigarette packs, as they have a very high longitudinal strength. In addition, such films do have a distinct tendency to split, although this has no restriction on the above applications.

Breathable films made from highly filled ($CaCO_3$) polyethylene are also oriented only in the machine direction. Defined hollow spaces up to the surface are thereby produced so that the required water vapor permeability is attained. Such film types are used in the hygiene sector as well as for the breathable layer in the building industry.

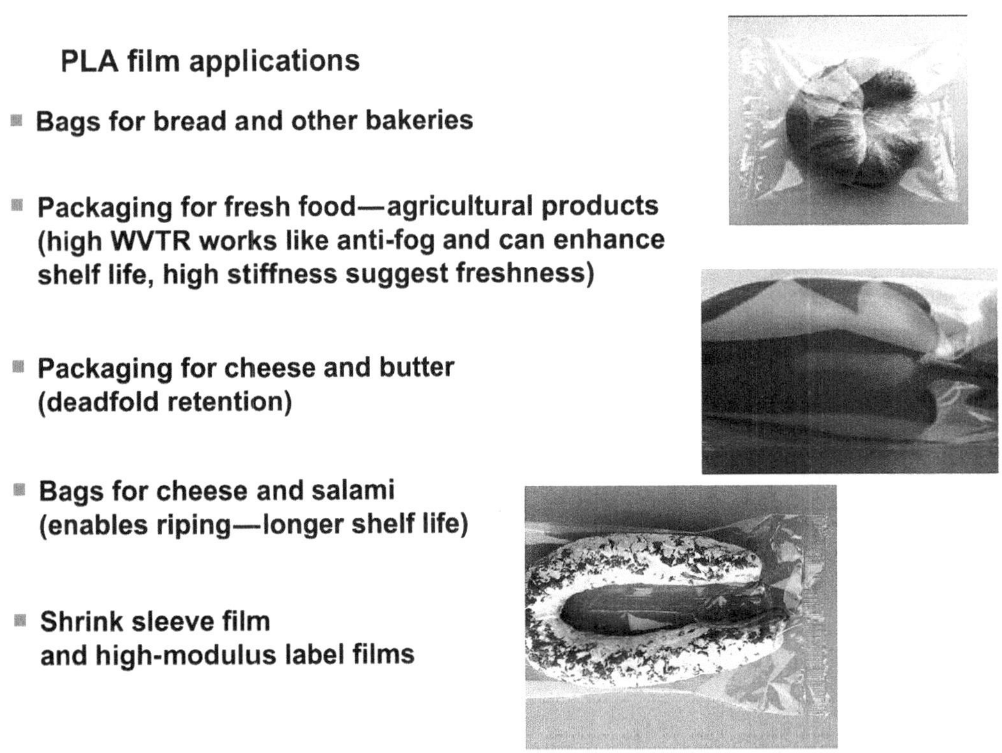

Figure 12.32 PLA film applications.

Also various special film types from longitudinally oriented polyamide (MOPA) are common. A three-layer structure PA/EVOH/PA is used to improve the barrier properties.

12.3.3 Film Oriented in Transverse Direction

A relatively large and growing market segment is represented by films oriented in the transverse direction. These are applied solely as shrink films where the demands are such that the films only shrink in the transverse direction while the machine direction shrink is not required. Such films are, to a large extent, used as sleeves and this anisotropic shrink behavior is required in order that the container-contours appear clearly and the desired print is attained (Fig. 12.33).

Shrink values of up to 80% in the transverse direction can be attained, whereas full body sleeves can also be attained for containers with strong contours. Polyvinylchloride, PS, PET-G, and PP materials are used where, in terms of shrinkage, the different characteristics of these materials become apparent (Fig. 12.34).

12.4 Trends for Oriented Films

Basically, oriented films are most suitable for meeting the trends in the packaging sector set by politics, society, and the industry. The stipulations, for example in Germany, namely packaging regulations, compel the industry to give consideration not only to the material and manufacturing costs but also to the disposal costs. This induces one to attain minimum packaging material and maximum protection with packaging. These goals can only be reached with high-strength materials to reduce thickness along with meeting the protection and barrier functions, plus operational properties that ensure high-speed packaging. With sophisticated orienting processes (eg, simultaneous stretching technology), a significant increase of strength can be attained for all plastics. Furthermore, future potential in terms of packaging can be further developed, for example, by the substitution of aluminum foil with transparent or metallized high-barrier stretched film. Another further large potential lies within the integration of many function layers in the production process of stretched films, so that complex processing steps can be waived [10]. For example, it was proved on a pilot line scale that all functions of a

Figure 12.33 Shrink film applications.

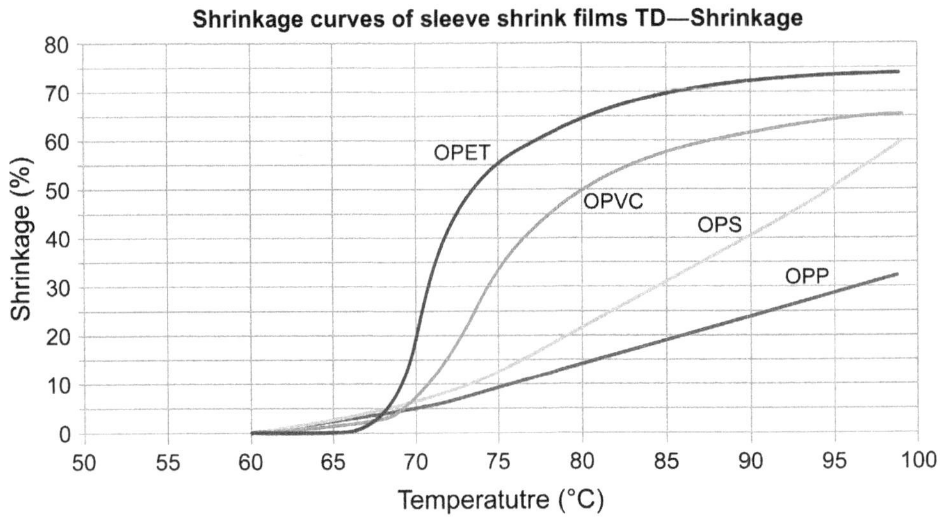

Figure 12.34 TD shrinkage versus temperature for four shrink films.

complex triplex laminate could be attained by a coextruded stretched film manufactured in one process step.

In view of the limited crude oil resources, coupled with ever increasing oil prices, plastics manufactured with crude oil bases are also subject to price increases. This, accordingly, gives a boost for alternative materials and thus the possibility of cost-efficient production on an industrial scale. The production of suitable stretchable films for packaging applications from such alternative raw materials is evident in the PLA example. One can predict that much research and development will be performed within this sector in the near future.

References

[1] Briston JH, Katan LL. Plastic films. 3rd ed. Harlow: Longman Scientific & Technical; 1989.

[2] Hansen F, Knappe W, Potente H. Handbuch der Kunststoff-Extrusionstechnik. Munich: Hanser Verlag; 1989.

[3] Tobita K, Miki T, Takeuchi N. Biaxially oriented film. In: Kanai T, Campbell GA, editors. Film processing. Munich: Carl Hanser Verlag; 1999.

[4] Breil J. Added value speciality films produced with sequential and simultaneous stretching lines. In: 18th Annual Maack World Congress, speciality plastic films, Switzerland, Zurich; October 29–30, 2002; 2002.

[5] Wellenhofer P. Monoaxiales und biaxiales Recken mit besonderer Berücksichtigung des biaxialen Flachfolienstreckens. Folien, Gewebe, Vliesstoffe aus Polypropylen. Dusseldorf: VDI-Verlag; 1979.

[6] Jabarin SA. Orientation and properties of polypropylene [Reprinted from Reinforced Plastics and Composites] Presented at the society of plastic engineers annual technical conference, May 1992, vol. 12. April, 1993

[7] Tsunashima K, Toyoda K, Yoshii T. Stretching conditions, orientation and physical properties of biaxially oriented film. In: Kanai T, Campbell GA, editors. Film processing. Munich: Carl Hanser Verlag; 1999.

[8] Goetz W. Polyamide resin for BOPA film. In: Polyamide 2005, Maack Conference, June, Dusseldorf; 2005.

[9] Nentwig J. Polystyrol, Kunststoff Folien, Herstellung—Eigenschaften—Anwendung. Munich: Carl Hanser Verlag; 1994.

[10] Breil J, Lund R. Development of new BOPP barrier films by coextrusion and simultaneous biaxial orientation. In: 2008 TAPPI PLACE conference, September, Portsmouth, VA; 2008.

13 Polymer Blending for Packaging Applications

Barry A. Morris
DuPont Performance Materials, Wilmington, DE, United States

13.1 Introduction

Blending of polymers is becoming increasingly important in packaging applications to enhance properties, improve processing or lower cost. Tailoring surface properties, such as coefficient of friction, adding color, promoting adhesion, increasing output, improving stability, and obtaining easy-opening features, are just a few of the attributes that can be achieved by blending.

The simplest blends can be made by mixing ingredients in the extruder used to convert the resin into a film or coating. For more complex blends, specialized screw designs or customized compounding equipment may be required to achieve the desired properties. These machines incorporate various mixing modes, such as flow rearrangement (distributive mixing) or high stress levels to breakup particles (dispersive mixing). Complex shear and elongational flow fields may also be used to obtain optimum mixing.

The final blend properties will depend not only on the flow and stress history, which is process dependent, but also on the thermodynamics and the polymers' thermal and rheological properties. Most polymer blends are immiscible, resulting in the minor component forming a separate dispersed phase or domain within the major component. The major component forms a continuous phase or matrix. The phase size and shape is known as the blend morphology. Blend morphology has a profound effect on the final properties and is the subject of much study. Morphology is influenced by many factors, including:

- interfacial tension (thermodynamics),
- dispersed to continuous phase viscosity ratio,
- elasticity of each phase,
- minor component concentration,
- mixing and melting order, and
- stress and flow history.

A basic understanding of these complex relationships between polymer properties and processing can aid in optimizing blends for a given application.

In this chapter, we will highlight some fundamentals of blending polymers for packaging applications. The literature is too broad for a detailed review of polymer blending and alloying. The goal is to familiarize the reader with those aspects of polymer blending that are important for inline mixing of resins and other ingredients during film converting. Some blend requirements, however, are too demanding for inline mixing and are best left to a resin manufacturer or compounder. Some of the technology that may be used to design and produce such blends is also reviewed.

13.2 Why Blend?

Even with the flexibility of controlling properties by introducing specific functional layers within a multilayer film, blending can be critical for the package function. Blending may be needed to make the polymer stable enough to extrude or have the right surface properties after extrusion. Blending may be a way to tailor specific properties into a layer, such as barrier or heat-seal performance. The resin manufacturer often adds polymer additives, such as antioxidants, catalyst killers, and various processing aids, to the polymer. But the film manufacturer may add these and other additives, such as slip, antiblock, antifog, antistatic agents, or processing aids such as fluoroelastomers for reducing sharkskin and melt fracture. Typically, these additives comprise less than 1% of the final composition. They are usually powders or liquids. Because of this and their low concentrations, they are typically first made into a masterbatch, a highly concentrated blend of the additive with a carrier resin. Masterbatches are typically produced by companies with specialized compounding equipment. The film manufacturer blends the masterbatch into

Multilayer Flexible Packaging. DOI: http://dx.doi.org/10.1016/B978-0-323-37100-1.00013-2

the resin at the extruder feed hopper used to make the film or sheet.

Pigments such as titanium dioxide (TiO_2) and various colorants are typically added as a masterbatch. Since the additive or pigment is well dispersed in the masterbatch, blending can be done in a single-screw extruder without special screw designs or compounding equipment.

Blending may be used to reduce the resin cost. For example, a metallocene polyethylene plastomer (mPE) may be diluted with standard linear low-density polyethylene (LLDPE) or low-density polyethylene (LDPE) to lower cost. Recycle may be incorporated back into the film structure. One form of recycle is inhouse scrap, which may be ground up and introduced into the extruder hopper as "regrind."

Blending may also help improve resin processability. Two grades of the same resin with differing flow properties (such as melt index) may be blended together to achieve the proper flow for a given process. LDPE is typically blended into LLDPE to reduce extruder pressure and torque, improve melt strength and bubble stability, which help increase output [1]. Other examples where blending improves processing include blending amorphous nylon (polyamide) into nylon 6 to increase extruder output and adding amorphous nylon or ionomer to ethylene vinyl alcohol (EVOH) to improve thermoformability [2,3].

A polymer deficient in one property is often blended with another to enhance that property. Blending in cyclic polyolefins (COCs), for example, can enhance LLDPE stiffness [4]. Soft polymers are often blended into harder polymers to improve the toughness. Examples include blending ethylene vinyl acetate (EVA) into LLDPE, mPE into LLDPE, and ethylene-propylene-diene rubber (EPDM) or mPE into polypropylene (PP) [5].

In the engineering polymer world, rubber is added to PA to improve low temperature toughness [6]. Adding polyethylene terephthalate (PET) to polycarbonate (PC) lowers cost and improves chemical resistance and processability. Blends of PC and acrylonitrile–butadiene–styrene copolymer (ABS) have lower cost than PC and higher heat deflection temperature and toughness than ABS [7].

Barrier properties may be enhanced by blending. High-density polyethylene (HDPE) is blended into LLDPE or LDPE to improve moisture barrier performance. Amorphous nylon, such as DuPont's Selar® PA, is blended into Nylon 6 to improve the oxygen barrier at high relative humidity [8]. DuPont invented a laminar barrier technology where Nylon is blended into HDPE forming large platelets that impede the flow of species trying to migrate through. The laminar morphology is accomplished by choosing specific resins and processing conditions [9–12].

Adhesion may be promoted with several resins. Adding EMA or EVA to PE can improve adhesion to certain inks. Anhydride modified polyolefins are blended with PE or EVA to improve adhesion to Nylon or EVOH in coextrusion. An acid-based additive has been developed for enhancing the adhesion of LDPE to aluminum foil in extrusion coating [13].

Additives may also help control adhesion during the heat seal process. "Contaminates" are often blended into sealant resins to achieve easy openability. Examples include blending polybutene-1 (PB) into LDPE, LLDPE, EVA or ionomers, and EVA/ionomer blends [14–17]. The blend morphology and phase compatibility is important for these applications. For example, in the blends containing PB, the PB is the minor phase. It forms spherical particles that are stretched into fibers and ribbons during the film fabrication process. The poor compatibility between the PB and matrix resin results in failure along these fibers and ribbons near the sealant interface, lowering the seal strength.

13.3 Blending Processes

Blending requires that the ingredients be brought together in the right proportions and then homogeneously mixed. In polymer blending, the former is important because there is little back-mixing in most continuous mixing devices used by the industry. This is particularly true of the single-screw extruder, which is the predominant device used by the film converting industry. For simple blends, such as those adding masterbatches, the pellets are premixed together and fed into the extruder hopper. In more sophisticated compounding devices, such as twin-screw extruders, ingredients can be added at various stages along the extruder.

Once the polymer enters the extruder or other mixing device, it is melted, and the mixing mechanism depends on the specific device employed. We will focus on mixing in single-screw extruders, introduce some more sophisticated devices, and compare and contrast them to single-screw extruders. Although specialty compounding devices, such as twin-screw extruders, kneaders and continuous mixers, have historically been used by the resin manufacturer or toll compounder and not by film converters, they are beginning to be used as inline compounders for some large-scale film operations [18].

13.3.1 Pellet Premixing

Pellet mixers fit into two general types, offline batch mixers and inline feeders/mixers. Batch mixers can be as simple as a cement mixer. The ingredients are weighed and poured into the mixer, blended, and then transported to the extruder hopper. Batch mixers generally are less expensive and occupy less space than inline mixers. They can be used to feed more than one extruder. Also, the ingredients' weight can be accurately measured, depending on the scale being used. On the other hand, they can be labor intensive, leave no automatic records and open up the chance for human weighing errors and the possibility for pellet segregation during transport. Human weighing errors can be eliminated with automatic weighing that use either gravimetric or volumetric feed systems, as described below.

Inline mixers generally are positioned above the extruder hopper. Each ingredient is fed by individual feeders. Fig. 13.1 shows a typical pellet blender. The ingredient feeders meter out a specific volume or mass over time, either by constant augur speed (volumetric feeders) or by an augur whose speed is controlled by the feed hopper loss in weight (gravimetric feeder). Volumetric feeders are less expensive but need to be manually calibrated for each ingredient since differences in density, bulk density, and compressibility affect the feed rate. Gravimetric feeders have become the most commonly used feeders. Their calibration is usually much simpler than volumetric feeders. For example, the feeder may measure the weight loss over 30 seconds at 10% maximum feeder speed to compute a feed factor that takes into account the variation due to bulk density, density, etc. The whole calibration process is usually done electronically and requires little operator interaction. Since the feed hopper weight is monitored by the computer control system, gravimetric feeders allow the individual feed rates to be recorded. Alarms can be set to ensure that each ingredient is being fed to the extruder. Care must be taken when the feed hopper is refilled. Various control schemes, such as temporarily going into volumetric mode, are used to ensure feed continuity as the weigh cell is recalibrated. Understanding how the gravimetric feeder handles refilling is helpful in ensuring a trouble-free operation.

Figure 13.1 Pellet blender. *Source: Courtesy of AEC.*

Inline mixers generally take up more space and are more expensive than batch mixers. They allow, however, for changing ingredient proportions during processing and can record the actual ingredient weights entering the extruder, which can be important for process control. They are also generally less labor intensive.

The ingredients in premixed pellet blends may segregate if they differ in density, size, or shape and are transported over a long distance. Keeping the transport distance as short as possible and avoiding mixing powders with pellets can minimize segregation.

Using properly sized feeders and augers is important to ensure ingredients are controlled to the correct proportions. An oversized feeder will likely have less precision and accuracy. There are also practical limits to pellet blending. A 1% pellet blend corresponds to blending one pellet in every 100. When possible, it is best to strive for ingredients equal to 10% or higher for greater control and accuracy.

Care must be taken to thoroughly clean the feeders, mixers, and transport lines when changing over products. One wrong pellet in a product can cause film imperfections and other quality issues.

The film producer incurs extra risk when blending since the blend properties cannot be directly measured or controlled, especially if the blend is a thin layer in a multilayer film. Premade blends from a resin manufacture or compounder can be tested for properties to ensure they meet specifications. The error associated with premixing is directly related to the precision and accuracy of the pellet premixing system, how well it is maintained, and procedures that have been adopted to ensure the ingredients never run dry. Gravimetric pellet mixers typically have precisions ranging from 0.1% to 1%, depending on hopper, feeder, and auger size.

13.3.2 Melt Blending

Once the ingredients have been fed in the correct proportions into the mixing device hopper, typically an extruder, the polymers, and/or the additives must be homogeneously blended together. This requires that the polymers be in the molten state. The mixing device melts the polymers, provides a means for mixing, and generates pressure for subsequent operations such as:

- making film when in-line mixing and
- pelletizing when compounding.

Mixing is described as either distributive or dispersive and is illustrated in Fig. 13.2. In distributive mixing, the polymer is rearranged by deformation. Separating flow with rearrangement and kneading are two examples of distributive mixing. In dispersive mixing, particles are broken up and dispersed within the polymer matrix. Shear stress is important for overcoming material yield stress. Dispersing pigment particles in a masterbatch is an example of dispersive mixing.

The single-screw extruder is the most commonly used device for film production. It efficiently melts the polymer and generates pressure for extruding the polymer through a flat or annular die. It is suitable for many blending applications but has its limitations, many of which can be overcome by optimizing the screw design. Single-screw extruders rely on the difference in friction between the solid polymer pellets and the barrel and screw surfaces to propel the pellets forward. Slippery ingredients may impede the pellet flow and cause surging and other unwanted effects. Single-screw extruders are typically flood fed, meaning the throughput is determined by the extruder screw speed, not how fast the ingredients are fed to the feed hopper. In more sophisticated compounding devices, such as twin-screw extruders, the throughput is decoupled from the screw speed. The output in these devices is determined by the feed rate, and the screw speed can be increased or decreased to change the mixing intensity and energy input. Single-screw extruders are also generally not well equipped to handle powder (due to potential problems with segregation in the feed hopper and nonuniform melting) and liquid feeds.

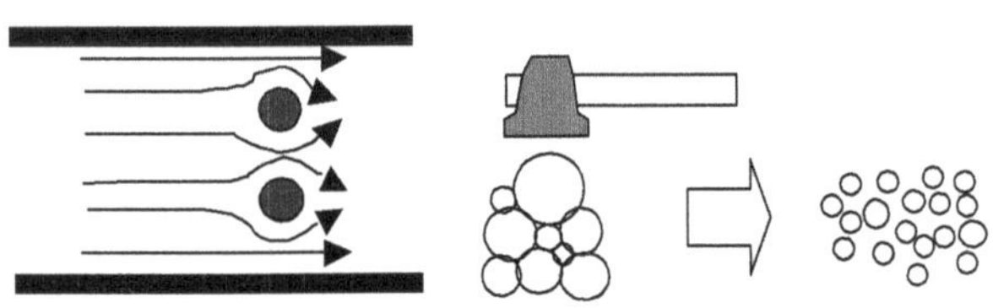

Distributive mixing—flow is divided, separated, and redistributed

Dispersive mixing—particles are broken up by application of stress

Figure 13.2 Distributive and dispersive mixing.

The melting mechanism for pellet blends with different melt points in a single-screw extruder is not well understood. The standard melting model for uniformly melting pellets envisions a solid bed compressed against the trailing edge of the screw channel accompanied by an ever-lengthening melt pool that results from the frictional heat near the barrel surface [19–21] (Fig. 13.23). Disruptions in the solid bed can result in poor melt quality; unmelted particles may exit the extruder, plugging screen packs, or appearing in the final film as "gel" particles [22]. A disruption in the solid bed can cause surging. Phase inversions and other aspects of blending polymers that have different melt temperatures may cause disruptions in the solid bed and are not well documented in the literature. The low melting polymer may also act as a lubricant, impeding the melting of the high melting component since viscous heat generation from the friction between the polymer and barrel wall is a primary mechanism for polymer melting.

Conversely, Morris and Kirschbaum [23] found that adding 1–2% of a low melting ethylene copolymer to polylactic acid lowered power consumption by up to 30%, which may lead to higher output on extrusion lines limited by power. Baird et al. [24] used a highly instrumented extruder to show that additive melting altered the frictional characteristics in the solids conveying section as well as the melting behavior in the melting zone. This resulted in less viscous energy dissipation and a more energy-efficient process.

In a standard screw, the polymer melt is subject to nonuniform temperature and flow histories. There are circulatory flow patterns in the metering section of the screw (Fig. 13.3).

Figure 13.3 Single-screw extruder circulatory flow patterns in the metering section. *Source: Adapted from Middleman [25].*

These patterns aid distributive mixing and, as they are nonuniform, they may lead to uneven mixing. The shear stress and flow rates vary across the flow channel (Fig. 13.4). As a consequence, the temperature varies across the flow channel. The temperature nonuniformity exiting the extruder can cause film thickness variability. Mixing elements incorporated into the screw design or at the extruder exit (such as static mixers) help homogenize the melt temperature for proper control of film thickness. These same mixing devices also help blend polymers. Mixing elements that have been utilized over the years include pins, restriction rings, and more specialized designs such as the pineapple, Dulmage, Saxton, and Maddock mixers [26].

Some mixing elements are characterized as distributive mixing elements—they achieve mixing by

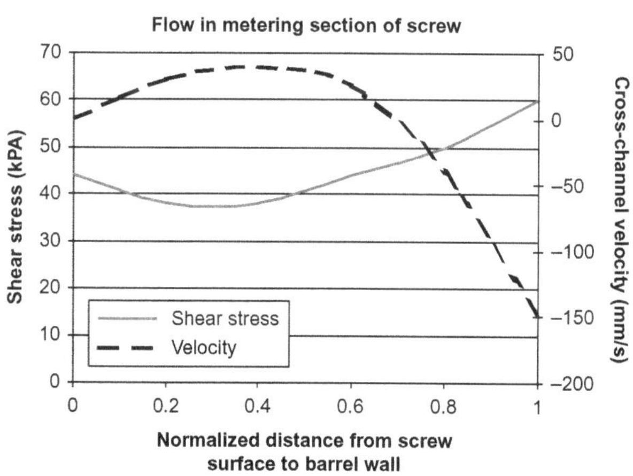

Figure 13.4 Single-screw extruder metering section stress and velocity distribution for polyethylene in the cross-channel direction. Calculated with commercial software from Polydynamics, Inc.

disrupting the flow. One way to do this is by inserting pins (Fig. 13.5), blisters, or other elements that impede the flow. Pins can cause problems, however, as stagnant flow zones are created behind the pins, setting up the potential for gel formation with some thermally sensitive polymers. More sophisticated designs, such as the Saxton mixer [27] (Fig. 13.6), eliminate the dead zones while dividing up the flow. Other elements are dispersive in nature, such as the well-known Maddock mixer (Fig. 13.7). In this design, the polymer flows down a fluted section and over a small clearance between the screw and barrel, which introduces high shear stresses to break up agglomerates. Its actual effectiveness at dispersive mixing is suspect, particularly compared to

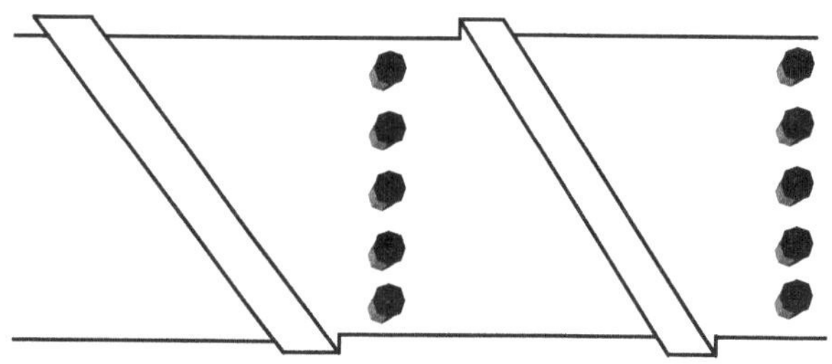

Figure 13.5 Pin mixing section.

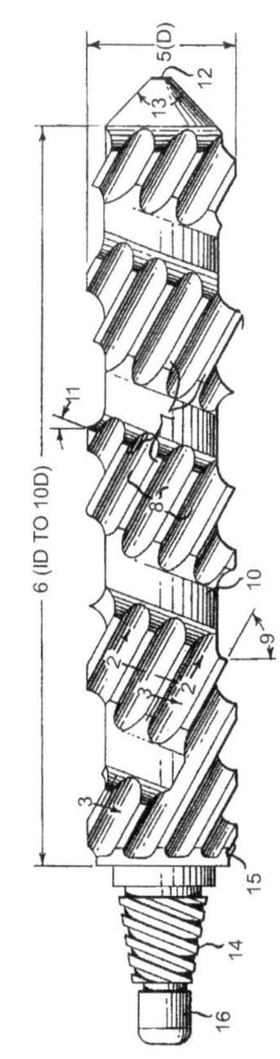

Figure 13.6 Saxton [27] mixing section.

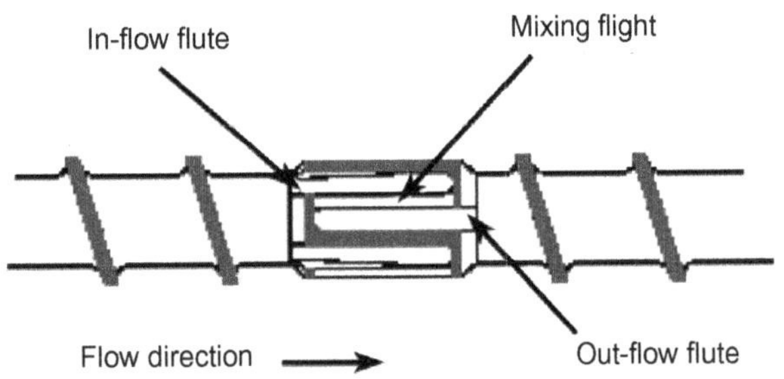

Figure 13.7 Maddock screw element. *Source: Adapted from Spalding and Hyun [22]). Courtesy of the Society of Plastics Engineers.*

specialized compounding devices, such as twin-screw extruders, but it is often used for multipurpose mixing.

In recent years, several new high-performance screw designs have been developed that try to improve the mixing efficiency while increasing output. These screws generally work by having the polymer flow through a series of relatively tight clearances. The solid material is both entrapped and given time to melt or subject to high shear or elongational flow fields that aid in melting and mixing [26,28]. The high-performance section replaces the whole metering section. Some of the more commonly used high-performance screws include the energy transfer [29], variable barrier energy transfer [30,31], double wave [32,33], stratablend [34], unimix [35], and CRD [36] screws. An example of a variable barrier energy transfer screw element is show in Fig. 13.8.

Although optimizing the screw design allows many polymer blends to be made on single-screw extruders, in some cases, specialty compounding devices may be necessary. This is especially true when intensive dispersive mixing, reaction, or devolatilization is required. There are several compounder designs on the market. For general purpose compounding, corotating twin-screw extruders are often employed [37,38]. Kneaders provide good distributive mixing and provide the ability to easily introduce liquid feeds and minimize temperature buildup, which is important for thermally sensitive polymers. Counter-rotating twin-screw extruders are often used for reactive extrusion and devolatilization since they can have long L/Ds (length to diameter ratios). The Farrel continuous mixer operates like a continuous Banbury mixer and is useful for making masterbatches with high filler loadings. Planetary mixers and other multiscrew devices are also available for specialty applications.

Of the specialty compounding devices, the corotating twin-screw extruder is the most often used. Corotating twin-screw extruders offer considerable versatility in design and operation. As described earlier, feeding is independent of extruder rpm, which allows the mixing and energy intensity to be independently varied. Multiple feeding ports are common, and screw designs can be changed from job to job to tailor the process. The screws are built from individual elements, which can be changed to suit the application. These elements come in several geometries with different mixing intensities. They provide good distributive and dispersive mixing and impart both shear and elongational flow fields. Fig. 13.9 shows some twin-screw modular elements.

Some twin-screw extruders have screws that are self-wiping, which results in a narrow residence time distribution for better mixing control. One drawback is that they are poor melt pumps. Because of this, temperatures can be difficult to control at the extruder exit. Twin-screw extruders are not used alone for inline compounding on a film line. They are adapted for better melt pumping by coupling the twin-screw extruder with a single-screw extruder or gear pump [18].

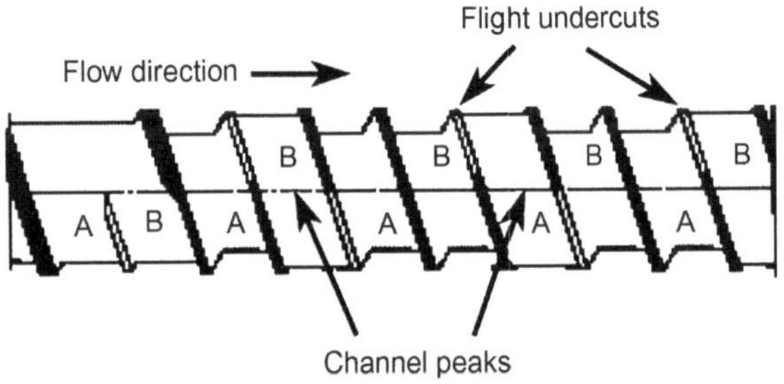

Figure 13.8 Variable barrier energy transfer section. *Source: Adapted from Hogen et al. [30] Courtesy of the Society of Plastic Engineers.*

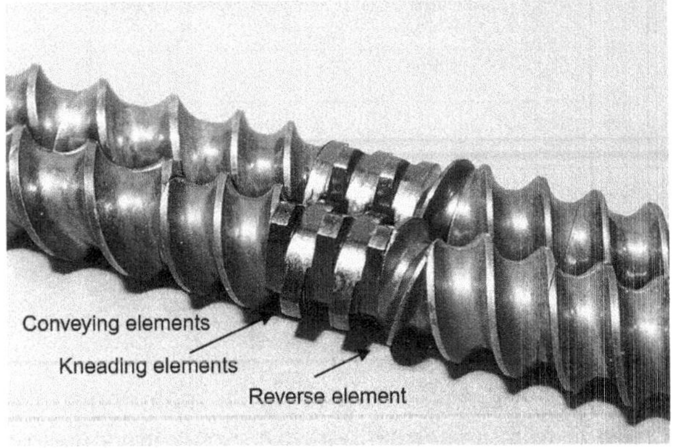

Figure 13.9 Twin-screw extruder modular elements.

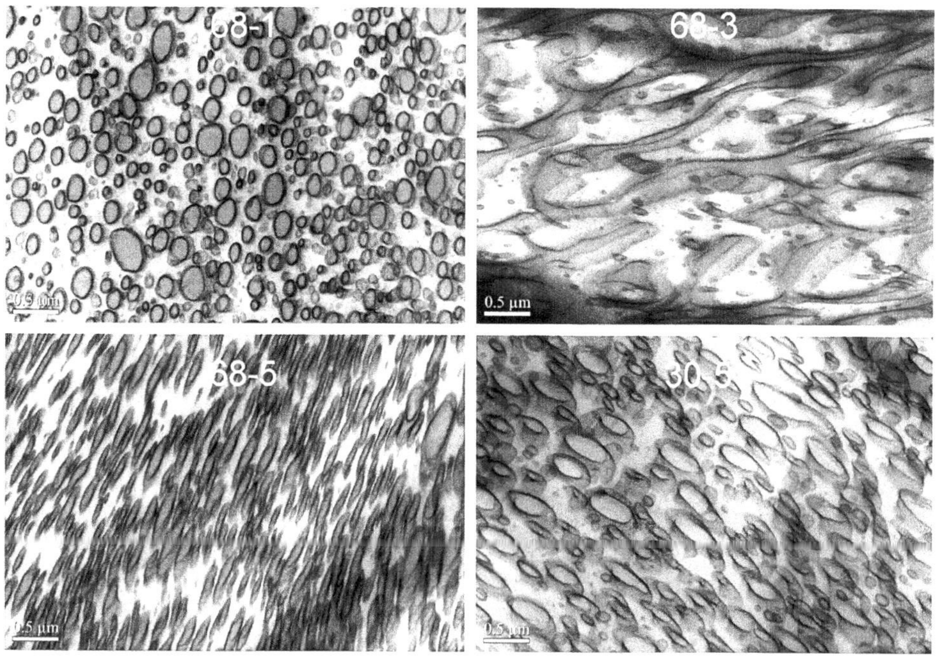

Figure 13.10 Transmission electron micrographs (TEM's) of immiscible polyethylene–styrene polymer blends of varying chemistries. *Source: Courtesy of Barbara Wood and I-Hwa Lee, DuPont. Further immiscible blend TEM examples can be found in Wood [39,40].*

Twin-screw extruders and other specialty compounding devices are considerably more expensive than single-screw extruders.

13.4 Physics of Blending

Polymer blend properties are influenced by specific interactions between the molecules (thermodynamics) and their response to deformation (rheology). The thermodynamics determine whether the blend forms a single phase (miscible) or multiple phases (immiscible). Miscible blends typically follow the rule of mixtures, namely, the blend properties that are directly proportional to the component ratio. Same polymer blends with different grades, such as two LDPE grades, are typically miscible. Most different polymer blends are not miscible. There are a few commercially important miscible blends marketed today. One is Sabic Noryl which is a polyphenylene oxide and high-impact polystyrene (PS) blend.

How do we determine whether a blend is miscible? One method is to look at transparency, either by microscopy or light scattering. An immiscible blend forms separate domains within the polymer matrix that may diffract light if they are large enough. Another technique is to measure the glass transition temperature (Tg) using differential scanning calorimetry or other thermal analyses. A miscible blend will have a single Tg, typically between that of the components. Both techniques use a fairly large sample or probe size, which can at times be misleading. Sometimes blends appear to be miscible because the probe size is too large. Other techniques with smaller probe sizes are X-ray and neutron scattering and various spectroscopy techniques, such as infrared and nuclear magnetic resonance.

Occasionally, we use the word "compatibility" to describe the degree to which polymers interact. Miscibility is maximum compatibility. Compatibility is a subjective term and is not well defined. Miscibility has a specific definition—two polymers are miscible if they form a single phase over their entire composition range at a given temperature.

Immiscible blends, by definition, form multiple phases. In the simplest case, a two-component blend, the minor phase forms domains within a continuous major component matrix. The domain sizes, shapes, and distribution are known as the morphology. Fig. 13.10 shows some immiscible blend morphologies. The morphology is influenced by the concentration, thermodynamics, component rheology, and the flow and stress history during mixing and processing.

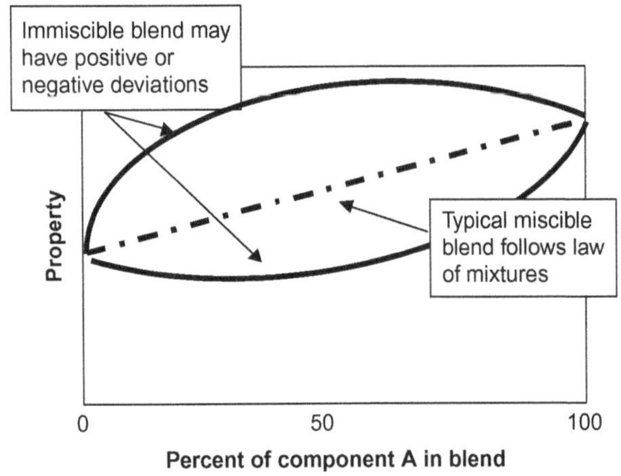

Figure 13.11 Properties versus percentage of polymer A in polymer blends.

The morphology is critically important to the final blend properties, which often do not follow the rule of mixtures (Fig. 13.11). In many cases, blends are designed to create a specific morphology to achieve certain property gains. An example is super-tough polyamide (PA) or Nylon, which is a PA66/rubber blend [6]. The rubber must achieve a certain domain size in order to stop cracks from propagating during impact. Another example is Selar® RB laminar technology developed by DuPont. Here Nylon is dispersed in HDPE so that the Nylon phase forms platelets parallel to the surface. The platelets create a tortuous path for diffusion, resulting in improved barrier performance [8].

Once a specific morphology has formed, it may change with further processing. The domains may coalesce or be stretched through orientation (Fig. 13.10). Compatibilizers are frequently used to stabilize the blend morphology.

There is a third blend system, namely melt-miscible blends that are miscible in the melt state but phase separate in the solid state. An example is polyoxymethylene/polylactide (POM/PLA) [41]. This phenomenon may be important for ease of processing in order to obtain a fine dispersion of one component into another during melt blending.

13.4.1 Thermodynamics

When two materials are brought together, there must be a decrease in the Gibbs free energy for them to transform into a single material or miscible blend. This can be expressed mathematically as:

$$\Delta G_m < 0 \text{ for miscibility} \quad (13.1)$$

The free energy includes enthalpy and entropy contributions:

$$\Delta G_m = \Delta H_m - T\Delta S_m \quad (13.2)$$

where:

ΔG_m	=	Gibbs-free energy of mixing
ΔH_m	=	enthalpy of mixing
T	=	temperature
ΔS_m	=	entropy of mixing

For most polymer blends, ΔH_m is positive and ΔS_m is nearly zero. Thus, it is rare that polymer blends are miscible.

Coleman et al. [42] derived the following relationship for ΔG_m based on the work of Flory and Huggins:

$$\frac{\Delta G_m}{RT} = \left[\frac{\Phi_A}{N_A}\ln \Phi_A + \frac{\Phi_B}{N_B}\ln \Phi_B\right] + \chi\Phi_A\Phi_B\left(\frac{\Delta G_H}{RT}\right) \quad (13.3)$$

where:

ΔG_m	=	Gibbs-free energy of mixing
R	=	ideal gas constant
T	=	temperature
Φ_A	=	polymer A volume fraction
Φ_B	=	polymer B volume fraction
N_A	=	polymer A degree of polymerization
N_B	=	polymer B degree of polymerization
χ	=	interaction parameter
ΔG_H	=	free energy of specific action between polymers, including hydrogen bonding

In Eq. (13.3) the first term on the right side arises from combinatory entropy. Since N, which is related to molecular weight (MW), is large for polymers, this term is nearly zero. The second term also arises from entropy and is always positive. The interaction parameter, χ, is defined by:

$$\chi = \frac{V_{\text{ref}}}{RT}[\delta_A - \delta_B]^2 \quad (13.4)$$

where:

V_{ref}	=	reference volume
δ_A	=	polymer A solubility parameter
δ_B	=	polymer B solubility parameter

The final term, $\Delta G_H/RT$, is negative when specific interactions are present.

From Eqs. (13.3) and (13.4), we see that we can improve compatibility by matching solubility parameters and achieve miscibility only when we incorporate specific interactions. Non-polar polymer blends, such as PP–PE blends, have no specific interactions beyond weak dispersive forces. As we shall see, even though their solubility parameters are nearly equal, they are not miscible.

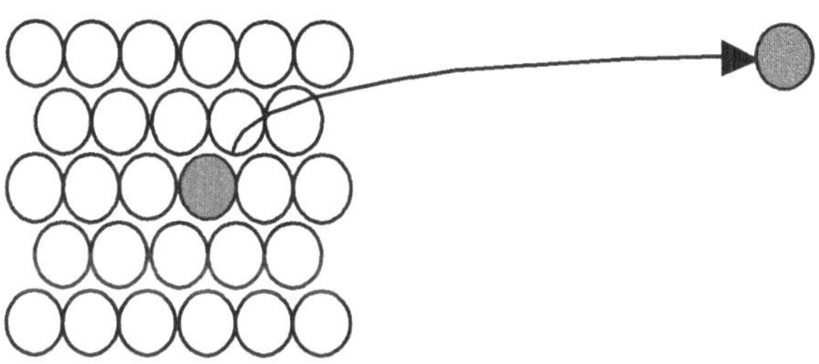

Cohesive energy density is the energy needed to remove a molecule away from its environment and is the square of the **solubility parameter**.

Figure 13.12 Cohesive energy density illustration.

In Eq. (13.4), we introduced the solubility parameter. We now discuss the origin of solubility parameters and how they can be helpful in understanding polymer blends. The energy per unit volume required to remove a molecule from a liquid or solid is known as the cohesive energy density. This is illustrated in Fig. 13.12. The cohesive energy density is a function of the forces that hold the material together. The solubility parameter is the square root of the cohesive energy density. It contains contributions from both non polar (dispersive forces) and polar (dipole–dipole and hydrogen-bonding) interactions

Comparing solubility parameters is a way to quantify the "like dissolves like" principle of chemistry. For two polymers to be miscible, the difference in solubility parameters should be between 0.1 and 3 Hildebrands, depending on their interaction strength. Coleman et al. [42] defined a critical solubility parameter difference, $\Delta\delta_c$, below which the polymers may be miscible. As shown in Table 13.1, the value of $\Delta\delta_c$ depends on what interactive forces are present.

Table 13.2 lists solubility parameter values for some polymers used in packaging films. Returning to our PE–PP blend example, we see that the difference in their solubility parameters is less than 1 ($\Delta\delta < 1$). However, since only nonpolar dispersive forces are present, the critical solubility parameter difference ($\Delta\delta_c$) is less than 0.1. Thus, these polymers are not miscible.

Table 13.1 Critical Solubility Parameter Difference Upper Limit

Specific Interactions Involved	Polymer Blend Examples	$\Delta\delta_{critical}$ Hildebrands
Dispersive forces only	Polybutadiene–polyethylene (PBD–PE)	<0.1
Dipole–dipole	Polymethylmethacrylate–polyethylene oxide (PMMA–PEO)	0.5
Weak	Polyvinyl chloride–butadiene acrylonitrile copolymer (PVC–BAN)	1.0
Weak to moderate	Polystyrene acrylonitrile–polymethylmethacrylate (SAN–PMMA)	1.5
Moderate	Polycarbonate–polyesters	2.0
Moderate to strong	Nylon–polyethylene oxide (Nylon–PEO)	2.5
Strong	Polyvinyl phenol–polyvinyl acetate (PVPh–PVAc)	3.0
Very strong	Polymethacrylic acid–polyethylene oxide (PMMA–PEO)	>3.0

From Coleman, Graf, and Painter [42].

Table 13.2 Solubility Parameters for Some Polymers Used in Packaging

Polymer	Solubility Parameter $cal^{1/2}/cm^{3/2}$ (Hildebrands)	Source
PE (polyethylene)	7.7–8.4	Van Krevelen [43]
PP (polypropylene)	8.2–9.2	Van Krevelen
PS (polystyrene)	8.5–9.3	Van Krevelen
PVC (polyvinyl chloride)	9.4–10.8	Van Krevelen
PVDC (polyvinylidene chloride)	9.9–12.2	Van Krevelen
PVOH (polyvinyl alcohol)	12.6–14.3	Van Krevelen
EVOH (44 mol% ethylene)(ethylene vinyl alcohol)	17	Evalca literature [117]
EVOH (32 mol% ethylene)	19	Evalca literature
Nylon 6	12.6	Evalca literature
EVA (9% VA) (ethylene vinyl acetate)	8.1	DuPont calculation [118]
EVA (25% VA)	8.2	DuPont calculation
EMA (20% MA) (ethylene methyl acrylate)	8.3	DuPont calculation
PET (polyethylene terephthalate)	10.3	Wu [49]

It should be noted that there is some controversy around solubility parameters in the literature. They are exact for polymers with only physical interactions. There are errors associated with trying to extend the concept to polymer systems involving hydrogen bonding and other polar interactions and with the indirect methods used experimentally to measure them. These errors typically result in a range that is not very useful for predicting miscibility, hence matching solubility parameters is not a necessary and sufficient condition for polymer miscibility. They are useful guides for compatibility, however and, since most polymer blends are not miscible, this may be their greatest strength. The solubility parameter difference has been related to the interphase thickness between immiscible polymers. Immiscible polymers with a solubility parameter difference about 0.5 Hildebrands or less may still build up enough strength at the interface for good mechanical properties [119].

Returning to our PP–PE blends example, PE is often used to modify PP properties. For example, ethylene–propylene rubber (EPR) is blended with PP to improve PP impact toughness. The EPR forms a phase with sufficient adhesion to the PP that the properties are enhanced, as would be predicted by their close solubility parameters.

LDPE–LLDPE blends are perhaps the most commercially important blends used in flexible packaging applications. A distinguishing difference between LDPE and LLDPE is long chain branching (LCB) in LDPE, which contributes to its relative processing ease. Typically, about 10–30% LDPE is added to LLDPE to improve melt strength, bubble stability, and other film production in general. Hussein et al. [44–47] found that LDPE and LLDPE are not miscible over their entire composition range. Miscibility is favored for LLDPE-rich blends. The miscibility range increases by lowering the LLDPE MW and by increasing the short-chain branch length (replacing the butene comonomer with octene). At the same MW and branch content, a Ziegler–Natta LLDPE is more miscible with LDPE than metallocene LLDPE (m-LLDPE), which has a narrow MW and comonomer distribution. At the same MW distribution and MW, an m-LLDPE with higher branch content is more miscible with LDPE than an m-LLDPE with lower branch content. Comonomer type does not have an effect on m-LLDPE–LDPE miscibility.

One can add specific interactions to promote miscibility. Coleman et al. [42] have developed software that uses group contribution theory to predict polymer solubility parameters that, in many, cases, agrees well with experimental results. This is illustrated in Fig. 13.13.

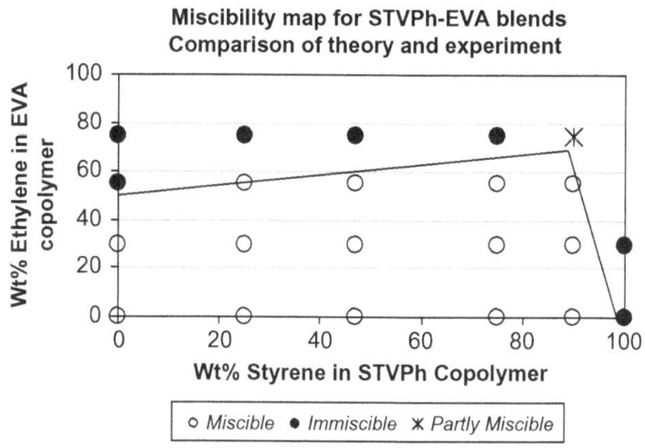

Figure 13.13 Example of using specific interactions between molecules to achieve miscible blends. *Taken from Coleman et al. [42].*

Here, vinyl acetate (VA) and styrene (St) contents are varied in an EVA copolymer–styrene vinyl phenol (EVA-StVPh) copolymer blend. Increasing the VA and the vinyl phenol content (VPh) promotes polar interactions (ΔG_H) that drive ΔG_m below zero (Eq. (13.3)). The experimental data are represented by the open and closed circles: the open circles represent miscible blends and the closed circles immiscible blends. The line demarking the region of miscibility comes from Eq. (13.3), using the software to predict the solubility parameters with specific interactions. The miscibility region depends on the comonomer percent in each polymer.

A practical note about compatibility and miscibility concerns masterbatches. The carrier resin in the masterbatch should be compatible and, ideally miscible, with the resin in which it is being blended. This way, the carrier resin will not harm the letdown resin properties. Poor compatibility can lead to poor optical properties, reduced barrier performance and, for sealants, poor seal performance. Benkreira and Britton [48] found that better dispersion occurs when the carrier resin is lower in viscosity and has a lower melting point than the host polymer, for reasons that will become clearer in the next section.

Solubility parameters may also prove useful for predicting miscibility of low MW additives in polymers.

13.4.2 Morphology Development in Immiscible Blends

To understand how morphology develops in a polymer blend, we will first look at how a single droplet suspended in a fluid is broken up during flow. The droplet is held together by interfacial tension, which arises from a nonuniform force distribution acting on the molecules at the interface (Fig. 13.14).

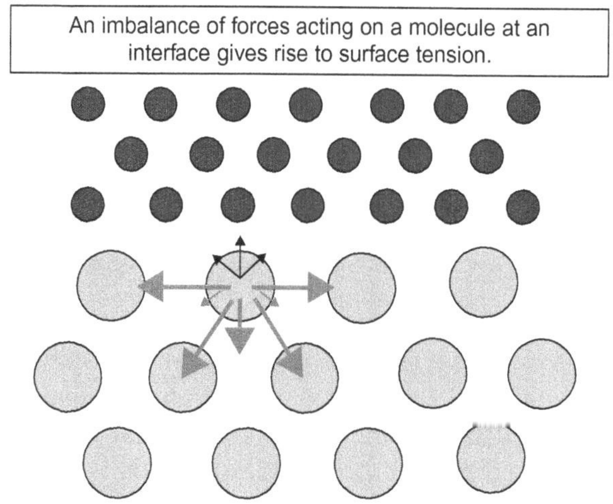

Figure 13.14 Illustration of the origins of interfacial tension.

Inside a material, a molecule is bound to its neighbors by attractive forces related to the cohesive energy density. At an interface, however, molecules are only partly surrounded by their own kind. The material across the interface may exert different attractive forces. The difference in these attractive forces gives rise to the interfacial tension. The more alike the materials are, the lower the interfacial tension and the smaller the driving force holding the droplet together. If the droplet has a radius a, then this holding force, $F_{\text{interfacial}}$, is proportional to Γ/a, where Γ is the interfacial tension.

We can relate interfacial tension to solubility parameters as introduced earlier. Wu [49] shows that the interfacial tension is related to the surface tensions of the two polymers:

$$\Gamma_{12} = \Gamma_1 + \Gamma_2 - 2\varphi\sqrt{\Gamma_1\Gamma_2} \quad (13.5)$$

where:

Γ_{12}	=	interfacial tension between polymers 1 and 2
Γ_1	=	polymer 1 surface tension
Γ_2	=	polymer 2 surface tension
φ	=	interaction parameter

This is known as the Good and Girifalco equation. Wu tabulates the interaction parameter, φ, for several polymer combinations. They range from 0.79 to 0.98 for the polymers considered. The surface tension can be related to the solubility parameter:

$$\Gamma_1 = 0.2575 \frac{\delta_1}{\sqrt[3]{\rho_1}} \tag{13.6}$$

where:

Γ_1	=	surface tension (dynes/cm),
ρ_1	=	polymer 1 density (g/mL), and
δ_1	=	polymer 1 solubility parameter (cal/mL)$^{1/2}$

Combining Eqs. (13.5) and (13.6), recognizing polymer densities are around 1 g/mL and assuming $\varphi = 1$, we find that the interfacial tension is directly proportional to the square of the difference in solubility parameters:

$$\Gamma_{12} \cong 0.26(\delta_1 - \delta_2)^2 \tag{13.7}$$

As shown in Fig. 13.15, the flow field exerts a drag force that acts to break up the droplet. For shear flow, the drag force is equal to the viscosity of the continuous phase, η_c, times the shear rate, $\dot{\gamma}$, acting over the area of the particle ($\approx a^2$):

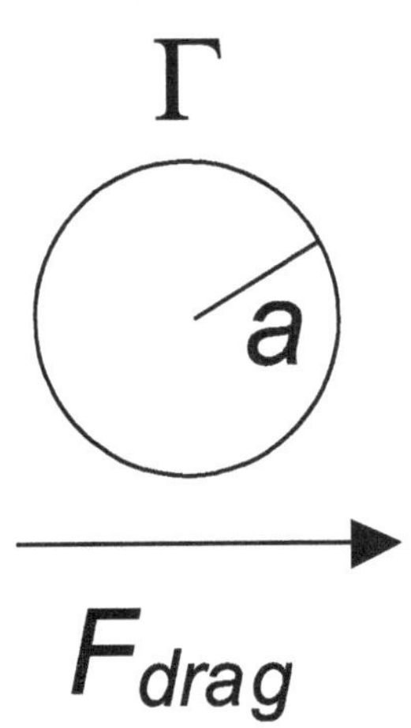

Figure 13.15 Balance of drag and interfacial forces on a spherical droplet in shear flow.

$$F_{\text{drag}} \sim \eta_c \dot{\gamma} a^2$$

We define the dimensionless capillary number, *Ca*, as the ratio of the drag force to the interfacial force:

$$Ca = \frac{F_{\text{drag}}}{F_{\text{interfacial}}} = \frac{\eta_c \dot{\gamma} D_d}{2\Gamma_{12}} \tag{13.8}$$

where:

D_d	=	droplet diameter (=2*a*)

When *Ca* exceeds a critical value, Ca_{critical}, the droplet breaks up because the drag force exceeds the force holding the droplet together. This hydrodynamic instability was first proposed by Taylor [50]. Taylor found that Ca_{critical} for a Newtonian droplet imbedded in a Newtonian fluid is a function of the viscosity ratio (droplet viscosity to the continuous phase viscosity ratio): η_d/η_c. This has since been confirmed by other investigators and is illustrated in Fig. 13.16.

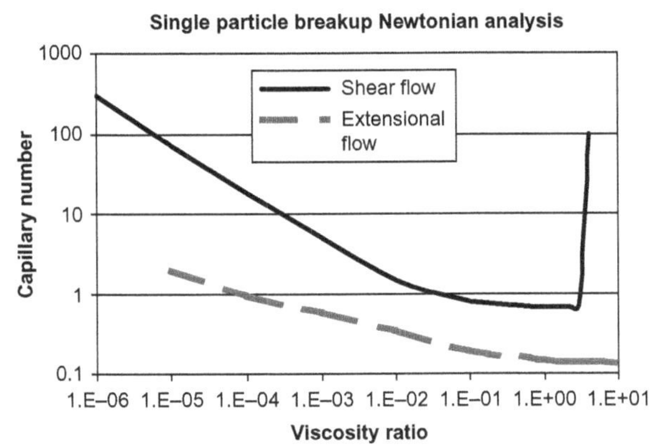

Figure 13.16 Critical Capillary Number for the breakup of Newtonian droplet in shear and elongational flow. *Source: Data taken from Grace [51].*

In a shear flow, Ca_{critical} reaches a minimum when the viscosity ratio is 1. When the viscosity ratio exceeds about 3.5, the droplet cannot be broken up in a shear flow, as indicated by the rapid rise in Ca_{critical}. The bottom curve in Fig. 13.16 shows the relationship for Ca_{critical} in elongational flow. Elongational flow is more effective for breaking up droplets than shear flow; Ca_{critical} is lower and droplets can be broken up even at high viscosity ratios.

For transient flow, a different mechanism for drop breakup has been proposed by Tomotika [52], based on Rayleigh's instability theory. Here, the droplet becomes an elongated ellipse or cylinder that, upon cessation of flow, breaks up into small droplets due to capillary disturbances, provided that the wavelength of these disturbances is greater than $2\pi a$ [53,54].

The single droplet analysis gives us considerable insight into dispersion in polymer blends (Fig. 13.17). Typically, we want the minor

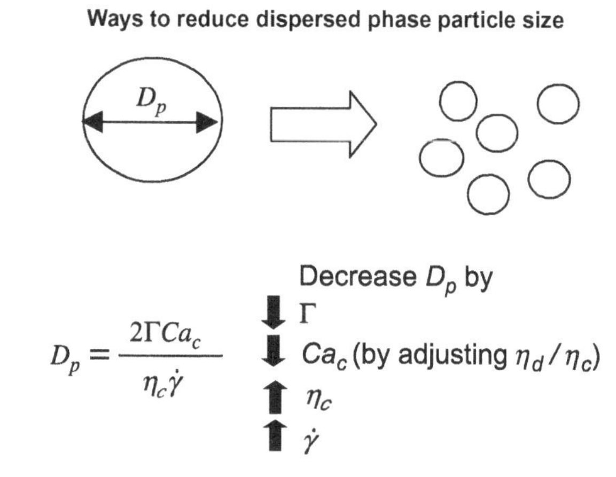

Figure 13.17 Learnings from single droplet mechanics.

component domain size (D_d) of the blend to be small. For example, for good clarity, the dispersed phase should be less than the wavelength of light, about 0.3 μm. For toughening, small soft rubber domains help prevent cracks from propagating. Eq. (13.8) shows we can decrease the droplet size by:

- increasing the continuous phase viscosity,
- increasing the shear rate, and
- decreasing the interfacial tension.

We can further reduce the droplet size by matching the polymer viscosities (a good guideline is to choose a viscosity ratio between 0.01 and 2) and using a mixing device that imparts elongational flow. The interfacial tension can be minimized by reducing the difference in solubility parameters (Eq. (13.7)) or by introducing specific interactions.

Compatibilizers are sometimes used to reduce the interfacial tension between polymers. Block or random copolymers often make good compatibilizers since they can be designed to contain two functionalities, each compatible with one of the polymers being blended. An example is using styrene–ethylene–butadiene–styrene copolymer to compatibilize HDPE and PS blends [55]. Another approach is to add functional groups to the compatibilizer that react with one of the polymers. An example is using an ionomer to compatibilize PA (polyamide or nylon) and PE blends. The acid groups in the ionomer react with the PA amine end groups, and the ethylene backbone is compatible with PE [56]. Another example is to use maleated polyolefins to compatibilize nylon or EVOH containing recycle streams [57].

Although the breakup of a single Newtonian droplet in a Newtonian fluid is well understood, there are several difficulties in extending the analysis to polymer blends. The first is that polymer melts are typically non-Newtonian in their flow behavior (non-Newtonian fluids have viscosities that vary with shear rate and often exhibit elastic effects, such as normal stress differences and extrudate swell). Although several droplet breakup studies for non-Newtonian fluids have recently been published, this phenomenon is still not as well understood as the Newtonian case. Table 13.3 summarizes many of these studies. One aspect of non-Newtonian behavior is the polymer melt elasticity, characterized by the first and second normal stress differences (see chapter 6 "Rheology of Molten Polymers" for an introduction to melt elasticity, or consult a rheology text such as Dealy and Wissbrun [58] for a broader discussion). Some investigators have used the dynamic storage modulus, G', obtained from dynamic mechanical analysis, as a way to characterize elasticity. G' is proportional to the first normal stress difference but only at low shear rates. At higher shear rates found when processing polymers, G' may underestimate the polymer elasticity. In general, it has been found that when the droplet is more elastic than the continuous phase, it is more difficult to disperse. An elastic force aids in holding the droplet together, resulting in larger Ca_{critical} values and larger droplet sizes. Conversely, when the continuous phase is more elastic than the droplet, it is easier to break up the droplet because the matrix resin elasticity adds to the drag force on the droplet to break it up. Quantification and modeling of this behavior in flow regimes typical of polymer processing is still in its infancy.

Experimental studies with non-Newtonian fluids have also revealed different droplet breakup mechanisms. A Newtonian droplet immersed in a Newtonian fluid breaks up via Taylor [50] and Tomotika [52] instabilities (Fig. 13.18). Under some conditions, non-Newtonian fluids also break up in this manner, but other mechanisms have also been observed. In some cases, the droplet flattens in the flow direction and becomes elongated perpendicular to the flow [65,73]. The ends of highly elongated particles find themselves in different planes with respect to flow and are torn apart by the velocity differences. At very high viscosity ratios, greater than the 3.5 cutoff for Newtonian

Table 13.3 Viscoelastic Droplet Deformation in a Viscoelastic Matrix Under Shear or Elongational Flow Studies

Source	Expt or Theo	Droplet	Matrix	Flow	Result
Flumerflet [59]	Expt	Newtonian	Viscoelastic	Shear	Minimum droplet size and critical shear rate for breakup increases with matrix elasticity
Tavgac [60]	Expt	Viscoelastic	Viscoelastic	Shear	Effect of elastic matrix depends on the viscosity ratio: when ratio is small, matrix elasticity stabilizes the droplet when ratio is high elasticity helps break up the droplets
Gauthier et al. [61]	Expt	Viscoelastic	Newtonian	Shear	Small viscosity ratio: similar to Newtonian drop in Newtonian matrix. High viscosity ratio: $Ca_c > Ca_{c\text{-Newtonian}}$
Parabodh and Stroeve (in Utracki and Shi) [54]	Expt	Viscoelastic	Newtonian	Shear	For viscosity ratio < 0.5, droplet viscoelasticity has a stabilizing effect
					For viscosity ratio > 0.5 droplet viscoelasticity has a destabilizing effect
DeBruijn [62]	Expt	Viscoelastic	Newtonian	Shear	$Ca_{critical}$ for elastic droplet is slightly higher than for Newtonian droplet
Elmendorp and Maalcke [63]	Expt	Viscoelastic	Viscoelastic	Shear	Viscoelastic drop deformation in Newtonian matrix decreases with increasing drop elasticity
					Newtonian drop deformation in viscoelastic matrix increases with increasing matrix elasticity
					Problems with quantifying the behavior due to fluid shear thinning
Varanasi et al. [64]	Expt	Viscoelastic	Newtonian	Shear	At any viscosity ratio $Ca_{critical}$ increases with increasing droplet elasticity
Levitt et al. [65]	Expt	Viscoelastic	Viscoelastic	Shear	Studied PP droplets in PS matrix at different viscosity and elasticity ratios Shear flow at 1 s^{-1}. For high elastic matrix found that the droplet widened in the direction perpendicular to flow. The width of the flattened drops depended on the differences in second normal stress differences between the phases
Han and Funatsu [66]	Expt	Viscoelastic	Viscoelastic	Elong	Viscoelastic drops are more stable than Newtonian drops in both Newtonian and viscoelastic matrices
Milliken and Leal [67]	Expt	Viscoelastic	Newtonian	Elong	Viscoelastic drops with viscosity ratios less than 1 have smaller deformation and $Ca_{critical}$ than Newtonian drops. For viscosity ratio > 1 the viscoelastic drop deformation behavior is similar to Newtonian drops
Delaby et al. [68]	Expt	Viscoelastic	Viscoelastic	Elong	For negligible interfacial tension, viscoelastic drops deform less than the surrounding media when the viscosity ratio is less than 1 and more when the viscosity ratio is greater than 1

Miejer and Janssen [69]	Expt	Not disclosed	Not disclosed	Elong	At small viscosity ratios, the droplet deformation in planar elongational flow resembles that for the matrix
Chin and Han [70]	Expt	Viscoelastic	Viscoelastic	Elong	Higher droplet elasticity results in less deformation compared to the matrix
Shanker et al. [71]	Expt	Viscoelastic	Viscoelastic	Elong	Higher droplet elasticity results in less deformation compared to the matrix
Mighri et al. [72]	Expt	Viscoelastic	Viscoelastic	Elong	Used Boger fluids which have elasticity and nonshear thinning viscosity behavior
					For a given elastic matrix fluid, increasing the droplet elasticity decreased droplet deformation relative to the surrounding media. Droplet deformation increases with increasing matrix elasticity. Defined k' as first normal stress difference ratio divided by the viscosity ratio. When $k' < 0.2$, the matrix elasticity has a greater effect on deformation than the drop elasticity. The opposite is true with $k' > 0.2$. The study was conducted over a fairly narrow viscosity ratio range (0.5–1.1); shear rates were not disclosed
Mighri et al. [75]	Expt	Viscoelastic	Viscoelastic	Shear	Used Boger fluids, which have elasticity and constant viscosity versus shear rate behavior
					Shear rates not disclosed; viscosity ratios ranged from 0.2 to 1.1. Defined k' as first normal stress difference ratio divided by the viscosity ratio. For high matrix elasticity ($k' < 0.37$), the elastic drop deformation in an elastic matrix was higher than for the Newtonian case with same viscosity ratio and interfacial tension. In some cases, droplet widening in the direction perpendicular to flow was observed. For $k' > 0.37$, elastic drops deform less than the Newtonian case. The critical shear rate for droplet breakup increases with increasing k'. For $k' < 4$, $Ca_{critical}$ increases rapidly with k'; for $k' > 4$, $Ca_{critical}$ levels off at 1.75
Mighri and Huneault [73]	Expt	Viscoelastic	Viscoelastic	Shear	Studied model systems and PS/PE system with viscosity ratio between 5 and 20 under relatively high shear rate ($1-20\ s^{-1}$) in shear flow. In PS/PE system, the PS drop widened in the direction perpendicular to flow which contributed to droplet breakup (the ends of the highly elongated droplet are in different flow planes). At high shear, a second breakup mechanism different from the Newtonian case was observed: attrition from the surface. An EPR/PP system had similar breakup mechanisms as the PS/PE system despite a $10\times$ difference in interfacial tension

(*Continued*)

Table 13.3 Viscoelastic Droplet Deformation in a Viscoelastic Matrix Under Shear or Elongational Flow Studies—*Cont'd*

Source	Expt or Theo	Droplet	Matrix	Flow	Result
Van Oene [76]	Theo	Viscoelastic	Viscoelastic	Shear	Developed an expression for the dynamic interfacial energy to account for viscoelasticity: $$\Gamma_{12} = \Gamma_{12}0 + D_p/12 * [(N_{11} - N_{22})_d - (N_{11} - N_{22})_m]$$ where: $\Gamma_{12}0$ = the interfacial tension in quiescent flow; $(N_{11} - N_{22})$ = the first normal stress difference. Droplet elasticity greater than the matrix stabilizes the droplet; matrix elasticity greater than the droplet destabilizes the droplet
Greco [120]	Theo	Viscoelastic	Viscoelastic	Shear and Elong	Assumed the materials are simple second-order fluids, which includes elastic effects for slow flows. Used a perturbation method to analyze the first normal stress difference effect on droplet shape for small drop deformations
Maffettone and Greco [121]	Theo	Viscoelastic	Viscoelastic	Shear and Elong	Developed a phenomenological model for the dynamics of a drop immersed in an immiscible fluid. Assumed the drop is ellipsoidal. Either or both fluids may have elasticity. Found that elastic drops deform less than Newtonian drops. Elastic drops in Newtonian matrix under shear are forbidden to break up at lower viscosity values than the Newtonian case. In elongational flow, the drop breakup is easier when the matrix is elastic and more difficult when the droplet is elastic
Lerdwijit-jarud et al. [122]	Expt	Viscoelastic	Newtonian	Shear	Studied deformation and breakup of viscoelastic drops (Boger fluids) in a nearly Newtonian matrix. Viscosity ratio controlled to around 1.0. Shear rates up to 5 s^{-1} used
					The deformation decreases for isolated drops at constant *Ca* as the drop elasticity increases. They did not observe droplet widening in the transverse to flow direction and suggest widening is influenced by factors other than first normal stress difference (such as second normal stress difference, shear thinning or a shift of the critical condition to lower *Ca* when the matrix is elastic). $Ca_{critical}$ increases with increasing droplet elasticity. For a 10% blend, the steady-state *Ca* is less than $Ca_{critical}$ for an isolated drop (they suggest this may be due to flow differences in concentrated suspensions). The steady-state *Ca* was found to increase monotonically with the dispersed phase first normal stress difference

Expt: experimental; theo: theoretical; elong, elongation.

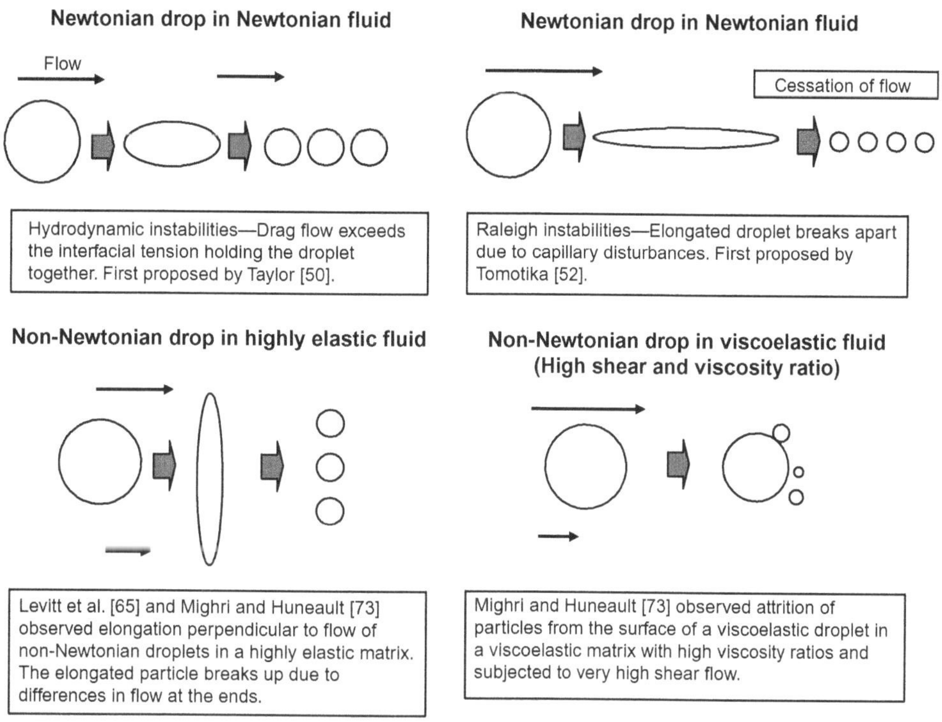

Figure 13.18 Single droplet breakup mechanisms in shear flow.

droplets, Mighri and Huneault [73] observed breakup via attrition at the droplet surface.

As the droplet concentration increases, the breakup mechanism becomes more complex. Utracki and Shi [54] and Macosko [74] found that when the polymer concentration exceeds 0.5–1%, the droplet size after shearing was much greater than that predicted by the single-drop experiments. As the concentration increases, the probability that droplets will collide and coalesce increases. The coalescence kinetics are not well understood and are thought to be critically important to the final blend morphology. Utracki and Shi [54] observed that the same factors that enhance droplet breakup, namely high shear rates and reduced droplet viscosity, favor coalescence. Recently, several investigators [74, 123, 124] have proposed that compatibilizers act to inhibit coalescence by providing a protective shell around the droplet (Fig. 13.19). The shell acts to repulse other droplets. Thus, dispersion stabilization may be more important to the droplet size than reducing interfacial tension. Indeed, Mighri and Huneault [73] found that an EPR–PP droplet–matrix system with very low interfacial tension broke up in a similar manner as a PS–PE system with an order of magnitude greater interfacial tension. They suggest that interfacial instabilities may not

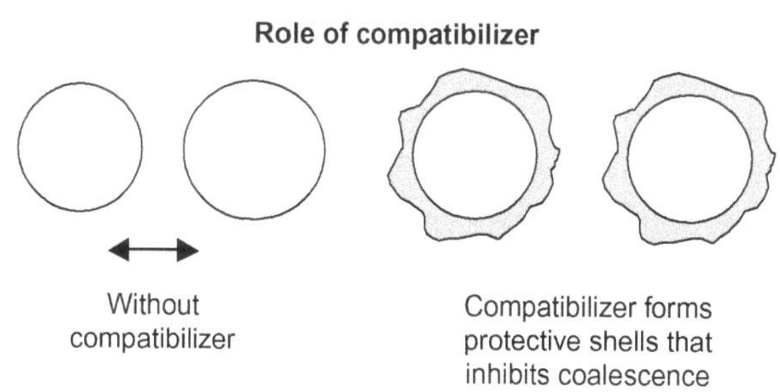

Figure 13.19 Role of compatibilizers in reducing coalescence in polymer.

be required for non-Newtonian droplets to break apart.

As the minor component concentration is increased, the morphology may change. At high enough concentrations, the minor component becomes the continuous phase. Fig. 13.20 shows the different morphologies possible for HDPE/PS blends studied by Bourry and Favis [55]. At low PS levels, the PS forms droplets in the HDPE matrix. As the PS concentration is increased, fibers are formed. At about 70% PS, the morphology becomes cocontinuous; the HDPE and PS phases form an interpenetrating network structure where both phases are continuous. Higher PS levels results

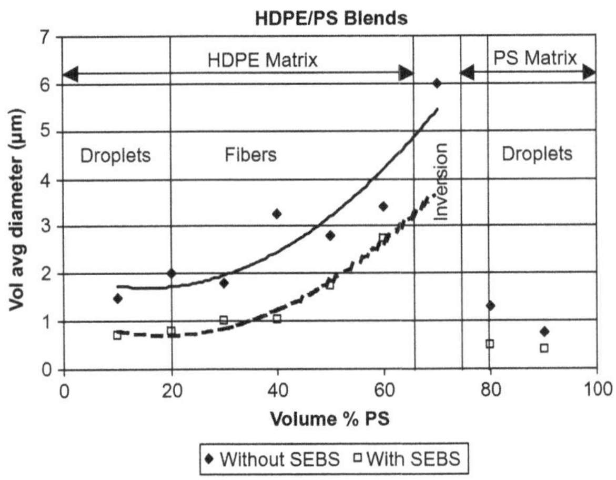

Figure 13.20 Blend morphology as a function of concentration. PS-HDPE blends from Baurry and Favis [55].

in HDPE droplets in a PS matrix. Note how the SEBS compatibilizer decreases the domain size.

Another concern in applying single droplet studies directly to polymer blends is uncertainties about the flow field in the mixing device and how to calculate the appropriate viscosity ratio. Polymer mixing devices, such as extruders, have complex flow fields involving both shear and elongational flow. The flow fields are typically nonuniform; the shear rate varies across the screw channel (Fig. 13.4). Further complicating matters are temperature changes along the extruder. Given these complexities, Lyngaae-Jorgensen [77] proposes measuring the Ca_{critical} versus viscosity ratio curve for the given device.

One must also consider the fact that the polymer melt viscosity varies with temperature and shear rate. Generally, there are two schools of thought on calculating the viscosity ratio. Some argue that the zero shear viscosities should be used. Others suggest using the viscosity at a representative shear rate for the process. However, it is stress, not shear rate, that drives dispersion and droplet breakup. Stress is determined by the flow field and is continuous across an interface, whereas the shear rate is discontinuous. A better method is to plot the viscosity as a function of stress (stress is equal to the viscosity times the shear rate) and compare the blend components at a representative stress for the process. This is illustrated in Fig. 13.21. Here, polymers A and B are the minor and major components, respectively. At a constant shear rate equal to 100 s^{-1}, the viscosity ratio is 1.6. But, at constant stress, the ratio is 2.6, very close to the 3.5 cutoff for shear flow.

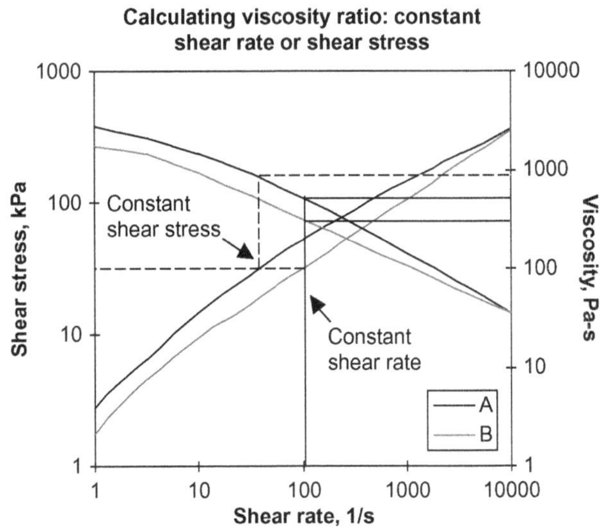

	Viscosity, A	Pa-s B	Ratio A/B
Constant shear rate	520	320	1.6
Constant shear stress	840	320	2.6

Figure 13.21 Calculating viscosity ratio.

The polymer temperature changes along the extruder as the polymer becomes molten and is conveyed to the exit. Huneault et al. [78] showed that, for a PS/HDPE blend, the viscosity ratio varied by six orders of magnitude depending on the temperature (Fig. 13.22).

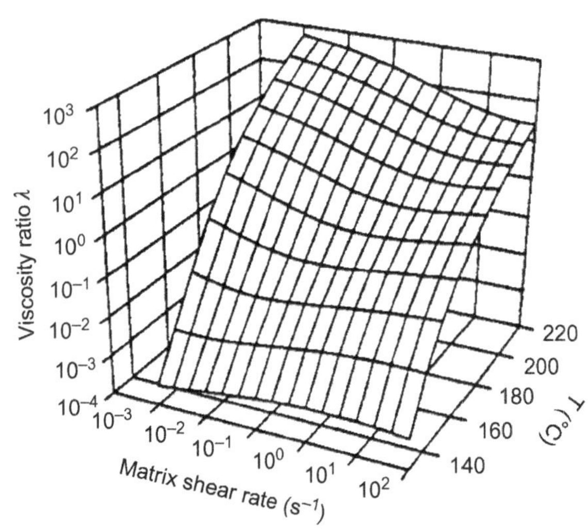

Figure 13.22 HDPE-PS blend viscosity ratio variation as a function of shear rate and temperature. *Source: Taken from Huneault et al. [78]—Courtesy of the Society of Plastics Engineers.*

Computing the viscosity ratio just at the final extrusion temperature may be misleading. They found that, for this blend, the mixing was much better than what they would have expected just by looking at the viscosity ratio at the final temperature (200°C). Notice in Fig. 13.22 that the viscosity ratio is sometimes less than one and other times greater than one. This indicates that, for the two polymers, the viscosity temperature dependence is not the same, another complication in computing the viscosity ratio.

So far, we have assumed that the polymers are both molten at the time they are mixed. This is generally not the case as mixing can begin during melting. Ghosh et al. [79] studied the softening/mixing of two similar amorphous polymers and found that laminar sheets or striations form that subsequently break up into droplets. Lindt and Ghosh [80] found that the striations come from the single-screw extruder melting process. A thin molten layer forms between the pellet solid bed and the barrel or screw surface caused by frictional heat and heat transferred from the barrel (see Fig. 13.23 for the solid bed melting model). The high stress and deformation rates in this region transform the molten polymer into laminar morphologies as the pellets melt. The domain size decreases from a few millimeters (pellets) to about 50–100 μm due to the melting process. They made measurements as well as calculations that confirmed these striations and abrupt domain thickness changes for blends of both rheologically similar and dissimilar resins. Huang et al. [81] studied blends of two semicrystalline polymers (PP and PA6) by sampling along the length of an extruder. They observed striations followed by droplet formation, the nature of which depended on the viscosity ratio.

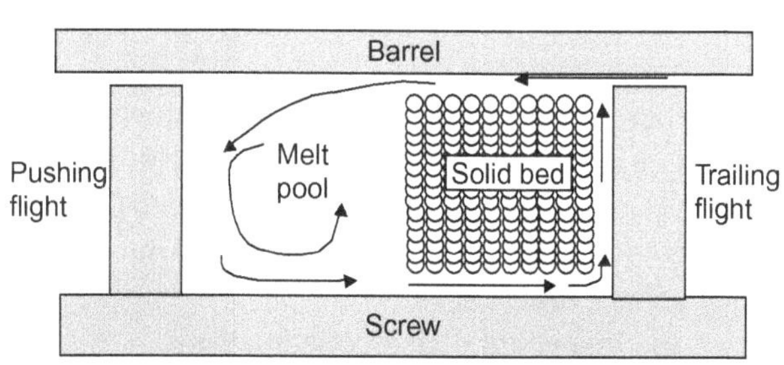

Figure 13.23 Single-screw extrusion solid bed melting model. Melting above and below the solid bed causes the minor component to form thin layers or striations in a two-component polymer blend. *Source: Adapted from: Lindt and Ghosh [80].*

Benkreira et al. [82] found that, when mixing a masterbatch into a host polymer, most of the mixing occurred during melting and little thereafter. They proposed a laminar mixing model where the striation stretching during melting leads to a reduction in dimensions. Scott and Macosko [83] describe how component melting and softening in a batch mixer leads to the domain size reduction. They found sheets and ribbons are formed first, which become unstable due to flow and interfacial tension effects. Holes form, then a lace-like structure, followed by irregular shaped particles and finally spheres (Fig. 13.24).

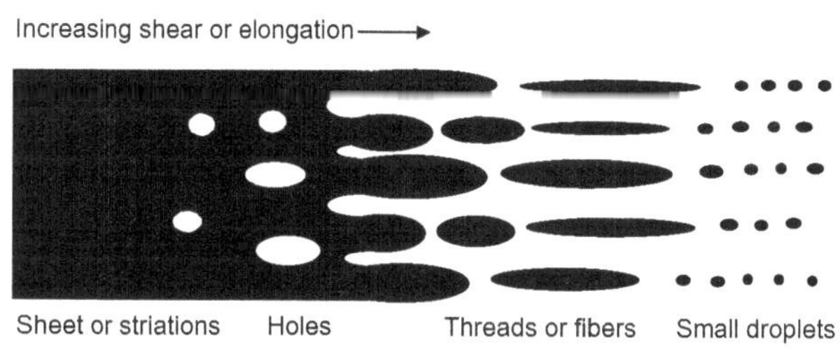

Figure 13.24 Morphology development during melting and subsequent shear in an extruder. Striations or sheets form during melting. Shear and stress cause holes to form, followed by threads and particle breakup. This process leads to large changes in the minor phase morphology.

Breakup is driven by interfacial instabilities. Burch and Scott [84] found similar behavior for miscible polymers during the initial mixing stages and proposed instabilities due to dynamic interfacial tension as the cause. Willemse et al. [85] found that sheets or striations formed during single-screw extrusion and that the final droplet size depended on the striation dimensions at the start of breakup rather than the capillary number. The sheet to droplet breakup during shear is very effective at dispersing the minor phase, more so than the elongated droplet to subdroplet mechanism found in single drop experiments.

In a single-screw extruder, Tyagi and Ghosh [86] followed a PP–EVA blend as it developed its morphology using freeze experiments pioneered by Maddock [19]. In these experiments, after the extruder has reached steady-state, the screw is stopped and the polymer is quickly quenched. The screw is pulled, and samples from the screw channels are analyzed. Tyagi and Gosh found that striations formed in the feed zone. The dimensions quickly diminished along the channel length. In the compression zone, the shear and elongation increases, causing the striations to break up. The droplets that form are an order of magnitude smaller in size than the striation thickness before breakup. The breakup into droplets requires a step-up in shear and stretching. They conclude that, for an extruder to be an effective mixer, periodic flow reorientations should occur. This concept is used in high-performance screw designs where the polymers flows through tight clearances.

Li et al. [87] found that the minor phase melts faster during mixing when a compatibilizer is present. They attribute this to faster heat transfer between the phases due to their intimate contact. Faster melting can affect the final blend morphology.

Shih et al. [88] studied the solid–melt transition for many polymer blends in a batch mixer equipped with a glass window. They found that blends go through four stages from solid to melt:

- Elastic solid pellets: Below the glass transition temperature (Tg) or melting point (Tm) the solid pellets were observed sliding in the mixer. Torque was low.
- Deformed solid pellets: As the temperature increased, some components began to soften and deform. The torque started to rise.
- Transition material: which took on several different forms including
  - a fluid with suspended solid particles,
  - fractured or semi-fluid material, and
  - and a dough-like material.

  This was the transition zone between solid and liquid. Different blends took on different transition phases. The temperature increased sharply and then remained constant near the polymer melt temperature (Tm) if it was crystalline. Torque rose sharply
- Viscoelastic material: The typical liquid appearance was observed. Temperature rose to about 20–50°C above the matrix resin Tm or Tg, and the torque decreased.

During the transition phase, they observed several phase transitions that aided in the mixing process. Shih [89] and others [90,91] found that phase inversions are to be expected when the minor component melts first. The molten minor component surrounds the still solid major component. As the major component melts, it becomes the continuous phase. This inversion is accompanied by a spike in the mixer torque and dramatically reduces the minor component phase size. They also describe how phase inversion can explain some unique morphologies, such as major component regions imbedded within the minor component domains.

The phase inversion onset is a function of the volume fraction ratio and viscosity ratio. A simple expression is given by [55,92]

$$\frac{\eta_1}{\eta_2} \sim \frac{\Phi_1}{\Phi_2} \tag{13.9}$$

where:

Φ	=	volume fraction

Two polymers with the same viscosity will undergo a phase inversion when the minor component reaches about 50% of the blend composition. If the minor component viscosity is substantially less than the major component, the minor component can become the continuous phase. This is the case when the minor component melts first. As the major component melts and its viscosity decreases, then conditions no longer favor the minor component as the continuous phase and the phase inversion occurs.

13.5 Morphology Development in Blown Film

Several morphologies have been described for immiscible blends in blown film, including elongated particles, such as fibrils and ribbons. Forming elongated morphologies is a two-step process:

- Sphere-like dispersed domains are formed during extrusion (in the absence of blowing or drawing). Shear flow fields in the extruder and die dominate the morphology. The domain size is a function of:
  - the dispersed phase concentration,
  - the viscosity ratio of the dispersed phase to the continuous phase,
  - interfacial tension,
  - continuous phase viscosity, and
  - shear stress.

- Stretching and orienting the spherical particles in the elongational flow fields at the extruder exit (blowing and drawing) form elongated structures. The morphology is influenced by:
  - initial domain size,
  - polymer elasticity,
  - minor component percentage, and
  - draw ratio and other film blowing parameters.

Fig. 13.25 shows the elongated morphology in a blown film. Here polybutene-1 (PB) has been added to a polyolefin to create a peelable sealant. Pirtle et al. [15] suggest that too much PB phase orientation in a peel-seal blend can produce unwanted stringy seals. Thus, morphology control is important for packaging applications. We will now examine some factors that influence the morphology in more detail.

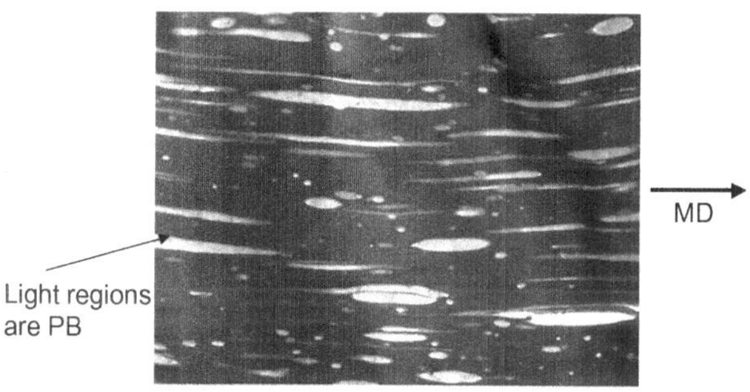

Figure 13.25 Elongated second phase (PB-1 in a polyolefin matrix) found in blown film. *Source: Photograph courtesy of DuPont.*

13.5.1 Viscosity Ratio

A viscosity ratio near one typically gives the smallest particle domains. These are difficult to elongate; the smaller the particle size, the harder it is to deform the particle. David et al. [93–95] studied PE–styrene block copolymer blends. They found that the dispersed phase viscosity has to be less than the matrix to obtain elongated morphologies. For viscosity ratios greater than one, the dispersed phase cannot be deformed enough to create the elongated morphologies.

13.5.2 Interfacial Tension

As described by Eq. (13.8), the lower the interfacial tension, the smaller the particle diameter (in the extruder and die). Smaller particles are more difficult to elongate. Adding compatibilizers to the blend tends to reduce the particle size and make elongated morphologies (such as laminar structures) more difficult to obtain [96–98].

13.5.3 Minor Phase Concentration in Blend

Increasing the dispersed phase volume fraction (ϕ_d) generally gives larger particle sizes, resulting in more fibril or laminar structures. The larger particle sizes may be due to coalescence, which becomes more significant as the concentration increases. Lopez-Barron et al. [98] studied PA6–LDPE blends where LDPE was the continuous phase. The PA6 domain size increased with increasing concentration for uncompatibilized blends and was constant for a compatibilized system. The compatibilizer reduces interfacial tension and acts to reduce coalescence. David et al. [93] found cocontinuous fibrils formed as ϕ_d increased. This was determined experimentally by extracting films with a solvent that selectively dissolves the minor component. As the volume fraction was increased, more minor phase was extracted, suggesting cocontinuous morphologies were forming.

David et al. [93,95] found that the onset of cocontinuous fibril morphology occurred at a lower volume fraction than predicted by Eq. (13.9). They attribute this to elongational effects.

13.5.4 Polymer Elasticity (non-Newtonian behavior)

Getlichermann and David [95] found that a viscosity ratio less than one was not sufficient to create elongated morphologies. If the dispersed phase was Newtonian in behavior, elongated morphologies were not observed after blowing and drawing. The polymer's non-Newtonian (elastic) behavior helps to stabilize the "threads" that forms during elongation, allowing more elongation without the threads breaking up into small particles.

They suggested that dispersed phase tension thickening is one attribute that should help stabilize the thread and help create elongated morphologies. Tension thickening refers to elongational viscosity measurements, which are difficult to conduct. Linear polymers, such as LLDPE, have transient elongational viscosity that decreases with time (tension thinning), whereas the elongational viscosity of polymers with long chain branching (such as LDPE) tends to increase with time (tension thickening).

13.5.5 Extruder RPM

Eq. (13.8) predicts that, as extruder speed is increased, the increased shear rate should reduce the droplet size. As discussed previously, smaller particles formed in the extruder and die are more difficult to elongate after they exit the die. Lee and Kim [97] found just the opposite for a LDPE–EVOH blend. They attributed this to the shorter extruder residence time at higher rpm. The EVOH, which melts at temperatures 50°C higher than the LDPE, has less time to melt fully and be dispersed. This gives rise to a larger particle size at the extruder exit and a greater tendency toward elongated laminar morphology.

13.5.6 Extruder Temperature

The extruder and die temperatures affect the morphology through their impact on the viscosity of both components. As the temperature is increased, η_c decreases. As described by Eq. (13.8), this allows less stress to be applied to breaking up the droplets, giving larger particle sizes. Also, the polymers' viscosity temperature dependence may differ, altering the viscosity ratio.

13.5.7 Shear Stress in Extruder, Adaptor, and Die

During pressure-driven flow in an extruder, adapter tubes, and die, there is a shear stress distribution across the flow channel. The highest shear stresses occur at the wall. The wall also experiences the longest residence times since the flow rate at the wall is low. Thus, there may be a morphology distribution as a result of the stress differences. This was found to be true by Lee and Kim [97] in LDPE–EVOH blends, and it has also been documented for blends involving PB-1 and LDPE [99].

Lopez-Barron et al. [98] observed a core-skin effect with uncompatibilzed blends of PA6 and LDPE, with the skin having small elongated domains and the core having larger particles. They attribute this to a discontinuity at the die wall at the die exit. The polymer near the die wall in the die flow channel has zero or nearly zero velocity (no slip boundary condition) and accelerates almost instantaneous to the free surface velocity at the die exit. This creates an intensive stretching that may lead to the observations. Their compatibilized blends showed no such effect.

13.5.8 Screw Design

For laminar morphologies, the literature teaches not to overmix the blend [9]. Careful screw design may be needed to accomplish this.

13.5.9 Draw Ratio

Before drawing (or blowing) (see Fig. 13.26 for blown film process description), the minor component morphology is typically spherical as it exits the extruder [93–98,100]. It is the elongational flow fields imposed by the drawing and blowing process that give rise to the highly elongated structures. The degree of elongation depends on the process. David [93] found that the convergence/divergence in a capillary geometry only gave rise to elongated ellipsoids. A flat die gave rise to fibrils in the machine direction (MD) and a blown film die gave fibrils and ribbons with two-dimensional orientation in the machine and transverse direction (TD) as well as cocontinuous fibrils.

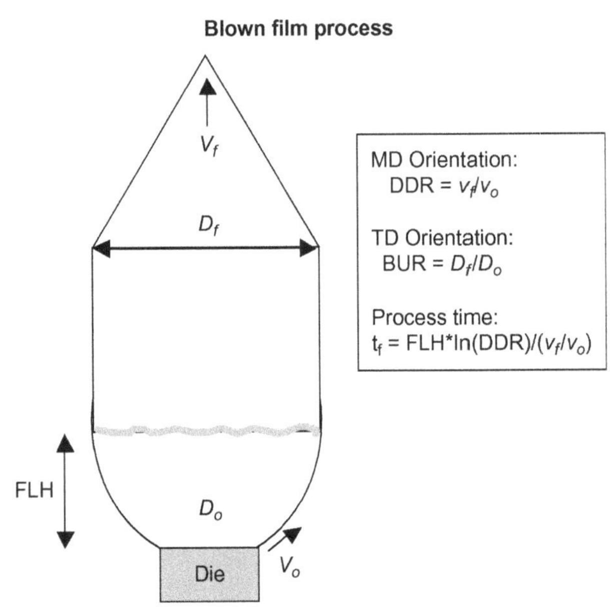

Figure 13.26 Blown film process.

The draw ratio in the MD is characterized by the draw down ratio (DDR), which is defined as the

haul-off speed divided by the polymer melt velocity as it exits the die. The blow-up ratio (BUR) characterizes the draw ratio in the TD or hoop dimension. BUR is defined as the final bubble diameter divided by the die diameter. The DDR is inversely proportional to BUR times the final film thickness divided by the die gap (Eq. (13.10)):

$$\mathrm{DDR} = \frac{V_\mathrm{f}}{V_0} \approx \frac{\text{Die Gap}}{\text{BUR} \times \text{Film Thickness}} \tag{13.10}$$

where:

V_f	=	haul-off speed
V_0	=	velocity of polymer at die exit

Thus, DDR decreases with increasing BUR and film thickness and decreasing die gap. (In Eq. (13.10), we have ignored the density change between solid and melt.)

David [93] found that the fibril diameter is not a straightforward function of draw ratio. Without any drawing, the particles were spherical with a 0.5 μm diameter. At a 5 and 20 DDR, the fibril diameters were 0.1 and 0.2 μm, respectively.

Robledo-Ortiz et al. [100] showed that the stretching force increases with increasing draw ratio and decreasing frost line height. The minor phase domain elongation (PA6 in their case) increased in the MD with increasing stretching force. The elongation in the hoop and radial (thickness) dimensions were balanced.

13.5.10 Frost Line Height and Process Time

Robledo-Ortiz et al. [100] showed that stretching force increased strongly with decreasing frost line height. The PA6 MD elongation in their PA6-LDPE blends increased accordingly. The frost line height is a component of the process time. Process time in the blown film process is defined as the time it takes the polymer to begin to freeze once it has exited the die. It is proportional to the frost line height and inversely related to the haul-off speed (Eq. (13.11)):

$$t_\mathrm{f} = \frac{FLH}{V_\mathrm{f} - V_0} \ln\left(\frac{V_\mathrm{f}}{V_0}\right) \tag{13.11}$$

where:

t_f	=	process time
FLH	=	frost line height
V_f	=	haul-off speed
V_0	=	polymer velocity at die exit

The effect of process time on the blend morphology has not been fully addressed in the literature. It is only within the last few years that this parameter has been accepted as an important scale-up parameter for blown film. It encompasses critical aspects of the film blowing process, including the cooling and crystallization time. More important for this discussion is that the process time is inversely proportional to the elongation rate. Stress is related to the draw ratio and to the draw or elongation rate. Therefore, process time is inversely related to the stress imposed on the process [101].

13.6 Dispersion of Rigid Particles and Nanocomposites

Dispersing inorganic and other particles in polymers is not well studied. As White [37] explains, this may be because breaking up agglomerates is hard to quantify, particularly since most fillers have a particle size distribution. Generally, it is thought that high stresses are needed to achieve dispersive mixing. Such stresses are typically not found in a single-screw extruder. Hence, most mineral-filled polymers are manufactured in specialized compounding equipment as described earlier. In packaging applications, blending fillers is generally limited to letting down masterbatches. Here, the filler, pigment, or additive has been predispersed and simple distributive mixing is all that is required.

One noteworthy aspect of dispersing rigid particles into polymers is particle attrition. Fillers, such as glass fibers, mica, and clay, may have long aspect ratios that are important for the blend properties. Too much dispersive mixing can reduce their aspect ratio, degrading performance. Filler abrading the mixing device surfaces can also cause excessive wear. Adding mineral filler in a downstream stage where the polymer is molten rather than to the first stage with the solid pellets can significantly reduce abrasion and attrition.

A key question when dealing with pigments is how do you know if good mixing has been achieved? One commonly used method is to compare a sample against a visual standard. Spectroscopic techniques have also been used, as well as microscopic image analyses [82]. Various indices, such as striation thickness, variance in minor component concentration, and segregation scales, have been used, as discussed by McKelvy [102]. More recently, computer image analysis (red-green-blue correlations) has been used. Translating such color analysis to a quantitative distributive mixing index has been difficult. Recently, Alemaskin et al. [103,104] have made progress in developing such an index by employing Shannon entropy. They have used both computer simulation and experiments to verify the method. They simulated mixing two ABS colors in a conventional single-screw extruder metering section using a numerical particle tracer analysis. The results were compared with extrusion experiments under similar conditions. Here, the extruder was stopped, rapidly quenched, and the screw pulled to obtain samples along the extruder length. The color homogeneity evolution was measured using digital computer imaging. Color homogeneity indices based on entropic considerations were computed for both the simulations and digital images, and they qualitatively agreed. Such an approach can potentially be used for scale-up, control, and process design optimization.

A polymer filled with nanoparticles is a special case. These blends are typically with inorganic particles that have nanometer dimensions. At levels between 3 and 5 wt%, they have been found to be effective in improving:

- stiffness,
- mechanical strength,
- barrier,
- electrical conductivity, and
- flame retardance.

Still in its infancy, two nanofillers have received the most attention, carbon nanotubes and clay. Carbon nanotubes have been found to impart good electrical properties for shielding and other potential applications. Because of their cost, except for some specialty electronics applications, carbon nanotube composites are not suited for packaging applications.

Of greater interest are clay nanocomposites [105–110, 125]. Most work to date has been done with montmorillonite clay, initially at Toyota. In its natural state, each montmorillonite particle is an agglomeration of many layers of nanosized platelets. The platelet length and width range from a few tenths of a micron to about 1.5 μm. Their thickness is only about 1 nm and is the reason they are considered nanomaterials. Their high aspect ratios give them their unique properties.

The platelets are hydrophilic and are held together at a distance of about 3.5 Å by attractive forces. Clay suppliers add surfactants to the clay to promote interplatelet expansion to about 20 Å. To obtain the best properties, the platelets must be completely separated from each other. This complete dispersion into the polymer matrix is known as exfoliation and is illustrated in Fig. 13.27. There are two typical ways to accomplish this. One is to introduce the clay

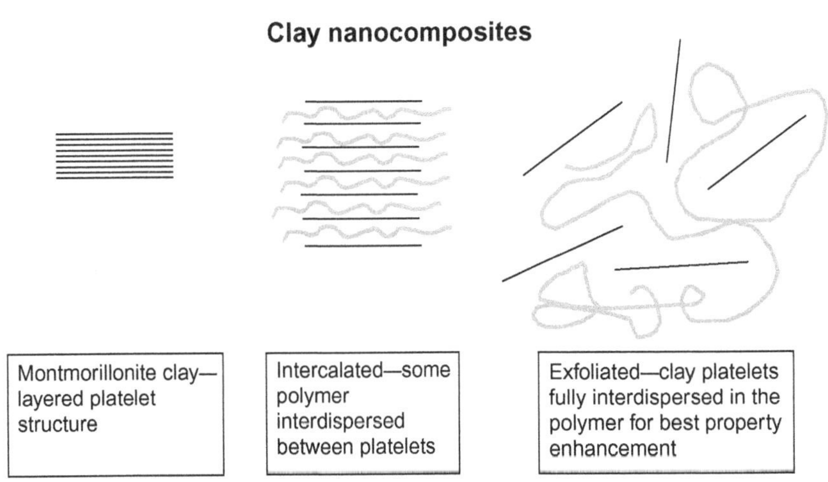

Figure 13.27 Illustration of states of clay nanocomposites.

during the polymerization process. The monomer is absorbed into the spaces between platelets. As the monomer polymerizes the platelets separate. This process is only amenable to certain polymers where the clay can be introduced during polymerization. Most often it has been applied to PA.

The second is to compound the clay particle with an already polymerized polymer and rely on the stress generated during mixing to separate the platelets. The surfactant selection is critically important; it must help turn the hydrophilic environment between the platelets into one that is compatible with the monomer or polymer. In some cases, compatibilizers are used to aid in the exfoliation. For example, in PP–montmorillonite nanocomposites, maleic anhydride grafted PP is often added as a compatibilizer. Nevertheless, so far it has proven difficult fully to exfoliate montmorillonite in a compounding process and is the focus of continued R&D. The first approach of polymerizing in the presence of the clay has been more successful.

Adding nanosized particles to the polymer restrains the molecule's movement, more so than in conventional fillers because of the extremely high surface area that is generated. The restrained motion enhances stiffness and heat deflection temperature. Because the individual platelets are small, toughness and optical properties do not suffer. Of particular importance for packaging applications is the potential for increased barrier performance, especially oxygen barrier. The orientation of the platelets is flexible, and their longer dimensions line up in the extrusion and orientation direction. The resulting platelet network structure creates a more tortuous path for gas molecules to diffuse through. Reductions in gas permeability of 60–80% for nylon or EVOH nanocomposites have been reported [97,105].

Nanoparticles have been used to change or stabilize the morphology of immiscible polymer blends. For example, Bose [111] found multiwalled carbon nanotubes (MWCNT)–compatibilized cocontinuous blends of PA6 and ABS. Tiawai and Paul [112] found that organo nanoclay particles stabilize and reduce the dispersed phase size of PP–PS blends. The nanoparticle shape (aspect ratio) has been found to influence which phase the particle ends up in and its ability to stabilize the interface [113].

13.7 Rheology of Polymer Blends

Utracki [114] provides a good review of the rheology of polymer–polymer blends. Miscible blends are easier to characterize than immiscible ones. Often a log-mean rule is used to estimate the miscible blend viscosity, although there may be some positive deviation:

$$\log(\eta_{\text{blend}}) = X_A \log(\eta_A) + X_B \log(\eta_B) \qquad (13.12)$$

where:

X_A	=	weight fraction of polymer A in the blend
X_B	=	weight fraction of polymer B in the blend

Eq. (13.10) is useful for estimating the viscosity or melt flow index (MI) for two grades of the same polymer. For example, a 70–30 wt%, LDPE1–LDPE2 blend where the LDPE1 MI is 4 g/10 min and LDPE2 MI is 12 g/10 min, yields:

$$\log(\text{MI}_{\text{blend}}) = 0.7\log(4) + 0.3\log(12)$$

$$\text{MI}_{\text{blend}} = 5.6\frac{g}{10\ \text{min}}$$

For immiscible blends, Lyngaae-Jorgensen [77] points out that steady-state data, such as capillary rheometry data, look superficially like homogeneous polymers. Unless they have been compatibilized [114], however, their behavior may be complex. They typically do not follow time–temperature superposition and mixing rules since the individual phases may behave differently. Their transient behavior and extrudate swell may also be complicated. The immiscible blend morphology is the driving force for this complex behavior.

Reactive polymer blends are a special case. Typically, the reactive blend viscosity increases due to cross-link formation that effectively increases the polymer MW and often times the interfacial contact area. An example is blending maleic anhydride modified PE with PA. The PA viscosity can increase by an order of magnitude. The anhydride reacts with the amine end groups on the PA, effectively reducing the interfacial tension between the PE and PA. This results in a smaller domain size of the dispersed phase (as given by the Capillary number in Eq. (13.8).

Smaller domain size increases the interfacial area and the opportunity for crosslinking.

Somc additives can cause gels and other problems when blended with a polymer. For example, the acid groups in acid copolymers and ionomers can react with a coupling agent used in some TiO_2 grades, causing gel and other problems. On the other hand, adding wax, tackifiers, and other low MW additives to polymers will reduce their viscosity.

How rigid fillers effect viscosity can also be complex. For very dilute solutions of rigid spheres in a Newtonian fluid, Albert Einstein derived the following result:

$$\eta = \eta_f(1 + 2.5\phi) \tag{13.13}$$

where:

η	=	viscosity of the suspension
η_f	=	viscosity of the suspending fluid, and
ϕ	=	volume fraction of the rigid spheres

The assumptions for this equation breakdown quickly for filled polymers as:

- the filler loading is not typically dilute,
- the filler is not spherical, and
- the polymer matrix is non-Newtonian.

Dealy and Wissburn [115] describe the following complexities:

- non-Newtonian effects caused by particle concentration and aspect ratio,
- buoyancy effects, and
- particle migration and agglomeration.

A number of equations have been proposed, including the following by Maron and Pierce [116]:

$$\eta = \frac{\eta_f}{[1-(\phi/A)]^2} \tag{13.14}$$

Here, A is a constant that is a function of the filler. Dealy and Wissbrun propose that it can be thought of as the maximum filler packing fraction. For example, the value for A for spheres is 0.68, close to the theoretical maximum packing fraction of 0.637. The value for A decreases with the increasing filler aspect ratio (0.44 and 0.16 for L/D's equal to 8 and 30, respectively). The value of A for filler with a nonuniform particle size distribution is greater than that for a uniform particle size. A nonuniform filler size distribution, therefore, can reduce the concentrated suspension viscosity.

There are other complications. One is the viscosity to choose when dealing with polymers, since η_f varies with shear rate. Dealy and Wissbrun point out that plotting the viscosity versus shear stress rather than shear rate, as noted earlier for polymer–polymer blends, gives the best results. At low shear rates, the filled polymer may exhibit a yield stress, below which the material behaves like a solid (and does not flow). The yield stress increases with filler concentration.

13.8 Conclusion

Blending is an important aspect of polymer/property design for many packaging applications. Achieving a consistent quality blend with the desired properties requires proper attention to both process and product design. The polymer blend properties are a complex function of the stress history imposed by the process and the rheological and thermodynamic properties of the components. Based on experience and science, we can propose several guidelines for reducing these complexities to practice:

- Careful control of the blend components fed to the mixing device feed hopper is important since most continuous mixers, such as extruders, do not have very good back mixing.
- All things being equal, target 10% or more for the minor component for better accuracy and control.
- Single-screw extruders can be used for many blending applications where only simple distributive mixing is required. For more demanding applications, high-performance screws or ultimately specialty compounding devices, such as twin-screw extruders, may be required.
- When using a masterbatch concentrate, make sure the masterbatch carrier resin is compatible with the resin it is being blended into.
- Very few polymer blends are miscible. For miscible blends, the viscosity can be approximated using the log mean rule (Eq. (13.12)).
- The immiscible blend morphology influences the properties, which generally deviate from the mixing rule. Some examples include using small domains of a soft resin to improve

toughness without detracting from optical clarity and laminar morphologies to help improve barrier performance.

- The immiscible blend morphology is influenced by:
  - the component concentrations,
  - stress history imposed by the compounding device, and
  - the viscosity and mutual attractiveness (thermodynamics) of the components.
- The minor component domain size of an immiscible blend can be reduced by:
  - increasing the stress (by increasing the matrix phase viscosity and the shear rate or rate of elongation during mixing),
  - matching phase viscosities a viscosity ratio in the range of 0.01 to 2 is preferred, particularly for shear flow),
  - lowering the interfacial tension by matching solubility parameters, and
  - minimizing coalescence.
- Compatibilizers reduce interfacial tension and coalescence thus reducing domain size. They also reduce the minor phase melt time, which can affect domain size.
- Elongational flow is generally more efficient than shear flow for dispersing the minor component of an immiscible blend. The flow in single-screw extruders is typically dominated by shear flows unless special mixing elements are used. Both shear and elongational flow can be found in twin screw extruders.
- Most mixing in a single-screw extruder occurs as the polymers melt, forming sheets or striations that break up into holes and droplets as the shear and elongation are increased along the extruder. Increased shear and stretching in the compression zone or high-performance metering zones (such as energy transfer screws) promote sheet breakup.
- For immiscible blends, orientation during the blown film process often elongates the minor component domains within the matrix resin.
- Dispersive mixing may be required to break up agglomerates when blending mineral fillers or pigments with polymers. For this reason, these blends are often made as masterbatches using specialized compounding equipment and let down using a single-screw extruder by the film manufacturer.

References

[1] Ghijsels A, Ente JJSM, Raadsen J. Melt strength behavior of polyethylene blends. Int Polym Process 1992;7(1):44–50.

[2] Fetel AI. Nylon-containing ionomer improves the performance of ethylene vinyl alcohol [EVOH) copolymers in multilayer structures, TAPPI Polymers, Laminations & Coatings Conference proceedings; 1997. p. 503–05.

[3] Fetel AI. Amorphous nylon improves the performance of EVOH copolymers in multilayer structures. Wixom (MI): Future-Pak; 1977.

[4] Jester R. Heat seal characteristics of cyclic olefin copolymer/polyethylene blends. Boston, MA: TAPPI PLACE Conference; 2002. p. 1308–34.

[5] Qiu G, Zhang P, Pan J, Xie Y, Zhoa S. Mechanical properties and morphology of polypropylene toughened with metallocine polymerized polyethylene elastomers. Polym Plastics Technol Eng 2003;42(1):33–44.

[6] Epstein BN. Tough thermoplastic nylon compositions, US patent 4174358; 1979.

[7] Encyclopedia of polymer science and technology, 2nd ed. New York: Wiley & Sons; 1987.

[8] DuPont, DuPont Selar® PA blends with nylon 6. http://www2.dupont.com/Selar/en_US/assets/downloads/selar_pa3426_nylon_blends.pdf [accessed 06.04.2014].

[9] Subramanian PM. Permeability barriers by controlled morphology of polymer blends. Poly Eng Sci 1985;25:483.

[10] Kamal MR, Garmabi H, Hozhabr S, Arghyris L. The development of laminar morphology during extrusion of polymer blends. Can Polym Eng Sci 1995;35:41–51.

[11] Garmabi H, Kamal MR. Improved barrier and mechanical properties of laminar polymer blends. J Plast Film Sheeting 1999;15 (2):120–30.

[12] Huang H-X, Huang Y-F, Yang S-L. Developing laminar morphology in higher-viscosity ratio polyblends by controlling flow fields in a single-screw extruder. Polym Int 2005;54:65–9.

[13] Morris BA. Improving the adhesion of LDPE to aluminum foil through blending, SPE ANTEC Tech Papers; 2004. p. 1101–1105.

[14] Soutar AM. Polymer combinations for peel/seal applications. TAPPI Seminar Notes, Coextrusion 1986:31–43.

[15] Pirtle S, Mergenhagen L, Wooster J, Cotton R. The effects of resin selection and extrusion variables on peelable seal performance, TAPPI PLACE conference, 2004.

[16] Hwo CC. Polybutylene blends as easy open seal coats for flexible packaging and lidding. J Plast Film Sheeting 1987;3(4):245–60.

[17] Mergenhagen LK, Pope TJ, Timper CA. Polyolefin plastomer-based peelable seal formulations. TAPPI Polym Laminations Coating Conference, vol. 1. 1996. p. 63–71.

[18] Martin C. Evolution of direct extrusion using high speed twin screw extruders. SPE ANTEC Technical Papers; 2002.

[19] Maddock BH. A visual analysis of flow and mixing in extruder screws. SPE J 1959;383–9.

[20] Tadmor Z, Duvdevani I, Klein I. Melting in plasticating extruders theory and experiments. Poly Eng Sci 1967;7:198.

[21] Mount E. A review of melting mechanisms in single screw extrusion, SPE ANTEC Conference; 2005. p. 304–310.

[22] Spalding MA, Hyun KS. Troubleshooting mixing problems in single-screw extruders, SPE ANTEC Tech Papers; 2003. p. 229.

[23] Morris B A, Kirschbaum D. Modification of PLA for Improved Processing and Properties, Bioplastics Processing Conference, Charlotte, NC; 2007.

[24] Baird JC, Christiano JP, Morris BA. An extrusion study: Examination of the improved processing characteristics of a PLA impact modified blend, SPE ANTEC Conference, Chicago, IL; 2009.

[25] Middleman S. Fundamentals of polymer processing. New York: McGraw-Hill Book Company; 1977. p. 154.

[26] Rauwendaal C. Polymer extrusion. New York: Hanser; 1986.

[27] Saxton R. Extruder mixer screw, US patent 3006029; 1961.

[28] Schut J. Single-screw compounding is learning new tricks. Plast Technol July 1999. Available at: http://www.plasticstechnology.com/articles/199907fa2.html [accessed 07.04.2014].

[29] Somers SA, Spalding MA, Dooley J, Hyun KS. An experimental study of the flows in an energy transfer screw, SPE ANTEC Tech. Papers; 2002. p. 307.

[30] Hogan TA, Spalding MA, Kim EK, Barr R, Meyers J. Performance analysis of a variable barrier energy transfer screw, SPE ANTEC Tech papers; 2003. p. 92.

[31] Myers JA, Barr RA. Improved screw design for maximum conductive melting, SPE ANTEC Tech. Papers; 2002. p. 154.

[32] Kruder GA, Calland WN. Fundamentals of wave screw extrusion, SPE ANTEC Tech. Papers; 1990. p. 74.

[33] Fan P, Vlachopoulos J, Smith N, Sheth H. Computer simulation of melt flow in wave screws, SPE ANTEC Tech. Papers; 1998. p. 97–101.

[34] Somers SA, Spalding MA, Hughes KR, Frankland JD. Performance of a Stratablend mixing screw for single-screw extrusion, SPE ANTEC Tech. Papers; 1998. p. 272.

[35] Myers JA. Feedscrew for injection molding and extrusion, US Patent 5,342,125; 1994.

[36] Rauwendaal C, Ponzielli G. Performance characteristics of elongational mixing screws, SPE ANTEC Tech. Papers; 2004. p. 178–182.

[37] White JL. Twin screw extrusion, technology and principles. Munich: Hanser Publications; 1990.

[38] Curry JE. Practical aspects of processing blends. In: Folkes MJ, Hope PS, editors. Polymer blends and alloys. London: Blackie Academic & Professional; 1993.

[39] Wood BA. Transmission electron microscopy of detection of rubber modifiers, 161st Technical Meeting of the Rubber Division of the American Chemical Society, Savannah, GA, April 29–May 1, Paper 45; 2002.

[40] Wood BA. Transmission electron microscopy of polymer blends. In: Finlayson K, editor. Advances in polymer blends and alloys technology, vol. 3. Lancaster, PA: Technomic Publishing Co., Inc; 1992.

[41] Ohme H, Kumazawa S, Kumaki J. Resin composition and molded article, film and fiber each comprising the same, US Patent application 2007/0260019A1; 2007.

[42] Coleman M, Graf J, Painter P. Specific interactions and the miscibility of polymer blends. Lancaster, PA: Techonomic Publishing; 1991.
[43] Van Krevelen DW. Properties of polymers: their estimation and correlation with chemical structure. Amsterdam: Elsevier Scientific Publishing Co.; 1976.
[44] Hussein IA, Williams MC. Rheological study of heterogeneities in melt blends of ZN-LLDPE and LDPE: influence of Mw and comonomer type, and implications for miscibility. Rheol Acta 2004;43(6):602–14.
[45] Hussein IA, Williams MC. Rheological study of the miscibility of LLDPE//LDPE blends and the influence of Tmix. Can Polym Eng Sci 2001;41(4):696–701.
[46] Hameed T, Hussein IA. Rheological study of the influences of Mw and comonomer type on the miscibility of m-LLDPE and LDPE blends. Polymer 2002;43(25):6911–29.
[47] Hussein IA, Hameed T, Sharkh B, Mezghani K. Miscibility of hexane-LLDPE and LDPE blends: influence of branch content and composition distribution. Polymer 2003;44:4665–72.
[48] Benkreira H, Britton R. The optimization of masterbatch formulations for use in single screw machines. Intern Polym Process 1994;9:205–10.
[49] Wu S. Polymer interface and adhesion. New York: Marcel Dekker, Inc; 1982.
[50] Taylor GI. The formation of emulsions in definable fields of flow. Proc Roy Soc, Ser A 1934;146:501.
[51] Grace HP. Dispersion in high viscosity immiscible fluid systems and application of static mixers as dispersion devices in such systems. Chem Eng Commun 1982;14:225–77.
[52] Tomotika S. On the stability of a cylindrical thread of a viscous liquid surrounded by another viscous fluid. Proc Roy Soc, Ser A 1935;150:322.
[53] Kang J, Bigio D, Smith T. Study of breakup mechanisms in cavity flow, SPE ANTEC Conference; 1995. p. 1988–1993.
[54] Utracki LA, Shi Z. Development of polymer blend morphology during compounding in a twin-screw extruder. Part I: Droplet dispersion and coalescence – a review. Poly Eng Sci 1992;32:1824–33.
[55] Bourry D, Favis B. Co-continuity and phase inversion in HDPE/PS blends: the role of interfacial modification, SPE ANTEC Conference; 1995. p. 2001.
[56] Sinthavathavorn W, Nithitanakul M, Magaraphan R, Grady BP. Blends of polyamide 6 with low-density polyethylene compatibilized with ethylene-methacrylic acid based copolymer ionomers: effect of neutralizing cations. J Appl Polym Sci 2008;107:3090–8.
[57] Torradas J, Pimentel RO, Dean DM. Additives to improve regrind utilization and recycling of high barrier blow molded containers, SPE ANTEC Conference; 2013.
[58] Dealy J, Wissbrun K. Melt rheology and its role in plastics processing: theory and applications. London: Chapman & Hall; 1995.
[59] Flumerfelt RW. Drop break-up in simple shear fields of viscoelastic fluids. Ind Eng Chem Fundam 1972;11:312–18.
[60] Tavgav T. Drop deformation and breakup in simple shear fields, PhD Thesis [Chemical Engineering). Houston, Texas: University of Houston; 1972.
[61] Gauthier F, Goldsmith H, Mason S. Particle motions in non-Newtonian media. II: Poiseuille flow. Trans Soc Rheol 1971;15:297–330.
[62] De Bruijn R. Deformation and breakup of drops in simple shear flow, PhD Thesis. Eindhoven, The Netherlands: Eindhoven University of Technology; 1989.
[63] Elmendorf J, Maalcke R. A study on polymer blending microrheology: Part I. Poly Engr Sci 1985;25:1041–7.
[64] Varanasi P, Ryan M, Stroeve P. Experimental study on the breakup of model viscoelastic drops in uniform shear flow. Ind Eng Chem Res 1994;33:1858–66.
[65] Levitt CL, Macosko CW, Pearson S. Influence of normal stress difference on polymer drop deformation. Poly Eng Sci 1996;36:1647–55.
[66] Han CD, Funatsu K. An experimental study of droplet deformation and break-up in pressure-driven flows through converging and diverging channels. J Rheol 1978;22:113–33.
[67] Milliken W, Leal L. Deformation and breakup of viscoelastic drops in planar extensional flows. J Non-Newtonian Fluid Mech 1991;40: 255–379.
[68] Delaby I, Ernst B, Muller R. Droplet deformation in polymer blends during uniaxial elongational flow: influence of viscosity ratio for

large capillary numbers. J Rheol 1994;38: 1705–20.

[69] Meijer HEH, Janssen JMH. Mixing of immiscible liquids. In: Manas-Zloczower, Tadmor Z, editors. Mixing and compounding of polymers. Germany: Hanser Publishers; 1994. p. 85–147.

[70] Chin H, Han C. Studies on droplet deformation and breakup: droplet deformation in extensional flow. J Rheol 1979;23:557–90.

[71] Shanker R, Spradling R, Pham H, Namhata S, Bosnyak C. Effect of viscoelasticity on transient drop dispersion, SPE ANTEC Conference; 1996. p. 2596–2600.

[72] Mighri F, Ajji A, Carreau P. Influence of elastic properties on drop deformation in elongational flow. J Rheol 1997;41:1183–201.

[73] Mighri F, Huneault MA. Drop deformation and break-up mechanisms in viscoelastic model fluid systems and polymer blends. Can J Chem Eng 2002;80:1028–35.

[74] Macasko C. Morphology development and control in immiscible polymer blends. Macromol Symp 2000;149:171–84.

[75] Mighri F, Carreau P, Ajji A. Influence of elastic properties on drop deformation and breakup in shear flow. J Rheol 1998;42: 1477–90.

[76] Van Oene HJ. Modes of dispersion of viscoelastic fluids in flow. J Coll Inter Sci 1978;40:448.

[77] Lyngaae-Jorgensen J. Rheology of polymer blends. In: Folkes MJ, Hope PS, editors. Polymer blends and alloys. London: Blackie Academic & Professional; 1993.

[78] Huneault M, Champagne M, Daigneault L, Dumoulin M. Blend morphology development during compounding in a twin-screw extruder, SPE ANTEC Tech Papers; 1995. p. 2020.

[79] Ghosh AK, Ranganathan S, Lindt JT, Lorek S. Blend morphology development in a single screw extruder, SPE ANTEC Tech. Papers; 1991. p. 232–236.

[80] Lindt JT, Ghosh AK. Fluid mechanics of the formation of polymer blends. Part I: Formation of laminar structures. Polym Eng Sci 1992;32:1802–13.

[81] Huang H-X, Huang Y-F, Li X-J. Detecting blend morphology development during melt blending along an extruder. Polym Testing 2007;26:770–8.

[82] Benkreira H, Shales R, Edwards M. Mixing on melting in single screw extrusion. Int Polym Process 1992;7:126–31.

[83] Scott C, Macosko C. Morphology development during the initial stages of polymer-polymer blending. Polymer 1995;36(3):461–70.

[84] Burch H, Scott C. Evolution of structure in the softening/melting regime of miscible polymer mixing. Polym Eng Sci 2001;41(6): 1038–48.

[85] Willemse RC, Ramaker E, van Dam J, Posthuma de Beor A. Morphology development in immiscible polymer blends: initial blend morphology and phase dimensions. Polymer 1999;40:6651–9.

[86] Tyagi S, Ghosh AK. Morphology development during blending of immiscible polymers in screw extruders. Polym Eng Sci 2002;42:1309–21.

[87] Li H, Hu G-H, Sousa JA. Morphology development of immiscible polymer blends during melt blending: effects of interfacial agents on the liquid-solid interfacial heat transfer. J Polym Sci Part B: Polm Phys 1999;37:3368–84.

[88] Shih C, Tynan D, Denelsbeck D. Rheological properties of multicomponent polymer systems undergoing melting or softening during compounding. Poly Eng Sci 1991;31(23): 1670–3.

[89] Shih C. Fundamentals of polymer compounding: The phase inversion mechanism during mixing of polymer blends. Adv Polym Tech 1992;11(3):223–6.

[90] Ratnagiri R, Scott C. Phase inversion during compounding of polycaprolactone/polyethylene blends, SPE ANTEC Conference; 1996. p. 59.

[91] Sundararaj U, Macosko C, Shih C-K. Evidence for inversion of phase continuity during morphology development in polymer blending. Poly Eng Sci 1996;36(13): 1769–81.

[92] Steinmann S, Gronski W, Friedrich C. Cocontinuous polymer blends: influence of viscosity and elasticity ratios of the constituent polymers on phase inversion. Polymer 2001;42:6619–29.

[93] David C, Trojan M, Jacobs R, Piens M. Morphology of polyethylene blends: I. From spheres to fibrils and extended co-continuous phases in blends of polyethylene with styrene-isoprene-styrene triblock copolymers. Polymer 1991;32(3):510.

[94] David C, Getlichermann M, Trojan M. Morphology of polyethylene blends. II: Formation of a range of morphologies in blends of polyethylene with increasing content of styrene-isoprene-styrene triblock copolymers during extrusion-blowing. Poly Eng Sci 1992;32(1):6.

[95] Getlichermann M, David C. Morphology of polyethylene blends: III. Blend films of polyethylene with styrene-butadiene or styrene-isoprene copolymers obtained by extrusion-blowing. Polymer 1994;35(12):2542.

[96] Ronzalez-Nunez R, Favis B, Carreau R. Factors influencing the formation of elongated morphologies in immiscible polymer blends during melt processing. Poly Eng Sci 1993;33(13):851.

[97] Lee S-Y, Kim S-C. Morphology and oxygen barrier properties of LLDPE/EVOH blends. Intern Polym Process 1996;XI:238.

[98] Lopez-Barron CR, Robledo-Ortiz JR, Rodrigue D, Gonzalez-Nunez R. Film processability, morphology, and properties of polyamide-6/low density polyethylene blends. J Plast Film Sheeting 2007;23:149–69.

[99] Basell product literature, 2001, Polybutene-1 for use in seal-peel applications, customer presentation aid.

[100] Robledo-Ortiz JR, Ramiriz-Arreola D, Gonzalez-Nunez R, Rodrigue D. Effect of freeze-line position and stretching force on the morphology of LDPE-PA6 blown films. J Plast Film Sheeting 2006;22:287–314.

[101] Morris BA. The effect of coextrusion on bubble kinematics, temperature distribution and property development in the blown film process, SPE ANTEC Tech Papers, 1998.

[102] McKelvy JM. Polymer processing. New York: John Wiley; 1967.

[103] Alemaskin K, Camesasca M, Manas-Zloczower I, et al. Entropic mixing characterization in a single screw extruder, SPE ANTEC Conference; 2004. p. 167–172.

[104] Alemaski K, Manas-Zloczower I, Kaufman M. Color mixing in single screw extruder: simulation & experimental validation, SPE ANTEC Conference; 2005. p. 255–259.

[105] Lan T. 2007, Nanocomposite materials for packaging applications, SPE ANTEC Conference, Online: http://www.nanocor.com/tech_papers/Antec-Nanocor-Tie%20Lan-5-07.pdf [accessed 15.10.2012].

[106] Leaversuch R. Nanocomposites broaden roles in automotive, barrier packaging. Plast Technol 2001:64.

[107] Sherman L. Chasing nanocomposites. Plast Technol 2004:56.

[108] Manias E, Strawhecker K, Wu L, Kuppa V. Concurrent enhancement of various materials properties in polymer/clay nanocomposites. Proc Am Chem Soc Composites 2001;16:236–46.

[109] Dahman S. Melt processed polymer-clay nanocomposites: properties and applications. SPE ANTEC conference 2000;58(2):2407–11.

[110] Akkapeddi K, Socci E, Kraft T, Facinelli J, Worley D. A new family of barrier nylons based on nanocomposites and oxygen scavengers. SPE ANTEC Conference 2003;61(3):3845–8.

[111] Bose S, Bhattacharyya AR, Bondre AP, Kulkarni AR, Potschke P. Rheology, electrical conductivity, and the phase behavior of cocontinuous PA6/ABS blends with MWNT: correlating the aspect ratio of MWNT with the percolation threshold. J Polym Sci, Part B Polym Phys 2008;46:1619–31.

[112] Tiawari RR, Paul DR. Effect of organoclay on the morphology, phase stability and mechanical properites of polypropylene/polystyrene blends. Polymer 2011;52:1141–54.

[113] Goldel A, Marmur A, Kasaliwal GR, Potschke P, Heinrich G. Shape-dependent localization of carbon nanotubes and carbon black in an immiscible polymer blend during melt mixing. Macromolecules 2011;44:6094–102.

[114] Utracki LA. Polymer alloys and blends: thermodynamics and rheology. New York: Hanser Publishers; 1989.

[115] Dealy J, Wissbrun K. Melt rheology and its role in plastics processing: theory and applications. London: Chapman & Hall; 1995. p. 392–400.

[116] Marion S, Pierce P. Application of Ree-Eyring generalized flow theory to suspensions of spherical particles. J Coll Sci 1956;11:80.

[117] Kuraray, Eval properties. Available at: http://www.eval-americas.com/us/eval-properties.aspx [accessed 08.04.2014].

[118] McBride, E. 1993, DuPont internal memo.

[119] Walsh DJ. 2005, personal communication.

[120] Greco F. Drop deformation for non-Newtonian fluids in slow flows. J Non-Newtonian Fluid Mech 2002;107:111–31.
[121] Maffettone P, Greco F. Ellipsoidal drop model for single drop dynamics with non-Newtonian fluids. J Rheol 2004;48: 83–100.
[122] Lerdwijitjarud W, Sirivat A, Larson R. Influence of dispersed-phase elasticity on steady-state deformation and break-up of droplets in simple shearing flow of immiscible polymer blends. J Rheol 2004;48:843–62.
[123] Uttandaraman S, Macosko CW. Drop breakup and coalescence in polymer blends: the effects of concentration and compatibilization. Macromolecules 1995;28:2647–57.
[124] Milner S, Xi H. How copolymers promote mixing of immiscible homopolymers. J Rheol 1996;40:663–87.
[125] Krook M, Hedenqvist M. Polymer-clay nanocomposites: one way to improve the barrier properties of polymers used in packaging. TAPPI PLACE conference; 2002. p. 194–7.

14 Water- and Solvent-Based Coating Technology

Edgar B. Gutoff[1] *and Edward D. Cohen*[2]

[1]Consulting Chemical Engineer, Brookline, MA, United States; [2]Consulting Services Inc., Fountain Hills, AZ, United States

14.1 Introduction

The coating process is an essential technology in the successful development and commercialization of flexible packaging films. Many available packaging substrates by themselves are not able to meet the specific requirements for a functional packaging film. Additional functional layers are often coated on the substrate to meet the requirements for the desired oxygen and moisture barrier, for adhesion, for preventing static-caused defects, and for printability, aesthetics, and durability. Multiple layers can be applied to both substrate sides. Recent advances in coating technology have resulted in the ability to simultaneously coat 12 or more layers. With slot die coating, two or three layers can be simultaneously applied with readily available commercial hardware. The hardware gets more involved and complicated with additional layers. For simultaneously coating more—many more—than two or three layers, slide coating (or precision curtain coating, which in many ways is similar) is the preferred method. Applying multiple layers simultaneously is much more efficient than multiple single-layer coatings, and also allows very thin individual layers to be coated. Slide coating is not used for single layers as it is more expensive and more technically complex than slot die coating, and at relatively high speeds slot die coaters can coat thinner individual layers.

This chapter discusses several aspects of the coating process and aims to provide a broad understanding of the many different coating applicators that are used. In addition, the major web coating process elements are discussed. The aim is to familiarize the reader with the many different coating processes available, so that coating technology can be applied successfully when developing products.

14.1.1 Coating Uniformity

Coating uniformity can be described by several methods and it is often given as a range, such as a ±10% uniformity. This specifies that all the data are expected to lie within 10% of the mean value. However, the range is not a precise value and depends on how many data points are involved. As more data accumulates, one would expect occasional points to lie further from the mean. To be more precise, statistical definitions should be used. If many data points are involved (an "infinite" number), we may ask what range will include 99% of the data, or 95%, or 99.7%—as is done in statistical process control. We believe that the commonly used range would include about 99% of the data, and thus is approximately the same as the 99% confidence limits. The coating uniformity we give in later sections is thus the approximate 99% confidence limits, and in statistical terms would be ±2.58 true standard deviations from the mean. When the standard deviation is estimated from a data sample, the number would be higher than 2.58 depending on how many samples are used to estimate the standard deviation, and one would have to use the two-sided *t*-distribution to get this number.

14.2 Coating Structures

A typical coated structure is a flexible substrate with functional layers coated on either side or on both sides. The substrate can be a single polymer such as polyethylene (PE), biaxially oriented polypropylene, polyethylene terephthalate, or polyvinyl chloride. There can be a single coated layer or multiple layers depending on the product requirements. The layers may function as oxygen, light or moisture barriers, or may be release layers to which

Multilayer Flexible Packaging. DOI: http://dx.doi.org/10.1016/B978-0-323-37100-1.00014-4

adhesives will not stick. In addition, adhesives are often coated to form laminated structures. Also, conductive antistatic layers may be coated to prevent charge buildup.

The specific coating solution composition depends on the coating's function. However, all coatings have the following general ingredients:

- A binder, which is the nonvolatile solid in the dried coating that binds the functional ingredients—such as pigments, clay, or other additives—together to form a film. The binder is usually composed of one or more polymers. The polymer may be dissolved in the coating solution or dispersed as a latex.
- A solvent to dissolve the binder and some of the ingredients, and to suspend and disperse all the insoluble components.
- Dispersed functional particles such as pigments for color and opacity, silver halides for photographic activity, or iron and chromium oxides for magnetic activity.
- Additives such as surfactants or dispersants added as coating aids, plasticizers for flexibility, protective colloids (usually the binder), UV absorbers, biocides to prevent bacterial growth, cross-linking or hardening agents for toughness and insolubility, and conductive materials such as cuprous chloride or some carbon blacks to prevent static buildup.
- Particles on the coating surface, such as silica or hard latex spheres, to control reflectivity and to give surface roughness to prevent "blocking" or adhesion between adjacent layers, such as that occurs between glass microscope slides.

The final solution properties, most importantly the viscosity and surface tension may affect the coating process, and are discussed in sections 14.4.3 and 14.4.4.

14.3 Web-Coating Machine Hardware and Functions

The coating machine used to produce the coated product is complex and has many separate elements that must function well while interacting with each other. The machines are typically quite versatile. They can continuously produce different product structures at high productivity and low cost. They can produce the more common continuous coatings such as barrier layers, release layers, adhesives, or antistatic layers. They may also produce intermittent coatings, such as in the printing process and in coating anodes and cathodes for thin film batteries.

In general, there is an ideal coater arrangement for any given product. However, most coating machines produce many different products with different coating thicknesses, and therefore the machine usually is a compromise made for the several applications. The coating machine has several process elements that are required to continuously produce the coated product.

The first element, while essential, is not part of the coating machine itself. It is the kettles and mixing equipment used to mix the fluids that are to be applied to the substrate. The final coating solution contains solvents, binders, pigments, colorants, cross-linkers, surfactants, etc., which need to be dissolved, dispersed, and mixed and treated to prepare a uniform homogeneous solution or dispersion. Usually, this is done in temperature-controlled kettles with high or low shear mixers. Scales and filtration equipment are part of this system. High shear mixers are needed to disperse particles and to obtain the desired particle size. After the solutions are mixed, they are filtered to remove larger particles, and then either sent to storage or transferred to the solution feed equipment, which is a part of the main coating machine. The solution preparation equipment is not integrated into the coating machine; it is a separate batch operation.

Coating itself is a continuous process and the individual elements must be integrated and function at the same speed. While all coating machines have the same common elements, the exact component design depends on the specific product and productivity needs. Each unit is essentially a semi-custom unit. Coating machine line speeds can range from about 0.01 m/s (2 fpm) for some medical test strips up to 15 m/s (3000 fpm) or even faster for clay coatings on paper, depending on the exact design. Coating widths can vary from about 10 cm (4 in) to 3.7 m (12 ft) or even wider, depending on need. Fig. 14.1 shows the components of a typical coating machine.

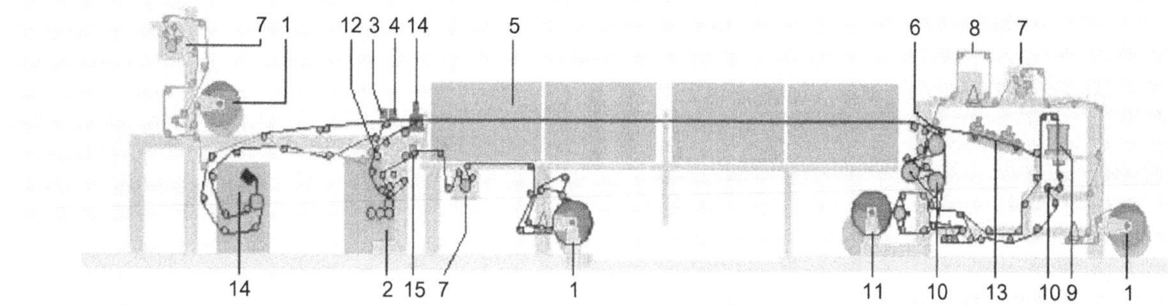

1. Unwinder
2. Coating unit
3. Infrared lamp (short wave)
4. Infrared lamp (medium wave)
5. Air floatation dryer
6. Dry laminating station
7. Corona station
8. Edge guide unit
9. Remoisturizing station
10. Cooling station
11. Rewinder
12. Wet laminating station
13. Ultra violet curing station (UV)
14. Curtain coater
15. Web cleaning station

Figure 14.1 Coating machine with three unwinds and two laminating stations.
Source: Polytype Converting SA.

The web transport system conveys the uncoated substrate from the unwind stand through the coating station and the dryer to the rewind stand, while maintaining the appropriate speed and tension control. It has several components. The unwind stand holds the uncoated substrate roll and feeds it into the coating applicator and then onto the dryer. Drive rolls, which may be pinch rolls or S-wrap rolls, or other pull rolls with large wrap angles to avoid slippage, or vacuum rolls or tables, pull the substrate through the coater at the desired speed and tension. At the rewind stand, the coated product is wound into rolls. There are dancer rolls or force transducers to measure and provide feedback to control the tension in the substrate, and edge guide systems to keep the substrate running straight in the coating machine.

The feed system maintains the coating solution at the proper temperature and pumps it at the desired rate to the coating applicator through filters to remove unwanted solid particles and through a debubbling system.

The coating solution is applied to the moving web with a coating applicator. There are many different applicators and configurations that can be used for this purpose, and these details will be discussed later. The coating station holds the coating applicator in a fixed position and permits the operator to precisely and accurately set the coating parameters such as gaps, roll speeds, etc. These settings must be maintained during the coating run. The coating station should not induce chatter in the coating.

The dryer dries the wet coating by supplying heat to evaporate the solvent and circulating air to remove the solvent. Most web dryers are impingement devices that blow hot air against the wet coating. In flotation dryers, hot air also blows against the back, uncoated side. This not only supplies additional heat but also supports the moving substrate. Additional heat may be supplied by infrared heaters. Infrared heaters are frequently used to increase the dryer capacity without increasing the dryer length.

The coating line may also contain a section between the unwind and the coating applicator to surface treat the substrate before it is coated. Flame treatment, corona treatment, and gas plasma treatment may used to improve the substrate wettability. This improves the coating quality and increases adhesion to the substrate by increasing the substrate surface energy and by removing or cross-linking a weak boundary layer on the surface. Web cleaners and static control devices may also be included in this section.

After the dryer and before the rewind, there may be a posttreatment section where coatings may be cross-linked with ultraviolet light or an electron beam, moisture may be added by passing the coated substrate through a steam chamber, or another sheet may be laminated to the coated substrate.

In addition to production coating machines, laboratory pilot coater machines are also available. These pilot coaters contain the same basic components as the manufacturing lines; however, they are narrower, often 15–60 cm (6–24 in) wide, and run at relatively low speeds, 0.05–0.25 m/s (10–50 ft/min). They can produce commercial quality coatings and can be used for research and development, scale-up, product development, and small volume production. All of the components—unwind, coater, web transport, dryer, rewind—and all of the control systems are contained in a steel and glass enclosure as shown in Fig. 14.2.

Figure 14.2 Yasui Seiki CAD laboratory coater. *Source: Reprinted with permission of Yasui Seiki Co.*

There are two laboratory coater designs that are used to develop product—batch sheet coaters and continuous coaters. Sheet coaters are used to develop new products where many different solutions need to be coated and only small samples are available. These methods use devices such as drawdown blades, dies, or wire-wound rods (Mayer rods) to spread a uniform solution layer across a substrate. Most applicators can provide wet coating thicknesses from 5 to 1300 μm (0.2–50 mil). Spray coaters may also be used to coat sheets. The coated substrates or sheets may be dried by ambient air or in a lab oven.

Laboratory automated sheet coaters are available. In these coaters the sheet coating is mechanically controlled. The coated sheets may be fed directly into a dryer, with the temperature and residence time controlled. Compared to hand coaters, these give better reproducibility with much higher productivity and lower overall costs. A typical laboratory sheet coater is shown in Fig. 14.3.

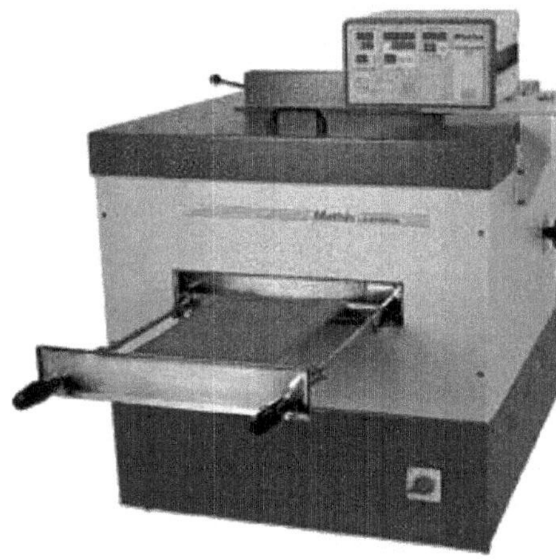

Figure 14.3 Laboratory sheet coater. *Source: Werner Mathis, USA.*

14.4 Coating Applicators

14.4.1 Classification of Applicators

There are many coating methods that are routinely used to coat various products. The methods range from the Mayer rod (wire-wound rod) coater, which has been in use for over 100 years, to relatively new methods such as the multilayer precision curtain coater and the multilayer slot die coater. Table 14.1 lists the major methods that are in current use. Each method has several configurations for the web path in the coating applicator, application point, blade design or doctoring devices, and the number and size of rolls used. As a result, there are about 1000 specific configurations that are in routine use.

It is convenient to classify these methods according to the mechanism that is used to control the wet coating weight or thickness. These four basic classes are:

- self-metered
- doctored
- premetered
- hybrid

Table 14.1 Major Coating Methods

Forward Roll	Reverse Roll
Reverse roll hot melt	Reverse roll precision
Forward gravure	Reverse gravure
Offset gravure	Mayer rod
Dip	Dip and scrape
Dip and squeeze	Floating knife
Kiss	Slide, Multilayer
Meniscus	Slot Die
Knife/metering bar	Knife-over roll
Knife-over blanket	Extrusion hot melt
Doctor blade	Air knife, metering mode
Air knife, squeegee mode	Curtain
Saturator	Spray
Rotary screen	Trailing blade
Coextrusion	Flexographic
Comma direct/indirect	Microgravure

In the *self-metered methods* the coating bead conditions control the final coating weight. In this classification, there are two separate modes that are in operation: the free meniscus and free meniscus plus film splitting. In the free meniscus mode the conditions in the meniscus, or coating bead, control the solution volume that is applied to the substrate. The principle variables are viscosity and line speed. Increases in these will increase the coating weight. Examples are:

- dip coating
- meniscus or bead-roll coating
- forward roll coating

In the free meniscus plus film splitting mode the coating is applied by the free meniscus mode and then additional roll(s) are added to create a nip or coating solution bead. The film is split in this nip and part of the coating remains on the substrate and part is returned to the coating pan. A reverse roll coater is an example of this method. This is a very popular method and, with precision bearings, can give high-quality coatings with excellent uniformity (±2%).

In the *doctored coating methods* a two-part applicator is used. The continuous liquid film is applied by any suitable method such as with an applicator roll, as in the previous methods. A doctoring device is then used to remove excess coating and thus control the final coating weight. This method can give good coating quality and stability with a many different coatings. The disadvantage is that the coating solution that recirculates back to the original feed system can lead to dirt, bubbles, and aging problems. Examples are air-knife coaters, blade and knife coaters, wire-wound or Mayer rod coaters, and dip and scrape coaters.

In the premetered *methods*, all of the coating solution fed to the coating die is applied to the substrate. Assuming a good coating is obtained, the coating thickness is just the flow rate per unit width divided by the substrate speed. Examples are slide coaters, slot die coaters, extrusion coaters, and precision curtain coaters. The advantages of these methods are:

- Ease of controlling coverage
- Excellent coating uniformity (±2%, except in extrusion coaters) and excellent coating quality
- No recirculation
- *Multilayer coating capability.* Over 12 layers for slide and curtain coating and up to 3 layers for slot die coating. Five or more layers can be made in an extrusion die, but normally a single-layer die is used preceded by a combining adapter
- Coverage is not dependent on viscosity

The *hybrid methods* combine features of more than one method to control coverage. Gravure coating is the most significant hybrid method. In gravure coating the cells or indentations on the gravure roll surface are filled in the coating pan, and then a doctor blade is used to remove the excess coating solution that is then recirculated. The majority of the solution in the cells transfers to the substrate. The coverage is controlled by the cell volume and the fraction that transfers. Thus it is at least partially premetered. However, the doctor blade and the force applied to press the gravure cylinder against the substrate on the impression role also play very important roles.

The coating methods can also be classified according to the number of layers that can be simultaneously applied. However, this limits the understanding of how the methods function, and most methods apply one layer at a time. Only the slide, precision curtain, extrusion, and slot die coaters can apply multiple layers simultaneously.

14.4.2 Role of Substrate

The substrate used in the coating process is a key element for a successful web-coating process. The substrate carries the coated layers. The effect of the substrate on the coating process needs to be understood, optimized, and controlled to insure high quality and minimum cost. *Web* and *support* are other terms that also refer to the substrate.

The major areas of technical importance for the substrate are:

- substrate structure
- surface properties treatments and coatability
- mechanical properties
- defects arising from the substrate
- substrate flatness
- substrate specifications
- substrate manufacturing
- substrate roll unwinding and winding
- substrate web handling in the coating machine
- storage

The substrate structure is critical because there are many substrates that are used in the coating process. They all have different properties that may require different operating conditions for all of the other elements in the coating process. In addition, there are many new substrates that are being introduced and often they require operating conditions that are not attainable with existing process hardware.

Several major properties categorize and describe substrates. The first is porosity. The structure may open (porous) or closed. An open porous structure has many cells penetrating into the substrate. On coating, the coated solution partially or fully fills the pores and the balance, if any, remains on the surface. This causes the coated layer to vary in thickness from top to bottom. In order to get a uniform thickness, a filler or sealer layer is coated which fills or covers the pores. A doctoring coating process is used to give a flat surface. Examples of the open structure are:

- paper
- fabric
- sintered metals, which may be porous or closed

In the second porosity class with closed structures, there are no pores and the surface is uniform and flat, unless it has been deliberately embossed. Examples of these substrates are:

- PET or Polyethylene terephthalate. Trade names include Mylar and Cronar.
- PEN, Polyethylene naphthalate
- PE, Polyethylene
- PP, Polypropylene
- PS, Polystyrene
- PVC, Polyvinylchloride
- PVF, Polyvinylfluoride. Trade name is Tedlar.
- PVDF, Polyvinylidene fluoride
- PI, Polyimide. Kapton is an example.
- Cellophane

Foamed plastics may be open or closed. Open foamed structures may be used as filters in air handling equipment or as absorbent sheets. Closed foam sheets may be used as insulating layers in clothing or blankets, or as a packing material. An extreme example is bubble wrap.

The coating applicator selected depends on the surface nature. A slot die applicator will give excellent results with a closed, flat surface because the coating will not penetrate and will be flat and uniform. The same applicator with an open surface will not give a flat coating, because some will penetrate into the pores. For an open structure, a knife or blade coater is normally preferred because the blade is a fixed distance above the substrate and should give a flat top coating while insuring that the pores are filled.

The substrate porosity influences the way the coating dries. For an open structure the coated thickness that must be dried varies from the deepest penetration into the pores to the surface (this may be the substrate thickness). This long diffusion path may greatly increase the drying time. A variation in pore size from roll to roll can lead to a wide variation in drying time. With a closed flat substrate, the coating thickness is uniform and the drying time is shorter than with an open structure at the same coverage. Thus the coating will dry faster and be more uniform. The tendency of the solvent to wet the substrate (or conversely, the ease that the substrate releases the solvent on it) also affects the drying rate in open structures.

With embossed plastic substrates, the embossed depth is usually much less than the pore depth which in open structures may be the substrate thickness. Thus the coating thickness variations tend to be less than with an open structure, even when coated with a doctoring applicator. With embossed structures, the increase in drying time compared to a flat surface is often too small to be noticed.

A second important substrate characteristic is its thickness and stiffness. Stiffness varies with substrate thickness cubed and with the modulus of elasticity. Therefore different substrates at the same thickness can have different stiffness. Thus polyester films are much stiffer than PE films at the same thickness. Substrates can be grouped into three basic categories depending on stiffness:

- *Very stiff and difficult to bend.* These substrates are usually metals that are thicker than foils, such as aluminum for lithographic printing plates, steel and tin plate for a variety of uses, and copper for electronic circuit boards, and also thick plastic webs.
- *Average stiffness.* These are the majority of the plastic and paper webs that are used in the web-coating process. They normally range in thickness from about 25 to 250 μm (1–10 mil).
- *Low stiffness.* Composed of thin substrates under about 15 μm (0.5 mil), depending on the specific plastic, and aluminum and copper foils. These distort and wrinkle very easily in the coating machine. Thin substrates are used in applications like thin film batteries to give more battery power in a given volume. There is also an economic incentive to use a thinner substrate and save on raw materials costs.

Web stiffness is an important property during coating application, transport, drying, and winding. Thicker substrates do not bend as easily and therefore require larger diameter rolls when there is a significant wrap angle. Thicker substrates require higher tension levels and therefore more power to transport through the coater from the unwind to the rewind.

On the other hand, thin substrates deform and wrinkle easily. Also, they require low tension levels. Although they can wrap around rolls easily, the low tensions they require may not be enough to drive idler rolls without scratching. In addition, in flotation drying the substrate can contact the opposing nozzles. This effect is sometimes called parachuting because of the substrate appearance.

A coating line is usually designed and optimized for a particular substrate and it may not have the capability to run widely differing product without modification.

A third major substrate property is its wettability and the coating adhesion to the substrate surface. The coating solution must wet the substrate surface and spread uniformly in order to get good coating uniformity. To get good wetting the substrate surface tension should be a little greater, perhaps by 5 dyne/cm, then that of the substrate. The coating surface tension can be measured using a Wilhelmy slide or the du Noüy ring balance [1]. The substrate surface tension or surface energy (these terms are equal, as 1 dyne/cm equals 1 erg/cm^2) can be estimated with “dyne pens.” These are a felt-tip pens filled with known surface tension liquids. The substrate “dyne level” equals the highest surface tension liquid that will still spread on the substrate. This is the highest surface tension that the coating fluid can have and still spread. If this condition is not met, then the coating solution surface tension must be adjusted down with surfactants. Surfactants are rarely needed for organic solvents that inherently have low surface tensions, and almost always are needed in aqueous systems. In addition, the substrate surface tension or energy can be increased by flame treatment, electrical discharge (corona discharge), or plasma treatment. Fig. 14.4 gives more details of the wetting requirements.

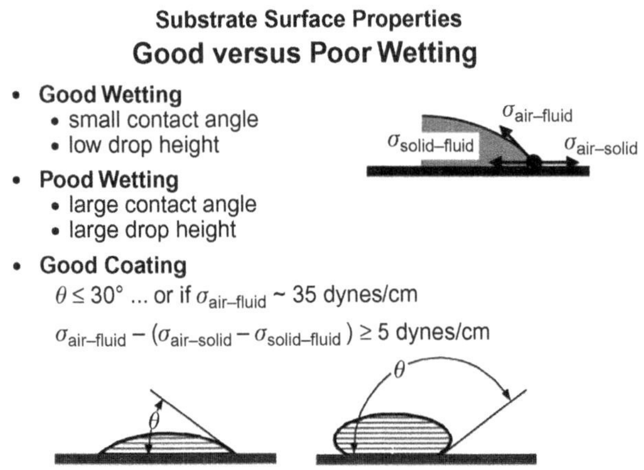

Figure 14.4 Wetting requirements. The upper sketch shows that for the fluid to spread, the force to the right (the surface tension of the solid) should exceed the sum of forces to the left, the solid–liquid interfacial tension plus the liquid surface tension times the cosine of the contact angle. When the fluid spreads the contact angle tends toward zero and the cosine tends toward unity.

For good adhesion, the coating fluid must wet the surface and also adhere to it. A weak "boundary layer" on the substrate surface is the usual cause of poor adhesion. Surface treatments such as flame, corona, or plasma treatment not only increase the surface tension or energy, they also greatly improve adhesion by either removing the weak boundary layer or cross-linking [2]. In addition, special precoat layers may be coated on the substrate to improve adhesion. These are sometimes called *subbing* layers.

14.4.3 Coatability Limits

In all coating methods, when one tries to coat too fast air gets entrained under the coating and the coating quality suffers greatly. This is the coatability high-speed limit. Air also tends to be entrained when one tries to coat too thin a layer. The low coverage and the high-speed coatability limits are very similar. There is also a low speed, or high coverage, coatability limit. When one tries to coat too slowly or to apply too thick a coating, the fluid may run back and some of it may not be carried off.

Different coating methods may have other limitations. Thus, in forward roll coating the ribbing defect appears when a dimensionless group called the capillary number (equal to the viscosity times the coating speed divided by the surface tension), exceeds a critical value that varies with the coating gap to the roll diameter ratio. In reverse roll coating, quality coatings can be made free of ribbing and another defect called *cascade* even at very high capillary numbers, in a region determined by the gap and the metering roll to the applicator roll speed ratio.

Other limitations may be present. For example, in curtain coating a certain minimum flow rate per unit width is needed to maintain the curtain integrity, independent of coating speed but dependent on the fluid properties. The minimum thickness that can be curtain-coated at any speed is then equal to the minimum flow rate per unit width required for the curtain integrity, divided by the coating speed. As a result, for reasonably thin coatings fairly high speeds are needed. Thus curtain coating is inherently a high-speed coating process.

And in slot die coating, there appears to be a low speed limit below which coatings can be made but only with certain coating defects.

14.4.4 Description of Coating Methods

14.4.4.1 Self-Metered Methods

Dip coaters. Dip coating, along with brush coating and wire-wound rod coating, is one of the oldest coating methods. The web passes under an applicator roll that is partially submerged in a pan of the coating fluid as shown in Fig. 14.5. The web

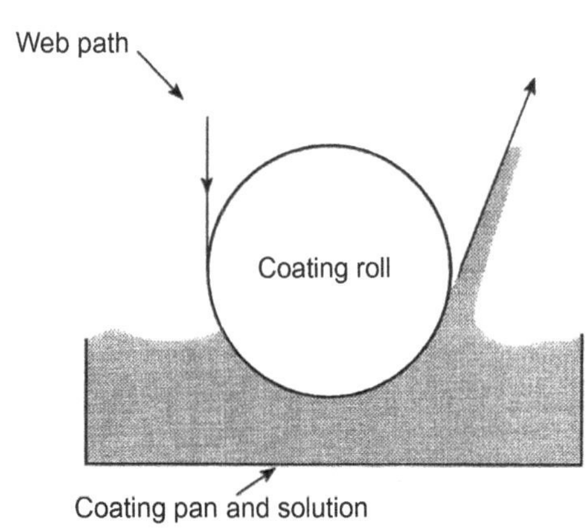

Figure 14.5 Dip coater. *Source: From Ref. [7]. Reprinted with permission of John Wiley & Sons, Inc.*

is actually dipped into the coating solution. The wet thickness is determined by:

- fluid properties
- coating speed
- withdrawal angle

The thickness increases with viscosity and coating speed. Dip coating has a wide operating range, is easy to run, and the hardware is inexpensive.

Operating parameters for dip coating are:

- viscosity: 20–2000 cP or mPa s
- wet thickness: 10–200 μm (0.4–8 mil)
- line speed: 0.5–7.5 m/s (100–1500 ft/min)
- coverage uniformity: ±10%

Dip coating is very commonly used for coating continuous objects that are not flat, such as fibers, and also for irregularly shaped discrete objects. Drops of coating at the bottom of dip coated articles may be removed by applying electrostatic forces while the article is moved along a conveyor.

Meniscus or bead-roll coaters. In the meniscus or bead-roll coater the web passes under a back-up roll that is just above the liquid level in a pan. The pan is raised until the solution touches the web and then is lowered just a little so that the solution still maintains contact with the moving web but only due to surface tension. The coverage is determined by:

- viscosity
- solution surface tension
- coating speed

Different configurations are available, including a two roll design, where an applicator roll dipping into the solution picks up the fluid and transfers a portion of it onto the web on a back-up roll with a gap just slightly greater than the thickness of the approaching fluid. Only surface tension keeps the substrate in contact with the fluid. Meniscus coating is limited in that the coating speed is very low, only about 0.1–0.2 m/s (20–40 ft/min) and a low-viscosity solution is needed. However, high-quality optical coatings may be produced. In the past, these coaters have also been used for adhesives and for photographic coatings.

Kiss coaters. In kiss coating, there is no back-up roll. The coating solution is applied to the unsupported substrate by a rotating applicator roll, called a kiss roll. There are many kiss coater designs. While the kiss roll can turn with or against (reverse) the web direction, it usually operates in the web direction. Kiss coaters are tension sensitive and are often used to apply excess coating prior to a metering device. Operating parameters for a kiss coater are:

- viscosity: 50–1000 cP or mPa s
- wet thickness: 5–75 μm (0.2–3 mil)
- line speed: 0.5–5.5 m/s (100–1100 ft/min)
- coverage uniformity: ±10%

Forward roll coaters. In forward roll coaters the substrate passes between an applicator roll and the backing roll, both of which are rotating in the same direction as the substrate. The applicator roll drags fluid into the nip, as shown in Fig. 14.6. The fluid exiting the nip splits into two directions, with some adhering to the substrate and the rest remaining on the applicator roll, to be returned to the coating pan.

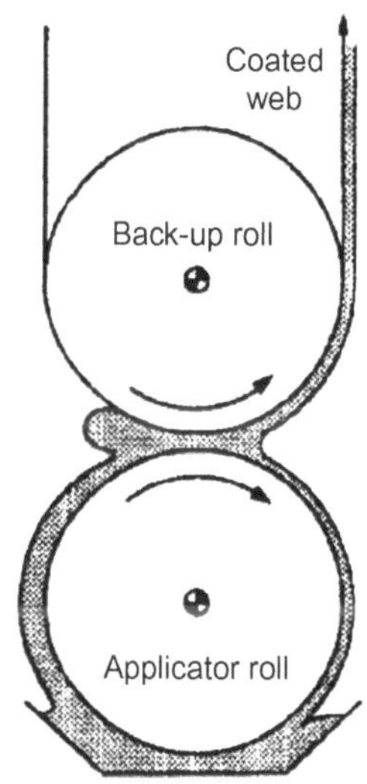

Figure 14.6 Two-roll, pan-fed forward roll coater. *Source: From Ref. [7]. Reprinted with permission of John Wiley & Sons, Inc.*

In pan-fed forward roll coaters, the applicator roll dips into the coating pan where it picks up the coating fluid. In fountain-fed coaters, an upward pointing slot die coater sends a fountain of fluid across a wide gap against the bottom of the applicator roll. In nip-fed coaters, fluid is fed directly into the nip and kept from flowing out by edge dams.

The forward roll coater is not widely used, especially compared with the popular reverse roll coater, because it gives a relatively poor quality coating and has less viscosity latitude and lower line speeds. It is also very susceptible to a defect called ribbing, where the coating thickness varies sinusoidally across the substrate. It appears as though a giant comb or rake were dragged down the wet coating in the machine direction. Ribbing occurs when the capillary number, the dimensionless ratio of viscous to surface forces, exceeds a certain value that depends on the gap-to-diameter ratio. The capillary number is:

$$C_a = \frac{\eta \cdot U}{\sigma}$$

where:

η	=	coating fluid viscosity
U	=	average surface speed of the two rolls
σ	=	surface tension

It is very difficult to avoid ribbing in forward roll coating. When the fluid is not self-leveling, a smoothing bar is often used to smooth out the ribs. It has been found that a fine wire or thread stretched across the gap exit and touching the liquid eliminates ribbing [3].

The range of operating parameters for forward roll coating is:

- viscosity: 20–2000 cP or mPa s
- wet thickness: 10–200 μm (0.4–8 mil)
- line speed: 0.5–7.5 m/s (100–1500 ft/min)
- coverage uniformity: ±10%

Reverse roll coaters. In reverse roll coaters the applicator roll is rotating in the direction opposite to the substrate that is carried on a backing roll, and a reverse-turning metering roll removes excess fluid from the applicator roll and returns it to the coating pan. The coating fluid is applied to the applicator roll by any of a number of techniques, such as having it rotate in the coating pan (pan fed), or by applying the fluid by a fountain from a slot die (fountain fed). The important action takes place in the gap between the applicator roll and the metering roll. The fluid remaining on the applicator roll after metering is completely transferred to the substrate traveling in the reverse direction on the backing roll. There are many possible configurations, one of which is shown in Fig. 14.7. The metered coating thickness on the applicator roll is a function of:

- the metering and applicator roll gap
- the metering roll to applicator roll speed ratio
- the applicator roll speed capillary number.

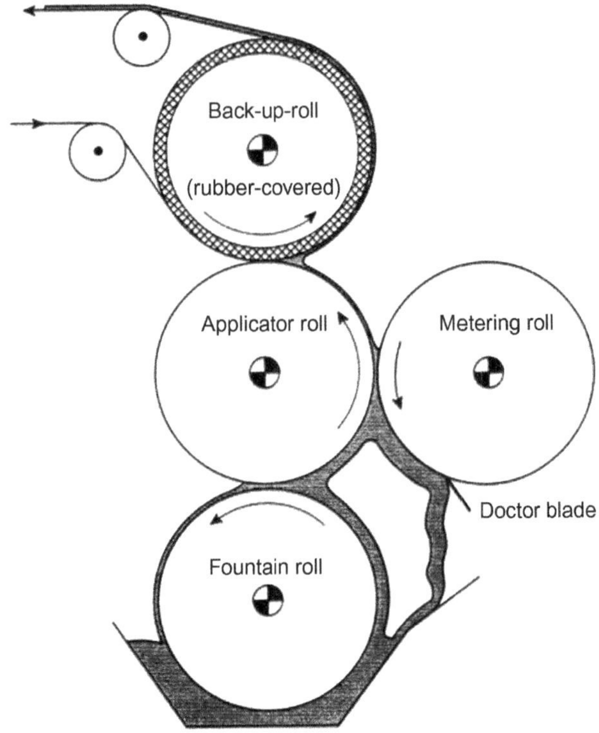

Figure 14.7 Four-roll, pan-fed reverse roll coater. *Source: From Ref. [7]. Reprinted with permission of John Wiley & Sons, Inc.*

The reverse roll coater is one of the most widely used coating methods. It is an extremely versatile coating method and gives a very uniform defect-free coverage over a very wide coating speed range. It can coat low to extremely high-viscosity fluids. The principal advantage is that reverse roll coating can be adjusted to give a stable, defect-free coating at high coating speeds. Using precision bearings, reverse roll coaters can give a coverage uniformity of ±2%.

Operating parameters for reverse roll coating are:

- viscosity: 200–50,000 cP or mPa s
- wet thickness: 14–450 μm (0.6–18 mil)
- line speed: 0.1–8.5 m/s (20–1700 ft/min)
- coverage uniformity: ±2–10%

A defect called cascade or seashore that is caused by air entrapment can appear under certain conditions in the metered flow on the applicator roll. Both ribbing and cascade can be avoided by adjusting the capillary number based on the applicator roll speed, the metering roll to applicator roll speed ratio, and the metering/applicator roll gap.

14.4.4.2 Doctored Methods

Mayer rod or wire-wound rod coaters. The Mayer rod coater has been in use for about 100 years and is still one of the most widely used methods. In it a wire-wound rod (Fig. 14.8) is used to doctor off excess fluid from the moving substrate. For increased life, the rod is slowly rotated. This evens the wear and prevents particles from getting caught under the rod and causing streaks. Normal rotation is in the reverse direction to the substrate travel. The wire-wound rod can be held against an unsupported substrate, as shown in Fig. 14.8, or against the substrate on a backing roll. When used against an unsupported substrate the web tension affects the coverage.

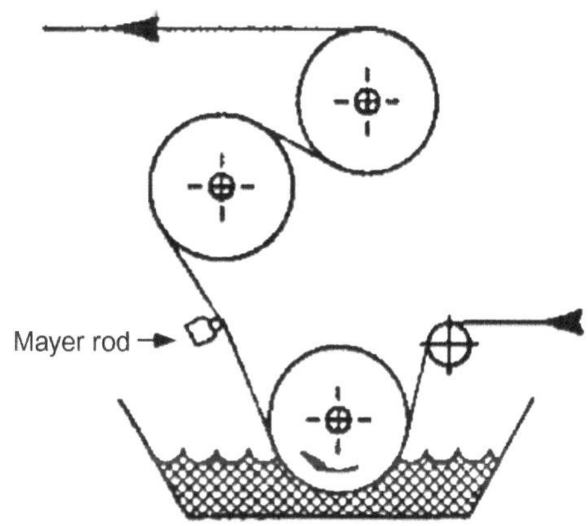

Figure 14.8 Mayer rod or wire-wound rod coater. *Source: From Ref. [7]. Reprinted with permission of John Wiley & Sons, Inc.*

The coverage is mainly controlled by the wire diameter wound onto the rod. If the wet coating occupies the area between the wire and the substrate, then it is relatively easy to show that the coated thickness should be 10.7% of the wire diameter. It is always less than this, sometimes as low as one-half. The actual coating thickness as a percentage of the wire diameter depends on:

- the fluid rheology
- the substrate speed
- the web tension when coating against unsupported substrate
- the rotation direction.

Different rod designs give different coating weights. Now rods machined to have the same profile as the wire-wound rods are also available. Mayer rod coating is very popular. The coating rods are compact and simple, but they wear rapidly when used with abrasive fluids. However, they are inexpensive and can be rapidly removed and replaced.

Because the wire wound around the central rod has an undulating surface, one would expect the coating to have a similar undulating surface. However, the down-web lines that one frequently sees are usually spaced at other than the wire diameter and are due to ribbing. If the solution is not self-leveling, a smoothing rod may be used to smooth out the surface.

Rod coaters are commonly used for low solids, low-viscosity coatings such as used to coat adhesives, optical coatings for windows, barrier layers on poly (vinylidene chloride), carbon paper, and silicone release papers. Dry coating weights range from 1.5 to 10 g/m^2, and speeds are as high as 5 m/s (1000 ft/min). Operating parameters for wire-wound rod coating are:

- viscosity: 50–1000 cP or mPa s
- wet thickness: 4–80 μm (0.2–3 mil)
- line speed: 0.05–5 m/s (10–1000 ft/min)
- coverage uniformity: ±7–10%

Air-knife coaters. The air-knife coater is a versatile coating process that is used for many products. However, the air blower power consumption is high, and newer premetered coaters can give higher qualities. Therefore air-knife coating is used less these days. Fig. 14.9 shows the major components of an air-knife coating process. They are:

- a coating pan and back-up roll to apply the coating solution to the moving substrate
- an air knife positioned after the pan regulates the final wet coating weight by doctoring the coating with a focused air jet

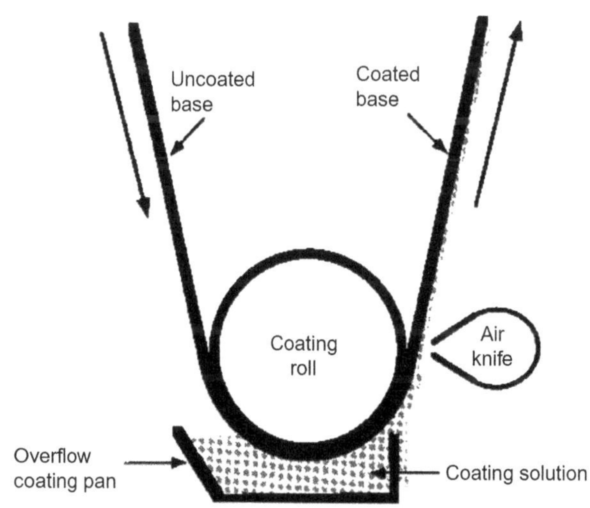

Figure 14.9 Air-knife coater. *Source: From Ref. [7]. Reprinted with permission of John Wiley & Sons, Inc.*

The excess solution is collected in an overflow pan and is either recirculated and reused or scrapped. The coating weight is determined by:

- the coating line speed
- the coating solution viscosity
- the air-knife characteristics

The air-knife characteristics are the volume, velocity, and air jet geometry impinging on the coated layer, including the air-knife angle of incidence relative to the substrate. Increasing fluid viscosity and line speed increase the thickness and increased air-knife pressure reduces the thickness.

The air knife can function in the precision or the squeegee mode. These give a very different coating and performance characteristics, although the same name is used for both processes. In the precision mode, the air knife uses lower pressures and doctors off some of the coating to control the coating weight and to level the surface to give a reasonable quality uniform coating. Good coating quality is usually obtained although the method is susceptible to streaks and chatter.

The operating parameters for air-knife precision coating are:

- viscosity: 1–50 cP or mPa s
- wet thickness: 1–200 μm (0.04–8 mil)
- line speed: 0.2–2 m/s (40–400 ft/min)
- coverage uniformity: ±5%

In the squeegee mode the air knife operates at much higher pressures and coating speeds than in the precision mode and effectively blows off the majority of the coating. This process is used for porous supports, such as paper, where the coating is absorbed into the voids. After the air knife, which effectively functions as a leveling device, the coating solids remain in the voids and in a thin surface layer.

The operating parameters for the air knife in the squeegee mode are:

- viscosity: 5–500 cP or mPa s
- wet thickness: depends on pore depth
- line speed, fpm: 0.6–10 m/s (125–2000 ft/min)
- coverage uniformity: ±5%

The advantage of the air-knife processes are:

- the low initial cost
- versatility for coating a variety of substrates and solutions
- ease of changing and maintaining the coating
- and good coating quality

The disadvantages are:

- the noise
- contamination problems created by the air stream
- the resulting spray
- solution viscosity limitations
- a somewhat restricted coating weight range
- and the high air blower operating cost

This method tends to give a uniform coating coverage, and the coating profile will follow the support profile. This is a precision method and can give good uniform coatings and has been widely used for both photographic and paper coatings.

Knife Coaters. The knife coater is a two-part applicator where the coating fluid is applied by any of several methods and then a stationary, rigid knife doctors off the excess coating. Knife coating, either against a backing roll or on unsupported substrate, is a simple, inexpensive, and effective coating method. These are simple devices, easy to operate, and need little maintenance.

Knives are usually held perpendicular to the substrate and are thick and rigid to prevent them from deforming due to the moving coating layer pressure. There are several possible knife configurations and shapes. The knife-over roll coater, Fig. 14.10, is probably the most common knife coater. It is simple and compact. The driven backup roll may be precision made and chrome-plated with a controlled gap between the substrate and the knife, or it may be rubber-covered with the knife pressing against the substrate. Here the pressure against the knife determines the coating weight, with higher pressures giving lower coating weights. Sometimes the knife is pressed against an

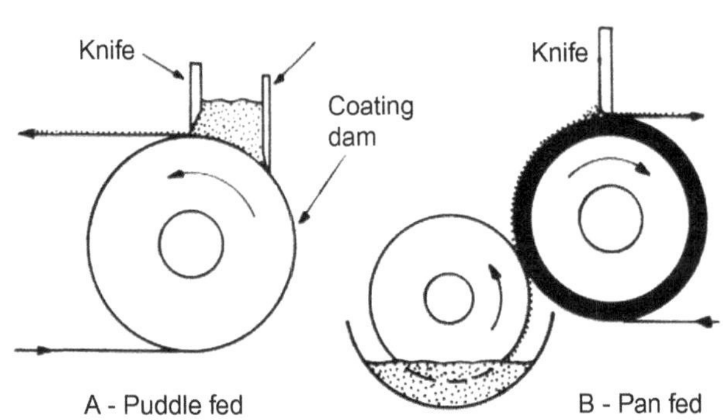

Figure 14.10 Knife-over roll coater. *From Weiss H. Coating and laminating machines. Milwaukee: Converting Technology Co.; 1977, with permission of Mrs. Sheila Weiss.*

unsupported substrate that is held taut by the substrate tension. Coating against a backing roll is more accurate than coating against an unsupported substrate, as it is independent of substrate tension. A full width endless belt may be used to support a weak substrate and pull it through the knife area without tearing, to overcome the knife drag.

The knife ends can be square, beveled, or rounded. The wet coverage is exactly one-half the gap when the end is square and parallel to the substrate, the upstream face is perpendicular to the substrate, and there is a fixed gap between the knife end and the substrate. In this case there is no pressure-driven flow and the flow is due only to the substrate dragging the fluid through the gap. The coated thickness will be greater than half the gap when there is converging flow up to the gap, especially when the converging angle is relatively low and the gap is tight. In this case strong hydrodynamic forces build up and tend to lift the knife. They also cause pressure-driven flow in addition to the drag flow, so more fluid flows under the knife and the coated thickness is greater than half the gap.

Diverging flow on the downstream side of the knife should be avoided as this very frequently causes ribbing. Thus, instead of using a round rod as a knife, one uses a bull nose knife, where the downstream quadrant of the knife has been machined away.

In knife coating any depressions in the support are filled in to give a smooth, flat surface. Knife coating tends to level rough surfaces rather than give a uniform coverage, a characteristic that can be desirable or not depending on the finished coating requirements. Streaks and scratches are hard to avoid, especially with high-viscosity liquids.

The operating parameters for knife coaters are:

- viscosity: 50–40,000 cP or mPa s
- wet thickness: 10–750 μm (0.4–30 mil)
- line speed: 1.7–25 m/s (350–5000 ft/min)
- coverage uniformity: ±10%

Blade coaters. Blade and knife coaters are very similar in their configurations, operation and variables. While knives are thick and rigid, coating blades are thin, only 0.2–0.5 mm (8–20 mil) thick, and can be rigid, or flexible spring steel. As in knife coating, blade coating gives a smooth surface and fills in surface depressions.

Flexible-blade coaters can be used with a horizontally moving substrate as shown in Fig. 14.11, with a downward moving substrate, or with an upward moving substrate as shown in Fig. 14.12. As with knife and many other coaters the fluid may be pan fed, fountain fed, or nip fed.

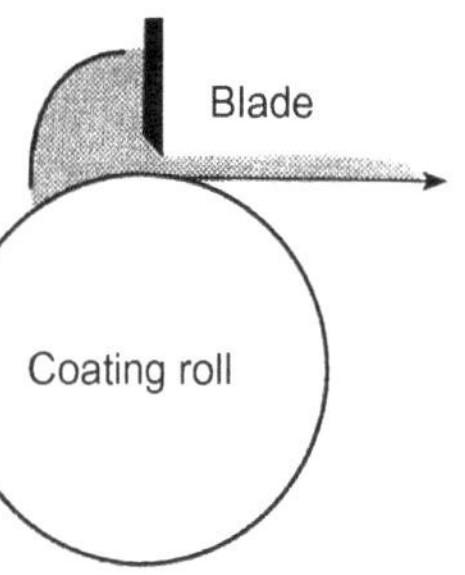

Figure 14.11 Rigid blade coater. *Source: From Ref. [7]. Reprinted with permission of John Wiley & Sons, Inc.*

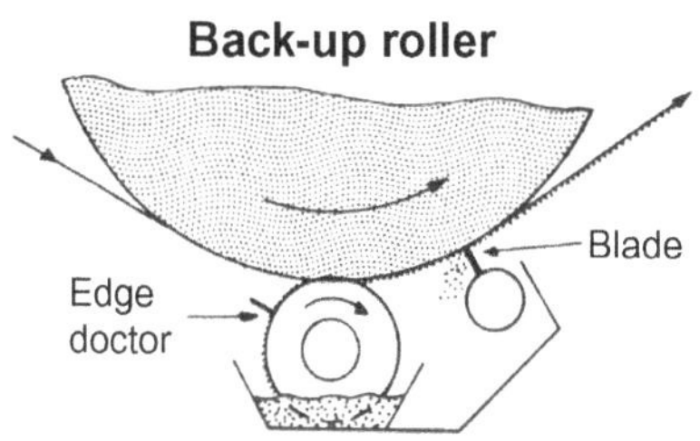

Figure 14.12 A blade coater against upward moving web. *Source: From Weiss H. Coating and laminating machines. Milwaukee: Converting Technology Co.; 1977, with permission of Mrs. Sheila Weiss.*

Blades, being thinner than knives, wear faster and have to be changed relatively often, perhaps two to four times a day. Blades are always pressed against the substrate, which is supported by a chrome-plated steel roll or a rubber-covered backing roll.

Blade coaters are commonly used on pigmented coatings. They have the unique feature of troweling in the low areas in a paper substrate, thus producing a coated surface that has excellent smoothness and printing qualities. The backing roll is usually covered with resilient material and is driven at the same speed as the web to stabilize the web and draw it past the blade. A replaceable blade is rigidly clamped at one end and the unsupported end is pressed against the substrate. The wet coverage is adjusted by varying blade thickness, the blade angle, and the force pushing the blade against the substrate. The force on the blade can be obtained by an inflatable rubber tube between the blade and a rigid member; or by rotating the blade holder to apply a greater or lesser force at the tip while keeping the angle the tip makes with the substrate approximately constant.

The wet coverage decreases rapidly as the force on the blade increases, since the force is concentrated at the blade tip. In knife coaters and in rigid blade coaters the metal does not deflect. However, with flexible blades further increases in force bend the blade more and a larger blade area presses the liquid against the substrate. With increasing loading a point is reached after which the blade tip lifts up and the coverage then begins to increase.

The beveled blade coater uses a rigid blade held at an angle of 40–55 degrees to the substrate. The end is parallel to the substrate and pressed against it. If initially the end is not parallel to the substrate, it will soon become parallel due to abrasion by the pigmented fluid. When the loading on the blade increases, the wet coverage decreases. With the same force but using a thicker blade, the pressure, or force per unit area, on the coating fluid between the blade and the substrate is less and the coverage then is more.

Both sides of a substrate can be coated simultaneously by using two flexible-blade coaters on opposite sides of the substrate, pressing against each other and also the substrate between them. The substrate usually travels vertically upward. The coating solution can be different on each side. The blades tend to be thinner and more flexible than the standard blades and the angle to the substrate is lower. The substrate has to have sufficient tensile strength to be pulled through the nip. Another configuration for two-sided coating is to use one flexible blade against the substrate on the roll, where the substrate moves vertically downward. On one side the coating fluid is nip fed between the substrate and the blade, and on the other side the fluid is nip fed between the substrate and the roll. Edge dams between the substrate and the blade, and between the substrate and the roll, keep the coating fluids contained.

The operating parameters for blade coaters are:

- viscosity: 50–40,000 cP or mPa s
- wet thickness: 10–750 μm (0.4–30 mil)
- line speed: 1.7–25 m/s (350–5000 ft/min)
- coverage uniformity: ±10%

14.4.4.3 Premetered Methods

Slot die coaters. In the slot die coater, the coating solution exits the slot between the die lips that are positioned very close to the substrate, forming of gap of 100–500 μm (4–20 mil). The fluid that exits the slot wets the die lips and forms a coating bead between the lips and the substrate. The coating fluid pumped to the die is deposited on the substrate. Fig. 14.13 shows a two-layer slot coating die. Although mainly used for coating single layers, slot coating can be used for two or even three simultaneous layers. A pump or a pressurized vessel feeds the fluid through a flow meter and control valve to the coating die. The die can operate at room or higher temperatures, if needed. The die may be maintained at the desired temperature by water flowing through internal channels in the die.

Figure 14.13 Two-layer slot coating die.
Source: Reprinted with permission of Cloeren, Inc.

A slight vacuum (up to 1000 Pa or 4 in. of water) stabilizes the coating bead by having the atmosphere press it against the substrate so that thinner coatings may be made. The maximum coating speed is increased slightly. A vacuum pump or a steam eductor provides the vacuum. Higher vacuums may be needed for higher viscosity liquids. There should be rubbing contact to seal the sides or ends of the rotating backing roll against air leaking into the vacuum box. The coating bead itself provides the seal where the coated substrate leaves the system. Where the substrate on the backing roll enters the vacuum box the gap must be as small as possible without the substrate surface touching anything stationary, to avoid scratching the substrate. The air in the vacuum chamber can and does resonate as in a musical instrument, such as in a giant tuba. These are just pressure fluctuations that, if large enough, can cause chatter marks at wide gaps. It is very difficult to eliminate this resonance, but one can reduce the amplitude of the pressure fluctuations by keeping the inward-leaking air flow as small as possible.

In premetered coating, when a uniform coating is made the necessary flow rate per unit width is just the desired wet coating thickness times the substrate speed. The viscosity does not enter into the calculation. Of course, if the viscosity varies from place to place in the die (due, perhaps, to temperature variations), then the flow rate per unit width and the coverage will also vary. In premetered coating, coverage control is simple and straightforward and in all the premetered methods using fixed or nonadjustable dies a uniformity of ±2%, even for very wide coatings, is obtained relatively easily.

The coating quality and the coating weight uniformity in the slot die are a function of die design and die geometry. Recent advances in understanding the fluid mechanics in the slot die have resulted in computer programs that help us design the die internals and lips to meet the specified requirements. Since the coating fluid rheology is either known or can be measured, the die can be designed to give uniform flow across the width with no adjustments. Because the viscosity is relatively low, the pressures within the die are also relatively low, and the die can be much less massive than an extrusion die that works with very high-viscosity fluids, and still withstand the spreading forces.

Typically, the substrate is supported by the backing roll in slot die coating. However, for very thin coatings, under about 15 μm (0.6 mil) wet thickness, the gap between the coating lips and the substrate becomes very tight, under about 100 μm (4 mil), and the system becomes difficult to control and operate. The bearing's run-out can become a significant fraction of the gap, and this can cause chatter marks as the beads fills and partially empties. Dirt can hang up in the gap to cause streaks. If the substrate contacts the coating die the substrate can tear, causing a shutdown. Coating against unsupported or tensioned substrate should be used for very thin coatings. In this case, web tension becomes an important variable.

The operating parameters for a slot die coater are:

- viscosity: 15–20,000 cP or mPa s
- wet thickness: 10–250 μm (0.4–10 mil)
- line speed: 0.02–8.5 m/s (5–1700 ft/min)
- coverage uniformity: ±2%

Extrusion coaters. The terms "extrusion coater" and "slot die coater" have no generally accepted definitions and are often used interchangeably. However, we consider them two different processes. In an extrusion coater a high-viscosity material (usually >50,000 cP), such as a molten polymer at elevated temperatures, is extruded from a coating die and onto a substrate where it is cooled and solidified. In the extrusion process the fluid exiting the slot does not wet the die lips. It forms a ribbon that is laid down onto the substrate. In slot die coating, in contrast, the low-viscosity fluid wets the die lips and fills the gap between the die and the substrate by forming a coating bead. In extrusion coating there usually is a relatively wide gap between the die lips and the substrate.

In coextruding multiple layers, in contrast to slot die coating that uses a die with multiple feed ports, multiple distribution channels, and multiple exit slots as in Fig. 14.13, normally only a single-layer extrusion die is used with one feed port. Attached to the feed port is a combining adapter [4] or feed-block with one rectangular chamber into which all the feed ports lead. The die feed port is now rectangular to match the combining adapter, unlike its usual round shape. This process works because different high viscosities layers do not mix and remain separate throughout the coating process.

As high pressures are required to push the high-viscosity fluid through the die, the die and support equipment are rigid and massive to prevent distortion from the high forces generated. The support equipment needed includes the extruder to melt and pressurize the thermoplastic polymer so it will flow through the extrusion die. The die is heated by electric heaters or by a heat transfer fluid circulating through channels in the die, permitting temperature control of the die. The die also contains adjusting bolts every 3–15 cm or so across the width that control the lip openings, and other bolts to control internal choker bars, to obtain uniform cross-web coverage. With manual die bolts, it is very difficult to get a completely flat profile. The preferred system is an online thickness gauge that measures cross-web profile and feeds a computer-controlled system to adjust the bolts. In this system the bolts themselves are electrically heated to cause them to expand or contract, and the computer controls the bolt temperatures. This is used to control the internal gap or the lip opening. A sophisticated computer program is used, as increasing the flow under one bolt decreases the flow under neighboring bolts.

An extrusion coating line is like a standard coating line. It has an unwind, a rewind, a web transport system, and a coating applicator—the extrusion die. Often there is a laminating station. There is no dryer because there is no solvent to remove. In its place there is equipment, often a chill roll, to cool the coated web down from the high coating temperature of perhaps 150–200°C (300–400°F). The driven chill roll is often chromium or nickel plated with a mirror or matte finish or with an embossed surface. The extruded film takes on the chill roll finish. The chill roll may be 60–90 cm (24–36 in.) in diameter with perhaps a 120-degree wrap angle and utilizes chilled water or water-ethylene glycol mixtures. It reduces the coated film temperature to about 65°C before it is stripped off the roll. Impingement air nozzles can be used to further cool the web. These only cool the web as there is no solvent to remove.

Extrusion coating is often used in food packaging where vapor and oxygen barriers are required and good heat sealability is desired. Necessary adhesives can be applied using extrusion coating. The expanding food packaging industry is the direct result of packaging improvements that can be attained by improving the surface and physical characteristics of a flexible web, by extrusion coating and by coextruding multilayered films. Operating parameters for extrusion coaters are:

- viscosity: 50,000–300,000 cP or mPa s. and higher
- wet thickness: 13–525 μm (0.5–21 mil)
- line speed: 0.6–9 m/s (120–1800 ft/min)
- coverage uniformity: ±5%

Slide coaters. The slide coater is a specialty coater that is only used to simultaneously coat multiple layers. We know that as many as 17 layers have been applied simultaneously, although often no more than 12 are needed. The slide coater is primarily used to coat color and black and white photographic films and papers. In color films and papers at least nine layers, three for each of the three primary colors, are coated simultaneously. The slide coater hardware was originally fabricated by the photographic product manufacturers and thus was not widely used. Recently, commercial sources for the hardware have become available.

A slide coater is shown in Fig. 14.14. It consists of rectangular plates with flow distribution channels machined into the plate with an individual coating solution feed port in each plate, except the top or cover plate. The plates are stacked to form a slide or an inclined plane and are then bolted together along the sides and back. The fluid is pumped to the plate, is distributed along the channel width, and then exits through the slots and onto the slide, and flows down the inclined plane in laminar flow without mixing. At the bottom of the slide the layers flow across the gap onto the upward moving substrate. In the gap the fluid layers form a coating bead. As with the slot die, a slight vacuum may be applied to stabilize the coating bead and permit a considerably thinner wet coating, and also to increase the maximum coating speed by perhaps 25% and reduce some coating defects.

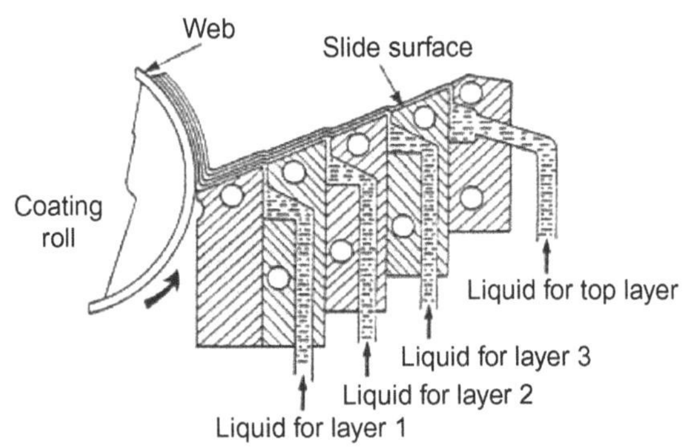

Figure 14.14 Slide coater. *Source: From Mercier JA, Torpey WA, Russell TA. US Patent 2,761,419; September 4, 1950.*

As with all of the premetered methods, the coating thickness is equal to the flow rate per unit width divided by the coating speed. With a good die design and flow control system, a very accurate coating coverage of ±2% can be obtained and the coating quality is excellent.

Coating defects that look like waves can result from interfacial instabilities in the fluid flowing down the slide. To avoid these interfacial waves, the physical properties of all the layers should be close to each other so that the system resembles a single layer. The important properties are density and viscosity. The densities are usually close to each other and rarely differ by even 10%, and often are not subject to adjustment. The individual layer viscosities should be reasonably close to each other and can often be adjusted by adding thickeners. Kobayashi [5] has suggested that to avoid interfacial waves, in any set of adjacent layers the ratio of the upper layer viscosity to the lower layer viscosity should be no more than about 1.5 or 2 (a ratio greater than 10 is also good) or less than 0.7. These critical ratios must be only approximate, as the layer thickness is also important. It is known that interfacial waves are less likely to form on the slide with thicker layers, and the layers will be thicker on the slide when one coats faster and thus require a higher flow rate. Thus, interfacial waves are less likely to form on the slide in production coaters than in pilot coaters at their much lower coating speeds.

The bottom layer should have the lowest viscosity to reduce drag forces and allow higher coating speeds. There is even one patent [6] that suggests using plain water as a desirable bottom or carrier layer to increase the coatability window. In one case with a viscosity of 50 cP for the second layer, the viscosity ratio is about 60, well over the 10 that Kobayashi [5] suggests, as noted in the previous paragraph.

In order for the layers to spread on the slide, the top layer surface tension should be lower than that for all the other layers. It does not matter what the other surface tensions are, as long the top layer has the lowest value. When surfactant is needed to get good wetting, as in aqueous coating, surfactant does not have to be present in the bottom layer as long as it is present elsewhere, as diffusion in these thin layers is so rapid that surfactant will be present in the bottom layer when needed.

The operating parameters for the slide coater are:

- viscosity: 1–500 cP or mPa s
- wet thickness: 25–250 μm (1–10 mil)
- line speed: 0.1–5 m/s (20–1000 ft/min)
- coverage uniformity: ±2%

Curtain coaters. In a curtain coater the coating applicator forms a falling sheet or curtain of the coating solution that falls onto the substrate passing under the curtain. There are two variations of this method that have very different properties: the standard curtain coater and the precision curtain coater.

Standard curtain coaters. In the standard curtain coater the curtain can be formed by fluid overflowing a weir or by a downward pointing die. The curtain is normally wider than the substrate and a catch basin is used to collect the overflow and return it to the feed system. This is not strictly a premetered system as all of the coating does not go onto the substrate. This technique has been in use since 1903 and is still widely used to coat adhesives, paints, food products, paperboard, plywood, irregular shapes, chocolate, thick coatings, and electronic coatings. Both continuous substrates and discrete sheets can be coated with this method.

It is a simple, inexpensive method of applying a coating with a moderate coating quality. One disadvantage is that a high flow rate of about 0.5–1.5 cm^3/(s cm width) is needed to maintain an intact, stable curtain. About double this minimum is desirable for a reliable coating process. Because of the high flow rate needed for a stable curtain, curtain coating is inherently a high-speed process. Water is the preferred solvent because of the environmental issues created by evaporation from the curtain. Wire edge guides or thin rods are used to prevent the edges from necking in due to surface tension.

When a weir is used to form the curtain, the weir usually is formed of many adjacent small *V*'s. The coating is delivered to the open weir uniformly across its width by a pipe having diffuser jet openings. The coating fluid flow rate controls the falling curtain thickness.

As with any recirculating system, air bubbles can be a problem.

The operating parameters for the standard curtain coater are:

- viscosity: 150–2000 cP or mPa s
- wet thickness: 25250 μm (1–10 mil)
- line speed: 1.5–6.5 m/s (300–1300 ft/min)
- coverage uniformity: ±5%

Precision curtain coaters. The precision curtain coater was developed to coat multilayer photographic films. A slide is used to generate the multilayer structure that then flows over an added die lip to form a curtain. Sophisticated wire edge guides or rods are used to prevent the curtain from necking in due to surface tension. For precision coating the curtain has to be completely uniform across the width. The curtain is narrower than the substrate, so there is no overflow and thus it is a true premetered coater.

The die's vertical distance above the substrate can be adjusted, and the falling curtain is protected from stray air movements by transparent enclosure sheets. This method is used only with continuous substrates. Similar to the standard curtain coater, this method has a high minimum flow rate to create a stable curtain, and this in turn necessitates a relatively high coating speed. There is a minimum speed set by the minimum flow rate per unit width for a stable curtain divided by the desired coating thickness. Below this speed the system will not function properly.

The precision coater is now used for other applications than coating photographic films. The method patents have expired and reasonably priced hardware is available. A good application of this method is coating single–layer, pressure-sensitive adhesives. Speeds up to 165 m/s (3300 ft/min) can be obtained with low defect levels.

The advantages of precision curtain coating are a high maximum coating speed, multilayer capability, and excellent coating quality. There is a very wide coating gap, easily allowing splices to pass, unlike the tight gaps in the other coating methods that usually require applicator withdrawal in order to allow splices to pass, and thus result in a yield loss. The wide gap also prevents the gap from catching dirt carried in by the substrate that leads to streaks and spots.

The operating parameters for the precision curtain coater are:

- viscosity: 5–500 cP or mPa s
- wet thickness: 5–500 μm (0.2–20 mil)
- line speed: 1.5–15 m/s (300–3000 ft/min)
- coverage uniformity: ±2%

14.4.4.4 Hybrid Methods

Gravure coaters. Gravure coating uses low-viscosity (10–5000 mPa s or cP) liquids to coat thin layers (1–25 μm wet coverage). It is used for long runs and it is used as a precision method to give thin coatings that other techniques cannot do. In gravure coating a chrome-plated cylinder, the gravure cylinder, is patterned with small depressions (cells) or reservoirs and normally picks up the coating fluid from a coating pan or reservoir. The excess fluid picked up is then doctored off, leaving fluid in the depressions. This fluid is then transferred to the substrate carried by the impression roll. The coating liquid can be applied to the gravure roll by other methods than by the pan-fed system. Also ceramic gravure cylinders are increasing in use. While they are more expensive than the conventional cylinders, they give an order of magnitude increase in cylinder life.

There are several process configurations used. Most use a backup or impression roll to create a nip and coating bead. In *direct gravure*, the coating solution transfers directly from the gravure roll to the substrate on the impression roll.

In *forward gravure* the gravure cylinder or the offset roll, if used, and the impression roll all run in the same direction. In *reverse gravure* the gravure cylinder or the offset roll, if used, and the substrate on the impression roll run the opposite direction. The sliding motion between the substrate and the reverse gravure cylinder requires that the gravure cylinder or offset roll force against the substrate on the impression roll must be relatively light. In *differential gravure* the gravure cylinder and the back-up roll run at different speeds. This is one way to control the coating coverage. In one gravure configuration there is no impression roll and the substrate is unsupported, similar to kiss coating. Here the web tension control is very important.

The fluid transferred from the gravure cylinder to the substrate or the offset cylinder depends on the fluid quantity in the engraved depressions, the cell pattern, and the fluid fraction that is transferred to the substrate on the impression roll. In forward gravure a high force is needed on the gravure cylinder against either the substrate on the impression roll or the rubber-covered offset roll to get a higher cell fluid fraction to transfer. The usual force is about 2000–20,000 N/m (12–120 lb/in). A different engraved cylinder is needed for each different coating weight. The fluid viscosity cannot be above about 5000 mPa s, and should also have a low surface tension. The coated fluid will tend to have the engraved roll pattern and so we would like it to tend to level. Reverse gravure and differential gravure with its sliding motion can improve leveling, or a smoothing bar may be used to remove the pattern.

There are three common cell patterns for gravure cylinders that are in routine use, as illustrated in Fig. 14.15. The pyramidal and quadrangular cells are similar, except that the quadrangular has a flat, not a pointed, bottom in order to empty easier. The trihelical pattern consists of continuous grooves spiraling around the roll, usually at a 45-degree angle. The volume factor is the term used to

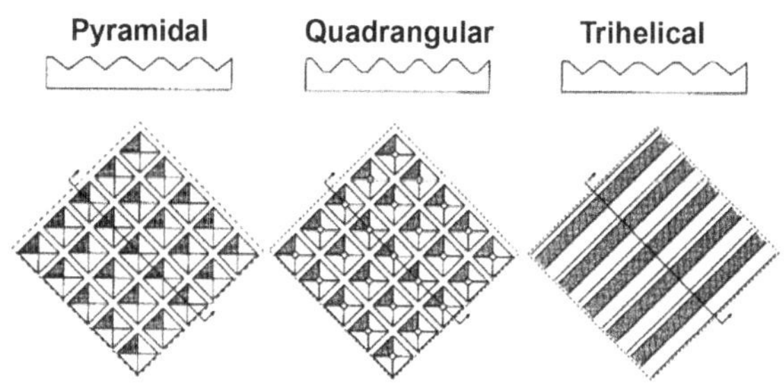

Figure 14.15 Gravure cell patterns. *Source: From Ref. [7]. Reprinted with permission of John Wiley & Sons, Inc.*

represent the total cell volume per unit area and therefore has the units of height. It typically ranges from 4 to 300 μm. The cell volume fraction that transfers varies greatly depending on the system. The cell pitch controls the substrate pickup stability. With high impression roll pressures one can expect about 58–75% cell volume transfer. The cell pitch or count is the number of cells per centimeter measured perpendicular to the pattern and usually ranges from 4 to 160 per cm (10–400 per inch). The cell pattern may be made by mechanical engraving, chemical etching, electromechanical engraving, or laser etching.

Excess solution is doctored off the gravure cylinder, usually using a 0.1–0.4 mm (4–16 mil) spring steel blade making a 55–70 degree angle with the incoming gravure roll surface. The blade is slowly oscillated about 6–50 mm (1/4–2 inch) to give even wear and to dislodge dirt that could cause streaks. Or one can use a reverse-angle doctor blade that makes an angle of 65–90 degrees with the exiting surface. This blade does not have to be loaded against the cylinder face as now the fluid presses the blade against the surface, and so the reverse blade can be made of softer materials such as bronze or plastic.

Using the standard doctor blade, a relatively soft blade, or one having a lower loading against an almost smooth cylinder with a shallow pattern, allows excess liquid to pass through, similar to the flexible-blade coater. A stiff, highly loaded blade against a cylinder having a large volume factor (cell volume per unit area) wipes the surface clean. In place of separate doctor blades to remove the excess coating solution, closed feed chambers with two doctor blades, one to seal the bottom and the other to doctor the top, may be used to both supply the fluid and doctor off the excess. These chambers are fairly small and are fed by what looks like a slot die. Air may be removed from the chamber on start-up by a suction pump.

Leveling the coating can be a problem. A large spacing between cells often results in printing the cell pattern rather than a uniform coating.

The operating parameters for gravure coaters are:

	Chamber	Direct	Offset
Viscosity, cP or mPas	10–200	1–500	50–1300
Wet thickness, μm	1–75	3–65	3–210
Line speed, m/s	0.1–12	0.1–12	0.05–5
ft/min	25–2300	25–2300	10–1000
Coverage uniformity	±2%	±2%	±2%

Microgravure coaters. The microgravure coater is a very useful gravure coating technique for applying a low coating weight on a thin gauge substrate. The method gives a uniform and smooth coated layer that is widely used for imaging, electronics, packaging, coated batteries, and other specialty applications.

The method is a reverse kiss gravure coater. A typical configuration is shown in Fig. 14.16. The gravure rotates in the opposite direction to the substrate and, as it is a kiss coater, there is no back-up roll. The method is unique in that it uses small diameter rolls, 20–50 mm (1–2 in) versus 150–300 mm (6–12 in) for conventional gravure, and in using the kiss principle so that there is no back-up roll. This results in a small stable bead

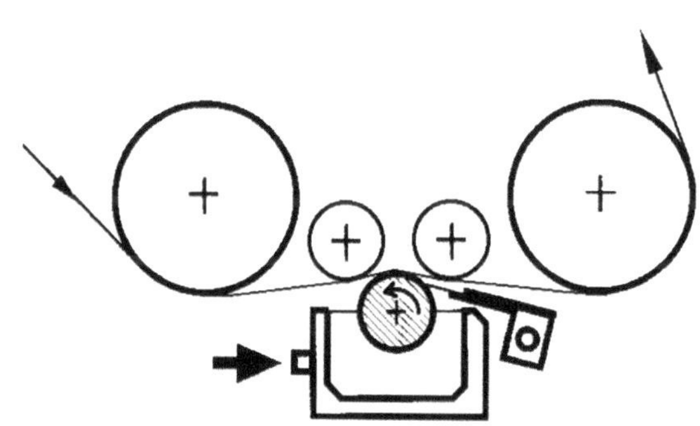

Figure 14.16 Microgravure coater. *Source: Reprinted with permission of Yasui-Seiki.*

that, combined with the reverse application, gives a very good quality and a low coating weight.

The operating parameters for the microgravure coater are:

- viscosity: 1–4000 cP or mPa s
- wet thickness: 8–80 μm (0.3–3 mil)
- line speed: 0.005–1.7 m/s (1–330 ft/min)
- coverage uniformity: ±2%

14.5 Effect of Solvent

There are many different solvents that are used in the coating process. The two principle categories are organic and aqueous. A solvent's primary function is to dissolve and disperse the formulation ingredients and maintain a stable solution or dispersion both during storage before coating and also during coating. Most often the coating solution contains insoluble particulates in addition to the soluble ingredients, and these must be dispersed and kept stable. Therefore, the primary criterion in selecting a solvent or solvent system is that it dissolves or disperses all of the formulation ingredients.

The other requirements for the solvent system involve:

- safety
- health
- environmental
- economic considerations

The solvent and its vapors should be nontoxic to the coating machine operators. Emissions and waste disposal must meet local and federal guidelines. Water as a solvent has several advantages. It is readily available, its vapors are not an explosion hazard, it is safe to use, and it is inexpensive. However, it may not dissolve the organic compounds in the formulation. Also, its high heat of vaporization requires more drying capacity. To increase aqueous solubility, co-solvents and surfactants may be used in the formulation. In addition, the polymeric binders can be synthesized using an emulsion polymerization process to give emulsions that are already in an aqueous formulation. Although pure water is environmentally benign, the final formulation may present disposal problems.

Organic solvents have excellent dissolving power and require less drying capacity than aqueous systems. However, they or their vapors can be hazardous to the operators' health, they may be expensive, and their vapors are not "green" as they are volatile organic compounds. Also, most organic vapors are explosive in a certain concentration ranges. We are mainly interested in the lower explosive limit or LEL. This is the minimum vapor or gas concentration in air, measured as volume percent, below which flame propagation (an explosion) cannot occur on contact with an ignition source. Below the LEL there is too little combustible fuel to sustain a flammable mixture. To avoid explosions, some companies operate with a nitrogen blanket in the dryer. Then all solvent concentrations are safe. However, the unwanted oxygen in the dryer has to be rigorously maintained below a safe level, normally 5%. Nitrogen is typically supplied as a cryogenic liquid, and the cold liquid nitrogen is used to condense out solvent from the dryer exhaust. The cleaned-up nitrogen is recycled to the dryer, along with make-up nitrogen that has been vaporized in the condenser while condensing solvent.

If there are safety and health hazards the coating process can be designed to minimize these hazards. However, this requires additional ventilation and may require special procedures and equipment. Toxic fumes usually need to be separated out, collected, and treated before disposal, and disposal may have to be in a hazardous waste site or by combustion.

For the coating applicator, the coating quality and operability are not a function of the solvent system. The formulation coatability in a coating applicator only depends on:

- solution or dispersion rheological properties
- quality level required
- desired coverage
- uniformity
- line speed

If the coating solution properties are the same for an aqueous and a solvent-based formulation, the product coated will be the same, although the product quality may greatly differ.

However, when selecting a coating method, the solvent evaporation rate, the vapor level in the coater air, the solvent emitted by the dryer, and explosion hazards need to be considered. Some coating methods are better suited to solvent coating because they emit less solvent. A slot die coater has only a small bead exposed to the coating room atmosphere and will emit less solvent than roll coaters that have large rotating wet surfaces and therefore higher evaporation rates. With curtain coating, the falling curtain also has a large surface for evaporation that will give high ambient solvent concentrations. Thus it is more suited to aqueous coating then to solvent coating.

It should be mentioned that in drying, due to water's relatively high diffusion rate, its small molecular size and the high heat of vaporization, most aqueous system drying takes place in what is known as the constant rate period, while for solvents most drying occurs in the falling rate period. This has no effect on the coating process, and is mainly of interest in calculating drying.

14.6 Hot Melt Coaters

Some formulations can be coated without using solvents. Among these are formulations with thermoplastic polymers and adhesives that have been heated and melted. Temperatures up to 200°C (392°F) are used to reduce the viscosity, as shown in Fig. 14.17. At the high temperature the hot melt viscosities can range from 9000 to 100,000 mPa s or cP. Formulations with these viscosities can be coated using reverse roll and slot die techniques, and extrusion dies may also be used. The applicators require heat sources to maintain the temperatures needed to keep the polymers fluid. Additives can be incorporated into the binder using high-temperature mixing equipment or extruders that can heat the binder and mix it with additives.

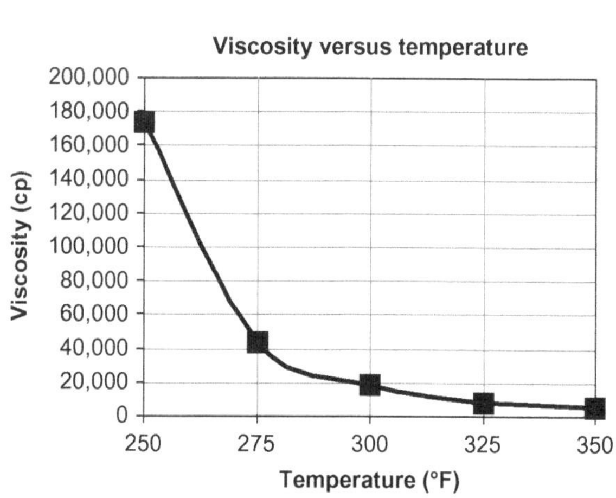

Figure 14.17 Viscosity–temperature relationship for a hot melt coater.

This method completely eliminates solvents and their safety, health, and environmental concerns. The coating machine can be simpler because solvent dryers are not needed. The dryers are replaced with cooling rolls or short air impingement cooling sections.

14.7 Selecting a Coating Method

A major requirement for the successfully manufacturing a coated product is that a defect-free high-quality product be produced with high productivity. This will insure a competitive product and competitive costs. To meet these requirements the appropriate coating method must be used in the scale-up and manufacturing process. This selection can be complex because there are many factors that go into choosing the best method.

All coating methods can coat a product at some conditions. However, those conditions may not insure a high-yield, defect-free product. Coating with a nonoptimal coater and coating conditions will produce product with some defects causing a resulting high yield loss. Both time and money are wasted by trying to make the product by a process that is not appropriate. Often a nonoptimal method is used because it is available at the site and must be used. Forcing a product into available facilities can, in the long run, be very expensive. It may be a better choice to modify the existing coating machine to use a new coating method while retaining the existing dryer and web handling equipment. Under other conditions, it is appropriate to install a completely new machine. One should mention that a coating process that works well at low speeds in the laboratory may not be appropriate for a manufacturing plant coating at high speeds. Conversely, a high-speed coating process may not be appropriate for laboratory trials.

Selecting a new coating method or a new machine may be needed when new products are being developed, when increased capacity is needed, and when quality and productivity improvements are needed. A study may also be carried out to see whether the current coating method is the best. It may be necessary that the best method be used to meet the requirements for the product. The procedure that should be used for this analysis consists of the following steps:

- establish the product requirements and the basic data
- evaluate the different coating methods' capabilities

- match the coating requirements with the coater capabilities
- select the best method
- use the best method

The first step is to establish the product requirements and to prepare a basic data document that quantifies the product requirements. There are two basic data categories. The independent parameters are those that are fixed and must be met, for they are the basic process needs. The second category is the dependent parameters that are controlled by and arise from the independent variables. The required product volume to be manufactured over 3–5 years is an independent variable. The coating speed and width are dependent variables that arise from the volume that needs to be coated in a given time.

In preparing the basic data, the best available quantitative data should be used. If there is some uncertainty about a particular parameter, then the best estimate should be included and changed as more data becomes available. At this stage, information should not be too narrow. Consider possible future needs. It is easier and cheaper to delete information as selection proceeds than to add an important factor later. The independent variables that need to be specified are as follows:

- dry coating weight
- coating weight uniformity
- number of layers
- quality level needed
- substrate: type, thickness, and surface treatments
- minimum coated width needed
- volume per unit time for the next 5 years
- drying and curing needs
- maximum mill cost needed
- solution application efficiency desired
- timing to market

The dependent parameters that need to be specified are as follows:

- wet layer thickness
- coating width (This is normally wider than the minimum coated width)
- web width
- line speed
- solvent system to dissolve ingredients
- solution rheology
- available hardware
- safety and environmental issues

The second step is to evaluate the different coating methods' capability. There are several sources that can be used for this data. There are published data in the literature, such as Cohen and Gutoff [7,8], Gutoff and Cohen [9], and Scriven [10]. Companies that manufacture coating applicators are happy to share this data. Contract coating companies also make this data available. And discussions with coating personnel can produce useable data.

The third step is to compare the basic product requirement data, Step 1, with the capabilities of the methods, Step 2 and then select the method that meets the criteria. There will be differences in the coating methods capabilities that meet all of the requirements. Therefore, the selected methods need to be analyzed to choose the best ones. A good way to do this analysis is to prepare a spreadsheet that has the coating methods being considered on the vertical axis and the requirements on the horizontal axis (Fig. 14.18). Each method is then rated for its ability to meet the specific requirements. A simple three-level rating system will work, where a 1 definitely meets a requirement, a 0 has the potential to meet a requirement, and a −1 does not meet a requirement. The values for all the requirements are then summed and the methods ranked. The top two or three are the methods that should be pursued. A wagon wheel chart can also be used. The spokes of the wheel are the requirements and they are quantified with the range of each requirement. One chart is used for the requirements and the operating range is colored in blue, and a second chart is used for the capabilities of the method and is colored in yellow. They are then overlaid and the method with the widest operating ranges where needs and requirements are in green can be determined (see Ref. [11]).

The fourth step is to select the best method to be used in scale-up and production. To do this, the best two or three methods from Step 3 are evaluated in pilot plant experiments. Statistical designs should be used to determine the product response to key coater variables and to evaluate product quality and performance. In addition, the operability window for the methods should be determined. A few experiments will be required to select the best methods and to check reproducibility. If the best coating methods are not available locally, then tests should be run using facilities at contract coaters or coater manufacturers.

Applicator Methods	Overall rank	Can batch coat?	Ctg. Wt. mg/cm^2	Quality	Coverage uniformity	Width (In)	Solvent	Volume ft^2/day	Availability	Experience	Capital needed
GOAL		Yes	0.3–1.5	Mod.	1–10%	0.5–15	All	20–1100	Yes	Yes	>200 K
Spray	**9**	1	1	1	1	1	1	1	1	1	0
Gravure	**8**	1	1	0	1	1	1	1	1	1	0
Hirano, comma	**4**	1	1	1	1	1	1	1	–1	–1	–1
USP 5 360,629	**4**	1	1	1	1	1	1	1	–1	–1	–1
Slot die	**3**	–1	1	0	0	0	1	1	1	1	–1
Extrusion coater	**2**	–1	1	0	0	1	1	1	–1	1	–1
Slide	**2**	–1	–1	1	1	1	0	1	0	1	–1
Curtain, precision	**2**	–1	–1	1	1	1	0	1	0	1	–1
Wire-wound rod	**1**	–1	–1	–1	–1	1	1	1	1	1	0
Printing, flexo, gravure	**0**	1	0	0	0	1	0	1	–1	–1	–1
Curtain, std.	**0**	–1	–1	1	1	1	0	1	0	–1	–1
Knife-over-roll	**0**	–1	0	–1	–1	1	1	1	1	0	–1
Reverse roll	**–1**	–1	–1	0	0	1	1	1	–1	–1	0
Forward roll	**–1**	–1	–1	0	0	1	1	1	–1	–1	0
Choinski patent	**–1**	1	1	0	0	0	0	0	–1	–1	–1
Air knife	**–2**	–1	–1	–1	–1	1	0	1	–1	1	0
Blade	**–3**	–1	–1	–1	–1	1	–1	1	–1	1	0
Porous roll	**–6**	–1	0	–1	–1	1	–1	–1	–1	–1	0
Screen	**–5**	1	–1	–1	–1	–1	–1	–1	–1	1	0
Rotary screen	**–6**	1	0	–1	–1	–1	–1	–1	–1	–1	0
USE CRITERIA: 1 = Meets Goal; 0 = Potential to Meet Goal; –1 = Does Not Meet Goal											

Figure 14.18 Coating methods comparison using a spreadsheet.

The fifth and final step is to use the best method in scale-up and production. Using contract coaters should be considered to scale up the product and to confirm the pilot plant findings on the operability window and on the effects of key variables. Contract coaters should also be considered for initial manufacture during construction of the production facilities. If during these tests the first method does not meet expectations, then the second method should be tried.

14.8 Drying and Solidification

The web-coating dryer removes the solvent used in the coating process without adversely affecting the coating quality and the product performance parameters. The dryer transfers heat into the liquid coating, thereby evaporating and removing the coating solvents. The dryer can also cross-link or harden the binder, vaporize chemicals other then the solvent, and equilibrate the dried coating with ambient air.

Most systems impinge hot air on the coating surface and, in flotation drying, on the back side too. Along with providing heat, the air also removes the solvent from the coating surface. Also, the heat transfer rate and thus the drying rate can be increased by increasing the air velocity impinging onto the coating surface and on the back side.

There are many different hot air convection nozzle configurations for delivering air to the film surface with many nozzle and jet designs. The two general configurations are single-sided dryers, and two-sided or flotation dryers. In single-sided dryers the air impinges on the coated side from nozzles, jets, slots, or orifices, and rolls are used to support the web in the dryer. In two-sided dryers, the air impinges on both sides, and also serves to float and support the web and aid in its transport through the dryer. Two-sided drying almost doubles the heat transfer rate compared with single-sided drying. (In two-sided drying the heat transfer coefficient is effectively doubled, but because the web and coating temperature are increased somewhat, the temperature driving force, ie, the air temperature minus the coating temperature, is decreased slightly. Thus, the heat transfer rate is less than doubled.) In addition, as the drying air supports the web, there are no rolls contacting the web. Thus scratches caused by the web sliding on a roll, and repeat marks and contamination caused by roll contact with the moving web are eliminated.

Typically, these dryers often have separate sections or zones where the air temperature, velocity, and solvent level can be controlled independently. The zoning improves quality by allowing mild treatment in the start of the drying process where the coating is very liquid and easily disturbed, and more severe conditions in the later zones where the coating is drier and is more resistant to deformation.

Drying air temperatures need to be controlled for several reasons. Higher drying temperatures result in more rapid drying. However, there are temperature limitations both for the substrate and for the coating. When plastic films are heated they can distort and stretch under the machine tension required to transport the film. Each plastic has its own temperature limit. The wet coating must not be heated above its boiling point or blisters will form on the surface. The as-coated coating boiling point may be near the solvent boiling point. However, as the coating solvent concentration decreases the boiling point rapidly rises, and thus the allowable temperature also rises. In addition, the coatings themselves may have a maximum temperature above which the coating degrades. Many photographic coatings, for example, should not be heated above about 45°C to prevent degradation.

In addition to supplying heat by convection using air impingement dryers, conduction heat transfer by running the coated web over a heated roll and radiation heat transfer using infrared or microwave radiation are also used. In conduction drying the hot roll is usually heated by steam. Air still has to be supplied to carry away the solvent vapors. Chilled rolls or drums are also used to rapidly cool hot extruded films. As the temperature falls, the plastic viscosity rapidly increases and the film solidifies. Chilled rolls can also be used at the end of a dryer to cool coated webs down to ambient conditions. Heat transfer by conduction is very efficient, but the large diameter, double-walled rolls are very expensive and have large inertias. This technique is used mainly at low speeds, and is not often used in newer systems.

When infrared and microwave radiation are used to supply heat, air must still be used to carry away the solvent. Radiation provides a high heat input over short distances. Infrared dryers are often used in conjunction with convection dryers. They are frequently positioned at the start of drying to rapidly bring the coating up to the desired temperature and evaporation rate. The higher initial evaporation rate gives a more rapid increase in viscosity than with an impingement dryer alone, thus allowing a more rapid increase in the impinging air velocity without damaging the wet coatings. Infrared heaters can also be placed between the air nozzles. This increases the drying rate and thus the production rate without increasing the dryer length, and without requiring additional air handling systems.

Microwave radiation is used only for aqueous coatings, since the heat is created by rapidly oscillating water molecules that are strong dipoles. Most other molecules have very low sensitivity to microwave radiation. This radiation is not often used as it is more expensive than the other heating methods.

14.8.1 Constant Rate and Falling Rate Drying

The drying process is divided into the constant rate period and the falling rate period. The requirement for constant rate drying is that the coating behaves as if it were a solvent pool and, as there is always enough solvent on the surface, the drying rate is then determined by the heat transfer rate to the surface and the mass transfer rate from the surface. These are determined by what takes place in the air, and not by diffusion in the coating. In the constant rate period the drying rate is constant only at equilibrium, not throughout the whole constant rate period. Because the heat transfer rate is determined by the temperature driving force from the air to the coating (and by the heat transfer coefficient), the coating has to remain below the drying air temperature for there to be a temperature driving force. In fact, in single-sided impingement drying of aqueous coatings, the equilibrium coating temperature in the constant rate period is the so-called wet bulb air temperature.

However, at the start of the dryer the equilibrium conditions have not yet been reached, but it is still in the constant rate period. There are cases where equilibration is never reached in the constant rate period and the drying rate continuously changes, yet it is still in the constant rate period. (One example is a photopolymer coating on 0.5 mm aluminum sheeting. The heat capacity of the "thick" metal is so great that the temperature can be accurately and easily calculated by just assuming that uncoated aluminum is being heated up in the dryer. And the drying rate always is changing in this constant rate period.)

After the constant rate period is over, the coating no longer resembles a solvent pool. The surface appears dull and dry. Now the drying rate is determined by how fast solvent can diffuse to the surface, and not by the conditions in the air. The diffusion rate to the surface is determined by the diffusivity and by the concentration driving force (dc/dx) in the coating surface. The diffusivity is a function of the coating temperature and of the solvent concentration in the coating. As the coating dries and the solvent concentration decreases, the diffusivity drops rapidly, the evaporation rate and the evaporative cooling decreases, and the temperature rises to approach the drying air temperature. At higher temperatures the diffusivity is higher. The drying rate continuously decreases in the falling rate period because the effect of the lower concentration on reducing the diffusivity is much greater than the effect of the increasing temperature on increasing the diffusivity.

14.8.2 Pollution Considerations in Drying

An important consideration in drying is that the solvent vaporized during drying is frequently a pollutant and must be removed from the exhaust air before it is discharged into the atmosphere, to insure that government standards are met. To do this the exhaust air is passed through a treatment facility where the solvent vapors and other pollutants are separated from the air by condensation, by adsorption onto activated carbon or other adsorbent, or by absorption into a circulating aqueous solution. The solvent can be recovered and reused or, if it is flammable, it can be burned. There are several commercially available systems to perform both functions. If the solvent is a pollutant, then the amount used should be reduced, and consideration should be given to using a nonpolluting solvent if possible.

14.8.2.1 Online Quality Control Systems

14.8.2.1.1 Introduction

Online quality control systems are an essential component of a web-coating machine [12]. They insure reproducible high quality, and an economically effective product can be manufactured. In these systems, various technologies are used to scan the applied wet coating to detect and characterize many physical defects that can occur such as contamination spots, ribbing, bubbles, streaks, chatter many more that are too numerous to describe. In addition, there are measurement systems that measure the applied coating thickness. These systems insure that the coating weight is within limits and that it is uniformly distributed across the substrate [13].

Utilizing these systems instead of the traditional manual and offline systems has several advantages when compared to the traditional offline quality control systems:

- Improved product quality and reproducibility
- Precise and reproducible defect characterization
- Real-time results
- Cost savings and reduced scrap by rapidly detecting defects
- Experiments to reduce defects can be immediately conducted online with rapid results; thereby, reducing the scrap product that is generated and providing data for process improvement
- The generated measured property data is stored for future retrieval and subsequent analysis
- Reduced loss for out of limits coating weight, better control of low coating weight products, and reduced loss in small volume coating runs
- Significant savings result in excellent Return on Investment

The system disadvantages are:

- High initial cost to purchase and install
- They are technically sophisticated and require precise operation trained personnel and appropriate maintenance
- Testing and machine optimization is required to insure that system functions as intended

14.8.2.1.2 Traditional Quality Control Systems

There are two quality control systems that have been used to monitor and control product quality:

- Online manual inspections system
- Offline laboratory characterization.

In the online manual inspection system there is a window in the coating line that is used by operating personnel to visually inspect the coated film for defects as it passes through the coating machine. These windows can be after the coating station and prior to winding the roll. Various lighting and filter configurations are used to increase the defect visibility. When defects are detected they are recorded on the run log sheet for use in removing defects, when the roll is converted into the useable sizes. This procedure was marginally effective at low coating speed. However; as line speeds increased it is not effective because the defects pass through the window and the human eye cannot capture and analyze them. In addition, it is very difficult for the inspector to be effective over long time periods and small defects cannot be detected. This procedure is better than no inspection, but; it is ineffective for the quality and cost requirements of current products.

In the offline laboratory characterization, samples are taken from the coated roll and then they are tested offline in the Quality Control laboratory for chemical and physical parameters using standard analytical techniques. The coating quality was determined by visually examining the coating to identify defects, their severity and mark their location. The advantage of this method is that it is inexpensive, it can detect defects and it is far more effective than the online manual inspection.

However, this method has several disadvantages:

- Because there is a finite time between the actual coating and characterization, large quantities of defect material can be produced before defects are detected. A 6 foot wide web being coated at 500 fpm produces 3000 sq. ft. per minute. A 20 minute test time results in 60,000 sq. ft. is coated before it is known if the product is acceptable and corrective action can be taken. This factor became more important as coating line speeds were increased.
- The human eye has limitations on its ongoing ability to see defects and cannot reproducibly detect them over long periods. In addition, there is variability between observers and their ability to see defects. Therefore, results for the same defect can vary depending on the observer.
- Increasing line speed will result in more defective material being produced before the quality is ascertained
- This procedure can often result in scrap material being manufactured before defects are detected.
- It is not effective for low volume coating campaigns because a significant percentage of the total volume can be coated before the quality data is available.

Therefore detecting defects offline has become more difficult and expensive and often quality product cannot be manufactured at a competitive quality and cost.

14.8.2.2 Online Defect Characterization

14.8.2.2.1 Basic Principles

Recently, the offline measurement deficiencies have been eliminated by utilizing new online measurement technology systems. There are two systems in use:

- Surface inspection
- Online coating weight measurement

They are installed in the coating line, either after the coating applicator and before the dryer or before the rolls are rewound. These online systems scan the entire coated film to measure the desired properties and determine if they are in Quality control limits. The acquired data can be statistically analyzed and is then stored in a computer database.

14.8.2.2.2 Surface Inspection System

Surface inspection systems:

- scan the substrate and characterize defects over the entire web
- store defect information for each roll
- display defect location in the coater control room

Advances in computer technology, data processing, and detection systems have resulted in several advanced surface inspection systems. These system advantages are:

- They are modular and can be configured to inspect many different product structures and defect types.
- They are very sensitive and reproducible, can classify defects according to their type and are cost effective.
- Defects can be detected in both coated film and uncoated substrates and these systems are more effective and reproducible than a human eye.
- Defects can be detected in real time; therefore, experiments can be performed to optimize quality and to reduce start-up losses.

Fig. 14.19 is an online surface inspection system for adhesive coating, label stock, and tapes [14–16].

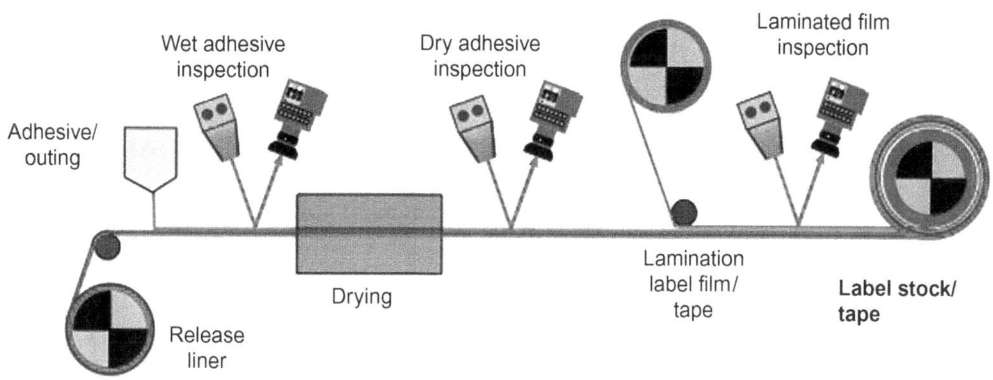

Figure 14.19 Surface inspection system. *Source: Dr Schenk Vision Systems.*

There are two basic inspection systems that are in use. A laser system uses a laser beam that is scanned across the entire web. The second system uses CCD Cameras to capture the defect image. Both single cameras and multiple cameras focused on the substrate can be used and interfaced with computer.

The laser system advantages are:

- Multiple optical channels to view substrate in several modes: Reflection, Transmission, Bright field, Dark field, and Polarity shift
- Can scan the entire substrate width
- Can detect many defects including scratches, pinholes, reticulation

The CCD system advantages are:

- lowest cost system
- excellent at bright-field detection >50 μm
- superior color/gray scale imaging

14.8.2.2.3 Defect Detection Capability

Figs. 14.20–14.23 are examples of defects in base film, laminated film, and coated film that were detected by surface inspection systems. Figs. 14.22 and 14.23 show an individual line scan and the images generated from these scans.

14.8.2.2.4 Online Coating Weight Measurement

Online coating weight measurement systems scan the substrate coating in the transverse direction, T_D and continuously, measure the coating weight to create a T_D coating weight profile, across the substrate. From these T_D profiles, the machine direction profile, $T_{M.}$ can be calculated. An accurate profile measurement is important because there can be peaks and valleys in these profiles. The film performance properties can vary depending on where the product is slit for use. Also, if the profile variations are immediately available, when there is nonuniformity it can be quickly corrected by adjusting the coating applicator. The measurements can also be used in a control loop to control coating weight.

The in-line coating weight gauges consist of a sensor system to measure the coating weight and a device to move the sensor across the substrate and back so that entire substrate is characterized. The computer system collects, stores, analyzes, and displays the data [17]. The system can be positioned in the coating line either before the dryer or at the end of the dryer. The gauge can measure a single point on the substrate or scan the entire substrate. The most accurate and informative system is to scan the entire substrate. However, lack of space to install scanning head and costs may inhibit using this system.

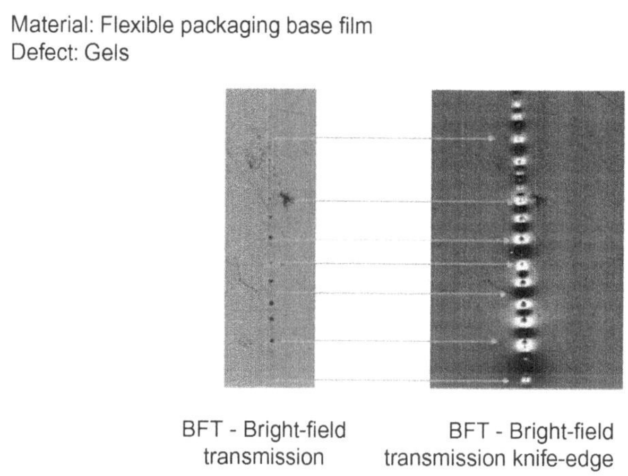

Figure 14.20 Base film inspection results. *Source: Dr Schenk Vision Systems.*

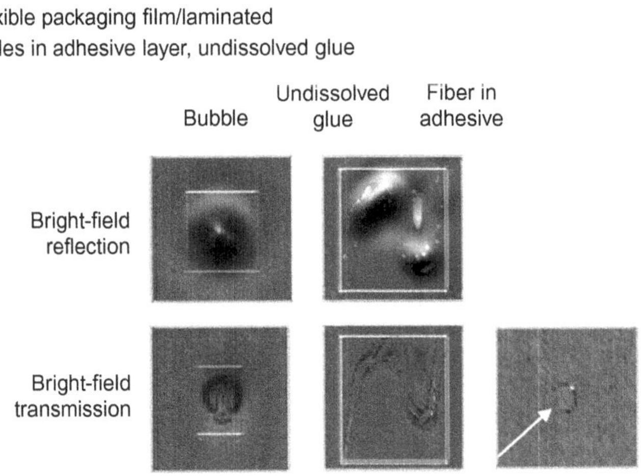

Figure 14.21 Laminated film inspection results. *Source: Dr Schenk Vision Systems.*

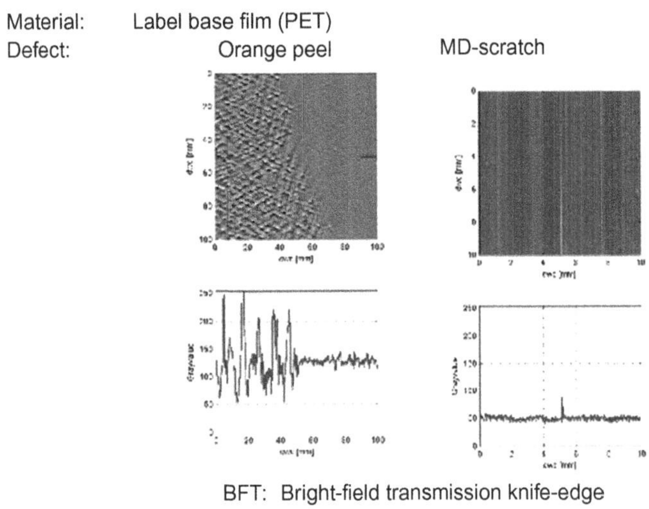

Figure 14.22 Base film inspection. *Source: Dr Schenk Vision Systems.*

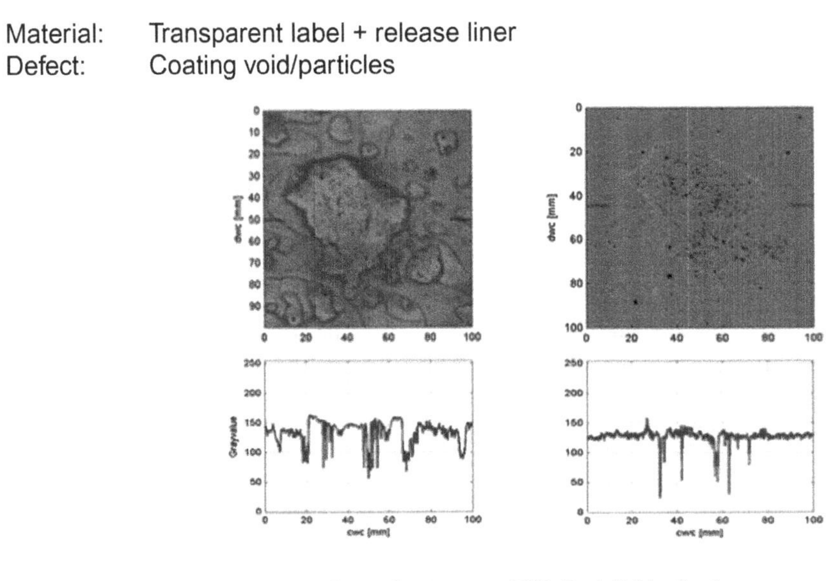

Figure 14.23 Adhesive coating inspection results. *Source: Dr Schenk Vision Systems.*

These systems are versatile and can measure several coated structure different properties such as:

- sheet thickness
- basis weight
- coat weight/coating thickness and
- barrier layer thickness

by selecting the appropriate sensor system. Table 14.2 lists the properties that these gauges measure and the sensor used for a specific measurement. Selecting the appropriate sensor depends on several factors:

- properties to be measured
- needed precision
- measurement reproducibility
- cost
- space available

The measurement systems used are direct coating weight gauges and total weight gauges. The direct weight gauge measures the coating weight only and is not influenced by the substrate. These gauges measure a unique component of the coating, such as metals, silver, silica, polymer, additives. The measured values will vary with the total thickness and thus is a reliable coating weight measurement. Calibration with a standard is required for accuracy and reproducibility. Infrared absorption, beta backscatter, and X-ray fluorescence are sensors that give direct measurements.

If a direct coating measurement on the substrate cannot be made, then a differential method is used. Total weight gauges that measure the total weight or thickness under the sensor head. In this method two sensors are used. The first sensor measures the substrate thickness weight before it is coated. The second sensor is located after the coating is applied and is in the same lane as the first sensor and measures the total substrate and coating weight. Subtracting the two measurements gives the coating weight. This measurement is not as accurate as the direct methods because the coating thickness is a small relative to the total measured thickness.

Table 14.2 Sensor System Technology and Measurement Capability

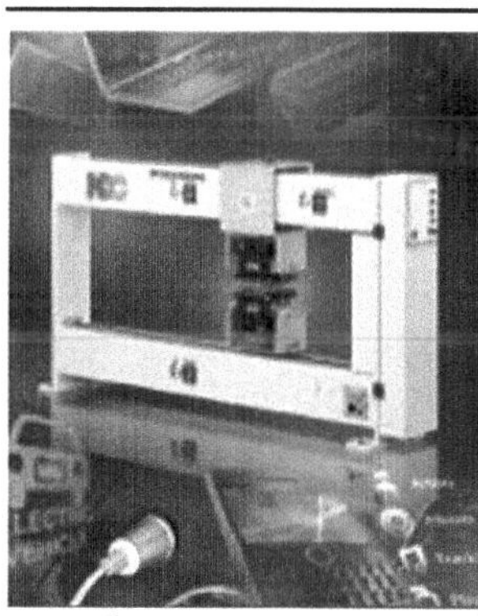

Sensor Technology	Measurements											
	Sheet thickness	Basis weight	Voided film thickness	Barrier layer thickness	Blown film thickness	Coatweight/Coating thickness	Moisture/water-based coating	Lamination thickness	Clear coatings on metalized	Calendared thickness	Basis weight	Wire balance, position, count
Gamma backscatter	■	■			■	■				■	■	
IR transmission	■	■	■	■	■	■	■	■				
IR backscatter						■	■	■				
IR forward scatter							■					
IR reflection									■			
X-ray transmission	■	■			■							■
X-ray backscatter	■	■								■	■	■
Optical micrometer	■									■		
Beta transmission	■	■			■	■				■	■	■
Laser	■									■		

This chart presents the typical sensor technology choices for the measurements indicated. Courtesy NDC Technologies.

Therefore, determining small differences between large numbers is not accurate. Beta transmission, gamma backscatter, and X-ray transmission are total weight sensors.

References

[1] Adamson AW. Physical chemistry of surfaces. 5th ed. New York, NY: Wiley-Interscience; 1990.

[2] Schonhorn H. Notes for "Adhesion and practical adhesion bonding tutorial." Chestnut Hill, MA: Schonhorn Consulting; 2006.

[3] Hasegawa T, Sorimachi K. Wavelength and depth of ribbing in roll coating and its elimination. AIChE J 1993;39:935–45.

[4] Mainstone KA. Extrusion. In: Satas D, editor. Web processing and converting technology and equipment. New York, NY: Van Nostrand Reinhold; 1984 [chapter 8]

[5] Kobayashi C. Stability analysis of film flow on an inclined plane. Indust Coating Res (Japan) 1992;2:65–88.

[6] Ishizuka S. Method of simultaneous multilayer application. European Patent Appl 88,117,484.1, Pub. No. 0 313 043; October 20, 1988.

[7] Cohen ED, Gutoff EB, editors. Modern coating and drying technology. Hoboken, NJ: Wiley-VCH; 1992.

[8] Cohen ED, Gutoff EB. Coating processes (survey). Kirk-Othmer encyclopedia of chemical technology, vol. 6. 4th ed. Hoboken, NJ: John Wiley & Sons; 2003.

[9] Gutoff EB, Cohen ED. Coating and drying defects: troubleshooting operating problems. 2nd ed. Hoboken, NJ: Wiley-Interscience; 2006.

[10] Scriven LE. "Bringing fundamentals to bear on choosing coating method" symposium on coating technologies for imaging materials. Springfield, VA: The Society for Imaging Science and Technology; 1991.

[11] Parodi A. Spin coating and alternative techniques for panel displays. Semiconductor Int January, 1996;101–3.

[12] Cohen ED. Are on-line defect detection systems effective in improving cost and quality? Converting Quart 2013; Quarter 1

[13] Cohen ED. Web coating defects: steps to improve coating weight uniformity. Converting Quart 2011; Quarter 2.

[14] Koenig J. Schenk Vision. How automatic metrology systems help to increase product quality while saving production costs and time. In: ICE 2015, Orlando, FL.

[15] Potter J. ISRA Vision. Solutions for all coating industries: advanced optical coating inspection for maximum quality. In: ICE 2015, Orlando, FL.

[16] Potts T. Dark field technologies, high speed, high resolution coating inspection. In: ICE 2015, Orlando, FL.

[17] Schnackertz T. NDC technologies, on-line coating weight measurement. In: ICE 2015, Orlando, FL.

15 Vacuum Metallizing for Flexible Packaging

***Charles A. Bishop*[1] *and Eldridge M. Mount, III*[2]**
[1]C.A. Bishop Consulting Ltd, United Kingdom; [2]EMMOUNT Technologies, Canandaigua, NY, United States

15.1 Introduction

There is a long history of roll-to-roll vacuum-deposited coatings with evidence that the earliest use, around 80 years ago, was for food packaging [1]. Since then the industry has grown enormously. Metallizing is still the most dominant process and with over 65% of metallized film used in packaging, the most dominant application. Of this 65% portion, approximately 56% is oriented polypropylene (OPP) and 37% polyethylene terephthalate (PET) [2]. What has changed over the last 70 years is the increase in variety of substrate materials as well as the functionality of the deposited coatings. Broadly speaking there are four main use areas:

- decorative or aesthetic
- barrier
- functional
- security.

It is possible that some coatings provide more than one function.

15.2 Decorative Coatings

Decorative coatings for packaging began with simply metallizing polymer films and papers, which gave the web a bright metallic appearance. This bright metallic web enabled products to be wrapped, boxed, or labeled using a material that aesthetically stood out from their competitors. As with all competitive products once one company produced a brighter package or label, others rushed to follow suit. As more products incorporated metallic effects to their packaging designs, there was a desire to differentiate products and so variants were produced. Instead of bright metallic silver reflecting material, colored metallic coatings were produced either by using a dyed substrate material or by overcoating the metallic coating with a transparent colored lacquer. Since then the most significant development was introducing embossed holographic substrate materials that can also be produced in silver or colored metallic versions.

Consumers scanning products on a shelf will generally have their attention drawn to packages that have some brightness to the design. Initially, this was simply a very white surface. Tests were carried out with packages with no white areas to almost all white designs, and people walking past and scanning the products always had their attention drawn to those products with a high white content. This all changed with introducing metallized webs. The products, with even a small metallized component included in the design, had a brighter appearance than the white packages and as they stood out more, they attracted the consumer's attention. Similarly, introducing metallized holographic substrates gave an even brighter appearance than the plain metallized webs [3]. In fact, holographic designs were assessed as being 3.5 times brighter than white [4]. As there are so many competing products, it is regarded as essential to attract the consumer's eye quickly as it has been shown that there is a higher chance that products that are seen first will be purchased. It has been demonstrated, particularly with holographic designs, that there is a considerable increase in market share after introducing holographic packaging [5–7].

There are different ways to make packages with a metallic look. The most common method is to metallize the whole substrate, such as paper or polymer, and then to print or laminate, with a selectively printed web, over the metal to complete the desired design. Where the metallic coating is overprinted and hidden and only a small patch exposed, one must metallize the whole surface which can be an expensive solution. An alternative approach, to achieve the same end result, is to use either a hot or cold stamping foil. In this process, the metallization is onto a polymer substrate that has been pre-coated with a release coating. Following metallization, an adhesive

Multilayer Flexible Packaging. DOI: http://dx.doi.org/10.1016/B978-0-323-37100-1.00015-6

is applied to the metal surface. When this material is pressed onto a surface, the adhesive sticks to that surface and the metal coating transfers from one surface to the other (Figs. 15.1 and 15.2). The metal separates from the original substrate because the release coating fails cohesively or metal releases from the coating. This process can be done either before or after printing and on the same machine. Depending on the system used, this can either be a hot or cold process (Figs. 15.1 and 15.2). If done cold, there tends to be a higher pressure required to transfer the metal. Hot transfer may be an easier process but in some cases the elevated temperature is a problem. An example of this would be in hot stamping a holographic foil where the temperature can degrade the embossing, reducing the embossed height by polymer relaxation, which can be seen as a lowered hologram brightness.

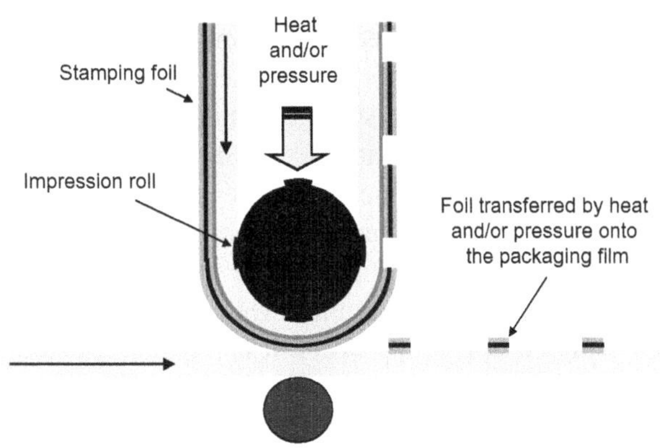

Figure 15.1 Hot or cold stamping foil transfer process.

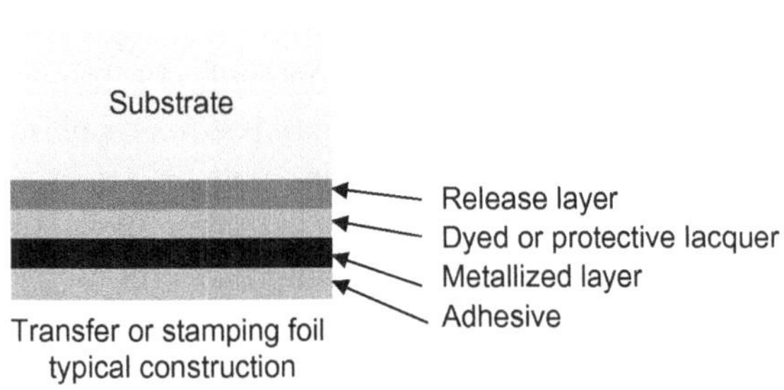

Figure 15.2 Typical stamping foil construction.

The metallized film quality largely depends on the substrate material quality. Metallized paper may appear dull unless the surface has been deliberately coated and smoothed to increase the specular reflectance (Fig. 15.3). Polymer films that contain a high filler levels may also have a rough surface and

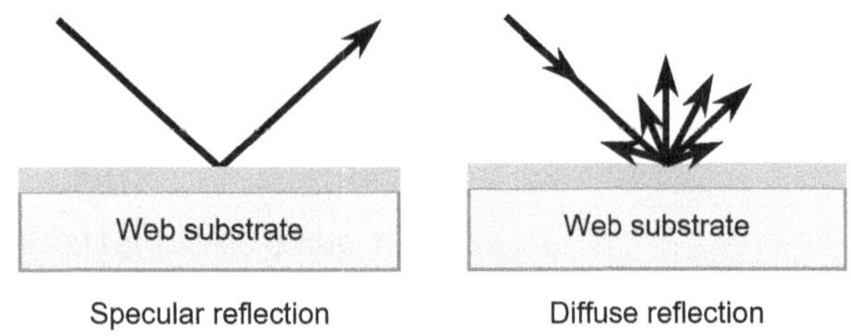

Figure 15.3 A schematic diagram showing the difference between specular and diffuse reflective coatings.

so scatter light giving a highly diffuse reflectance and lower specular reflectance. This has led to coextruded films that have an unfilled surface to give a very smooth surface and consequently a very highly reflective coating.

Holographic embossing can be done either before or after metallization. If the embossing is done before metallization, the coating is conformal with the surface but the deposition process can heat the embossing which can relax and lose some brightness. If the embossing is done after metallization, the embossing can stretch the coating and crack the metal and be more abrasive to the embossing shim. High quality, high value, high security, and lower volume holograms are often coated after embossing with the lower cost decorative holograms embossed after metallization.

A final way to add a metallic coating to packaging film, that also starts life as a metallized film, is by printing using metallized ink [8]. The metal pigment quality used to make the ink metallic can affect the reflectivity. The metallic pigments were used to be made from ball-milled aluminum particles that were known as "cornflakes" because of their crinkled shape [9]. This uneven surfaced pigment produced ink with a highly diffuse reflectivity. This pigment has been considerably improved with the change to using vacuum metallized pigment. Fig. 15.4 shows this process.

A roll with very high quality, smooth, flat surface is coated with a release layer and then metallized. After metallizing, the film passes through a stripping bath where a solvent interacts with the release coating and the metallic layer lifts off the surface. This metallic layer is fragile and breaks up into flakes. This slurry is filtered and the solvent exchanged to a suitable carrier for manufacturing ink. When the ink is printed, these flat metal flakes align parallel to the substrate surface. Each flake has a very high specular reflectance, and so although

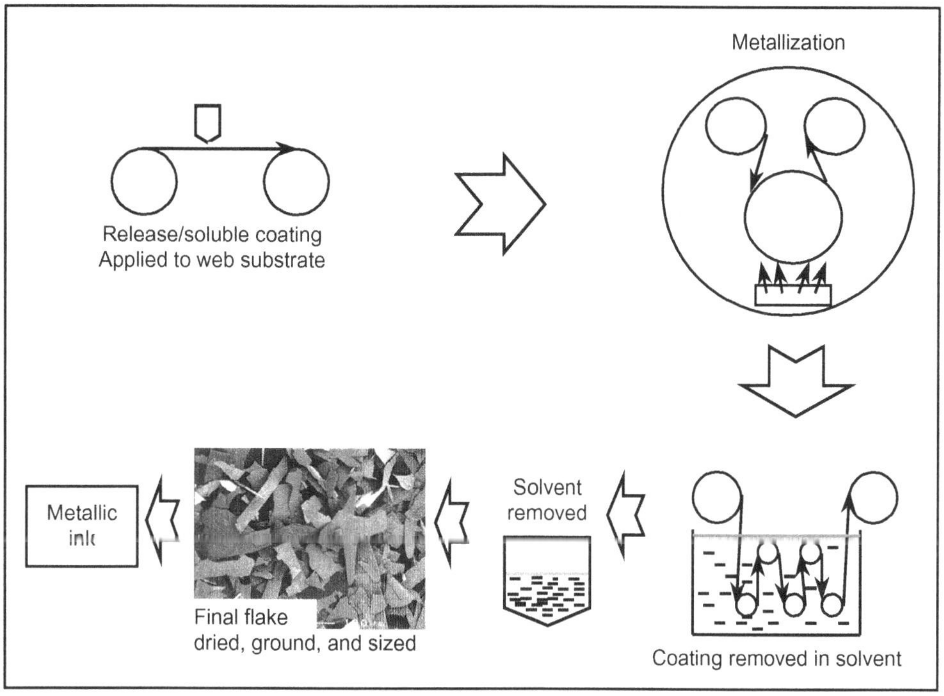

Figure 15.4 A schematic of the flake metallic pigment production process.

not all flakes are perfectly aligned, they are closely aligned enough that the specular reflectivity remains high.

15.3 Barrier Coatings

The "barrier coatings" category is very broad and includes coatings that reduce light, oxygen, water vapor, or other gas transmittance [10]. We shall look at these properties in turn.

15.3.1 Light Barrier

Many foods degrade by photocatalytic light reactions that degrade color, fats, flavors, or vitamins within the food. This shortens the food shelf life and makes the food unattractive, off flavor, and thus harder to sell.

15.3.2 Oxygen Barrier

Oxygen turns fats rancid and oxidizes vitamins such as vitamin C reducing their potency. It is not only foods that require barrier packaging but some high technical specification devices also need to have protection against oxidation. Examples would be organic light-emitting devices (OLEDs) and copper indium gallium diselenide solar cells.

15.3.3 Water Vapor Barrier

Water is absorbed or lost by foods turning them "stale" and a texture change losing their appeal. An example would be potato crisps or chips that are expected to be brittle and crispy. If they absorb too much moisture they lose this feature, which could lead to customer complaints.

Foods that are already moist, rather than absorb moisture, may actually lose moisture and the food dries out. Examples would be cheese, cookies, bread, or dried fruits. This moisture loss changes the food from what is expected and shortens the product shelf life.

OLEDs and solar cells are also adversely affected by moisture and so require a suitable moisture barrier along with an oxygen barrier.

15.3.4 Gas Barrier

There are instances where it is advantageous to exchange normal atmospheric gas with an alternative (called gas flushing) that can maintain some food property better than air. In these cases, the aim is to provide a barrier to the gas being flushed into the package, as well as a barrier to keep out the air and water vapor. The materials used in gas flush operations are known as modified atmosphere packaging films.

Barrier coatings provide a structure that resists gas or liquid diffusion through the material. Nature drives all systems towards thermal and concentration equilibrium, so a food packaged in a modified gas atmosphere, rich in specific gases; this gas will diffuse out and the surrounding atmosphere will diffuse in, until an equal concentration with the surrounding atmosphere is reached within the package.

Polymers have specific gas, moisture, and chemical barrier performance with some polymers performing better than others. Simplistically, it was observed that both glass and metal foil have a good barrier performance and it was thought that thin glass and metal layers on a polymer would produce an ideal barrier material. The metal or glass could be thin with the polymer acting as a support material. What became clear is that the barrier performance is never as good as predicted [11]. The glass or metal coating should be a perfect barrier but it is not.

Glass is a brittle material, cracks easily, and so is quite fragile. Very thin and more flexible coatings partially overcome this problem. Metal coatings are somewhat more flexible than glass or ceramic coatings, and they also do not achieve their predicted performance. In regards to the gas barrier properties, the main problem for these films is pinholes [12–14]. Pinholes are formed when dust or debris that is present on the polymer surface during metallization moves after metallization leaving behind an uncoated small area known as a pinhole or pinwindow. There may be other ways to produce pinholes such as pick-off, where high winding forces and high spots on the film's unmetallized surface along with poor metal adhesion lead to transferring some metal from the metallized surface to the unmetallized surface. Anywhere there is a pinhole, there is effectively no barrier at all.

Handling the film following vacuum coating can also introduce more defects. Dust/particulate debris can be moved during contact with rolls or as the web is rewound. Also if there is any film slipping, scratching happens as the debris is dragged across the surface. In making packages, the film may be folded and it is common for the web to be stretched. This folding or stretching as part of the packaging process can exceed the coating yield strength resulting in cracking and other stress-related defects.

The two main barrier coatings are opaque and transparent barrier coatings. Opaque barrier materials are primarily aluminum metallized coated polymer films. The barrier requirements may be a mixture of gas and moisture barrier as well as opacity for light barrier. The coating thickness will determine the opacity and may also affect the gas and moisture barrier.

Transparent barrier coatings, although they have been available for around 20 years, are in many respects still in their infancy compared to the aluminum metallized films. These transparent barrier coatings provide the same oxygen and moisture barrier along with transparency. This allows consumers to see the products they are buying and so may have some selling advantages. Also during high-speed manufacture and packaging, it makes scanning for metal contamination much easier than when using metallized films. The transparent barrier coatings are generally two to three times more expensive than aluminum metallized films. Much development work has reduced these coating costs [15–18]. There are several different techniques and materials all aimed at producing transparent barrier materials.

There are competing processes for depositing transparent barrier coatings. One early method used induction heated, thermal evaporation sources to deposit silica coatings [19]. Probably the simplest and most cost-effective process developed uses a standard aluminum metallizer, and introduces oxygen into the process just after the aluminum was deposited. This process relies on the very high reactivity of aluminum to oxidase the growing coating [18]. The process developed by system manufacturers is now a standard process offered on a metallizer. As the deposition rate is almost the same as for aluminum, the product costs are only slightly higher than for the metal [20]. Competing with this is a hollow cathode activated plasma used with resistance-heated deposition sources [21,22]. This plasma not only converts the aluminum to the oxide but also bombards the growing coating to increase the coating density. This densification process reduces the coating porosity and so improves the barrier performance. This process does require power to drive the hollow cathode sources and thus increases the product cost compared to the simple oxidation process.

The latest technology developed to produce a better or lower cost barrier coating is atomic layer deposition (ALD) [23]. The ALD process involves two gaseous precursor materials with the substrate being brought into contact with each one sequentially. When the first gas contacts the substrate surface, a monolayer is deposited on the surface. This precursor gas is removed and the second gas is

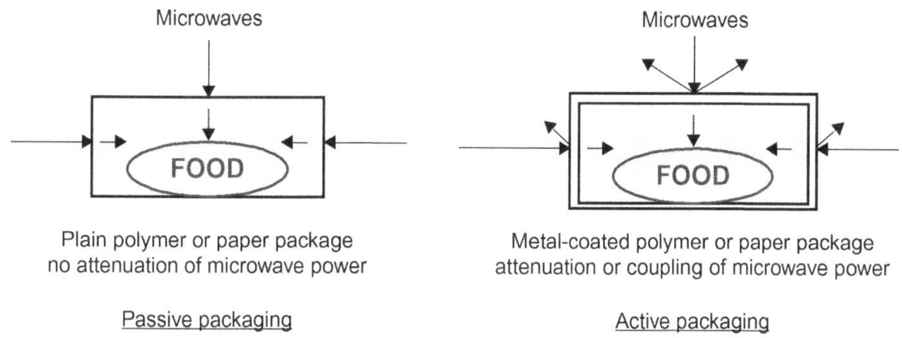

Figure 15.5 Comparison between passive and active microwave packaging.

brought into contact with the substrate surface, and it reacts with the first monolayer deposited to produce a monolayer of the desired coating. There are competing technologies that employ different ways to apply alternating layers of the two precursor materials. It remains to be seen which will deliver the best coating at the lowest cost. The advantage that ALD has to the other PVD techniques is that the gas deposits a monolayer over the whole surface and where there is debris, the particle does not shadow the substrate and so if it is moved later the pinhole left behind is considerably smaller. So for the same substrate material, ALD can either produce a better barrier performance for the same thickness coating or the same barrier performance with a thinner coating or combines both attributes.

The most flexible deposition process uses an electron beam deposition source which can deposit almost any material, and so has the flexibility to deposit a transparent barrier material on one substrate and then deposit an optically opaque barrier material onto the next roll. However, this is a much higher capital cost system than the simple aluminum metallizers and the coating cost are higher [17]. The process that is expected to deliver lower cost transparent coatings is plasma enhanced chemical vapor deposition (PECVD). In PECVD, a source gas is introduced into the vacuum system and the plasma technology breaks the gas down into the chemical components that deposit a silica coating. Although this process started out as a simple design, it has increased in complexity as the speed and barrier performance improved. The current process [16] requires a system that cost almost as much as an electron beam deposition system. A continued advantage is that the precursor liquid is inexpensive and so the material costs are lower than most other transparent barrier materials. Ultimately, the transparent barrier coating costs are controlled by the raw material costs as opposed to the processing costs.

15.4 Functional Coatings

In one sense, this group is a catch-all as it contains coatings as different as microwave susceptors, antistatic coatings for semiconductor packaging, some electromagnetic shielding coatings and some radio frequency identification devices, that can also fall under security coatings.

Microwave radiation can be used to heat foods directly, where the microwave energy is coupled into the food water molecules (Fig. 15.5), and vibrate them, generating heat. Fig. 15.6 shows a typical microwave structure. However, different foods contain different amounts of water and so will heat at different rates. Thus, in instant meals, that may contain different foods that all need to be heated at the same time, it is useful to control how the microwave power is coupled into different areas to allow all

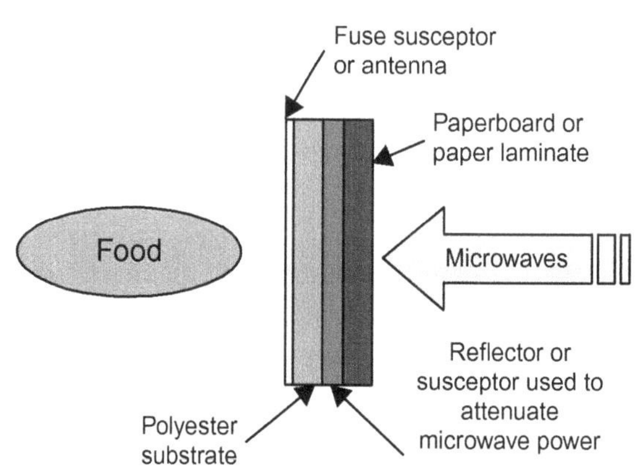

Figure 15.6 A simple packaging structure.

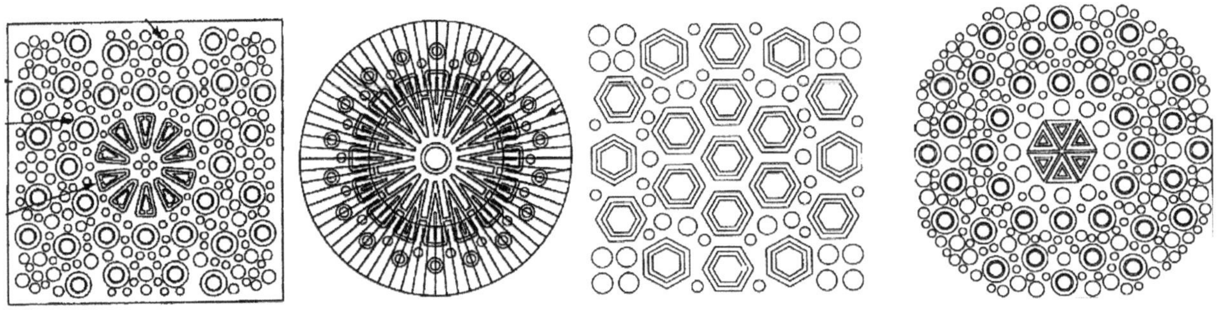

Figure 15.7 Different patterns used for microwave heating foods [24].

the food to arrive at the preferred cooked state at the same time. This can mean complex patterns are required as shown in Fig. 15.7.

For microwave cooking, different coatings used are:

- a continuous thin metal layer which acts as a microwave energy absorber,
- a patterned coating which acts as an antenna and focuses the power into the food to provide extra heating, and
- a continuous thick metal coating which acts as a reflector blocking out some microwave energy to portions of the container.

By combining these different coatings, complete meals can be cooked selectively.

A second application for microwave cooking is where a heated surface is required to crisp or brown the food in contact with the surface. One difficulty is to limit the device temperature to prevent charring. The microwaves are absorbed within the thin conducting metal coating, and this absorbed power causes the metallized surface to heat. The different thermal expansion coefficients between the substrate and metal coating control the ultimate surface temperature. As the polymer expands faster than the metal, it puts the metal coating into tension. Then, if the differential expansion is large enough, the metal coating cracks. As it cracks, it loses conductivity and this limits the microwave heating as the microwaves are no longer absorbed by the coating.

One problem seen in the early packages that relied on this feature was that the cracking could be variable and so there would some uncracked areas, that overheated, and other areas that failed early, and underheated the food giving incomplete cooking. To correct this problem, a pattern of demetallized areas was incorporated into the metallized surface design [25,26]. These patterns concentrated the stress so that the stress concentrators would act as a fuse (Fig. 15.8). In this way, the metal would crack across these stress concentrators, in a very uniform manner, and the film would reproducibly reach a specific temperature where the fuses would crack and limit the temperature.

Antistatic or electromagnetic screening applications are applications where the applied coating may be part of the same total structure or may be for separate applications. Packaging materials for electronic components, such as semiconductor chips, can be quite complex structures. There may be a problem using polymer films for packaging, as polymers in contact with other materials can build up a significant triboelectric (static) charge. This static charge can reach several thousand volts, and if this high voltage is discharged across the semiconductor or electronic device, it can damage the electronics. As polymers are insulators, once the charge has been generated it is very slow to dissipate. Thus, an applied conductive coating will dissipate any charge. As charge can build up on both the package inside and on the outside, this conductive coating may be applied to both surfaces (Fig. 15.9).

Furthermore, the antistatic bags may be subject to an electrostatic discharge from outside and so it can also help to have a conductive coating to protect the packaged products against a static discharge. A metal coating can provide a 97% shielding for a 1000 V static pulse. In addition, it is preferable for the coatings to be transparent, so that the bar code on the packaged goods can be read or the product visually identified, without having to open the package and expose the products to potential contamination or damage. Fig. 15.10 shows three basic structures for antistatic packaging materials.

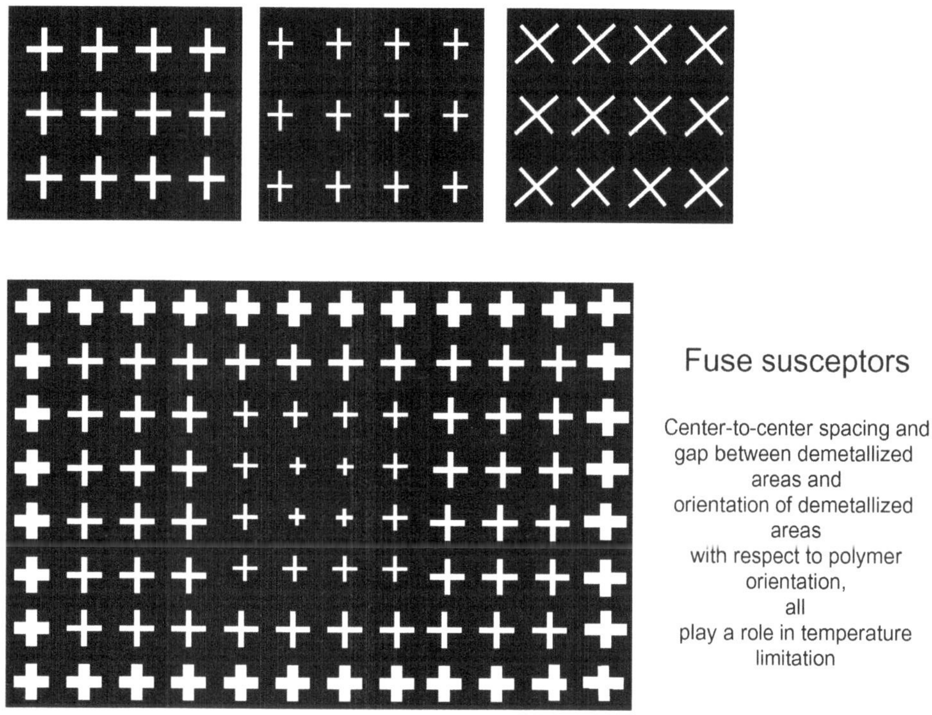

Figure 15.8 Various fused susceptor designs. The white areas are demetallized.

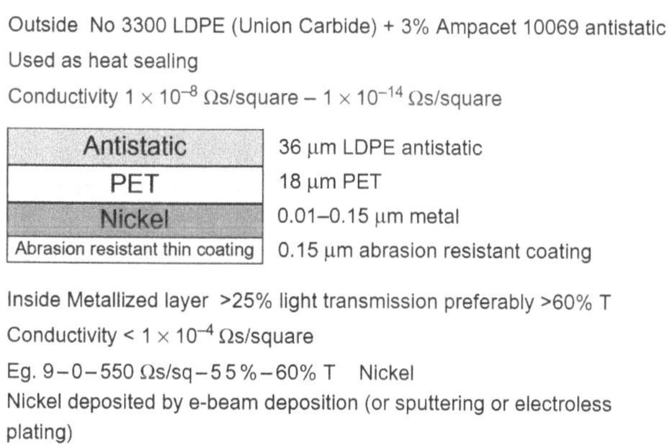

Figure 15.9 A typical antistatic coating as per US Patent 4,151,344.

Having several antistatic films, coatings, and metallized layers in the product designs allows one to use the material as either metal-in or metal-out to form the bags. This is self-explanatory with the metal coatings either facing the bag inside or outside. The metal-out structure needs to be protected from damage to the thin metal coatings by overcoating or lamination. The metal-in bags can be slower to dissipate charge than metal-out bags where the metal coating is closer to the surface than for metal-in bags.

There are higher specification bags with metal coatings on both sides allowing for better static protection. The classification of the products to be

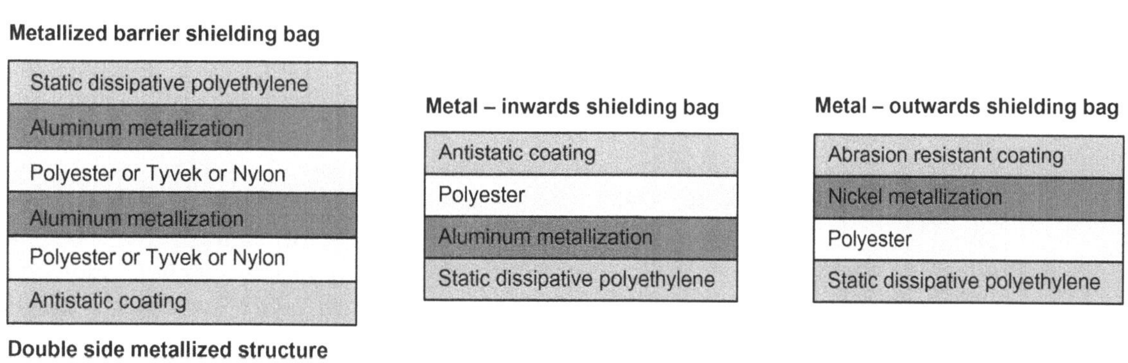

Figure 15.10 Basic metallized antistatic structures.

packaged, as well as the materials suitable for packaging them, has to meet various standards that are periodically revised [27].

Other variations are where the material is required to have a high moisture barrier performance as well as a high corrosion resistance for long shelf-life products. Generally, this requires much thicker metal coatings that are opaque. Also, as semiconductors are being developed to work at lower power levels, they tend to have a lower damage threshold and so the static protection has to be improved. This also includes protection against induced currents, due to electromagnetic interference or radio frequency interference (RFI), and this also requires thicker metal coatings. These metallized structures still have less metal than foil packaging, and so have a lower environmental impact.

15.5 Security Applications

The security packaging application familiar to most people are the holographic security labels, as these are visible and well-publicized security applications. They are by no means the only vacuum-deposited security devices available. There are tamper-evident coatings that fracture if lifted, many radio frequency identification labels (RFID), holographic tear tapes, customs and tax stamps, and various color shifting or optical variable devices [28].

As counterfeit products make up 5–7% of world trade, it is no surprise that companies have had to develop ways to verify legitimate products. High margin products are prime targets and often have the newest defensive devices included in the packaging design. As counterfeit protection spreads and becomes cheaper to make, it means that the high security protection technology becomes compromised, or easy to defeat. Then it is typically applied to lower value products with a lower protection level. This is what has happened to holographic security devices.

Holograms were once:

- very difficult to produce
- were expensive to produce
- commanded a high price
- there were limited sources.

Now almost anybody can produce holograms, and as the consumers find it difficult to discriminate between good and bad ones, it means that their value as a single security device is limited. The complexity that can be included in a hologram means that they can still be used as forensic security devices. This allows a "wallpaper" background hologram to be used as part of the general packaging aesthetics, but within the overall design at one, or more, specific locations, different high-level security features can be added to the design. In this way, extra security is added, as simple counterfeiting would not be able to replicate the entire design. Simple holographic labels can also have other security features such as optical variable or interference coatings added, so that the combined design becomes much more difficult to reproduce. This increases the difficulty to counterfeit and therefore increased security.

The basic design process is the same for security holograms as for decorative holograms. A holographic pattern (an origination) is designed and produced. This pattern is then "written" or exposed onto a light sensitive film with a photopolymer coating. The film is then "developed" which produces a plate with a surface structure or topography due to the removal of unreacted photopolymer from cross-linked and insoluble photopolymer. This surface structure is metallized to make it conductive and a thick metal plate grown by electroplating the metallized surface. This electrodeposited plate is known as a shim, and from this shim replicas are produced that are used to press, into a suitable surface, to reproduce the origination surface structure. This origination structure, when suitably illuminated, will reproduce the original image or artwork. This embossed surface is then aluminized. This material can then be laminated and printed to make a complete packaging material with the hologram as an integral part of the packaging material. Rudolf van Renesse [29] edited a book that details optical security designs and devices.

Another vacuum-coating security, that finds its way onto packaging, is labeling materials for tamper evidence. Tamper-evident labels indicate if the package has been opened. Unfortunately, the criminals who tamper with products rarely use the correct way to enter the package to adulterate the products and so often the tamper-evident labels are only of limited use. The vacuum-deposited labels usually include more than one vacuum-deposited layer; one layer is fragile, such that it breaks relatively easily [30]. The total optical layer design produces a particular color to the label. When the frangible layer breaks, there becomes an additional air layer, which disrupts the optical design, and the color is lost indicating the label has been used or tampered with.

A large growth area is RFID tagging for products. Vacuum-metallized coatings can be used for the antenna part of these devices. This does require the vacuum coating to be patterned to form a suitable shape to form the antenna. The conductivity required does depend on the device and the range that the tag is to be interrogated from. The greater the distance the better the coating layer conductivity required. What this means is, that for many applications, a conducting printing ink has sufficient conductivity to work well. Therefore, there is little or no justification for using the more expensive vacuum deposition production process.

15.6 Metallization Basics

The basic metallization process comprises a large vacuum coater that contains the substrate to be coated, the deposition sources, and the feed system to supply material to the deposition sources (Figs. 15.11 and 15.12).

The aluminum is supplied as wire on reels, and each reel is controlled to feed the wire onto a resistance-heated intermetallic boat. As the wire touches the boat, it melts and forms a molten pool where it evaporates. The vacuum level is such that there is no gas to cause any collisions. Therefore, the vapor cloud arrives uninterrupted at the cooled deposition drum. The vapor condenses on the film as it passes through the deposition zone supported by the deposition drum [31].

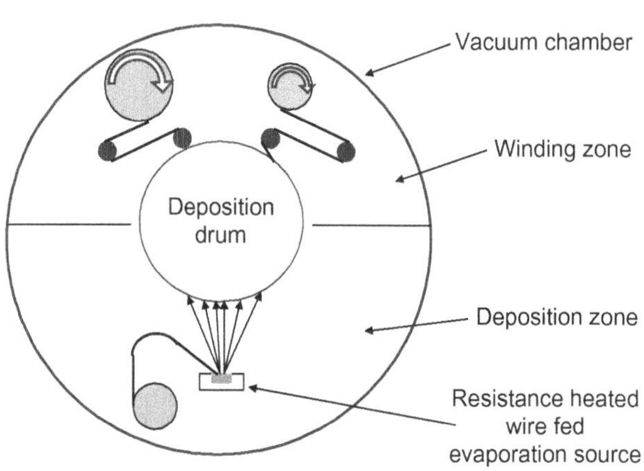

Figure 15.11 Basic aluminum vacuum metallizer system.

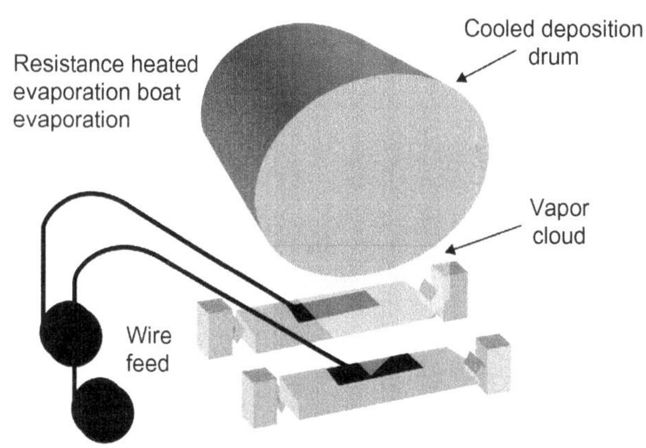

Figure 15.12 Resistance heated evaporation boats in an aluminum vacuum metallizer system.

Many transparent barrier coatings are deposited by different deposition processes, such as induction-heated evaporation, electron beam evaporation, or PECVD. Induction-heated evaporation sources (Fig. 15.13) use induction heating instead of resistance heating to evaporate the silica source material to produce silica coatings. The induction-heated source–based machines are similar in cost to standard resistance-heated source aluminum evaporators.

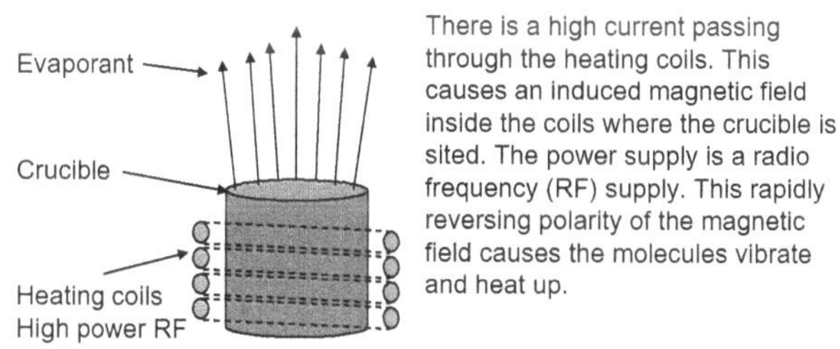

Figure 15.13 An induction heated evaporation source.

Electron beam evaporation is much more versatile as it can be used to evaporate many different materials. In this process, a hot filament is used to provide electrons that can be accelerated, bent, and focused onto a crucible where the electron beam energy is used to heat the source material. This evaporates the material, which condenses onto the substrate. The electron beam source (Fig. 15.14) can be a very large source, which can coat up to 1.25 m wide material per gun by sweeping the beam across the coating width. Alternatively, a series of smaller electron guns can be used across the film width, in exactly the same way as the resistance-heated sources (Fig. 15.15). The larger, sweeping e-beam source–based systems are anywhere from double

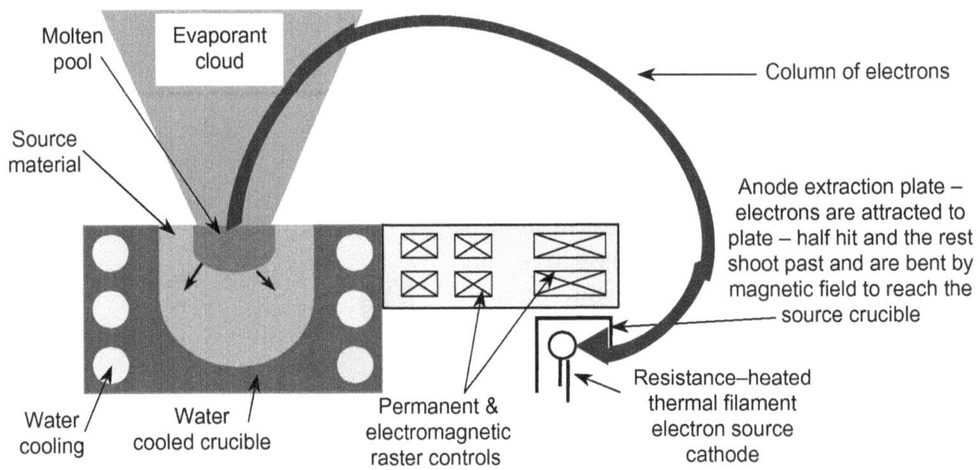

Figure 15.14 A small electron beam gun.

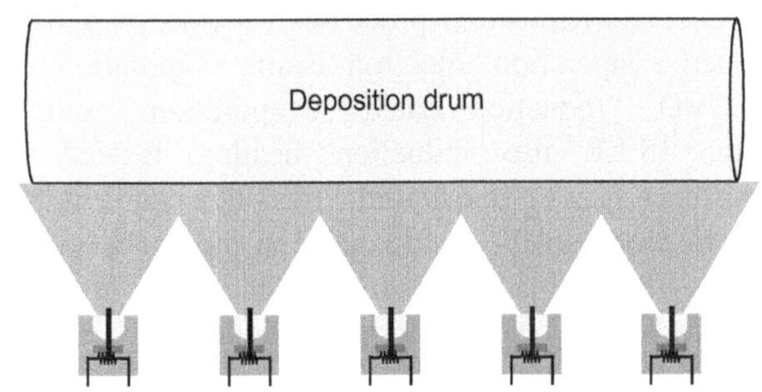

Figure 15.15 A series of electron beam deposition sources showing them arranged for overlapping deposition.

to triple the cost of an equivalent resistance-heated source machine. This means that it is preferable to deposit materials that are not as easily evaporated, such as silica as a transparent barrier material, using resistance-heated sources.

The other major deposition process, specifically developed for the depositing transparent barrier coatings, is PECVD. In this process, a gas or liquid precursor chemical is fed into the vacuum chamber where a plasma decomposes the precursor chemical, and one or more by-products are condensed onto the web to make the coating (Fig. 15.16).

The transparent barrier coatings use a precursor that decomposes to provide silicon or a silica sub-oxide that can be combined with oxygen to produce the desired silica coating. The normally diffuse plasma is densified using magnetic confinement, and this can increase the deposition rate. This process has the advantage that the precursor cost is low compared to some e-beam and induction-heated precursor materials. The deposition rate tends to be slower than by electron beam deposition. Thus, there is competition

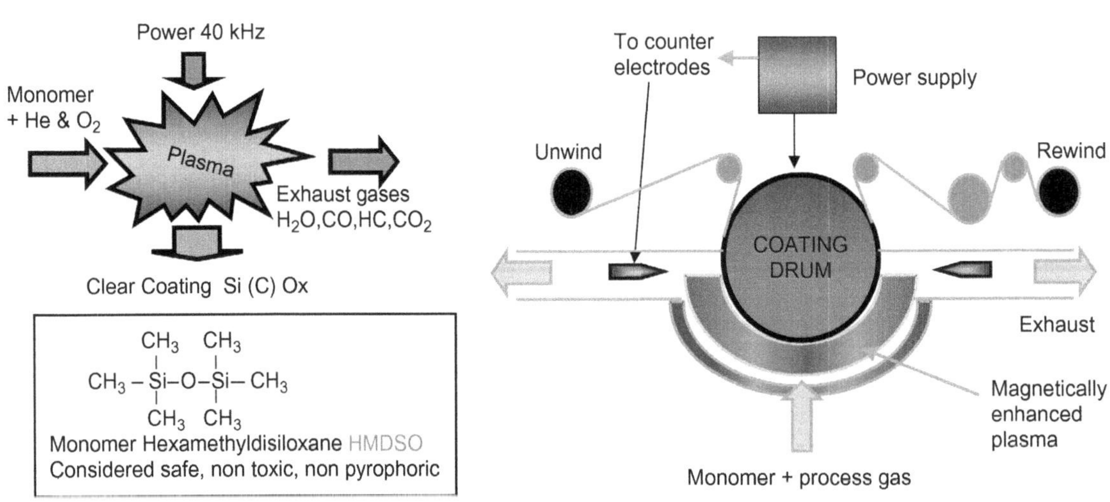

Figure 15.16 PECVD process using HMDSO as the precursor.

between the two processes to produce the most cost-effective (cheapest) silica barrier coating. The electron beam deposition process system cost is higher than for the PECVD process and the deposition rate is higher. It is possible to increase the PECVD process deposition rate with multiple deposition sources but this adds to the system complexity and cost.

The PECVD precursor material most widely used is hexamethyldisiloxane (HMDSO). This material also decomposes to carbon and some of this carbon can be advantageously included in the coating to further improve the barrier performance.

Common to all deposition processes is how incoming web substrate quality and any surface pretreatment impact the final coating quality. There are many factors that affect the deposition process quality, and it all starts with the incoming substrate material quality. Most adhesion problems observed in vacuum-coated products are attributed to problems with the incoming material. This can include fillers or additives as well as residual products from the polymerization process that can be present on polymer film surfaces.

These surface contaminants will usually form a weak boundary layer, cause the substrate surface energy to be low, and any layer deposited on the surface will have poor adhesion due to the weak layer between the coating and the substrate surface.

Surface treatment is always good policy. This can be flame, corona, atmospheric plasma, or vacuum plasma treatment [32]. It is important for any surface treatment to be done correctly, as it is possible to make the adhesion worse if it is done incorrectly. The surface treatment goal is, ideally, to increase oxygen at the surface to improve the metal bonding [33]. Also, surface treatment can be used to remove any low molecular weight material or other contaminants, as well. The atmospheric surface treatments have a common problem, that after they have been used to increase the polymer film surface energy, the film is then rewound and stored for some time. Once the film is rewound, the treated surface comes into contact with the untreated opposite (back side) surface that will still be contaminated. Once in contact with the untreated surface in the roll, any low molecular weight material may transfer from the untreated surface onto the higher energy treated surface, contaminating the treated surface. This gives vacuum plasma treatment a distinct advantage.

If the surface treatment power (or residence time) is plotted against surface energy, it can be seen that the surface energy (dyne level) will increase. This increasing surface energy will eventually reach a plateau. However, if the adhesion is plotted against treatment power, there is no plateau. Instead, the adhesion increases through a maximum adhesion level and then begins to decrease. If the treatment power is increased past the peak adhesion level, it can sometimes fall below the original low surface energy. Therefore, it is important for any surface treatment to be optimized so that the adhesion is close to, or preferably at, its maximum. This is best done by plotting adhesion versus Watt density of the treatment process.

At this point, it is worth pointing out that high adhesion is not closely related to surface energy. High adhesion comes from the coating being bonded onto polymer chains that are long enough to be entangled in the bulk polymer [34]. If you pulled on this coating bonded to the polymer chain, it would try to pull out from the bulk polymer. The entangled chain will not easily pull out and so the polymer will initially distort, stretching and absorbing energy until a point is reached where a bond will break. This type of bonding will result in very high bond strength. What happens with surface energy is different. If the surface energy is low when the coating is depositing, the coating will not wet the surface and the growth will be as a series of hemispheric crystals. If the surface energy is raised, the coating will wet the surface and the crystals will want to spread out and so the resultant coating will be smoother than when the surface energy is low (Fig. 15.18). Each crystal will cover a larger surface area. There is a low adhesion state that relates to proximity such as van der Waals force. In changing the surface energy, it is possible to change this proximity bonding.

Separate to this change in surface energy, the plasma treatment can cause chain scission and thus enables the coating to bond directly onto the ends of polymer chains that are entangled in the bulk polymer. It is this that can enable a large change in adhesion levels. Too much plasma treatment and the increased chain scission can cause several breaks in the same polymer chain resulting in more short-length polymer chains. The coating can still bond to these, but there are less polymer chains that are entangled into the bulk polymer and so the total adhesion falls rapidly.

It is important to optimize the surface treatment for each film type, because different polymers react differently to treatment and formulations are different from different suppliers. Different manufacturers, of what is nominally the same film, will have different proprietary coatings or additive packages with different chemical compositions. Therefore, what may be an optimized process for one film is likely not an optimum for others.

Treatment levels which lower a desired property, such as adhesion, are said to be "overtreated." What happens is that the overtreatment damages the polymer surface by continuing to break polymer backbone bonds. This results in lower molecular weight (decreased chain length) surfaces with the result that the surface becomes carbonized and eventually powdery. The chemical composition of the overtreated surface may not change much, and so the surface energy remains high.

Many coatings applied by vapor deposition are expected to be barrier coatings, and there are many factors that can affect the barrier performance. Fig. 15.17 shows how theory predicts a perfect barrier for a coating with no defects. The right half of the figure shows potential barrier problems. If you consider that the coating may only be 10–50 nm, the depiction of the debris is inaccurate, in that the particle is shown as very small relative to the coating thickness. Typically, the debris would be microns in diameter and much bigger than the coating thickness. This is why it is relatively easy to move the debris after metallization, leave a pinhole, and possibly scratch the surface as well.

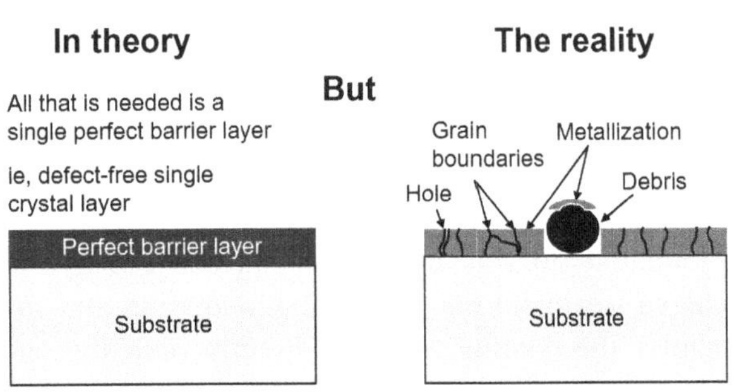

Figure 15.17 Theoretical and real barrier coating comparison.

The other defects such as grain boundaries and channels in the metal layer, generally relate to the substrate polymer, surface treatment, and deposition process. If the substrate has a low surface energy, the depositing metal will not easily wet the surface and the deposit may be poorly formed. If the substrate has been surface treated to increase the surface energy, the metal will wet the surface better and allow better nucleation and crystal growth of the deposit. This improved wetting gives some advantages to the coating quality and can often times appear as a higher optical density at constant deposition rate.

The improved surface wetting leads to a more continuous coating at a lower coating thickness. It also reduces the number of holes in the coating. These holes are formed where several growing crystals impinge against each other as they grow on the surface. If you imaging pushing three balls together, there will always be a gap in the center. This happens on the surface when several growing crystals impinge, but with continued deposition these holes are progressively filled in by continued crystal growth. The smaller and a fewer number of holes, which remain at the end of the deposition, will result in improved barrier performance.

Fig. 15.18 shows how substrate surface energy level impacts nucleation and growth due to improved wetting. Improved wetting will improve not just the adhesion, but also reduce the number of holes in the coating and improve the barrier performance. The nucleation and growth process is not perfect, and there will still be some defects as well as some grain boundaries, which also provide a diffusion path through the coating.

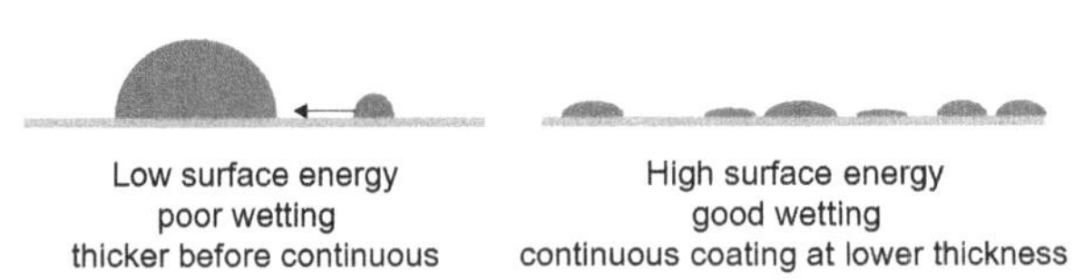

Figure 15.18 Diagram showing how surface energy can affect the wetting and hence coating quality.

Fig. 15.19 shows how the coating growth would progress if it were to be continued to a few microns thick [35,36]. The material that deposits on the substrate surface starts to nucleate and grow with any crystal orientation, and the individual crystal plane spacings will be different. This means that the planes with the greatest separation grow more quickly and these will eventually shadow slower growing crystals. The result of the shadowing is the columnar growth as shown above. Most metallized coating are very thin and so this gross columnar coating is never observed. However, what are seen are the roots of this columnar growth. This can have grain boundaries, which directly connect the atmosphere to the substrate and so present a very direct diffusion path through the coating.

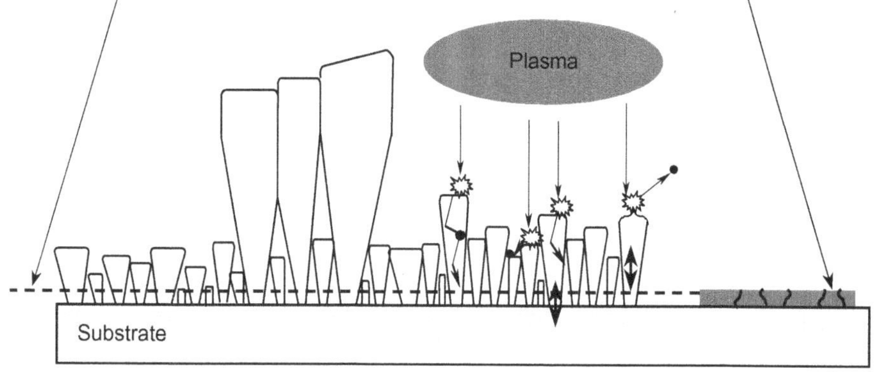

Figure 15.19 Nucleation and growth of a thin film coating.

Some transparent barrier coatings are sometimes improved by the deposition process or by plasma bombardment during the coating growth [37]. This ionic bombardment from the plasma compacts the coating as well as removing any poorly bonded material from the surface. This improves adhesion, as well as improves the coating density, which can minimize the number and size of grain boundaries. Fewer and smaller grain boundaries give fewer defects through the coating reducing this diffusion path. This may not be without side effects as increasing the coating density can also increase the stress in the film, which can lead to substrate curl and handling difficulties.

The substrate surface cleanliness and quality becomes much more important when making ultra-barrier coatings used for packaging the OLEDs or solar cells. In this case, any surface bumps or dents can lead to subsequent coating cracking as shown in Fig. 15.20.

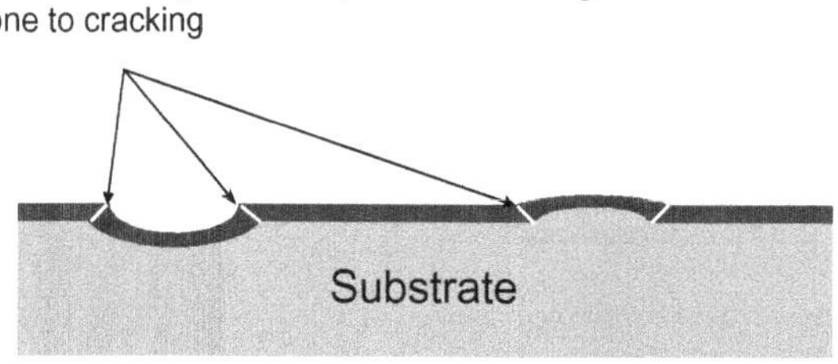

Figure 15.20 A schematic of how bumps and dents lead to stress raisers and subsequent cracking of the coating.

If we look at all these potential defects, we can rank them. Pinholes are the most detrimental and to make significant barrier improvements, one must remove surface debris on the substrate surface before coating. Next would be creating sufficient surface energy to increase wetting of the coating, improve adhesion, and minimize holes between impinging crystals. After this, densifying the coating will improve the barrier further. Finally, it may also be worth considering looking at the substrate surface material to see if there are any others that give better barrier performance.

Reducing surface debris amount and size reduces pinholes. This is achieved by substrate surface cleaning. Plasma treatments do not remove large debris from the surface and give only a molecular cleaning. One must not use a physical cleaning method that may remove the debris and damage the surface during the process. Two surface cleaning methods that work well are: a tack roll in contact with the web surface, and which the debris sticks to, and so can be removed [38–40], and an ultrasonically pulsed gas with electrostatic neutralization and vacuum to remove the particles. The ultrasonic pulsing flutters the web, which shakes off the dust, with the vacuum extraction removing the debris as it leaves the surface. The electrostatic neutralization helps prevent the dust being attracted back onto the surface. The tack roll has more recently been developed so that it can be used inside roll-to-roll vacuum coaters. The current limitation with including the tacky roll inside the vacuum system is that the high tack roll cannot be refreshed automatically and so only rolls that are short enough than the high tack roll does not become clogged can be cleaned. This limits using this technique inside a vacuum to short specialized rolls that are typically less than 10-km long, such are used for the ultra-barrier products.

This is only a partial solution for ultra-barriers, which need a much higher coating perfection than typical barrier coatings for food packaging. For ultra-barrier coatings, not only does the substrate need cleaning but any remaining very small debris as well as any inherent bumps or dents are covered by depositing a polymer coating inside the vacuum system immediately before the regular deposition process [41,42].

This polymer deposition process (Fig. 15.21) deposits a coating thick enough to cover all the defects. Also, the deposited polymer needs to have sufficient time in the liquid state to allow the coating to level out on the surface before it is cured. In this way, the new surface is expected to be the cleanest, flattest, and smoothest and most defect-free surface possible. This surface may still require a plasma treatment to increase wettability for the subsequent coating. Despite this high-quality surface preparation, the barrier performance of the coatings is still not as good as required for the high-tech applications. To improve the barrier further, a second polymer coating is sometimes added, after the metal or transparent ceramic barrier layer, to protect the barrier layer from subsequent damage that would reduce the barrier performance. It has been suggested that the polymer fills any holes by capillary action reducing the diffusion rate (Fig.15.22).

Even with these additional pre- and post-deposition layers, this still does not produce the necessary barrier performance for the highest specification barriers. To achieve this higher specification, additional alternating polymer and inorganic material layers are added. Even these multiple coating layers do not eliminate all defects, but what they do is significantly increase the tortuous path that any gas or liquid has to follow to pass from one side of the barrier coating to the other (Fig. 15.23).

The tortuous path idea has been utilized for many years for packaging films. It was recognized that pinhole defects were a big problem that was expensive and impossible to eliminate totally. At best pinholes can only be reduced, which still leaves a significant barrier problem. The low cost solution was to metallize both sides of the substrate. The fact that pinholes are randomly distributed means that, although there are pinholes on both sides, statistically the pinholes will rarely be directly aligned. The effect is that the diffusion path becomes tortuous and thus the diffusion rate is significantly reduced.

Another significant packaging material is paper and this too can be metallized. Approximately 66% of the paper is used for labeling [43,44]. Paper metallization allows gold and silver effects to be included

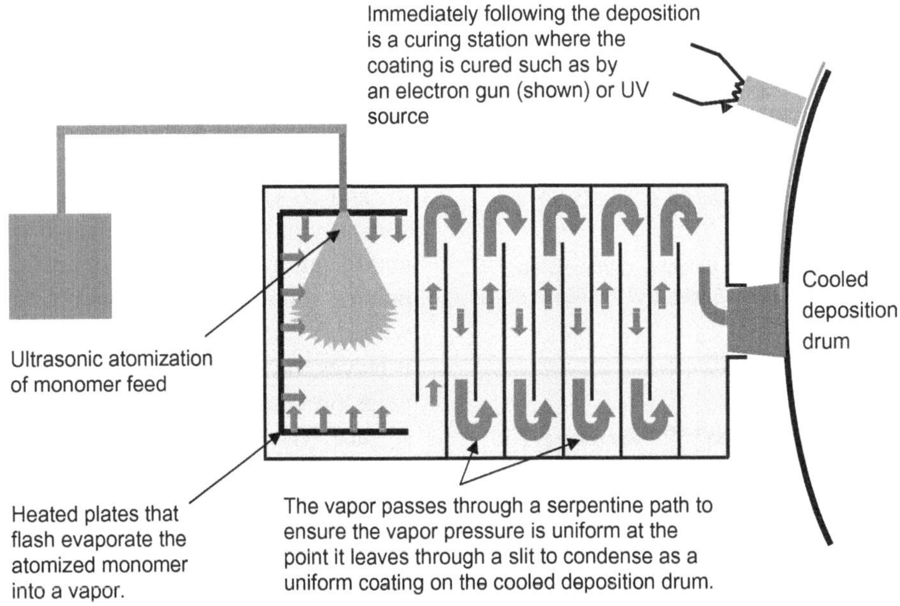

Figure 15.21 A schematic of in-vacuum polymer deposition.

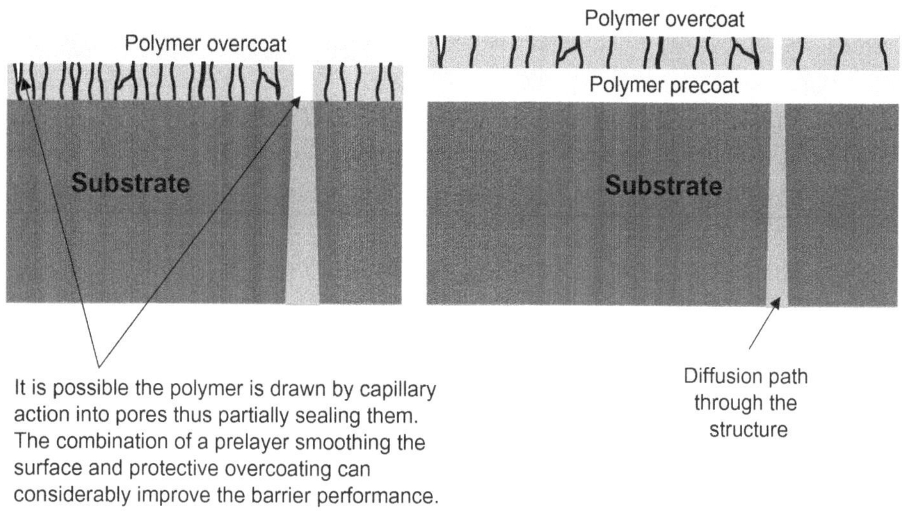

Figure 15.22 A schematic showing the benefits of minimizing surface defects.

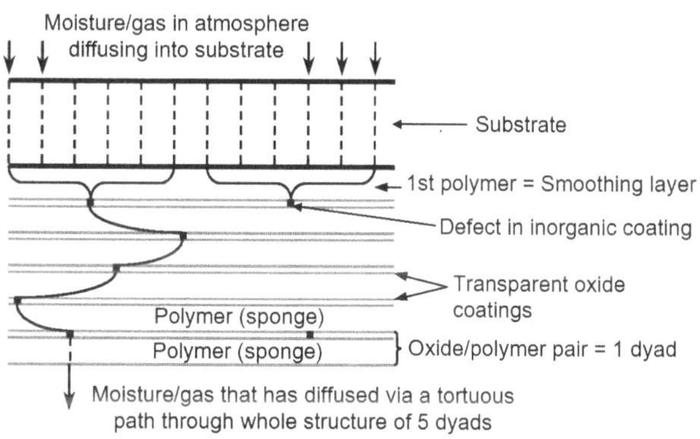

Figure 15.23 A schematic showing how a tortuous path improves barrier performance.

in the label design. Paper is a very different substrate than polymer films, particularly with regard to water content. Paper can contain as much as 20% water, if stored, for a prolonged period, in high humidity conditions. This moisture content adds a huge vapor load to the pumping system. Vacuum systems that are designed to metallize paper usually have additional cryopumping, included specifically to pump the water. This increased cryopump capacity can reduce the paper water content to below 5%.

Care must be taken that the moisture level does not fall too much, as the paper can become brittle and very fragile resulting in tears or web breaks becoming a significant problem. Similarly, after metallization it is often required that the paper be rewound so that it can become rehydrated on exposure to the atmosphere. If this rewinding is not done, the paper can rehydrate on the roll and the resultant swelling can damage the roll.

The paper surface quality is also important to the metal reflectivity. Therefore, it is common for the paper to have one or more polymer coatings to prepare the metallization surface by both sealing and smoothing the surface. The surface may also be calendared before coating. The calendaring process smoothens and compresses the paper fibers and is a surface polishing process. Metallized paper has a high demand in products, where it is perceived that polymers can add an odor or taint that papers do not have. One very big user for metallized paper is the tobacco industry.

15.7 Pattern Metallization

Pattern metallization is the title given to the process where the aluminum deposition is masked, in some way, so that only part of the polymer web surface is metallized. There are several ways that this can be achieved. Either the polymer surface can be printed with a pattern before it is put into the vacuum coater or the aluminum is printed after it is deposited with a resist and then the exposed aluminum removed by etching.

In the first case, the printed pattern is a soluble material and the whole surface is metallized. Once metallized and removed from the vacuum system, the roll is rewound through a solvent and the printed pattern will dissolve away taking with it the metal deposited onto the printed pattern [45].

In the second process, the substrate is metallized. It is then printed with a protective pattern (resist), covering the metal in the areas where the metal is to remain, and then the roll is rewound through a sodium hydroxide solution, which will remove, or etch, the aluminum in the areas not protected.

There has always been a desire to reduce the process steps by bringing the patterning process into the vacuum system where the patterning and metallization are done at the same time. This is done for simple striping by rotating a metal mask that came into contact with the polymer substrate (Fig. 15.24) as it passes through the deposition zone [46]. This process could be prone to generating dust, as the accumulated aluminum on the mask flakes off, and also there is a limit to the fineness of the lines that could be produced.

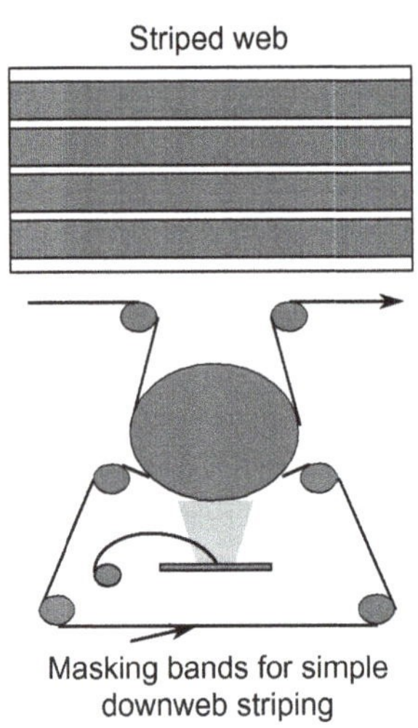

Figure 15.24 Masking process that uses masking bands to produce simple stripes.

The newer process (Fig. 15.25) is to print on the polymer substrate some low vapor pressure oil [47]. This oil is printed onto the areas where the aluminum is not required. The process is designed to print just enough oil such that the polymer substrate is virtually dry of oil once the substrate has passed through the deposition zone. What happens is that the aluminum as it deposits onto the oil does not stick and it re-evaporates. As it deposits on the oil surface, it will also heat and evaporate off some

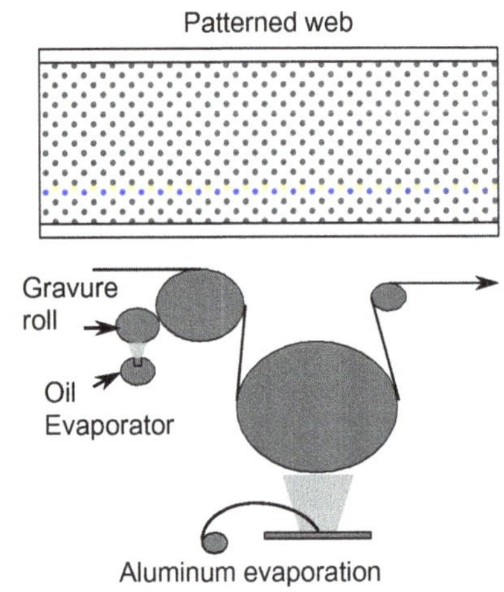

Figure 15.25 Schematic showing the much more versatile oil printing process for producing finer and more detailed patterns.

of the oil, and so eventually all the oil will be removed. If too little oil is printed on the surface, there will be some aluminum deposition, if too much oil is deposited, there will be residual oil and this will contaminate the roll as it gets rewound.

This pattern metallization process (Fig. 15.25) has been used for flexible electronic circuits such as RFID antenna, capacitor films, microwave susceptors (including fused susceptors), demetallized security holograms, and packaging films (Fig. 15.26).

In packaging films, there has always been competition between hot or cold stamped foil and pattern metallization. Both techniques are designed to only put the metal where it is desired, rather than metallizing the whole surface and overprinting with an opaque ink to hide the metallization in areas where it is not required.

15.8 Trends

There is a concern that there will not be enough food to keep the world population fed. The United Nations had a project to evaluate waste in the supply chain and how this might be addressed. In broad terms, the developed countries had significant waste from highly packaged, purchased but uneaten food that spoiled—in domestic storage. In the underdeveloped countries, there was significant food waste between the growers and the consumers. It was determined that this could be reduced in many cases by some simple packaging. So the

Figure 15.26 Various antenna shapes for microwave applications that could be produced by the oil printing pattern metallization technique.

conclusions included the desire for the developed world to reduce the drain on resources by reducing the unnecessary packaging. At the same time, for the underdeveloped world, they needed to increase the packaging of many fresh food items to prolong the storage lifetime to enable more to get to the consumers before they spoil. There was also a desire for more packaging to be fully recyclable and so they wanted to reduce mixed polymer structures that are hard or impossible to be recycled.

More locally, the packaging suppliers and converters have the driver that the cost of existing packaging is constantly under competitive or manufacturing cost pressure, which is passed on through the supply chain. The demand is for lower cost substrates, metallization, lamination, and printing. One way that some cost reduction has been achieved has been by downgauging (thinning the substrate). In downgauging, the substrate thickness is reduced with the expectation that the overall substrate performance is maintained. This is not without problems, as it is generally harder to handle thinner substrates during manufacturing and use. Also, managing the heat load through the deposition zone, during metallization, can be more difficult. The other aspect of down-gauging can be reduced layers in packaging laminates.

A good example would be coffee packaging, where the initial packaging material was an aluminum foil laminate (Fig. 15.27—1). A competitive material was produced that replaced the aluminum foil with a metallized polymer film in a triple layer laminate (Fig. 15.27—2). This triple-layer packaging material was still thick at approximately 110 μm total thickness, so there was a demand for down-gauging. The triple-layer barrier laminate was later successfully replaced by a thinner, two-layer laminate with equivalent performance (Fig. 15.27—3). This down-gauging process continued and the double-layer laminate was further reduced from 85 μm down to 60 μm (Fig. 15.27—4). It is believed possible to further reduce the laminate thickness and still meet the required barrier performance [48,49]. The reason why further thickness reductions are not yet used is the

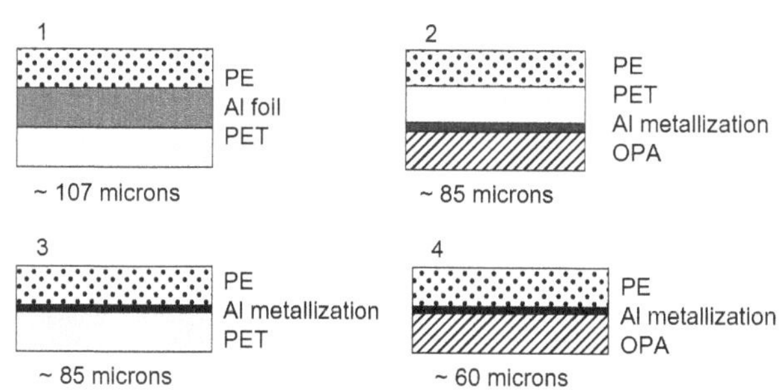

Figure 15.27 Down-gauging progression. 1–4, from the original foil packaging for coffee to the triple-layer replacement and then to the double layer and then the thinner double-layer current product.

mechanical performance of the thinner laminate. With the reduced overall laminate thickness comes a reduced stiffness making forming and filling the package more difficult. Thus, any future down-gauging will probably also include a further material change such as a change to oriented polypropylene. Some packaging machines use a horizontal form and fill process that requires the package to be stiffer than a vertical form and fill process and so it is likely that there will be many different products for this single application.

The other critical fitness for use criteria for down-gauged packages is heat-sealability. It has been found that the limitation to further down-gauging is most likely to be due to problems with the heat-seal integrity and not with the metallized or deposited transparent barrier coating performance [50]. Metallized BOPP thickness has been successfully reduced from 20-μm BOPP with a 40 nm metallized coating to a 12-μm BOPP with the same metallized coating thickness.

Down-gauging is not exclusive to polymer films but also applies to paper where it is used for high barrier applications. Similar to the polymer versions, the paper laminates started with aluminum foil as the high barrier layer that got substituted with a metallized polymer film. Typically the approximately 10-μm aluminum foil is replaced by a metallized polymer film at least 12 μm thick. In this case, the initial total thickness is increased but the laminate weight and cost is reduced [51] (Fig. 15.28).

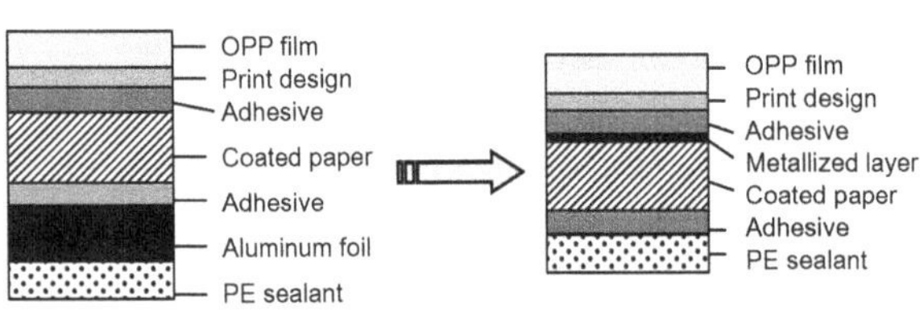

Figure 15.28 Aluminum foil/paper label with the replacement label that uses a metallized polymer [51].

Because polymer films can be produced as white films, and not just transparent films, it is also possible not only to replace the aluminum foil but also to replace the paper with polymer film too while maintaining the overall product ascetics.

The trend of replacing aluminum foils has been established for a number of years [52]. The metallized films used as replacements are not true replacements in that they rarely match the barrier performance of the aluminum foil. However, it is often the case that the foil has a barrier performance beyond what is required for the product shelf life, and so the metallized film can meet the specification although technically it does not match the ultimate performance of the foil it replaces [53]. Where the thinner metallized films can often outperform, the foil is where the package is to be folded. The aluminum foil will crack or begin to perforate (pinhole) under dead fold conditions. As the technical understanding increases, on how to produce more perfect barrier layers that do really approach the barrier performance of foil, then the environmental pressures will continue the drive to replace foils with metallized films. This is driven by the fact that laminations with metallized films are lighter in weight and take less energy to produce and transport which lowers the package total energy requirements.

A further target for developing these multilayer films relates to recycling. The above examples show that many laminates have mixed polymer content. While it is possible to recycle these films, the value of the recycled polymer is lower than if each of the layers were to be based around the same polymer and could be recycled as a single polymer type.

Recycling is also impacted by the adhesives, lacquers, and inks that are used. Some materials may need to be heavily cross-linked to make them effective, but this may make recycling more difficult. Thus, in the future some material choices may appear to be taking a step backwards, in performance, but may enable either a lower carbon footprint or a lower recycling cost or penalty for the packaging.

An example would be the lidding material used as shown in Fig. 15.29. The product described in Fig. 15.29 has an amorphous PET (APET) tray and the lidding has a structure comprising a co-extruded, heat-sealable PET film adhesive laminated to a transparent aluminum oxide–coated PET film. This

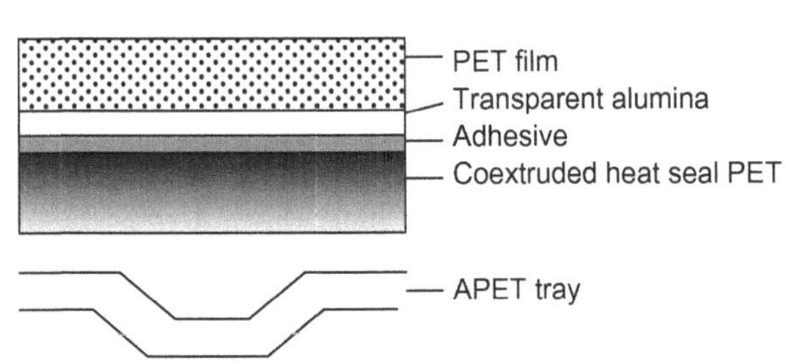

Figure 15.29 PET heat-seal laminate barrier-lidding material.

allows the lidding to be transparent and the whole package to be recycled as a single polymer [18]. Using a co-extruded PET allows the APET layer, which can be heat-sealed, while retaining the film dimensional stability, by using a co extruded crystalline PET layer. In this way, the amorphous polymer replaces an alternative polymer heat-seal coating.

Currently, many mixed polymer structures are used to reduce packaging costs. Polyethylene (PE) film does not have PET's mechanical performance but has lower cost. To use a thicker single PET film would be very expensive, whereas using a PE/PET laminate can give the best of both materials and reduce cost. As the costs of recycling and disposal continue to rise, and become a more significant part of the total cost in the packaging process, the number of materials will be simplified. To continue to achieve the expected performance, we will have to use every possible technique to continue to improve the performance, using fewer materials.

We know that the film surface quality is critical in producing good barrier films and that in theory we only need the polymer carrier substrate and a perfect coating to achieve a perfect barrier. This requires the polymer substrate to be perfect, including the cleanliness, chemistry of the film production, minimization of any additional filler or additives, as well as any optimized pretreatments. Currently, many products made only control one or two of these factors. This means that there is room for improvements by controlling the remaining factors. Cost currently stops this from being done. It is always more expensive to clean film or to produce a smoother surface with fewer residues or additives. As the costs change to meet the environmental regulations, it is likely to become economically favorable to reevaluate some of these factors. This could mean having to consider paying higher quality substrates that would enable superior products to be achieved.

Metallized papers for packaging, like aluminum foils, are also in decline. The metallized paper volumes have remained constant, despite the metallized packaging market increasing [53]. Most metallized paper is for labels. Removing labels from glass bottles using high temperature liquids, which may also clean the bottles, is regarded as an easier process than removing polymer labels. With the increase in polymer bottles, either direct printing or polymer labeling is increasing. Although there are many niche markets, that will always prefer metallized paper or board, it would be expected that metallized paper volumes will not only continue flat but also will start to decline.

In calculating product shelf life, it is possible to separate out the gas diffusion rate through the barrier film but this is not the only source of the gas. There may be some gas captured within the package as it is filled and sealed. There are several ways to eliminate this gas, such as vacuum, modified gas atmosphere and scavenging, or a combination of these mechanisms.

Shelf life for some products may be adversely affected by oxygen and simply applying a barrier layer to the packaging may not achieve the required shelf life. In some cases, a scavenging chemical is added that preferentially getters, or absorbs, the gas, further extending the shelf life. The scavenger level used can be sufficient to remove any gas left in the package, after it was sealed, plus sufficient material to getter all the gas diffusing through the barrier material over the desired lifetime. Recently, there has been work to change these scavengers from powders, included in a porous sachet, to incorporate them in the structure as part of the vacuum-deposition process becoming a more integral part of the lamination [54]. An example would be adding palladium towards the end of the electron beam deposition of silica.

15.9 Summary

Good quality vacuum-deposited coatings start by using high quality substrates. Ideal barrier coatings are only achieved through minimizing the film surface contamination, optimizing the surface chemistry and maximizing the surface smoothness. By having a clean, smooth surface and then maximizing the surface energy before vacuum coating, the deposition material will wet the surface and produce a continuous, low defect, coating at the thinnest coating thickness. This care of the surface and increased wetting will also help maximize the coating adhesion. If aluminum coatings are allowed to age and oxidase before use, the metal surface will have a tougher oxide surface and this will help protect the metal against any damage from subsequent handling.

Using the above recommendation will provide the user with a high quality barrier material that can then be a core layer that the packaging can be designed around.

References

[1] Mattox DM, Mattox VH. 50 years of vacuum coating technology. Society of Vacuum Coaters. ISBN 978-1-878068-27-9; 2007.

[2] Llewellyn W. Metallized papers and films—Developments in the global market. Proceedings AIMCAL Technical Conference; 2006.

[3] Beeching D. Coors: Summer silver adds sales sparkle. Proceedings Holo-pack-Holo-print; 1998; Paper 4.

[4] Williams K. Holography adding to the competitive edge. Proceedings Holo-pack-Holo-print; 1998; Paper 8.

[5] Keser O. Three years of holographic authentication of Caykur tea. Proceedings Holo-pack-Holo-print; 1998; Paper 3.

[6] Kontnik LT. Costs, benefit and prices. Proceedings Holo-pack-Holo-print; 1998; Paper 9.

[7] Collins D. Holograms and metallized packaging. Pira International Ltd; 2003; pp. 35–39; ISBN 1-85802-930-9.

[8] Yates M. Metallic printing inks—a complimentary metallic effect. Section XIV Proceedings Metallized Papers & Films—Europe; 2002. AWA Alexander Watson Associates BV.

[9] Seubert J, Fetz A. PVD Aluminium pigments: superior brilliance for coatings and graphic arts. Paints & Coatings Industry 40–48; 2000.

[10] Mount EM. III Aroma and moisture barrier properties of plain, coated and metallized OPP films. Tech. Proceedings SME Barrier Technology for the Food Packaging Industry. 1997.

[11] Decker W, Henry B. Basic principles of thin film barrier coatings. 45th SVC Ann. Tech Conf.; 2002. pp. 492–502.

[12] Bishop CA. Polymer web surface cleanliness Proc 45th SVC Ann. Tech Conf.; 2002. pp. 52–57.

[13] Jamieson EHH, Windle AH. Structure and oxygen-barrier properties of metallized polymer films. J. Matls. Sci. 1983;18: 64–80.

[14] Bishop CA. Tullo G. How items such as dust, oligomers & slip agents can affect the polymer surface quality and be potential problems in high-tech roll-to-roll vacuum deposition applications. Proc. 49th SVC Ann. Tech Conf.; 2006. pp. 648–653.

[15] Fayet P. Commercialisation of plasma deposited barrier coatings for liquid food packaging. Proc. 38th SVC Ann. Tech Conf.; 1995. pp. 15–16.

[16] Smith AW, et al. PECVD of SiO_x barrier films. Proc. 45th SVC Ann. Tech Conf.; 2002. pp. 525–529.

[17] Krug. Transparent barriers for food packaging. Proc. 33rd SVC Ann. Tech Conf.; 1990. pp. 163–169.

[18] Kelly R. Transparent barrier coatings for films. Section IX Proceedings Metallized Papers & Films—Europe; 2002. AWA Alexander Watson Associates BV.

[19] Miyabayashi T, et al. Recent advances on the world largest 122 inch vacuum web coater aimed at cast polypropylene (CPP) films. Proc. 3rd Intl. Web Coating Conf. 2–23, 1989.

[20] Struller C, et al. Characterization and conversion of aluminum oxide coated flexible barrier films. Proceedings AIMCA Tech. Conf.; 2013.

[21] Schiller S, et al. Deposition of Alumina Layers on Plastic Films Using Conventional Boat Evaporators. Proc. 39th Ann. Tech Conf.; SVC; 1996. pp. 404–412.

[22] Schiller N, et al. Vacuum roll-to-roll technologies for transparent barriers on polymer webs. Proceedings AIMCAL European Tech. Conf.; 2012.

[23] Dickey ER. Advances in Roll to Roll Atomic Layer Deposition. Proc. 55th Ann. Tech Conf.; SVC; 2012. pp. 73–80.

[24] Lai L, Zeng M. Microwavable container with heating elements having energy collecting loops. US Patent 5,698,127 1996.

[25] McCormick JA. Walters GJ. The fuse suceptor. Proc. 38th Ann. Tech Conf.; SVC; 1995. pp. 47–51.

[26] US Patent 5,412,187 Fused microwave conductive structure. 1984 McCormick J.A. & Walters G.J.

[27] ESD association standard for the protection of electrostatic discharge susceptible items. Packaging materials for ESD sensitive items. ANSI/ESD S541-2003. An American National Standard Approved March 25, 2003.

[28] Bishop CA. The use of vacuum deposited coatings for security applications. Proc. 44th SVC Ann. Tech Conf.; 2001. pp. 425–430.

[29] van Renesse RL. ed. Optical document security. 2nd ed. Artech House. ISBN 0-89006-982-4.

[30] Alcan International Ltd. US Patent 4,837,061 June 6, 1989.

[31] Bishop CA. Vacuum deposition onto webs, films and foils. William Andrew Publishing; 2007ISBN 0-8155-1535-9
[32] Chan C-M. Polymer surface modification and characterisation. Hanser Publishing N.Y.; 1993.
[33] Mount EM III. Metallizing technical reference. 3rd edn. Association of Industrial Metallizers, Coaters and Laminators AIMCAL 2001.
[34] http://www.stevenabbott.co.uk/Practical Adhesion
[35] Movchan BA, Demchishin AV. Study of the structure and properties of thick vacuum condensates of nickel, titanium, tungsten, aluminium oxide and zirconium oxide. Fiz Metal Metalloved 1986;28(4):653−60.
[36] Thornton JA. High rate thick film growth. Ann Rev Mater Sci 1977;7:239−2260.
[37] Messier R, et al. Revised structure zone model for thin film structure. J Vac Sci Technol A 1984;2(2):500−3.
[38] Hamilton S. Yield improvement in metallizing lines by using contact cleaning technology 51st Ann. Tech Conf.; Proc. Society of Vacuum Coaters; 2008. pp. 778−780.
[39] Hamilton S. Advances in contact cleaning for yield improvement, Proc. AIMCAL Fall Tech. & 22nd International Vac. Web Coating Conf.; 2008.
[40] "Web cleaning efficiency using contact cleaning roll (CCRs)" Downloaded from http://www.polymagtek.com/pdf/dirt_rpt.pdf
[41] Affinito JD. et al. A new hybrid deposition process, combining PML and PECVD, for high rate plasma polymerisation of low vapor pressure and solid monomer precursors. Proc. 42nd SVC Ann. Tech Conf.; 1999. pp. 102−106.
[42] Yalizis A, Mikhael M. UV versus electron beam radiation curing of vacuum deposited levelling coatings for high barrier applications. Proc. 47th SVC Ann. Tech Conf.; 2004. pp. 600−605.
[43] Gooseman M. Paper for metallization—speciality or commodity? Section VI Proceedings Metallized Papers & Films—Europe; 2002. AWA Alexander Watson Associates BV.
[44] Saillard A. Barrier metallized papers: aesthetics, ecological and protection. Section XIII Proceedings Metallized Papers & Films—Europe; 2002. AWA Alexander Watson Associates BV.
[45] Searl GA, Pizzanelli D. Hologram devices and method of manufacture. UK patent No. 2,136,352; December 3, 1982.
[46] Yamamori T. Margin generating mechanism in ULVAC's roll coater. Proc. 2nd Intl. Web Coating Conf.; 1988. pp. 58−76.
[47] Walters GJ. Pattern metallized films used for higher energy density capacitor components. Proc. 42nd SVC Ann. Tech Conf.; 1999. pp. 453−455.
[48] Hoekstra T. Metallized materials for Fast Moving Consumer Goods (FMCG)—packaging trends and needs. Section II Proceedings Vacuum Coating & Metallizing Industry Conference; 2004. AWA Alexander Watson Associates BV.
[49] Langowski HC. Barrier mechanism of vacuum coated polymers—where are the limits for downgauging of barrier substrates and for barrier properties of coated films? Proceedings ICE 2005—surface treatment of polymer films.
[50] Decker W. Metallized OPP as Alufoil replacement—US developments. Section III Proceedings Vacuum Coating & Metallizing Industry Conference; 2004. AWA Alexander Watson Associates BV.
[51] Blecken G. Markets, applications and trends for metallized PP. Section IV Proceedings Metallized Papers & Films—Europe; 2002. AWA Alexander Watson Associates BV.
[52] Jeffries E. Perfect foil. Flexible 2004;3(3):34−41.
[53] Collins D. Hot metal. Flexible 2003;5:16−23.
[54] Wanner T. Coating techniques for active packaging—oxygen absorbers, indicators and anti-microbial films. Proceedings ICE 2005—surface treatment of polymer films.

16 Web Handling and Winding

David R. Roisum
Finishing Technologies, Inc., Neenah, WI, United States

16.1 Web Handling Essentials

Web handling is the art and science of getting a web through a machine as fast as possible and with minimal waste and certain other troubles. These skills are essential to all processes from web manufacturing, to roll-to-roll converting and up to the final step of cutting the web to length when it is no longer a web. While elements of web handling have been known for more than a century, it was not until the mid-1980s when the WHRC (Web Handling Research Center) was formed that people across many industries recognized they had a common set of problems [1]. The most common problems are:

- Baggy webs
- Curl
- Length and width (dimensional control)
- Registration (position control)
- Web breaks
- Winding defects
- Wrinkling.

These include some of the largest causes of waste and delay in the web industries. These troubles include winding that is one of the half dozen or so major topic areas of web-handling study.

It was not until several technical leaders from different industries got together in the same room that we recognized that webs from the thinnest polyester to some of the heaviest aluminum sheet and carpeting; all suffered from wrinkles as just one example. The WHRC not only accelerated the pace of turning art into science, but they also helped make web handling a "household name" in tens of thousands of plants worldwide. At the time of this writing nearly 5000 people have been formally trained in just one of the multiday web-handling courses [2] and the total trained by all web instructors probably approaches 20,000.

Another important change in the web-handling world began in the mid-1990s, about a decade after the formation of the WHRC, when books were written to collect and organize our knowledge. Perhaps the most important book ever published is the *Ultimate Roll and Web Defect Troubleshooting* published by TAPPI Press [3]. This book is an encyclopedia describing almost all of the defects you will ever see in the web world. It gives pictures, paragraph descriptions, causes and cures as well as a detailed bibliography if more detail can be found elsewhere. No less important is that it codifies our language so that, for example, a wrinkle is not a wrinkle is not a wrinkle. There are six mutually distinct major wrinkle classes with about 20 different subclasses, or root causes if you like. Other examples include core crush, starring, and telescoping that are really a collection of distinctly different defects that happen to share the same outcome. A dictionary with aliases of thousands of words helps makes sure that we are all speaking of the same thing. Major areas covered in this troubleshooting encyclopedia include:

- Web handling
- Wrinkling
- Winding
- Coating
- Lamination
- Slitting
- Papermaking
- Extrusion
- and many others.

Without exaggeration, this is the one must-have book in our industries. Some of our best tools have recently been made much more accessible by the free and easy AbbottApps that will run on most any smart device that has an internet connection [4]. These apps can do a variety of web-handling calculations and diagnostics. However, there are also

Multilayer Flexible Packaging. DOI: http://dx.doi.org/10.1016/B978-0-323-37100-1.00016-8

many other apps in adhesion, heat sealing, packaging, and other areas. Also of note are the more than 100 short instructional YouTube clips on a variety of basic and advanced web-handling topics that are indexed under the Web201 app.

Aside from the WHRC and TAPPI already mentioned, other organizations that contributed greatly to web handling include AIMCAL (Association of Industrial Metallizers and Laminators) and ICE (International Converting Exhibition) which is the successor to CMM (Converting Machines and Materials). Magazines that included a large web-handling content include the *Converting Quarterly* (AIMCAL), the *Converting Magazine*, and the *Paper Film Foil Convertor* magazine. Blogs with multiple posts per week for nearly a decade now add to this ever-growing collection of knowledge. Now our problem in web handling is not too *little* information; it is often too *much* information. The simple solution here is to take formal training that greatly speeds up the process and gives a structure for past and future learning to build on.

16.2 Rollers

One easy way to recognize a web-handling machine is the presence of rollers. In fact, rollers are the most common element found no matter what the web is and what machine it is running on. The most important resource here is again a book by Roisum [5]. This textbook provides engineering tools used to design, size, apply, and maintain rollers. For example, if the *roller size (diameter)* is too big, you cost your company money because larger rollers' cost more. On the other hand, if the size (diameter) is too small, you cost your company a *lot* of money due to problems that can happen with excessive roller deflection including wrinkling. This vital roller-sizing step is now made easier with the Roller Bending Calculator by AbbottApps [4].

Rollers also must be adequately *aligned*. Here, the conventional machine building and maintenance frame-of-reference are level, square, and common centerline [6]. However, web handlers know that the really important references are in-plane bending and out-of-plane twisting. In-plane bending is far more risky in most cases because it causes four risks for every single roller that is misaligned as seen in Fig. 16.1. Because these misalignment risks are calculable [4,5], they are thus predictable to the

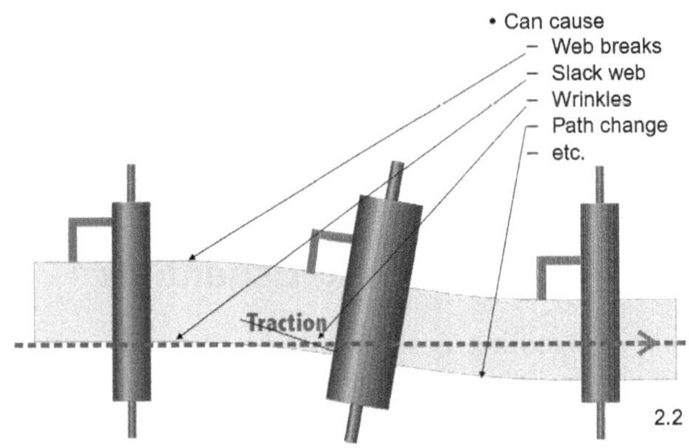

Figure 16.1 Four risks for every single misaligned roller.

point of suggesting guidelines for precisions [7]. Here, it may surprise that many *hairsbreadth tolerances* are often required, though the means to achieve those has been known and practiced for about a half century in the paper industry. Converting is now also starting to catch on to the need for mechanical precisions. What is important is the practical view of rollers. It does not matter how many rollers are done right. *All* that matters is how many that are *not* done right (or at least good enough). A single misbehaving roller can change the entire waste and delay character of a machine.

Another precision roller requirement is adequate *cylindricity*, which includes both out-of-round (sometimes called TIR for total indicator runout) and diameter variation across the width. Another requirement is that the supporting frame must be adequate to hold the roller's bearings in a sufficiently fixed location despite changes in applied loads. These and other geometries, again, may need to be on the order of a hair's thickness in tolerance, that is 0.005″ or 5 mils or 125 μm. Rollers may have metallic coatings on them for a variety of reasons including smoothness, roughness, or abrasion resistance. Similarly, rollers may be covered with an elastomer, sometimes called a rubber cover even though natural rubber is almost never used [8].

One last major topic is the concept of *traction*. Whether rollers are (motor) driven or undriven (a.k.a. idlers), they need to run precisely at web speed. There needs to be a sufficient combination of tension, wrap angle, and web-to-roller COF (coefficient of friction) to transmit the torque imposed by bearings (idler roller), motor (driven roller), and inertia during speed changes. Loss of traction is tantamount to loss of control and can disable guides, spreaders, and many other elements in addition to drives. Fortunately, traction capacity as a design calculation is well documented in books [5] and widely available through a free internet app [4]. Only one last traction complication needs be mentioned here. That is a loss of traction will occur in certain cases when *air entrainment* is present. Those air entrainment sensitive cases require all of the following to be true about the web: not narrow, not slow, and not rough. Again, the good news is this is a design calculation that can anticipate the risk as well as sizing a remedy such as roughening or grooving of the roller surface.

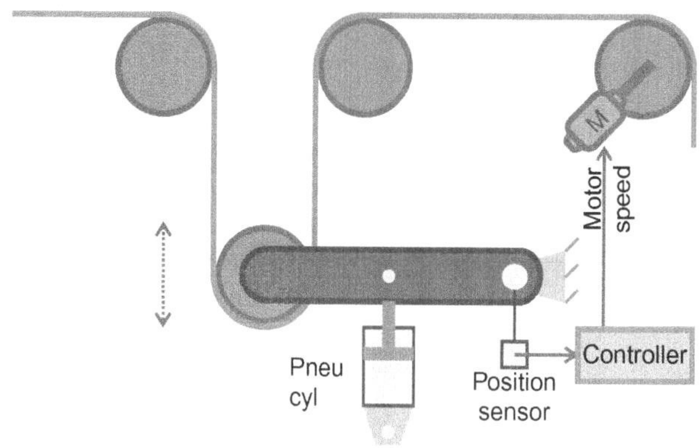

Figure 16.2 A pivoting dancer.

16.3 Tension Control

Tension is the first of the TNT (tension, nip, temperature, and path) controls that we will discuss because it is the most common. However, common does not always mean well understood or well executed. First is a definition. The most common units for web tension are force per unit (web) width. In the metric system, this would be newton per meter (N/m) or kilonewton per meter (kN/m). In the US system, the units are lb/in or pounds per lineal inch most commonly abbreviated as PLI. Occasionally, on some machines and in some industries (such as foil) and in some research, we nondimensionalize tension as force per unit width *per unit thickness*. This is engineering stress in pascal or PSI (pounds per square inch in the US system). Occasionally, especially on less engineered machines and machines where web width never varies, such as in some converting operations, the tension is given in machine units of total force in lbs or kN.

The most common instrument for measuring tension is a roller on a calibrated *load cell*. Very often that load cell might be used as feedback for a motor under tension control, though sometimes load cells are merely used as indicators. There are other drive motor control strategies such as dancers and speed control. *Dancers* are a roller on a pivoting arm with position readout as seen in Fig. 16.2. Dancers are a form of tension control that lacks tension *feedback* readout. Nonetheless, the dancer *setpoint* can and should be calibrated in the tension units above, even though it is seldom done for purely cultural/historical rather than practical reasons.

There are many guidelines for a "best" *tension setpoint* [9]. These guidelines tend to be somewhere between low-tension troubles, such as inability to hold a stable path, and high-tension troubles, such as web damage [10]. However, the result for many webs running through most machines is that they tend to run best between 5% and 25% of the web breaking (or yield) strength. Because the web world is quite varied, there are exceptions to this simple guideline such as foil and heavy construction materials that may run even lower than that range and creped tissue that may run higher than that.

We have just defined tension, gave some guidelines to its best size and merely mentioned *drive motor control strategies*. We will end with a brief discussion of web drives. This is easily one of the most complex areas of web handling; so much so that a working proficiency usually requires formal web training [2] as well as formal training in web drives [11] as well as a specialization or dedication for many years under something like an apprenticeship. It is thus well beyond the scope of this introduction to go into any details. Here, I will just offer a couple of concepts. The first is that is usually matters little whether the control strategy is dancer or load cell control, for example, because either can usually work well enough provided that all the details are mastered. Second, it usually matters little whether the motors are DC, AC, servo, or even seeming old-fashioned air brakes, for example, because any of these torque-developing devices can usually work well enough provided that all the details are mastered. As such, we see well-behaved and poorly behaved dancers and the same is true of load cells. As such, we see the old-fashioned air brake when well done outperform a servo that is not well understood by the control engineer. All that matters in the end is that the web tension setpoint be suitable, as discussed above, and that variation is held within suitable tolerances. An often and long-quoted *tension quality spec* would be to hold tension within 5% of setpoint at steady state and within 10% of setpoint during speed changes and perhaps 50% for the rare short violent events such as automatic roll changes on a turret winder [5]. These tension control quality specifications could be put into performance agreements when upgrading or buying new machinery [12].

16.4 Nip Control

After tension, the next most common web control is the nip between two rollers or, in the case of a winder, between a (wound) roll and a roll*er*. Nips might be used purely for transport, such as a driven pull roller section that has more traction capacity than a mere wrapped roller. However, the far more interesting applications are for processing such as calendering, coating, laminating, printing, and so on. In many cases, a nip pressure variation across the width can cause finished product variability or even product destruction such as in wrinkling. So in this introduction, we will focus on quality rather than on processing. Nip control quality is defined very much like tension control quality: a suitable load that is held to a suitable tolerance or variation across the width.

Let us first define units for nip load. They are not machine units such as force on the ends of the nip roller and certainly not cylinder pressure; even if it is often the norm in industries. The proper units for nip are the same as tension, kN/m or PLI, for example. However, just as we could account for thickness in tension to get pascal or PSI, we can also account for nip width to get pressure in precisely the same units of pascal or PSI. Nip control is about evenly divided into two different control schemes: *load control* as described above and *position control* where a small gap (or interference between two rollers if one is soft such as rubber covered) is maintained by precision adjustable stops. These two methods are shown in Fig. 16.3. Load control is common with laminating and winding. Position control is more common with coating and printing.

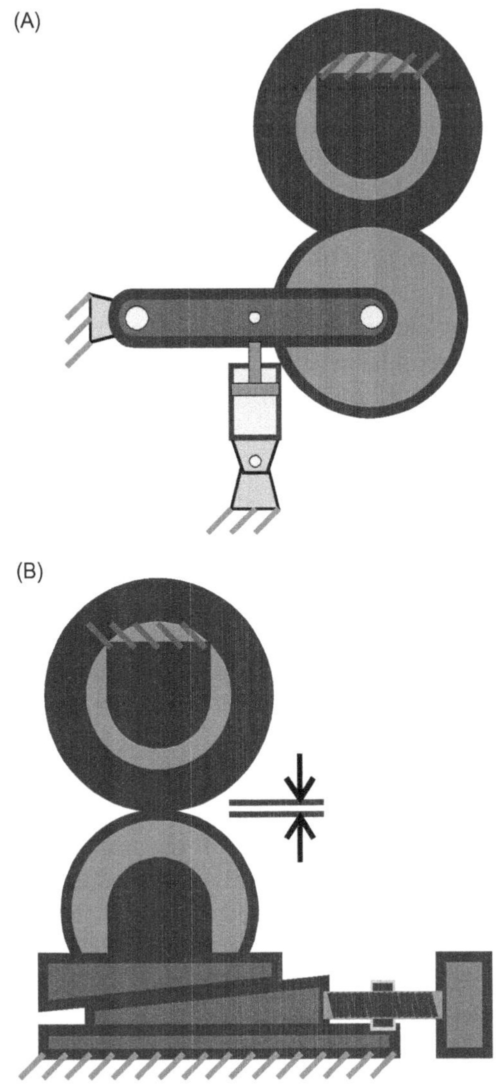

Figure 16.3 (A) Fixed load nip control. (B) Fixed position nip control.

To avoid going any deeper with nip mechanics, we will finish by talking about sources for nip variation and nip pressure quality control measurements. The pressure in a nip can vary across the width (and with time) due to many imperfections in the system such as:

- Manufacturing or maintenance
  - Roller misalignment
  - Roller radial runout
  - Roller diameter variations
  - Cover hardness variations
- Design shortcomings
  - Roller shell wall deflection
  - Roller deflection
  - Roller crown/load mismatch
  - Loading system variations.

It can be tedious and time consuming to measure each of these potential problems independently. Fortunately, we can detect most of these variations and have a good clue as to the cause by using *nip impression measurements* [13,14]. The most common nip impressions are made with something like carbonless paper, though there are many different products including one and two part films as well as embossed foil. The nip impression method may be static, in which case we read the uniformity or lack thereof of nip width across the width of the web or dynamic where we read the color of the nip impression as it might vary across the width. Both of these are summarized in Fig. 16.4. Nip impressions can be used for troubleshooting the possible causes of streaky products, for example, or preventative maintenance by doing the impressions on a schedule. Since these procedures are done near a nip, extreme attention to *safety* require it be done in a lock-out-tag-out condition where the sensor is placed and removed while the machine is locked out and the nip engaged only when the technicians are at a safe distance behind barriers or light curtains.

16.5 Temperature Control

Temperature control spans both web handling and web processing. In web *handling*, we might, for example, want to cool a hot web prior to winding so that further thermal shrinkage does not overstress the wound rolls. Though webs will cool in air, especially if the air velocities are high such as at an air knife, the most common way is to cool on a large hollow "chill roll" that has a circulating cooled liquid such as water or, in some extreme cases, glycol. However, the more common and more interesting cases for temperature control

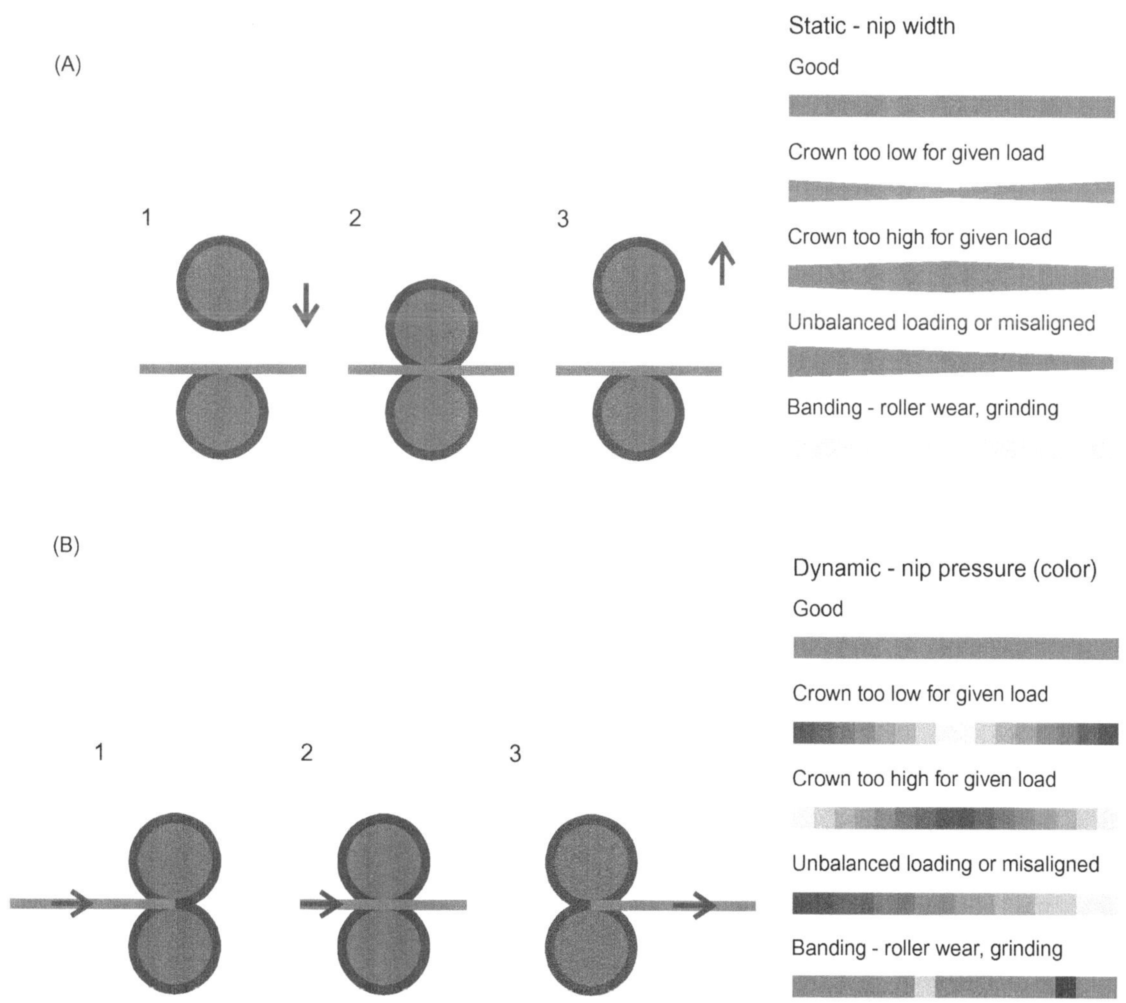

Figure 16.4 (A) Static nip impression. (B) Dynamic nip impression.

are web *processing*. This includes hot extrusion, either blown or cast. This includes hot lamination. The web handler should have a great respect for temperature, perhaps even more so than nip for the same reason. That is a small difference in web temperature can cause web profile problems such as *thickness variation* across the width that in turn is the leading causes of *bagginess*. Temperature uniformity starts with good thermal element design that is often optimized by finite element modeling. Maintenance and troubleshooting should use a good quality infrared camera to assess uniformity. While it is difficult to get a trustworthy reading on some shiny rollers, the web or winding roll can also be targets of inspection. Point infrared readouts and thermal strips applied to the web are other ways to get temperature information, though they are clumsy at best.

16.6 Guiding and Path Control

Every roller that touches the web steers the web. Since rollers are never perfectly cylindrical and are never perfectly aligned that means that the farther the web goes, the more scattered its path position will be. This is one of the many reasons for the first commandment of good web machine design is to minimize the rollers necessary to do the job [5]. However, even a perfect machine, if there were such a thing, would not result in a perfectly straight and controlled web path because the web itself can steer in different directions if it has even the slightest asymmetrical bagginess that is called *camber* in guiding language. Here, again how much the path moves depends on the path length going through the machine.

Allowing web path to vary through a fixed width converting process such as coating or slitting will cause an increase in *trim waste*; hence one reason to control path. Another is that customers often require good roll edge straightness or good *registration* such as color-to-color or print-to-edge. Commercial tolerances are often as tight as the thickness of a human hair, 5 mils or 125 μm for position, registration, and width. Good path control begins with good design and good webs. However, all of that may not be enough. We may need to correct the path. Either we could guide the web on the unwinder, to start the web off in a consistent position, or the winder, to give the customer had better roll edge quality. No less interesting is guiding in the middle of a process. While there are several guides, the most common in converting is a *displacement guide* shown in Fig. 16.5. Guiding is one of the oldest of web science and was thoroughly modeled at the PhD level in the mid-1960s [15]. The practical upshot of this is that little is left as a craft and the application rules are well known. They include:

- right angle entries and exit,
- the length of the vertical legs (about half web width),
- the proper position of the sensor (near the second moving roller),
- all aspects of the control algorithm.

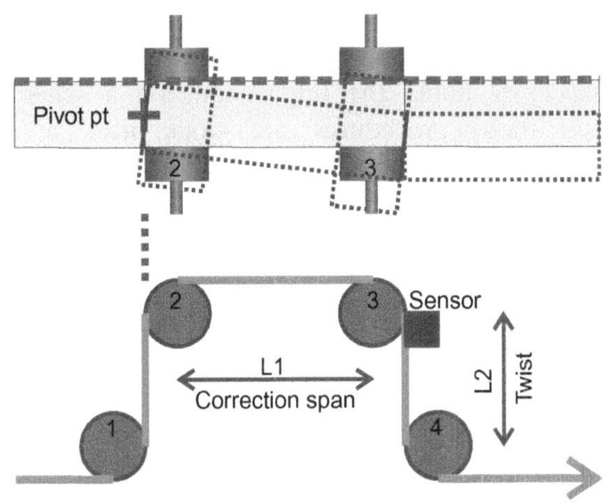

Figure 16.5 Displacement web guide.

While this compact guide shown as a "hat" orientation in Fig. 16.5 is the most common, there are eight different and yet entirely equivalent geometries that can be used.

16.7 Wrinkles and Spreading

Wrinkles, including bagginess, are the leading cause of waste and delay in the collective web industries. It should be no surprise than that there has been much study and hence much learned as to the causes and cures. We will define wrinkles for the purpose of troubleshooting as wherever and whenever the web is not dead flat. (Customers and quality assurance (QA) may have a looser definition of how much lack of flatness can be tolerated before rejection, but troubleshooters need not wait until that condition to do their work.) Also vital is to not lump wrinkle types together. Here, a wrinkle is not a wrinkle is not a wrinkle.

A way to break up wrinkles into major and subcases is shown in the following family tree:

- *Machine Direction (MD) Wrinkles*
  - Tension too high (stretchy materials)
  - Tension drop in span (stretchy materials)
  - Temperature increase (film, foil)
  - Moisture increase (nylon, paper)
  - Slender roller deflection
  - Excessive roller grooving width
  - Roller groove or bump
  - Improper spreading
- *Diagonal Shear Wrinkles*
  - Crooked machine
  - Crooked web
- *Baggy Webs*
- *Cross Direction (CD) Wrinkles*
  - Air buckles in wound rolls
  - Tunnel wrinkles in laminates
- *Corrugations*
- *Curl.*

There is very little in common between the subcases other than the web is not flat and sometimes spreading will help with different "root causes." A chapter in a book [3] and a module in a short course [2] provide details and explanations. The wrinkle troubleshooter app is a handy reminder and diagnostic tree aid [4].

The first step in wrinkle remediation is proper diagnostics of "root cause" such as outlined above. The second step would be to eliminate the root cause to the extent possible. For example, if your supplier is giving you baggy webs, then by all means complain and return the rolls. This is especially important in light of the fact that there is little that can be done effectively to run baggy webs aside from not making the situation worse with poor web machine mechanicals and controls such as described above [16]. Other causes of wrinkles such as excessively misaligned rollers are also most cost effectively treated by addressing the root cause, such as poor design or maintenance in this case. Unfortunately, there are cases where the root cause is the very process you hold dear, such as adding water (to nylon and paper) or adding heat (to film or foil). While we might try to "not coat" or "not heat" the web to avoid expansion wrinkles, for example, it probably will not get us very far with management and customer alike. Here and elsewhere, spreaders may be a last chance effort to salvage the situation.

We might anticipate where most spreaders might be needed even before a machine is built. This would often include immediately in front of any nip such as calendering, coating, laminating, printing, and winding. It may also be after expansive processes such as adding water or heat. It may also be immediately before (to flatten the web for consistent web width) and after (to separate lanes slightly) slitting the web into lanes. However, after we have a process the needs are clearly demonstrated wherever the web does not lay dead flat on a roller. So we might use the word "location" as being no less important for spreading than it is for real estate. One roller too soon and there is no need. One roller too late and it is, of course, too late.

Once we have identified the root cause and location of a particular wrinkle, we can choose the spreader most appropriate for that application. This subject is too large to cover here, but is well detailed in articles [17], books [5], and short courses [2]. The two most common spreaders are the modest *concave spreader* (Fig. 16.6) and the more capable and more complex *bowed roller* (Fig. 16.7). As you might suspect, these spreaders have been thoroughly modeled and understood by experts for decades so that there is no use "reinventing the wheel." I will just mention three limitations for spreading. The first was with bagginess already discussed, little can be done including spreading to make the situation much better. The second is that spreading cannot be used in some applications such as with narrow webs (perhaps less than 1 m) and with stiff webs (such as foil) to name just a couple. The third is that spreading is complex enough that many spreaders have the tendency to *make* wrinkles instead of the *tendency* to remove wrinkles. This is a reflection not of principle, but only of application. Some detail is not attended to or the user has unrealistic expectations of what can be done.

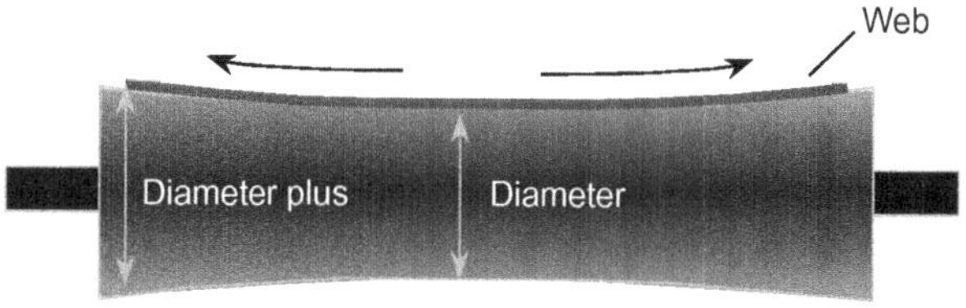

Figure 16.6 Concave spreader. *Courtesy: Davis-Standard.*

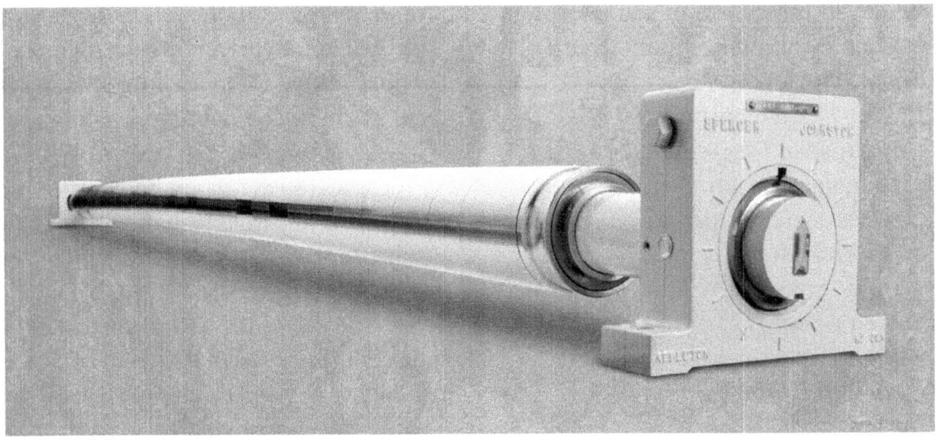

Figure 16.7 Bowed roller. *Courtesy: Spencer-Johnston.*

16.8 A Word About Slitting and Other Converting Processes

While web *handling* preserves the web as it goes down through a machine, web *processing* changes the web. As important as they are, the web handler must pass on some aspects of the most common converting machines such as printers. The same is true for other processes such as calendering, coating, and laminating. That does not mean that we cannot be useful on these machines, because *most* of the machine is still web handling. It is just that the coating head or the dryer or other process and any aspect of chemistry or application is often best left to others. There is one exception to this somewhat artificial delineation of handling and processing and that is the process of slitting. While strictly speaking slitting *does* change the web permanently, we have often taken that particular process "under our wings" so to speak. There are at least a hundred articles on slitting. It is also the subject of one- [18] and two-day [19] short courses.

16.9 Winding Is Essential

Winding is found at the end of web manufacturing. It is found at the end of roll-to-roll converting. It is found again when large rolls are cut into smaller shipping rolls, such as in a slitter-rewinder department. Some webs may be rewound yet again. Without exaggeration, winding enables the web manufacturing industries. The simple reason is that the wound roll is a very effective way to store webs and just about the only one aside from putting them in boxes. It should be no surprise than that winding is one of the best documented areas of web handling with nearly 1000 articles, columns, and conference papers on the subject [1]. There are also many books [20–23] and a short course [2]. This documentation reflects not only the essential nature of winding to webs of all kinds, not just flexible packaging, but also to the complexity of the subject.

In this section, we will merely introduce the subject of winding with the goal of providing a structure for further learning. We will not detail other essential areas of winding *machines* such as rollers, tension control, nip control, and drives. These were introduced earlier. We begin by describing the knobs that can make wound roll tightness, then we talk about classes and arrangements of winding machines, then we add a bit of applied theory so that many roll defects are made more understandable and then finish with a tour of some of the classic defects. Automation and roll changes and roll handling will not be covered here.

One way to keep complexity at a minimum and provide a foundation upon which to build the entire subject is to first be clear about the goal. That goal is suggested above. Winding is nothing more than a packaging operation; just a way to store webs without cutting or folding. As such the wound roll can be judged just like any other package based on the following three essential qualities:

- Stores the material expected by the customer (length, width, and web specifications)
- Protects the material from internal stresses due to winding as well as external stresses due to support, storage, and handling
- Looks pretty.

It may at first surprise some that looking pretty would be important, but it can be essential to being competitive in many markets. More than that, however, the great majority of wound roll defects can be detected visually and often *only* be detected visually. So whether the roll needs to look good to evidence good workmanship or look good as in no visible *functional* defects matters little.

However, trouble-free winding cannot necessarily be found by merely finding the right settings on the right winder. There are many webs that cannot be wound because they cannot withstand the stresses that are a necessary by-product. Some webs are not windable because some property, such as COF, is too extreme or the material is not dimensionally stable. There are many webs that cannot be wound because their quality is inadequate. Specifically, web profile thickness variation across the width will usually need to be less than 10% and perhaps even less than 1% in order to wind without defects. Winding is thus in a very real sense a team effort. Everyone from product designers, to web manufacturing, to winder designers, and maintenance and operation must all do their job. Indeed, even the customer must be part of this team by having reasonable expectations.

16.10 Winding Tightness and the TNT's

One of the many important duties of a winder is to *make a roll that is neither too tight nor too loose*. If a wound roll is too tight, the web risks defects such as:

- blocking,
- one of the types of core crush,
- stretching thicker lanes into bagginess,
- and many others.

If a wound roll is too loose, the roll itself risks defects where cylindricity is lost such as:

- getting a flat spot on the bottom during floor support,
- going egg-shaped during core support,
- two of the half dozen types of telescoping,
- and several others.

As much as we might wish otherwise, it is not uncommon to have both tight and loose defects in the same roll. This is often the result when caliper (gage, thickness) profile varies excessively across the width. A few defects, such as some starring and telescoping, are due to poor "roll structure." Good *roll structure*, on the other hand, means a relatively tight start at the core, a looser finish at the outside diameter (OD) and, no less important, a smooth transition from tight to loose as seen in Fig. 16.8.

Many defects that are not sensitive to winding tightness are maintenance or operational defects. Defects in this category include:

- rough roll edges
- wrong roll width
- offset core
- and many others.

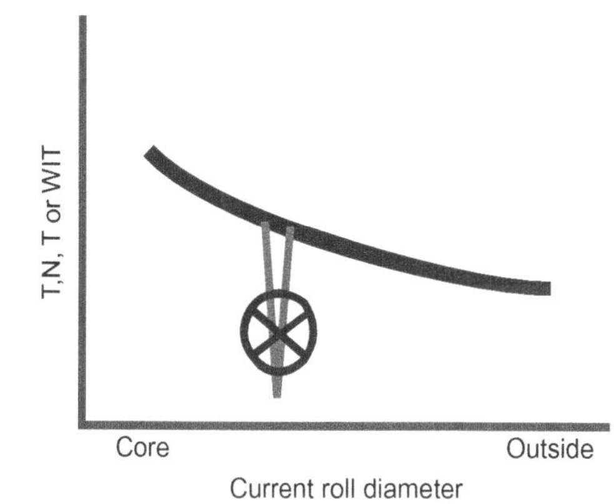

Figure 16.8 Three principles of good roll structure.

The one must-have book; the *Ultimate Roll and Web Defect Troubleshooting Guide* [3] documents all defects.

The knobs that make tightness are known as the *TNT's*. The first T, *Tension*, is the easiest. It is merely the web's tension as it enters the windup station and between 5% and 25% of web strength is often needed for good web handling up to the windup station. *Nip* is the lineal load between the winding roll and a roller; either of which can be stationary and the other is moving to accommodate roll growth. The last T has such a long proper name, center-surface wind torque differential, that it is often abbreviated to just *Torque*, even though it is most certainly *not* simple motor torque. It is motor torque *difference* between two motors; one connected to the roll and the other to the roll*er* or, in the case of two-drum winders, between two rollers. Note that few winders have all three of these knobs. A few have only one knob: tension. Most have only two: tension and nip.

The little "s" stands for *speed*. It is little (lower case) because only in some cases is winding tightness speed dependent. We know from web handling that these cases are where the following criteria are all *simultaneously* present:

- not narrow,
- not rough surfaces,
- not slow speeds.

These are often the case with flexible packaging materials. The units for the first three TNT's, including torque differential, are the same: lb/in, abbreviated PLI in the US system or N/m in the metric system. Quite simply, speed loosens the wound roll in the above case due to entraining air. One final and yet quite major contribution to tightness is certain *web material properties*, most particularly MD modulus. If the web is very elastic, such as polyethylene or polypropylene, even a small tension will add to enormous tightness and pressure inside the wound roll. If the web is stiff, such as polyester, it will be difficult to wind a tight roll even when applying maximum TNT winder settings. Adding any stiff material layer will make the entire composite stiff.

So what is this *tightness* that is created by the above winder knobs in conjunction with material properties? The answer depends on what you do for a living. If you are an operator, you might view tightness as the TNT settings on your machine. If you are a scientist, you might view tightness as the interlayer pressure and MD tension/compression in a layer measured in psi (lb/in^2) or pascal. If you are in QA, you might view tightness as the roll length for a given roll diameter. If you are in QA or troubleshooting, you might measure tightness using one of several roll hardness meters. Some common measures of tightness are:

- Process
  - TNT settings and machine
- Stresses–strains
  - Wound-in-tension
  - Radial stress distribution
  - Tangential stress distribution
  - Strains/displacements
- Geometry
  - Bulk
  - Density
  - Diameter and length
  - Diameter, width, and roll weight
- Fundamental measures
  - Axial press
  - Caliper in roll
  - Core torque
  - Density analyzers
  - Pressure in roll
  - Pull tab
  - WIT-WOT (Wound-In-Tension, Wound-Off-Tension)
- Nonfundamental measures (bill club hardness)
  - Firmness
  - Rhometer
  - Schmidt hammer
  - Smith needle
  - and many others.

These are thoroughly detailed in the last third of Good and Roisum [20] need not be detailed here. Fig. 16.9 is a picture of one of the many hardness meters.

At this point the reader could understandably get confused: we potentially have four knobs plus several material properties to make tightness and dozens of ways to measure tightness. We will describe those knobs in more detail in a later section, but here is one way to keep the idea of tightness simple. There is only one type of tightness. By this, I mean that whether the roll is made tight by tension or nip or some combination matters little for this purpose; it is merely tight. This means that if one knob is giving you trouble, such as tension causing web breaks as an example, you can reduce tension and make up for it by nip and the roll may be equally tight. The same applies with nip that can give you great grief, especially if caliper and machine are not dead level. Reduce the nip load and make up with tension. The only thing to ask then is which knobs are most powerful and which are most risky. Here, it is a bit challenging to give generalities because the web and winding

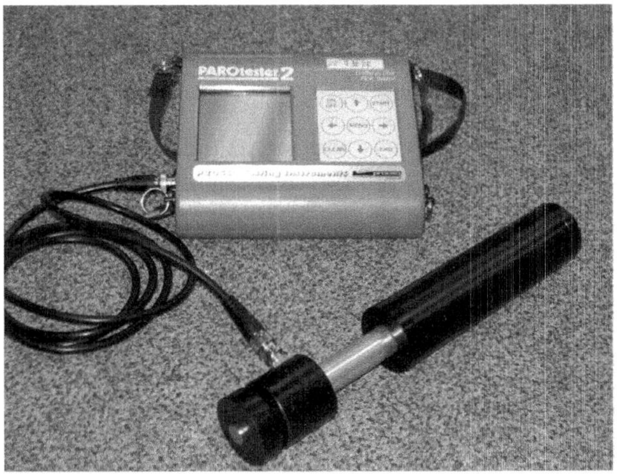

Figure 16.9 The parotester for wound roll hardness (tightness).

world are quite large. I will just say that nip is usually important, or quite important in the case of softer webs or those that entrain air. Nip is also the most risky. This means that, in a very real sense, *nip should be the "gold-plated knob"* [24].

Another way to keep things straight is to, at least initially, not dwell so much on the differences between the various measurements of tightness. They are all interchangeable in the sense that there is a monotonically increasing relationship between any two measures of tightness. If we increase the tension on the winding machine, the roll hardness will increase as well as the density and everything else. Merely choose the measure that makes most sense for what you do for a living. The only thing you need to make sure of is that that the measure you choose is easy and repeatable (trustworthy) for your application. Repeatable means, at the very least, calibrated and calibratable.

One last complication before we move on to machinery. That is, many if not most modern winders can *program "curves" for any or all of the TNT's*. Thus, for example, tension could be tapered from the core to the OD as the roll winds. We could do the same for the nip and the other knobs. This makes winders different and more complicated than many other processes because we can have set-point *functions* that vary with current wound roll diameter rather than set points that are constant for a run. One way to keep things simple is to not think in terms of old-school *taper* that was a convenience for the electrical engineers to program. Rather, the *two-point method* is much better for programming the starting and ending point settings in consistent units of lb/in or kN/m as seen in Fig. 16.10. That recognizes that the wound roll really acts like three entities: the bottom, the middle, and the top with regard to many defects and roll structuring.

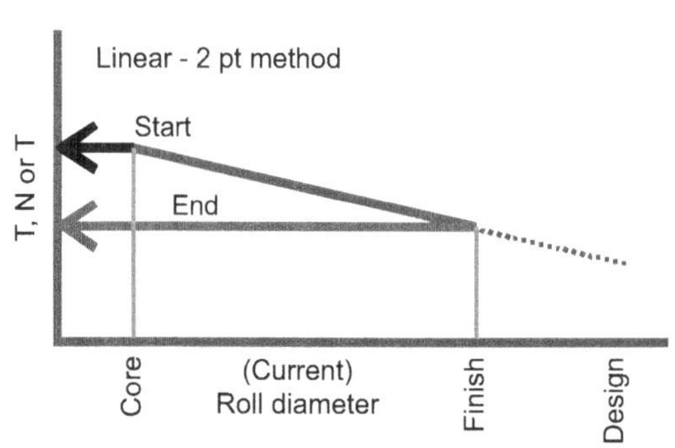

Figure 16.10 Two-point method for roll structure "curves."

A simple strategy is as follows. First, look at the bottom of the roll near the core. Determine if you have a tight or loose defect and adjust tightness using any or all of the TNT's accordingly. If there are no tightness-related defects near the core, then no move is needed. Then look at the OD and determine if you have tight or loose defects there and adjust tightness using any or all of the TNT's accordingly. Again, if you have no tight or loose defects near the outside, then no move is needed. Finally, connect the TNT's at the core to the TNT's at the OD with a simple straight line. This is the two-point method applied, one at a time, to all the tightness knobs. In the rare case of roll structure or global defects, such as some types of starring and telescoping, program from maximum tightness near the core to minimum tightness at the OD. This would be called maximum taper in old-school lingo.

16.11 Winding Classes and Arrangements

Winders fall into one of four *major classes* that, in effect, determine which "knobs" are available to make tightness in a wound roll as seen in Fig. 16.11. The simple *center-wind* has only one knob to make tightness: the web tension that enters the windup station that is a product of motor torque applied through the core. The tightness range that can be made on this winder is largely determined by the tensions that can be run reliably up to the windup plus material properties. Fig. 16.11 also shows generically compares the tensions that are available for the four winder classes. Because center-winds have only one knob (besides speed), they are thus limited and are for that reason uncommon in wide web manufacturing and converting. In many web industries, particularly converting, a nip roller might be added to the center-wind making it a *center-wind with lay-on roller*. This nip roller goes by several aliases such as lay-on roller, pack roller, rider roller, and so on. The nip adds to the tightness range that can be made, but only on the upper end, ie, the tightening end. You cannot make a wound roll looser that can be made with minimum center-wind tension by adding the nip knob. By similar extension, adding the third knob, the second T, the torque *difference* between two motors, you can increase the tightness even more.

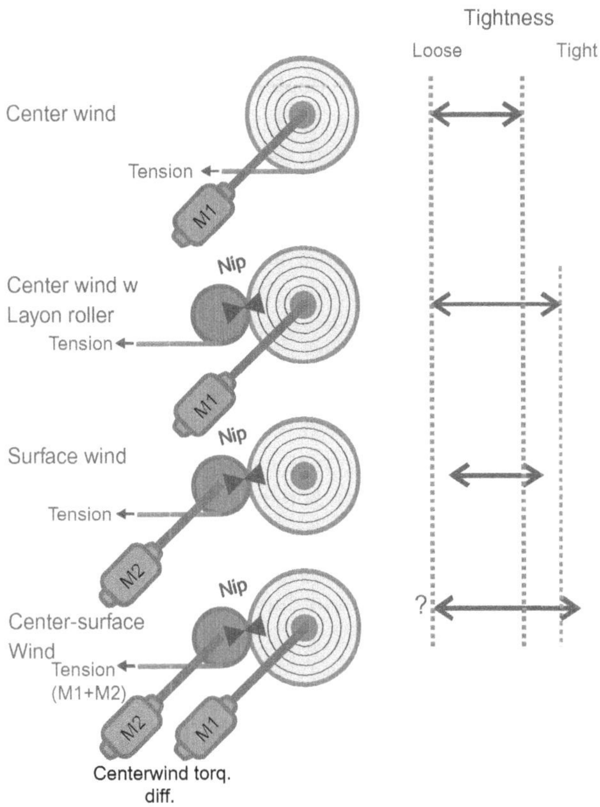

Figure 16.11 Classes of winders and the tightness range achievable.

This two-motor machine is called a *center-surface winder*. The torque knob can also loosen the roll, but probably not below the loosest obtained by a pure center-wind. The *surface winder*, common in nonwovens, paper, rubber, and textiles, but rare in flexible packaging, has the same two knobs as the center-wind with lay-on roller; (web) tension and (lay-on) roller nip. Yet, as again seen in Fig. 16.11, the tightness range is very much limited on both ends for reasons that will not be explained here.

Not only are winders classified by the knobs they have for adjusting wound roll tightness, they are also classified by *arrangement*. There are six common winder arrangements: two are continuous winders, two are off-line rewinders and two are specialty winders. These and other winders are well detailed by Gronewold [21]. Fig. 16.12a–f shows these most common arrangements schematically. The continuous winders do not slow or stop to make a roll change. As such, they are essential to most web manufacturing processes because starting and stopping a paper machine or extruder is nothing resembling graceful. Continuous winders are also essential to many high-speed roll-to-roll converting operations. The *turret winder*, shown in Fig. 16.12a, is by far the most common continuous winder in flexible packaging as well as most other materials. The turret can cut and transfer webs from one spindle to the next. (Analogous automation exists on turret *un*winds and is called flying splices.)

The turret winder is also one of the best candidates for the specialty *gap winding* arrangement as shown in Fig. 16.12f. Here, either the winding roll or a nearby roller is moved as the roll builds to maintain a small gap (part of an inch or a centimeter). This winding mode is potentially the most resistant to making wound roll wrinkles. It begins with good upstream spreading to help ensuring a wrinkle is not made upstream and simply transferred down into the roll. It may also work better if that last roller is very large and very slippery to kick in the flattening effect (poor man's spreader). Then the still flat web is so close to the winding roll that the web does not have time or distance to go from flat to wrinkled. Of course, the gap winder is extremely limited because without a lay-on roller you only have the tension knob to develop tightness. That may not be nearly enough, particularly for smooth webs wound at high speeds.

The turret is, in some ways, the most flexible winding machine because it can be made or operated as a pure center-wind, center-wind with lay-on roller, a center-surface winder if equipped with a motor on both roll and (lay-on) roller, as well as gap winder. That does not mean there are not limitations besides the prodigious cost and complexity of the machinery. The turret winder must support the wound roll by the core and transfer torque through the core. This can be problematic with large wound roll diameters as either corresponding interlayer pressure or torque near the core can cause various defects, such as the telescope. While the surface winders such as the *reel*, Fig. 16.12b, and the *two-drum*, Fig. 16.12d, do not impose torques through the core, they are not without analogous limitations for pressure. Nip load on the *outside* of the roll, especially if this is due to roll weight, can be no less problematic with large rolls and/or with delicate materials than core support can be on the *inside* of the roll. For these reasons and others, the reel and two-drum surface winders are uncommon in flexible packaging as we have already mentioned.

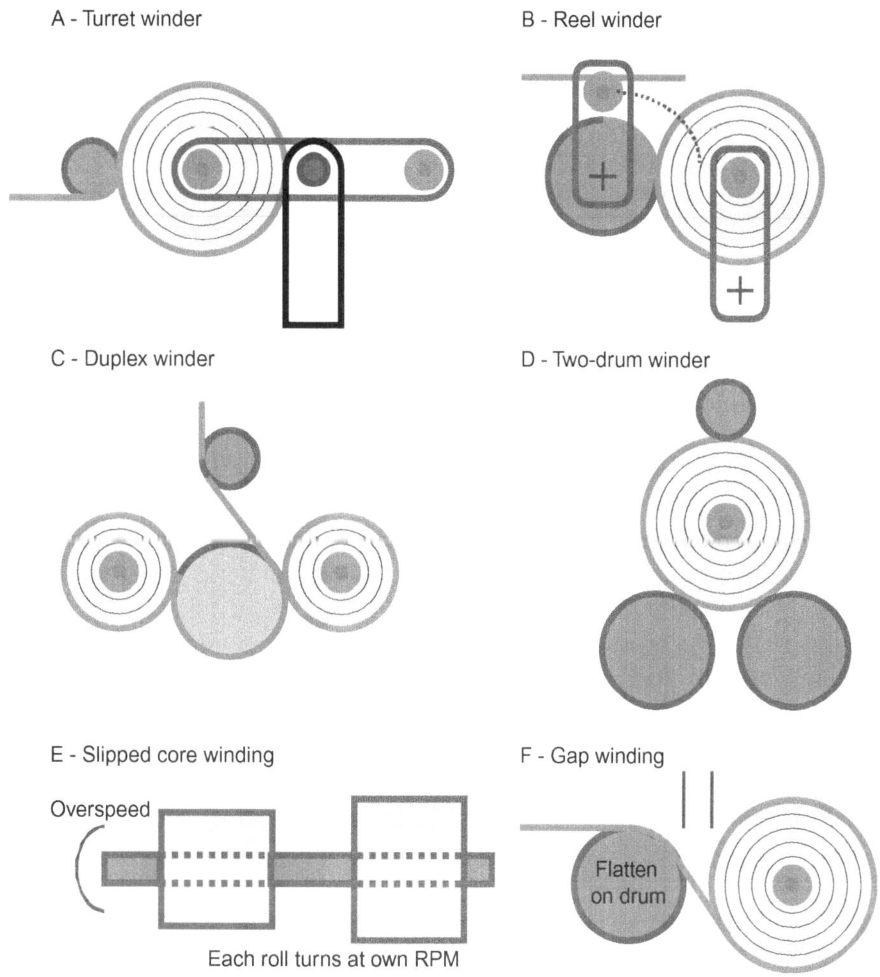

Figure 16.12 Common arrangements of winders.

One limitation common to all winders is that they all produce *upsets at the top and bottom of the roll.* This is true no matter what class or arrangement or operating strategy is used. To take the turret winder as one example, we note a severe tension upset is present during transfer. First, tension may be *unintentionally* changed as the turret indexes and increases the web path length. Whether there is feedforward and/or feedback to mitigate this to an extent does not change the fact that a tension upset of some size will exist for this cause. Also, *intentional* tension upsets will exist due to change of tension control strategy during this event for reasons that a drive engineer could best appreciate. Another possible intentional upset is the positive (flying knife) or negative (stationary knife) tension boost that may be programmed during the roll change to help the knife cut the material. Similar upsets occur with the nip roller during roll index. Similar upsets occur with all of the TNT's with all winders at roll change. Even slowing or speeding a winding roll will cause a tightness change. Thus all winders have upsets in tightness at the very bottom and very top of the roll of varying sizes and for various reasons. What may be more avoidable is upsets *during* the wind. A challenge here is a *winder stop* that will always produce a fault in a roll that could cause a defect such as starring or telescoping. Some web products on some winders cannot stop and be restarted without risk of defect at those layers. Thus our third roll structure principle described earlier, smooth transition between tight start and looser finish, is an idealism not always obtained.

Off-line winders are used for many reasons including cutting large master rolls into smaller diameter and/or width in order to ship rolls ordered by the end use customer. Continuous winders have great difficulties and limitations with both small cores and in-line slitting so that slitter rewinders will be with us forever for most web products. While the two-drum surface winder is used for some paper, rubber, nonwovens, and textiles, the *duplex rewinder* shown in Fig. 16.12c is by far the most commonly used for some dense webs and those whose caliper is insufficiently level across the width. The duplex winder gets its name from the fact that (usually) every other slit roll is wound on opposite sides of a central drum or roller. The duplex, despite its extreme cost and extreme complexity and extremely low productivity, has several distinct advantages compared with other off-line slitter rewinders. First, the separation of web path out of the plane means that in-plane separation via spreading is not usually needed. Second, the tightness of each winding roll can often be somewhat independently adjusted or even completely independently adjustable on some very high-end models because each winding roll is effectively a separate winding station in those cases.

Also, the shafted version of the duplex winder can often be run in the specialty *slipped core* winding mode as shown in Fig. 16.12e. Instead of an expandable shaft that turns all individual rolls at precisely the same revolutions per minute (RPM), each wound roll can turn independently. Here, an over-sped slipped core shaft transmits a somewhat similar torque to each roll. Fig. 16.13 schematically shows the problem it may help solve. Here, web caliper variations across the roll cause slight differences in the build-up diameter in each of the winding rolls in "locked core" mode. Lower relative caliper means a slightly smaller wound roll diameter there and vice versa. In the locked core mode, the small diameter rolls turn at a slower surface speed (even though at the same RPM). This

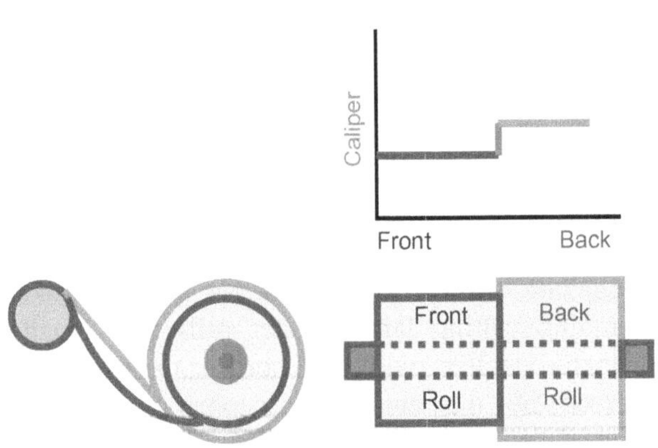

Figure 16.13 Reason for slipped core winding.

means that the small diameter rolls will have less, much less or, most likely, zero tension with stiff/dense webs. By allowing each roll to turn at its own RPM, *some* of this tension variation is mitigated. However, this is at best an imperfect solution to caliper variation because the shaft is essentially running in open-loop control. Differences in the COF of friction elements inside the shaft translate directly to tension variation. Worse yet, the most common caliper variation, within a single roll rather than at the junction of two rolls, are completely unaddressed by this costly and complex "remedy." Even so, most films are not windable on other winders and thus slipped core is what we might call the "lesser of evils" when comparing tightness variations due to caliper variation and slip core variability. The conclusion here is true of all winding: profile variation from upstream manufacturing will cause trouble, sometimes debilitating troubles, despite the best efforts of winder design and operation.

One criterion for off-line winders that differs distinctly from online winders and indeed most processes is the complex subject of productivity. With manufacturing and online winders, productivity mostly depends on speed (width and caliper) and is usually limited mostly by a single easily identified element such as drive speed rating or drying capacity to give two examples. With off-line winders, productivity is much more complicated. Fig. 16.14 shows the nominal "time cycle" of these winders in a speed versus time plot.

Here, the winder has repetitive, ie, deterministic elements, such as acceleration, run, deceleration, finished roll change for a single "set" and master roll change at the end of one, two or even a dozen sets. This nominal time cycle and resulting rewinder productivity can be easily calculated by a method described by Roisum [23] or, more conveniently, the Winder Productivity app by *AbbottApps* [4]. To this simplistic calculation, we must consider stochastic elements such as web breaks, slitter changes, operator inefficiency, and so on. The whole point of this exercise is not to buy even slightly less winder capacity than needed, else the plant is bottle-necked, or much more winder capacity than is needed, else the slitter-rewinder department runs less efficiently. The analogous consideration for continuous winders is to specify and maintain a maximum allowable missed wound roll transfer rate such as not less than 1 per 100 or perhaps even 1000 rolls. The reason is

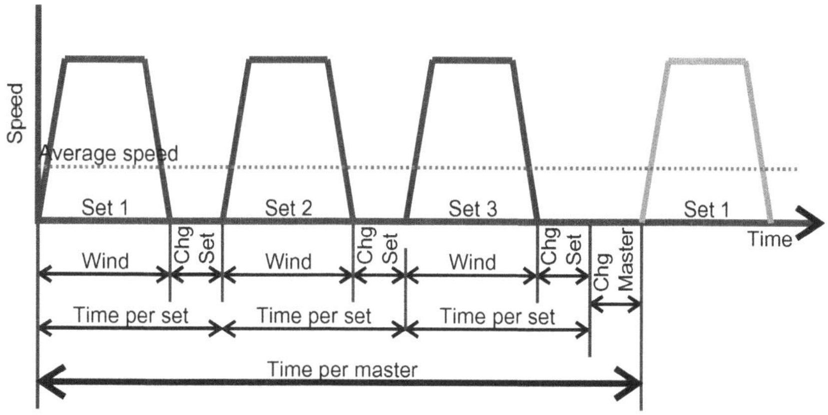

Figure 16.14 Simplified slitter rewinder time cycle.

quite obvious: the waste and delay of restarting the manufacturing or roll-to-roll process may take at least a half hour and, in some especially egregious cases, many hours to get back on-spec.

16.12 Theory Applied

In this section, we go beyond mere roll structure which, to be truthful, helps little beyond a generic proscription that at best helps with only a very few defects. What we need is a bit more information about the TNT knobs discussed earlier. More importantly, we need to understand how those contribute to the stresses within the wound roll that in turn helps explain many tightness sensitive defects.

Let us begin with *tension*, the one knob that all winders have and the one that is easiest to understand. The tension knob needs to accomplish two major goals. The first is to tighten the web in a range that it would like to run well in the drive span leading up to the windup. This is usually between 5% and 25% of web strength and most often near 10%. The second is to contribute to tightness in the wound roll. The big challenge is how to resolve these two, possibly disparate, goals. The place to start would be to center web tension between the high- and low-tension operating points and then observe the winding roll. If roll defects indicate more or less tightness would be best, move the tension knob in the appropriate direction until you are just short of bumping into *web* tightness/looseness defects. Finally, review the winding defects again and try to finish the job with the nip knob, or better yet torque if you have it. What we are trying to do is, when possible, to separate the knobs so that tension is mostly selected for web handling work, while nip (and torque) does the winding work. This makes even more sense in light of the fact that tension is not always a very impressive knob for wound roll tightening. Obviously, if you only have a center-wind winder, you will need to try to find the best compromise between tightness-related *web-handling* defects and tightness-related *winding* defects. Similar compromises might be needed if you have nip (and torque) but they are also limited. There are many circumstances where all of these best intentions may fall short. One could be if the drive is limited on the low end by control quality or at the high end by torque or power. Another more common situation is to have both tight and loose defects under the same winder settings; often because caliper variation is excessive. However, the most common situation of all is when the defect(s) you are trying to avoid is(are) not particularly tightness sensitive.

Nip, as we have already mentioned, is impressive not only by its relative strength for creating wound roll tightness, but it is also impressive by its ability to cause trouble; particularly if the caliper profile or winder is not level enough. Because of this, we want the knob to be repeatable and reliable and to be translatable between machines. This requires that nip be calibrated and calibratable in engineering units of lb/in or kN/m, a trait not always found on many winder makes and models. In contrast, tension is most often calibrated. Calibration aside, however, we should mention that the mechanism for creating tightness is quite different between materials with greatly differing properties. For "fuzzy" web such as some papers and all nonwovens and textiles, the nip-induced tightness is made by interlayer slippage. When this slippage is not gradually varying through the diameter or uniform across the width (such as due to caliper variation or excessive windup station deflection), shear is developed that can lead to a variety of defects. With smooth products, such as flexible packaging, wound at all but the lowest speeds, the nip operates in a different fashion. In that case, the nip excludes some of the air that creates wound roll looseness. The idea, however, is not to exclude *all* the air else one would literally have a solid (blocked) piece of plastic, for example. Rather, the idea is to meter in the appropriate amount of air so that the roll is appropriately tight. Pound for pound (or newton for newton), nip is perhaps 10× as powerful as tension for creating tightness so that again, nip should be the gold-plated knob.

(Center-surface winding) *torque* (differential) is the rarest knob. You can skip this paragraph if you do not have this knob because only very few windup stations have two motors. It is also one of the weakest knobs primarily because motors are often undersized for that need. For example, say a single-drive windup needs a 10 kW motor. Then a two-drive windup may then require (2) 20 kW motors or 4× as much total power because the motors are *intended* to fight each other to do their work. Fig. 16.15 shows differential torque changes tightness for the center-surface winder. As relatively more torque is put into the winding roll, perhaps even pulling against a braking roll*er*, the wound roll gets tighter. As more torque is put into the roll*er* the wound roll gets relatively looser. Note the surface wind and the center-wind with lay-on roller, both single-motor winders, fall inside this range of potential tightness. If you have the torque knob, its strategy is usually easiest given the fact that motors are often undersized. That is a strategy of redline tight or loose at the core as defects at the core indicate; to redline tight or loose at the OD, again as defects at the OD indicate. For a case where only type Ia or Ib telescoping exists, as one example, you might start with redline motoring torque on the winding shaft at the start to redline motoring torque on the roll*er* at the finish. In any case, web tension is unaffected and determined independently. In any case, the speed (or tension) of one drive is controlled as with a single-motor winder while the other motor shares a portion of the total torque or is set at a fixed difference.

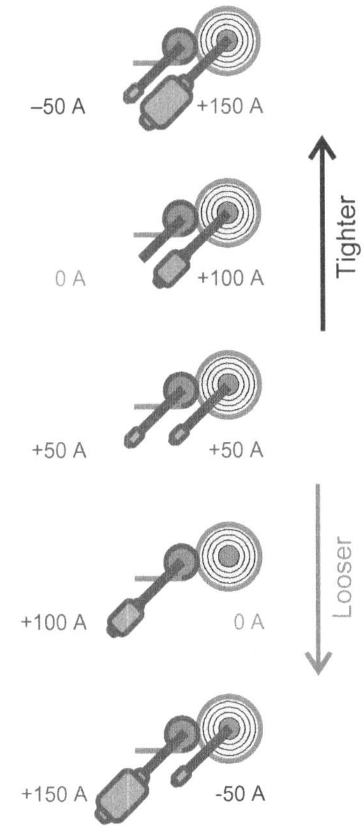

Figure 16.15 Tightening and loosening effect of the torque (difference) knob.

So how do we put all of these tightening knobs together in a way that might make sense of certain defects; most particularly why some defects favor the core and some favor the outside and why those that are found in the middle are not usually tightness sensitive aside from global defects such as starring and telescoping. The central idea of winding modeling science is that all wound rolls end up with two sets of stresses [2,23]. These are the "Z direction" pressure between the layers and the tension (or compression) in the "machine direction" (MD) of a layer. Fig. 16.16a shows both the interlayer pressure (the top curve) and the stress distribution (lower curve). Fig. 16.16b shows the tangential stress distribution. The trends are more or less the same for all webs, all winding machines and all winder settings; only the stress (magnitude, size) changes. There are two important things to note. That is that there are really three parts to every roll: inside, middle, and outside. This idea of treating the different parts of a roll separately has scientific foundation. For example, interlayer pressure goes from high at the core, to medium in the

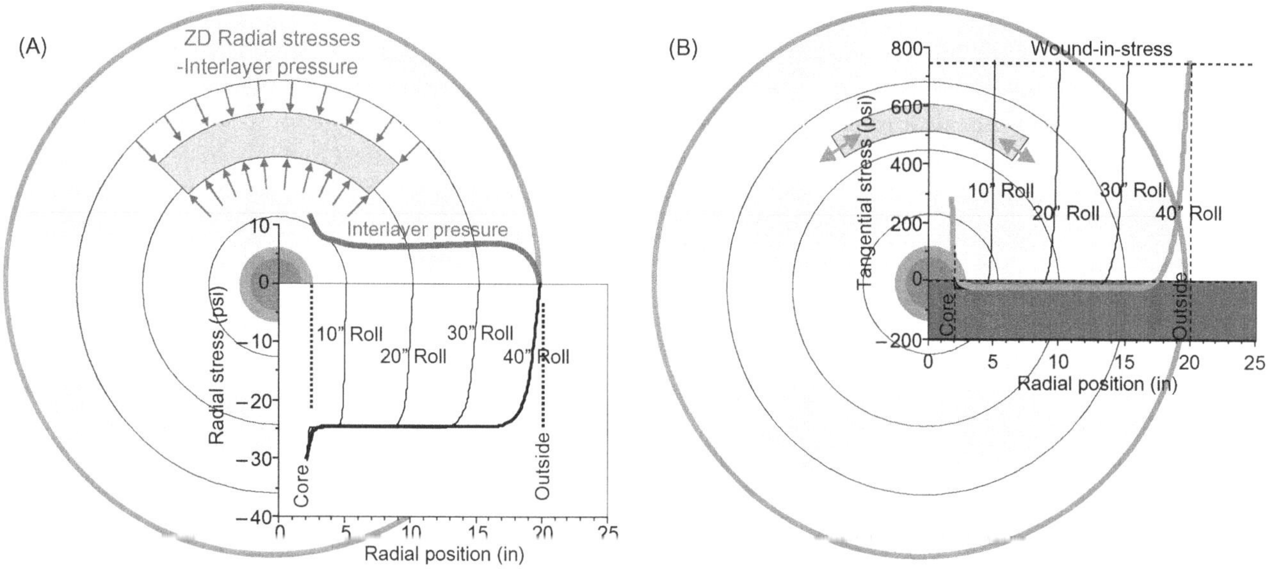

Figure 16.16 (a) Interlayer pressure and radial stress versus radial position inside all wound rolls. (b) Tangential stresses versus radial position inside all wound rolls.

middle to low (or zero) at the outside. Now we can understand why some defects, such as blocking (layers sticking together) and bulk loss (permanent loss of web thickness inside a roll) favor the core, or at least the bottom half of a roll. The pressure is highest there. Now we can see why other defects, such as flat tire and other out-of-round troubles, favor the outside. The pressure and thus resistance to handling deformation is less.

There is another complication that can exist in some uncommon situations. That is the wound roll may change with time, sometimes called aging. Fig. 16.17 shows one example known as *buckle*. As we will recall, air is entrained between film layers or other smooth webs. It is particularly troublesome for laminated packing films such as those containing foil. There is nothing to prevent some of that entrained air from escaping out the edges of the wound roll. The air has a very long path compared to its thickness that is perhaps on the order of a few percent of the thickness of a human hair. Thus, it takes some time for a noticeable fraction of the air to escape, usually hours to a day or so. This air had supported the layers. If enough air and thus support is removed, the roll may collapse on itself. The problem will show as CD-oriented buckles that favor the bottom half of the wound roll. Why did the web then buckle? It does it because there is MD compression in the bottom part of the roll coming from the winding physics just described in Fig. 16.16b. When the support is lost, the layers move inward and thus create even more compression. What may strike some as usual is that the layer-to-layer distance composed of web plus air only decreased by a fraction of a few percent (the percentage of air in a roll) of web thickness. That small numbers can create big problems is of no surprise to the web handler, however.

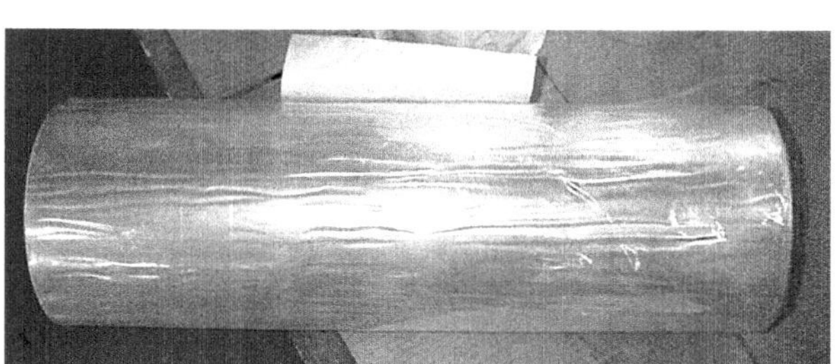

Figure 16.17 Buckle defect.

This air rearrangement loss with time also can cause other problems. That is the wound roll may visually look good after it is wound, but perhaps terrible a day later because it shows hills and valleys and other gage-related problems. There are some webs where the roll needs to be "quarantined" and re-inspected a day later to make sure that the customer does not receive an ugly roll.

In the above cases, what changed was not so much the web, but the air between the layers. Though not common (except for baggy webs that we will describe shortly), it is possible for the *web to change dimensionally* and cause troubles. Perhaps the classic is freshly extruded bi-axial oriented polypropylene (BOPP) and other films whose polymer constituents are not completely crystallized when it is cooled. Further crystallization in the next day or so causes the web to shrink. This can cause an enormous tightening, perhaps kicking off tight defects such as *blocking*. Another is winding a web that is hot. It will cool and shrink causing an increase in tightness. In the above cases, it is not unusual for the web to be more *tightened by thermal or chemical changes* than by the TNT's of winding. There are other odd chemistries where the web can shrink in thickness or expand in length. What should be noted is that the dimensional change is on the order of a few parts per thousand. Yet the effect of these small changes may be nothing like small. Better living through chemistry comes to mind here. DFM (design for manufacturability) comes to mind here.

One last time-dependent change we will bring up is not related to stresses, but rather to a very important but under-appreciated material property. That is web–web *COF*. If the web is too slippery, certain defects such as common *telescoping* may be much more likely. (It is also true that if the web is too grippy, more specifically tacky, other winding problems such as out-of-round rolls or winder vibration may be much more likely.) What can happen with many films, coatings, and inks is that they may get slipperier with time; typically over a few weeks. This means that you might be able to safely wind the roll without a (type Ia) telescope, but your customer might not be able to safely unwind that same roll without a (type Ib) telescope. Again, better living through chemistry and DFM comes to mind here; *after* you have played the TNT's to the fullest extent on the winder.

16.13 Other Common Wound Roll Defects

It is not the purpose of this section to completely review any particular defect, much less the more than 100 winding defects that are well described in a must-have book by Smith [3] or another winding book by Roisum [23] or references such as might be found by the free internet application, the Roisum Library by AbbottApps [4]. We will also not cover web or wound roll *wrinkling*, even though it is perhaps the largest single cause of waste, because that is also a very well covered in books [3] and articles [25] and the Wrinkle Troubleshooter by AbbottApps [4]. What we will do here is to simply give a few more of the more common and classic defects with a brief description of their mechanics.

We begin with the bottom of the roll and look at a defect that is not only important to some people, but also a defect that has something to teach *all* of us. Fig. 16.18 shows the *core crush*. While the appearance and physics may vary, the outcome is the same: the core gives way to pressure from some source. However, while the outcome is the same, how it got there and more importantly the list of remedies is quite distinct between different pressure sources. The reason is that core crush is not a single defect but rather three totally different defects that have the same outcome.

- Case I is where the core gives way due to winding pressures and happens during or very soon after winding. This case is a tight defect and is most common with low-modulus webs, such as many used in flexible packaging.
- Case III is where the core gives way due to handling pressures and happens during handling, never during winding. Case III, curiously, is a loose defect and is most common with high-modulus webs.

Figure 16.18 Three different causes of the outcome of core crush are I—pressure due to winding, II—thermal- or chemical-induced pressure, and III—external handling pressures.

- Case II is much rarer but may be found when the web shrinks chemically or thermally and those pressures cause the core to give way after a given time as we described above. Case II is a tight defect up and until when all of the TNT's are loosened to the extent possible. It is then no longer a winding defect because there is no winding solution. This is a process/product design problem.

In all tightness sensitive defects, one goes to the tightness knobs; first being mindful of which specific case one is dealing with. After that, one considers product redesign, the most palatable here will probably be stronger cores such as by increasing core wall thickness. Extremely sturdy and tight core plugs can help a lot in some applications. Using TNT first then considering product design is the same one two-step process you will often use for many tightness sensitive defects.

Web *bagginess* (camber, etc.) is an epidemic problem in nearly all web industries. Curiously, however, a web being made baggy in manufacturing is much less common than a web being made baggy during winding. The reason is that a flat web that has caliper variations will build a wound roll with gage bands (ridges) and valleys with corresponding stress risers in both pressure and tension at those gage bands. Fig. 16.19 shows this problem graphically. If the web yields even the tiniest bit, even just a few parts per thousand in length, the web may be stretched longer into bagginess on those gage band lanes. While this has been a concern for as long as webs have been made, little analytical progress had been made until recently where Land in her PhD thesis [26] and landmark paper by Thuer [27] were written that quantified long-standing ideas. In the thesis, caliper variations

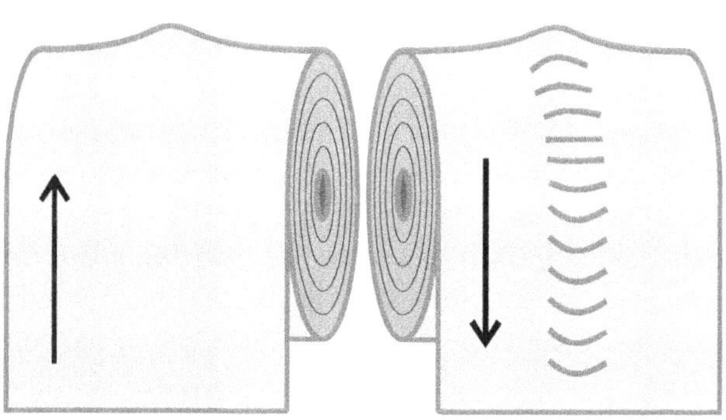

Figure 16.19 Baggy webs—the most common cause.

as low as 1% and roll diameter variations as low as 0.1%, both across the width of a single roll, were found to potentially cause rejectable bagginess. In Thuer's paper [27], she shows that a combination of average roll hardness and maximum roll hardness variation across a single roll could predict runability with great reliability and without necessarily even running it. Fig. 16.20 summarizes this hardness and hardness variation. This shows the same two-step philosophy. First turn to the winder, in this case loosen, then turn to product redesign, in this case more uniform caliper.

Not only is a caliper *ridge* a potential defect source (bagginess, etc.), a caliper *valley* can be a potential source of another extremely common defect, the *corrugation*. Here, the high–low–medium caliper profile across a narrow band causes a corresponding diameter variation in the wound roll. Fig. 16.21 shows an exaggerated schematic. Adding a nip causes the big diameter side to slow down and the little diameter side to speed up as the nip tries to enforce a common web speed even though the roll would want to go a common RPM. The effect can be a shear where the web collapses into the low spot. Again, we see the one two-step philosophy. First, loosen the roll, mainly with the nip. Then turn to product redesign, again to make the caliper more uniform. Not surprisingly, roll hardness and hardness variation across a narrow region is also a great predictor of corrugation just as it was for bagginess.

Starring is a collection of different defects with the same outcome; MD buckling of layers. Fig. 16.22 pictures how this may appear on the end of the roll as a star pattern with spikes or as wavy layers. Because we now know from winding modeling and measurement science that much of the roll is in MD compression, we should not be surprised that this might happen. When and why it happens though, is a bit more difficult to predict. While traditional teaching had been that buckling could be reduced by better roll structure, it turns out that right out-of-the box, most modern winders do *not* produce poor roll structuring as we described earlier; the exception being stopping a roll midway during winding. Rather, the most starring subcases are due to relatively low caliper. If, for example, a roll is starred on one end but not another, there is a good chance that the caliper was relatively lower on that end. Another extremely common case is on coated webs where the coating does not extend to the very edges. Another slightly different subcase is crushed cores. We

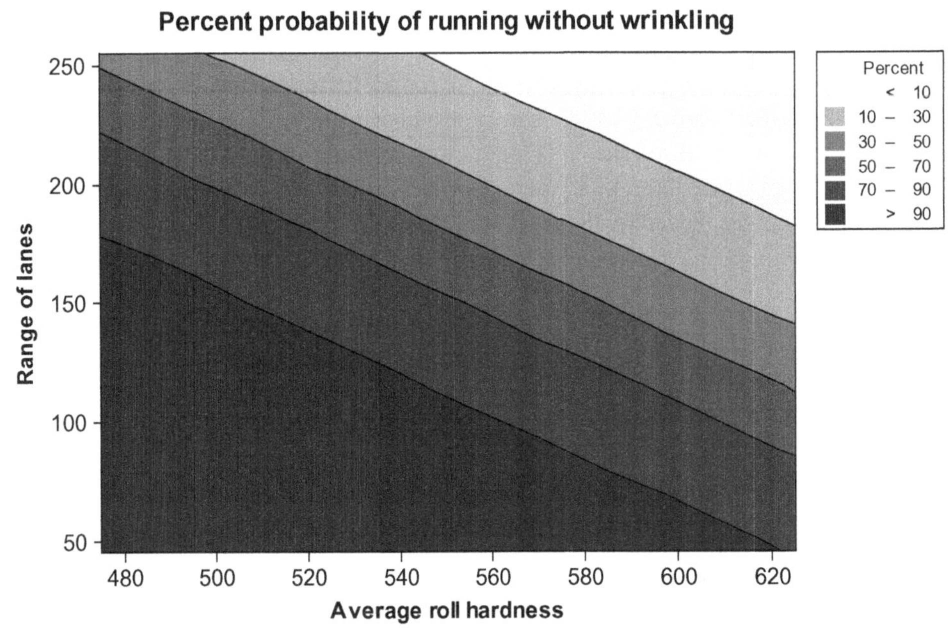

Figure 16.20 Screening wound rolls via the hardness measurement. *Courtesy: Amy Thuer of Avery Dennison.*

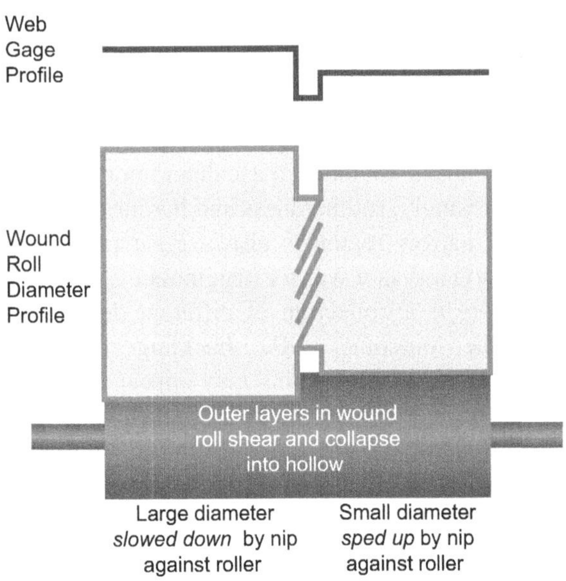

Figure 16.21 The corrugation and its cause.

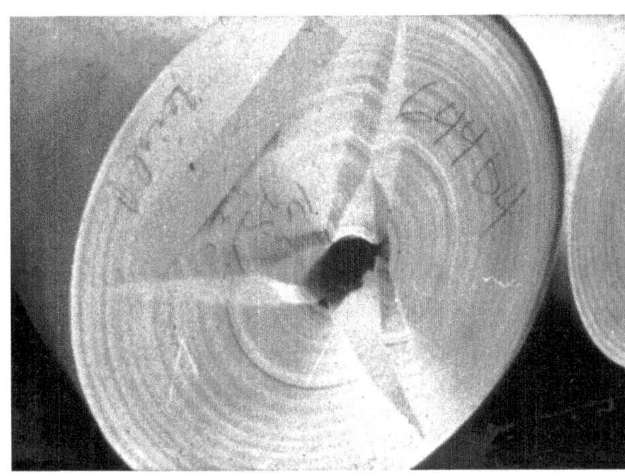

Figure 16.22 Starring.

discussed air buckles earlier as yet another subcase. If you examine the different subcases closely, you will find that all or nearly all involve some sort of loss of support. This inadequate support could be thin material, whether intentional or unintentional, crushed cores or loss of air as just some examples discussed earlier.

By now we should be familiar with the concept that a wrinkle is not a wrinkle is not a wrinkle. The same idea is true with core crush and starring that we have already discussed. This is also the case with telescoping. At least half dozen completely different subcases for *telescoping* that merely share the same outcome; a change in the roll's edge is described by Roisum [28]. Fig. 16.23 shows the most common which is torque-induced telescoping. This may occur during winding, case Ia, or unwinding, case Ib. With these two subcases, good roll structure or maximum taper is the best winder strategy. Case II is completely different, it happens during storage, particularly of pressure sensitive adhesive laminated materials. Here, the winding strategy is clearly *not* maximum taper; it is to run minimum tightness throughout the roll from bottom

Figure 16.23 The most common winding telescope.

but especially to top. The different telescope cases are described along with their causes and cures can be found in Roisum [23].

One whole class of winding defects is where the roll edge is not flat, but especially not smooth. You might view layers seen on the end of a wound roll as "tree rings" that show good seasons and bad seasons. Unfortunately, knowing the reason for the bad season or bad ring of material often requires complex diagnostics. The reason is that the wound roll end shows just about everything you did wrong from product design, to product manufacturing, to winder design, maintenance, and operation. One way to get started is to be extremely organized and use something like the diagnostic troubleshooting tree (Fig. 16.24).

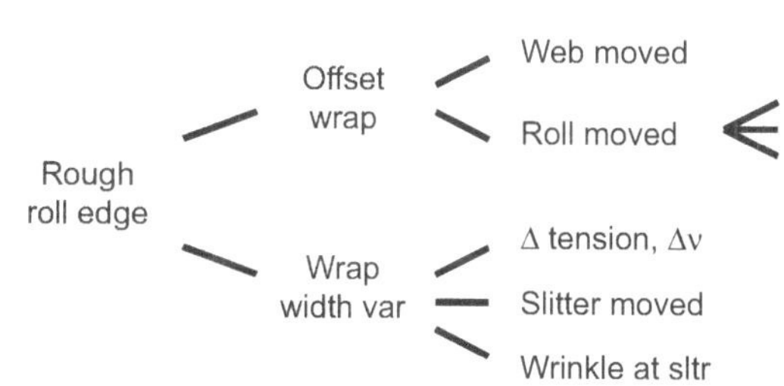

Figure 16.24 Rough roll edge troubleshooting tree.

References

[1] Roisum DR. A century of web handling literature. In: Proceedings of the 11th international conference on web handling; Oklahoma State University, Stillwater OK; June 2011.

[2] Roisum DR. Web101[SM]. Shortcourse.

[3] Smith RD. Ultimate roll and web defect troubleshooting manual. 3rd ed. Peachtree Corners, GA, TAPPI Press; 2014.

[4] Abbott S. AbbottApps, http://www.stevenabbott.co.uk/AbbottApps/.

[5] Roisum DR. The mechanics of rollers. TAPPI Press; 1996.

[6] Roisum DR. Roller alignment—mechanics. In: AIMCAL web handling conference, Prague; June 2012, and Myrtle Beach; October 2012.

[7] Roisum DR. Roller alignment—standards. In: International web handling conference. WHRC; 2013.

[8] Rubber Roller Group. Rubber roller handbook.

[9] Roisum DR. What is the best tension for my machine? In: TAPPI PLACE conference, Cincinnati; September 2006.

[10] Roisum DR. Optimization by integrating engineering and business models. In: Proceedings of the eighth international conference on web handling. Oklahoma State University; June 5–8, 2005.

[11] Klassen C. Drives for web-handling. AIMCAL online shortcourse.

[12] Roisum DR. Web machine buying guide. Lancaster (PA): Destech Publications; 2010.

[13] Genisot T. Nip impressions—a tool for troubleshooting and maintaining quality. Tappi J 1994;77(9):246–50.

[14] Kennedy WB. Nip impressions. Tappi J 1991;74(5):277–80.

[15] Shelton JJ. Lateral dynamics of a moving web [Ph.D. thesis]. WHRC: Oklahoma State University.

[16] Roisum DR. Baggy webs: making, measurement and mitigation thereof. In: Proceedings of the sixth international conference on web handling. Oklahoma State University; June 10–13, 2001 and other venues.

[17] Roisum DR. The mechanics of web spreading—parts I and II. Tappi J 1993;76(10):63–70; 1993;76(12):75–86.

[18] Schable R. Blade runners seminar. Shortcourse offered through Tidland until about 2005.

[19] Rumson D. Web slitting. Public and online shortcourse offered through AIMCAL from 2012 until the present.

[20] Good JK, Roisum DR. Winding machines, mechanics and measurements. Lancaster (PA): Destech Publications and TAPPI Press; 2007.

[21] Gronewold J. Anthology of winding. TAPPI Press; 2000.
[22] Gronewold J. Winders the complete guide. TAPPI Press; 1998.
[23] Roisum DR. The mechanics of winding. TAPPI Press; 1994.
[24] Roisum DR. Nips—the gold plated knob—YouTube, www.youtube.com/watch?v=wMQuyt6J68g.
[25] Roisum DR. The mechanics of wrinkling. Tappi J 1996;79(10):217–26.
[26] Land, C. Baggy paper webs [Ph.D. thesis]. Karlstad University, http://www.kau.se/en/paper-technology/baggy-webs.
[27] Thuer A. Baggy web screening. Given at the 2013 WHRC conference and AIMCAL Fall Tech conference, summarized by DR Roisum in AIMCAL Converting Quarterly 4th quarter 2013 and http://www.youtube.com/watch?v=n2vM__3K54Y.
[28] Roisum, DR. Diagnosing telescoping. AIMCAL 2014 and CEMA 2015.

PART IV

Multilayer Films – Descriptions, Performance Characteristics, Uses, Considerations, Properties

17 PE-Based Multilayer Film Structures

***Thomas I. Butler*[1] *and Barry A. Morris*[2]**

[1]Blown Film Technology, LLC, Lake Jackson, TX, United States; [2]DuPont Performance Materials, Wilmington, DE, United States

17.1 Introduction

Flexible packaging is used to deliver a product from the manufacturer or distributor to the retailer or ultimate consumer and to protect that product during shipping, display, and storage. Multilayer flexible packaging is the combining of two or more layers into a composite web or tube that provides functional, protective, or decorative properties. The introduction of new polymers, the development of new processing equipment technology, and the emergence of new packaging applications have resulted in significant growth rates in coextruded and laminated structures. Whatever the application or use, polymer materials are selected and the entire packaging structure is designed to meet the requirements specific to that particular application. These could include one or more of the following:

- Specific performance properties
- Reduced cost
- Reduced number of processes

The requirement for *specific performance properties* sometimes cannot be met by one polymer or even with polymer blends extruded in a monolayer film. Blending may not be desirable if the polymers are incompatible. Coextrusion with a high-strength or high-barrier polymer can allow significant down gauging, while maintaining or improving key properties. Heat seal polymers can be incorporated into a film structure to improve packaging line efficiency or speed.

Multilayer flexible packaging structures can *lower the cost* of many film structures by reducing the expensive polymer used, increasing the less costly polymers, and using recycled material or reducing film thickness. Competitive advantages can be achieved for many film structures ranging from the high-technology barrier food packaging films to the heavy-duty shipping bag market.

Coextruded multilayer flexible packaging structures can *reduce the number of process* operations required when several polymers are needed to obtain the desired properties [1]. Eliminating process steps saves labor, equipment overhead, and reduces turnaround time. The more operations that can be combined into a single process means more space available for other equipment and less scrap generated with multiple process steps [2]. Coextrusion can eliminate the use of solvent-based adhesives. This may provide some raw material cost savings and, with increasing regulations on solvent use and disposal, reduce the ancillary costs of solvent handling recovery and incineration.

The polymers available for extrusion have increased in recent years [3]. There are many polymers to choose from, each with their own set of attributes, which can be combined with other polymers to meet the packaging needs. These polymer attributes include:

- high barrier
- selected permeation rate
- adhesion
- high-strength sealing
- easy opening (peelable) sealing
- low-temperature sealing
- high hot tack sealing
- high tensile strength
- high impact strength
- high tear strength
- high modulus
- high temperature resistance
- low temperature impact resistance
- high clarity
- abrasion resistance
- chemical resistance
- low taste and odor
- high cling
- low slip
- stability

Multilayer Flexible Packaging. DOI: http://dx.doi.org/10.1016/B978-0-323-37100-1.00017-X

- compostable/biodegradable
- antistatic
- antifog
- pigmented
- thermoformable.

This list of polymer performance attributes will continue to grow as application requirements are identified. A critical factor in developing successful flexible packaging applications has always been a good understanding of the target application. The performance properties required by the application and economic comparisons should be evaluated against the many alternative structures. Performance requirements may include all user requirements in the chain of use.

For consumers, this may mean that the packaging:

- protects the product
- identifies the product
- is easy to open.

For retailers, the packaging may provide:

- eye-catching graphics that help sell the product
- the proper physical form for display purposes.

For the packager, flexible packaging may need to provide:

- high packaging speeds
- low scrap rates
- the functional requirements for protecting the product inside the package.

Specific performance requirements will vary greatly from one package to another but, in every case, meeting the performance requirements will help assure proper protection of the goods being packaged.

Polymer films may be manufactured by blown film or cast film extrusion or by extrusion coating of a polymer onto another substrate, such as paper or aluminum foil. Blown films are made by melting and pumping polymer through an annular die [4]. Cast films are made by melting and pumping polymer through a flat die. The extrusion coating process is similar to the cast film process except that the molten polymer is coated directly onto another material. The manufacturing process selected is governed by factors such as:

- job size
- packaging material to be made
- end-user packaging performance requirements
- equipment availability.

Cast film extrusion typically operates at much higher output rates than blown film so, for larger volume production, it has an advantage with high usage single-use films such as stretch film. Blown film extrusion typically runs at a lower rate and may result in film with improved physical properties. Blown film also allows for bubble size adjustment and thus the film width produced. This is a key advantage when many different film widths must be produced on the same machine. There are many existing coextrusion processes ranging from two-layer to eleven-layer capability.

The coextrusion process is used to combine multiple materials into a single film [5–7]. Both blown films and cast films may be coextruded in three, five, seven, nine, or more layers. The combination of multiple materials in a single film allows a cost-effective means of combining the performance properties of several polymers in a single film [8]. One example is the coextrusion of a barrier polymer such as ethylene vinyl alcohol (EVOH) or polyvinylidene chloride (PVDC) with a sealant resin such as linear low-density polyethylene (LLDPE), ethylene vinyl acetate (EVA), or a polyolefin plastomer (POP). Coextrusion is widely used in producing high-performance packaging films, such as those used to package foods. It is also widely used to produce industrial films, such as stretch film, and is increasingly used in producing industrial films.

As coextrusion technology has evolved over the last 40 years, the number of layers has increased [9]. Whereas 10–15 years ago, a five-layer line was state-of-the art, now it is common to see seven-layer and higher lines installed [10–12]. In addition to the advantages described earlier around combining different polymer materials, the extra layer capability gives the converter greater flexibility and control over their process [13,14]. For example, if a five-layer line was designed to produce five layers of equal thickness, it may be a challenge to produce an unbalanced structure such as a barrier cereal liner: (60% high-density polyethylene (HDPE)/5% tie/5% EVOH/5% tie/25% EVA) [9]. The line may have to be slowed down to achieve the desired HDPE thickness because of extruder output limitations. At low line speed, however, controlling the thin layer thickness can be difficult since the extruders may be oversized. Making the same structure on a seven- or nine-layer line is easier. The HDPE layer can be split into more than one layer and fed by multiple extruders, allowing for greater output and control over the process.

Another advantage of greater layer capability is the ability to split barrier layers into two or more layers [15]. For example, a simple nylon barrier film (polyamide (PA)/tie/LLDPE) may be split into (PA/tie/PA/tie/LLDPE). Separating the barrier layers insures barrier continuity—if a pinhole develops in one layer, the second layer still may be intact. Thin layer orientation and property nonlinearity with thickness suggests that two thin layers may have better barrier performance than a single layer of the same total thickness [16].

Polymer films may be stretched, or oriented, to impart improved properties useful for packaging applications. Biaxial oriented film is produced by a double bubble or tenter frame process. Machine direction orientation may be done between hot rotating rolls. A thick film or sheet is manufactured, typically 250–1000 μm (10–40 mil), and is subsequently oriented (stretched) in a semisolid state to many times its original dimensions [17]. The multiple step production is normally done in a continuous operation [18]. The sheet stretching or orientation may occur sequentially in the machine and transverse directions or the stretching may occur simultaneously. After orienting, the films are typically 12.5–25 μm (0.5–1.0 mil) thick. The film is typically supplied in roll form. Biaxially oriented polypropylene(OPP) is most often manufactured with a tenter frame process. Oriented PE films are usually manufactured using a double bubble process. Polyvinyl chloride films may also be oriented. Some oriented films are cross-linked further to enhance their performance. Compared to other films, oriented films typically provide improved optical properties, higher stiffness and increased shrinkage during packaging, which leads to improved package appearance. Coextruded barrier films may also be oriented, typically using a double bubble process. Applications include shrink bags and sausage casings.

Layer multiplication technology, developed at Dow Chemical in the 1980s [19–22] and used primarily for optical films [23,24], is starting to see application in packaging structures. The concept involves forming the layers in a coextrusion feedblock, dividing them vertically and stacking the layers on top of one another, as shown in Fig. 17.1.

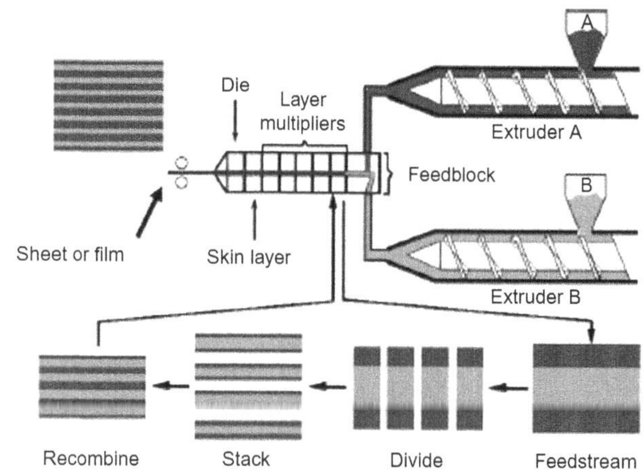

Figure 17.1 Layer multiplication technology schematic. *Source: Taken from Ref. [25].*

Each multiplier module doubles the number of layers. By adding several modules in series, hundreds of layers have been achieved. Improved strength and barrier properties have been claimed [26–31]. For example, researchers at Case Western Reserve show a 100 times or more increase in oxygen barrier of PE oxide when the layer dimensions reach the nanometer domain. They attribute this to changes in the crystalline structure [31]. Time will tell whether this can be applied to more conventional polymers used in packaging.

While originally developed for flat die coextruded cast sheet or film extrusion, recently Schirmer [32,33] and Dow Chemical [25] have introduced blown film versions. Zumbrunnen of Clemson University has been promoting technology based on chaotic advection that supposedly reproducibly creates unique morphologies, including microlayers of barrier and other polymers [34]. As some of the original Dow patents have expired, equipment suppliers have begun to promote their own versions in the market.

Lamination is used to combine two or more films into a single packaging structure [35]. It allows materials that cannot be coextruded to be combined. An example would be an aluminum foil and a PE sealant film lamination. More complicated laminations may include different polymer films, paper and foil. Laminations are usually either adhesive laminations or extrusion laminations. In adhesive laminations, the substrates are combined using an adhesive material [36]. In extrusion laminations, the substrates are adhered together using a molten polymer; often LDPE is used as the adhesive layer. Lamination can also protect the printing ink by placing it between layers, thus providing superior graphics to surface-printed packages. For example, glossy stand-up pouches have a reverse-printed outer layer laminated to structural and sealant materials. Laminations are also used to provide oxygen, moisture, or light barrier. The barrier functionality may be provided by foil or a barrier polymer such as EVOH or PVDC. Most high-value processed meat and cheese packages are laminations. This allows for combining various materials into the packaging structure and for superior graphic properties when using reverse printing. Since laminations are more costly than coextruded or monolayer films, laminations are generally reserved for use in higher value applications.

Metallization is used to apply a thin coating, typically aluminum, to a polymer film. This provides improved oxygen and water barrier properties as well as light barrier. The major use of metallized film is potato chip bags. Metallized films are also used for nuts and salty snacks. Metallized films may be coated to provide sealability or may be laminated to another polymer film to provide improved properties, such as seal integrity. Other coatings, whether to provide barrier properties or other functionality, may also be applied to polymer films used in flexible packaging.

17.2 Polymer Selection

Polymers are selected for the specific performance that they provide and are combined in the final package design to meet all the requirements for the specific application in which they are being used. Often, there are many different materials combinations or film constructions that will meet an application's minimum performance requirements [37]. In these cases, the packaging structure selected may be based on considerations such as availability from multiple suppliers and ability to provide differentiation over competitive packaging. For example, a box with an inner liner or a stand-up pouch may be used, each combination providing the minimum requirements for product protection and safety. One consumer goods company may select to use a box and inner liner. Another consumer goods company may elect to package their product in a stand-up pouch for the same product. Likewise, one manufacturer may choose to use a stand-up pouch and another manufacturer may choose to use a pillow pouch for the same product.

Polymers are chosen for individual layers to achieve specific performance properties. For example, polymers could be selected to contribute to the film's:

- tensile strength
- permeation resistance
- sealability
- adhesion
- optics
- formability
- machinability
- economics.

An individual layer can consist of virgin polymer, blends of polymers, regrind/recycled material, or high levels of additives [38]. Most thermoplastic polymers can be coextruded together. The PE family of resins, encompassing homopolymers and many different types of copolymers, is the largest volume thermoplastic used in flexible packaging. But many other polymers are used as well. A list of some of the most commonly used polymers in flexible packaging is found in Table 17.1.

Some of the key performance requirements for high-performance flexible packaging include the following:

- Barrier properties: to keep oxygen, water, light, flavor, aroma, or grease from entering the package or from leaving the package. Gas barrier properties may be characterized by measuring the oxygen and water vapor permeation through the packaging material.
- Selective permeability: to allow oxygen and carbon dioxide to permeate through the package at a calculated rate to extend the shelf life of fresh-cut produce. Oxygen, carbon dioxide, and water vapor permeability are frequently measured and specified.
- Abuse resistance: to prevent damage to the packaging material and its contents during shipping and storage. Abuse resistance may include

Table 17.1 Common Polymers Used for Flexible Packaging Applications

Polymer	Abbreviation	Density (g/cc)
Ethylene acrylic acid	EAA	0.925–0.950
Ethylene carbon monoxide	ECO	0.930
Ethylene ethyl acrylate	EEA	0.925–0.950
Ethylene metha acrylic acid (ionomer)	ION	0.940–0.950
Ethylene metha acrylic acid	EMAA	0.925–0.950
Ethylene methyl acrylate	EMA	0.930–0.950
Ethylene vinyl acetate	EVA	0.925–0.945
Ethylene vinyl alcohol	EVOH	1.14–1.16
Grafted maleic anhydride (grafted PE)	PE-g-MAH	0.91–0.940
High-density polyethylene	HDPE	0.940–0.965
High molecular-weight HDPE	HMW-HDPE	0.940–0.962
Linear low-density polyethylene	LLDPE	0.915–0.940
Low-density polyethylene	LDPE	0.915–0.925
Medium-density polyethylene	MDPE	0.925–0.940
Metallocene polyethylene	m-LLDPE	0.865–0.960
Polyolefin plastomer/elastomer[a]	POP/POE	0.856–0.915
Enhanced polyethylene	EPE	0.900–0.925
Polyamide (nylon)	PA	1.12–1.14
Polybutene-1	PB	0.909
Polycarbonate	PC	1.2
Polyethylene terephthalate	PET	1.3
Polypropylene	PP	0.89–0.902
Polystyrene	PS	1.04
Polyvinyl chloride	PVC	1.16
Polyvinylidene chloride	PVDC	1.7
Ultralow-density polyethylene	ULDPE	0.90–0.915

[a]POP is used in this chapter primarily to describe metallocene very LDPE resins, such as Dow Chemical's Affinity and ExxonMobil Exact resins. Some of the specialty ethylene–propylene copolymers such as Dow Versify and ExxonMobil Vistamaxx resins are also included in this category.

puncture resistance, tear strength, impact strength, and modulus. Some packages require good toughness at refrigerated or freezer temperatures.

- Sealability: to allow packages to be made at high packaging speeds and keep the product secure by preventing the package seams from failing. Sealability may be characterized by heat seal and hot tack strength, heat seal and hot tack initiation temperatures, seal through contamination performance, caulkability, and seal integrity. Hot tack refers to the seal strength while still in the molten state. It is critically important for packages where the product drops into the package while the seal is still partly molten, but also for horizontally filled packages involving gussets where the spring-back nature of the folded film creates an opening force. Caulkability refers to the ability of the sealant material to flow, filling in gaps around folds, wrinkles, or product contaminants.

- Machinability: to allow the packaging films to be easily run on high-speed automatic packaging equipment. Machinability is governed largely by film modulus, film thickness, seal properties, and coefficient of friction (COF).
- Consumer *appeal*: package appearance is an important factor driving product preference by consumers. Appeal is largely related to print quality and package gloss. Film thickness and modulus may also impact consumer appeal. Easy open and recloseable features also affect consumer appeal.

17.3 Mechanical Properties

Most published film data sheets are developed from monolayer films. A coextruded structure's mechanical strength may be estimated using the law of mixtures as shown in Eq. (17.1), that is, the summation of the tensile strength per unit layer thickness, multiplied by its thickness, divided by the total thickness.

$$M = \frac{(t_1M_1) + (t_2M_2) + \sum_{i=3}^{n} t_iM_i}{t_1 + t_2 + \sum_{i=3}^{n} t_i} \qquad (17.1)$$

M	=	estimated coextruded film mechanical property
t_i	=	polymer layer *i* thickness
M_i	=	polymer layer *i* mechanical property/ unit thickness
n	=	number of layers

Determining multilayer film properties with Eq. (17.1) does not account for any interactions (positive or negative) between layers or the influence of fabrication variables or orientation. Sometimes unfavorable interactions can lead to interlayer destruction, such as when a very ductile layer is adhered to a brittle layer, resulting in the film exhibiting the properties of the brittle layer [39]. Fig. 17.2 shows the effect of an LDPE core versus an HDPE core in a three-layer coextrusion structure (A–B–A) with increasing LLDPE polymer content in the skin layers on dart impact.

In the first case, the dart impact strength is a linear function of LLDPE content and can be approximated

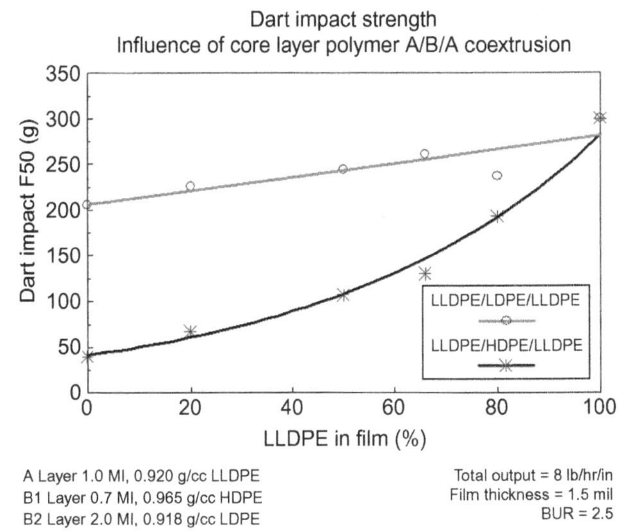

Figure 17.2 Influence of polymer in core layer.

by the law of mixtures. However, in the second case, dart impact is nonlinear and is disproportionately negatively influenced by the HDPE content. The LLDPE polymer is a highly elastic material, which allows it to absorb high levels of impact energy. The LDPE, while not as elastic as LLDPE, does not detract from the impact strength. HDPE is more brittle than the LLDPE polymer and tends to form localized stress concentration sites resulting in lower dart impact values for the film structure.

Another reason mechanical property data from monolayer film can be misleading is because some properties have a bias versus tested thickness, that is, there may not be a linear relationship between the mechanical property and film thickness. Also, the film making process itself can impart differences between coextruded films and monolayer films. For example, the stress history may differ due to quenching differences between mono- and multilayer films, resulting in differences in orientation and properties [40]. Therefore, the law of mixtures for coextruded films may be used for estimates and film properties should be measured.

Film stiffness affects:

- machinability on packaging equipment
- wrinkling
- final package hand or feel
- and, in some cases, the package function (eg, a stand-up pouch).

The flexibility or bending stiffness is an important film attribute. Some package forms, such as stand-up pouches, require high stiffness. The ability to handle the film during feeding and registration in high-speed packing lines depends on the stiffness. Consumer perception of product quality can also be impacted by the package stiffness. The bending stiffness of a monolayer film decreases with the cube of the film thickness; a small reduction in film thickness gives a large decrease in stiffness, which can be an issue when trying to down gauge a film structure. For multilayer films, the bending stiffness is a function of the tensile modulus, thickness and layer position. By manipulating these variables, the bending stiffness of a multilayer film can be maintained while down gauging the overall film thickness.

Morris and Vansant [41] showed that the outermost layers in a multilayer film often have the greatest impact on stiffness. Like an I-beam, separating stiff layers in a structure is an effective way to impart stiffness to the structure. Thus, using a stiff sealant layer can help increase stiffness and allow the total structure to be down gauged. Likewise, when laminating two stiff films, such as OPP, biaxially oriented polyethylene terephthalate (BOPET), or aluminum foil, increasing their separation builds in stiffness. The further they are separated with an adhesive or bulking layer, the greater is the bending stiffness. For this reason, an extrusion lamination process can be more effective in increasing stiffness than adhesive lamination; the extrusion process can lay down a thicker adhesive layer for separating the films.

17.4 Barrier Properties

Flexible packaging films are used to provide barrier to gases such as:

- oxygen
- nitrogen
- carbon dioxide
- water vapor.

Other applications may require taste and odor or aroma resistance properties. Permeation occurs in polymer films and this mechanism allows interaction with the environment. Molecules may permeate through a package in a three-step process. First, the molecules must dissolve into the film structure. Then the molecules will diffuse through the film layers. And, finally, the molccule will desorb on the opposite side. The molecule can move either from outside the film into the package or from inside to outside. This permeation happens due to a concentration or pressure gradient and is sensitive to temperature and, for many polymers, relative humidity (RH). The permeation rate of a material through a polymer is a function of the diffusivity at steady state times solubility in the polymer as shown in Eq. (17.2). Diffusivity is also a function of solubility especially at the low solubility levels.

$$P = DS \tag{17.2}$$

P	=	material permeability through a polymer (g/s-cm)
D	=	material diffusivity through a polymer (cm^2/s)
S	=	material solubility coefficient in a polymer (g/cm^3)

Permeation rates are usually measured at steady-state conditions. The adsorption of a material can significantly influence the permeation rate and, if the adsorption mechanisms are of sufficiently long duration, this may prevent steady-state conditions from being reached during the expected shelf life of the package. Thus, in these instances, the polymer location in a film structure could influence actual permeation rates during a package's shelf life. For most flexible package films, steady state is reached within a couple days.

The gas transmission rate at steady state through a given polymer is inversely proportional to the layer thickness. The permeation coefficient through a multilayer film structure may be estimated by summing the permeation resistance of each layer as shown in Eq. (17.3):

$$\frac{1}{P} = \frac{f_1}{P_1} + \frac{f_2}{P_2} + \cdots + \frac{f_n}{P_n} \tag{17.3}$$

P	=	coextruded film permeation coefficient
f_n	=	polymer n layer thickness ratio (layer n thickness divided by the total film thickness)
P_n	=	polymer n permeation coefficient

The actual transmission through a coextruded film is then calculated as shown in Eq. (17.4).

$$TM = \frac{P}{t_t} \quad (17.4)$$

TM	=	transmission through a film
t_t	=	coextruded film total thickness

Table 17.2 provides a *sample* calculation of moisture transmission rates based on permeability factors for a three-layer film typical for a cereal liner application using Eqs. (17.3) and (17.4). This apparent two-layer film is a typical cereal liner structure which is generally made on a three-layer blown film line with the HDPE layer split into two layers due to extruder size limitations.

Table 17.2 Sample Calculation of Moisture Transmission Rate

Basis		
At Steady-State Conditions		
HDPE	HDPE	EVA
0.9	0.8	0.3 mil
45%	40%	15%
P(HDPE) = 0.6 g-mil/100 in^2/day		
P(EVA) = 2.0 g-mil/100 in^2/day		
$1/P$ = [(0.45/0.6) + (0.4/0.6) + (0.15/2.0)] = 1.49		
P = 0.67 g-mil/100 in^2/day		
TM = 0.674/2.0 = 0.34 g/100 in^2/day moisture transmission		

For many high barrier films, the overall permeability rate is controlled by the highest barrier polymer. Permeability rates are sensitive to temperature and will increase as ambient temperature increases and follow the Arrhenius equation. Some polymer's permeability rates are also a function of RH and increase with increasing RH. Table 17.3 lists some oxygen permeability data for various polymers ranked from the lowest to the highest measured at 1.0 mil, 23°C, and 0% RH.

Table 17.4 lists the moisture transmission rate data for various polymers from the lowest permeation to the highest measured at 1.0 mil, 100°F (37.8°C), and 90% RH.

Table 17.3 Oxygen Permeability Coefficients

Oxygen Permeability Coefficients	
1. 0 mil, 73°F (23°C), 0% RH	
Polymer	**Oxygen Permeability Coefficient (cc-mil/100 in^2-day-atm)**
PVOH	0.03–0.06 [42]
EVOH	0.02
PVDC	0.15
PA	2.6
PET	3.5
PVC	5–20
PLA	30 [43,44]
PP	150
HDPE	150
EAA	200–500
EMAA	200–500
ION	200–500
PS	350
PB	385
LDPE	420
LLDPE	440
EPE	500–800
EVA	600–1000
EMA	600–1000
ULDPE	600–950
POP/POE	600–2000

Understanding the package environment during filling, processing, distribution, and storage is required to determine the permeation needs of a multilayer flexible package. Inadequate understanding of the product barrier requirements poses a design problem in predicting shelf life from gas transmission data. Packagers must rely on extensive shelf life testing of individual food products to evaluate barrier film performance requirements.

Taste and aroma barrier can be very important in some packaging applications. It is not possible to predict the polymer aroma and flavor barrier to other chemicals from common gas transmission data. The chemical component molecular size and solubility will determine the permeation rates in the polymer.

Table 17.4 Moisture Permeability Coefficients

MVTR (1.0 mil, 100°F (37.8°C), 90% RH)	
Polymer	**MVTR g-mil/100 in^2-day**
PVDC	0.10
PP	0.7
HDPE	0.4–0.8
LLDPE	0.8–1.2
ULDPE	1.2–1.5
LDPE	1.0–1.2
PB	1.0–1.2
EVA	1.0–5.5
EAA	1.0–1.6
EMAA	1.0–1.6
ION	1.0–1.6
EMA	1.0–9.0
POP	1.3–2.0
EPE	0.9–1.2
POE	2.0–3.0
PET	2.0–3.3
PVC	0.9–5.1
EVOH	2.0–4.5
PS	7.0–10.0
PA	10.0–20.0
PLA	40 (51)

Economics is a critical concern in all flexible packaging applications. Many factors are involved in determining the value a polymer is providing. For example, when comparing the cost of two barrier polymers, the comparison should be based on an equal transmission rate and package surface areas at the target application conditions (temperature and humidity). Use Eq. (17.5) to determine the best barrier polymer value:

$$\mathrm{TR} = P\frac{A}{100}\frac{1}{t} \tag{17.5}$$

TR	=	gas transmission rate through film (cc/[day-atm])
P	=	gas permeability coefficient through polymer (cc-mil/[100 in^2-day-atm])
t	=	polymer thickness (mil)
A	=	surface area exposed to a gas (in^2)

Table 17.5 provides the calculations required to compare two barrier polymers. First, the equivalent thickness of each polymer that will provide the same barrier property is determined using Eq. (17.5). Then, the cost of each polymer is determined based on the equivalent barrier properties. The data needed are:

- the required transmission rate (TR)
- permeability coefficient (P_1 and P_2)
- package surface area (A).

Oil resistance of a polymer is influenced by its polarity and crystallinity. Polar polymers, such as nylon and polyester, have good oil resistance. Among polyolefins and ethylene copolymers, PP and HDPE generally have the best oil resistance due to their high crystallinity. For PE, when the density is reduced by introducing comonomers, its oil resistance decreases. An exception is ionomers, which have excellent oil and grease resistance because of their polarity. Oil penetration generally increases with increasing temperature.

Table 17.5 Calculation of the Most Cost-Effective Barrier Polymer

Polymer	Thickness	Package Cost
1	$t_1 = P_1\frac{A}{100}\frac{1}{\mathrm{TR}}$	$C_{\mathrm{Polymer1}} = \frac{t_1}{1000}A\rho_{\mathrm{Polymer1}}\$_{\mathrm{Polymer1}}$
2	$t_2 = P_2\frac{A}{100}\frac{1}{\mathrm{TR}}$	$C_{\mathrm{Polymer2}} = \frac{t_2}{1000}A\rho_{\mathrm{Polymer2}}\$_{\mathrm{Polymer2}}$

t_n, *required polymer thickness (mil);* P_n, *permeability coefficient;* A, *package area (in^2);* ρ_n, *polymer density (lb/in^3); TR, transmission rate required to protect package contents;* C_n, *polymer cost per package;* $\$_n$, *polymer* n *price (\$/lb).*

17.5 Polymer Sealability

Heat sealability is a critical property for many packaging applications. Fig. 17.3 shows a typical plot of heat seal strength as a function of seal bar temperature.

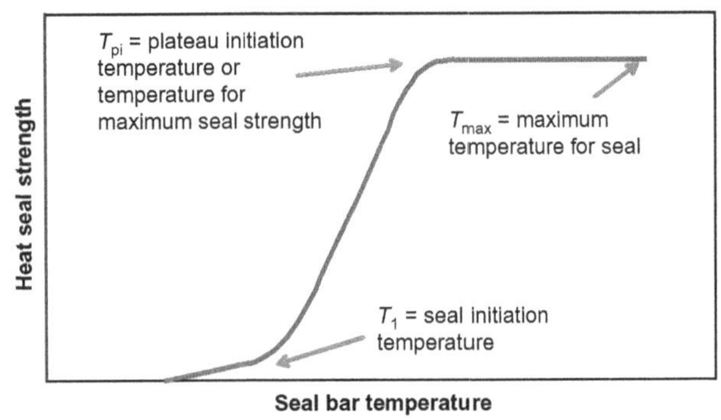

Figure 17.3 Heat seal strength versus seal bar temperature.

A polymer that exhibits low-temperature sealability and maintains seal integrity over a broad seal temperature, dwell time, and seal pressure can significantly increase packaging line speeds, improve efficiencies, and minimize seal failures. Such a polymer may be unacceptable, however, for packages requiring heat resistance as in applications such as retort, boil-in-bag, or microwave cooking.

Heat seal properties are influenced by a polymer's thermal and rheological properties as well as factors such as [45]:

- seal bar temperatures
- seal bar pressure
- dwell time
- sealing bar configuration
- package design.

Table 17.6 provides some heat seal data that compares the minimum seal bar temperature required to obtain a 3.5 N/cm (2.0 lb/in) seal strength for monolayer films at the prescribed conditions.

Ethylene copolymers have low melting temperatures and are often used as sealants; generally, the

Table 17.6 Polymer Thermal and Sealing Properties

Polymer	Melting Point °C (°F)	Vicat Softening Point °C (°F)	Fusion Seal Temperature+ °C (°F)
EMA (20%)	80 (176)	59 (138)	82 (189)
POP (0.902 g/cc)	94 (201)	80 (176)	88 (190)
EVA (12%)	90 (194)	81 (178)	91 (195)
ION	93 (200)	72 (161)	99 (210)
EAA (9% AA)	98 (208)	83 (181)	93 (200)
ULDPE (0.912 g/cc)	121 (250)	105 (221)	101 (215)
LDPE (0.920 g/cc)	110 (230)	92 (198)	107 (225)
LLDPE (0.920 g/cc)	122 (252)	110 (230)	118 (245)
HDPE (0.960 g/cc)	135 (275)	125 (257)	121 (250)
PP	168 (334)	133 (271)	149 (300)
PS	120 (248)*		
PVDC	160 (320)*		
EVOH	160 (320)		
PA	216 (420)		
PC		154 (310)	

*+Hot tack sealer, 50 μm (2.0 mil) film with 50 μm (2.0 mil) PET backing, 0.5 s dwell, 275 kPa (40 psi) pressure; * Tg.*

higher the copolymer content, the lower the seal initiation temperature. The fusion seal initiation temperature and the temperature for maximum seal strength of polyolefins and ethylene copolymers strongly correlate with their crystallinities [46]. The seal interface strength is due to molecular chain interpenetration and entanglement across the interface. Only after the polymer fully melts does enough penetration occur to ensure maximum seal strength. For LLDPE copolymers, the molecular weight and comonomer distribution influences the penetration rate: homogenous metallocene LLDPEs have a more rapid seal strength increase than do heterogeneous (Ziegler-Natta) copolymers [47]. The penetration needed to build strength depends on the polymer chemistry. Polar polymers, such as ionomers, require less penetration than nonpolar polymers such as LLDPE. Ionomers have longer range ionic forces present that help build strength, but these same forces slow their diffusion across the interface [48].

Thicker packaging structures require longer dwell times for a given seal bar temperature to reach the same heat seal strength as thinner structures. It takes longer for heat to transfer to the sealant interface in thicker films, and it also takes longer for the interface to cool down once the seal bars open. More strength can develop during the cooling phase than the heat-up phase for some structures [48].

Another critical sealability property is hot tack strength which is the seal's ability to maintain seal integrity while still hot and with an applied load. Hot tack is important in vertical form–fill–seal (VFFS) machines since the product weight puts a force on the bottom seal while it is still hot [49]. It can also be important in horizontal form–fill–seal (HFFS) applications, particularly in gusseted areas where the films are folded. A film's dead fold or "spring-back" nature exerts an opening force that must be countered by good hot tack strength to avoid channel leakers. Hot tack derives from two competing mechanisms. The first is polymer diffusion and penetration; as in the case for heat seal performance, strength at the interface is built by polymer interdiffusion and entanglement. The second is melt strength, which is related to polymer viscosity. Diffusion increases with increasing temperature whereas melt strength decreases. Thus, hot tack strength versus sealing temperature typically goes through a maximum. Fig. 17.4 compares hot tack for ionomer versus POP. Ionomer has a broad hot tack range and POP has a narrow hot tack range.

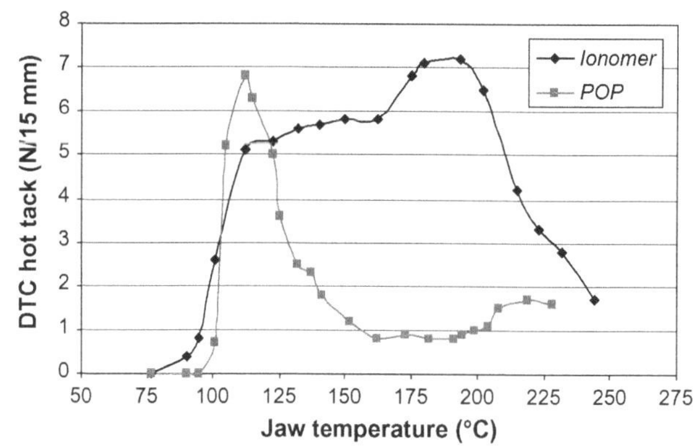

Figure 17.4 Hot tack performance. *Source: From de Garavilla [50].*

The maximum hot tack strength, the peak hot tack temperature, and the hot tack range can all be consequential in the end-use. The hot tack range versus temperature can be particularly important since it indicates how much flexibility packers may have in their packaging operation. Often, seal bar temperatures are not well controlled or the line speed is ramped up and down during the day, which affects the seal bar temperature. The molecular architecture and polymer chemistry play a role in the hot tack curve shape. Some POPs have very high maximum hot tack strengths near their melting points that tail off quickly at higher temperatures. Strain-induced crystallization in these homogeneous polymers has been proposed for the unusually high hot tack strength [51]. Poor hot tack at elevated temperatures is due to lower melt strength. Heterogeneous LLDPE and ULDPE tend to have broader hot tack curves. Broader yet are acid copolymers and, finally, ionomers, which have exceptionally high melt strength due to their unique ionic morphology [52].

Caulking or flow into cavities formed by wrinkles or gussets is sometimes required. Polymer flow is characterized by its viscosity, which is a function of temperature and shear rate. Higher sealing temperatures result in more flow and caulking. However, if the temperature is too high, the sealant may be squeezed out of the seal area, resulting in poor seal performance. Heat sealing is generally a very low shear process. The zero shear viscosity at the sealing temperature is a good starting point when comparing different polymers for their ability to caulk. Melt index, a flow measure typically used in the industry, is only a crude indicator for caulking since it does not account for a polymer's viscosity as a function of temperature and shear rate.

Seal bar pressure is another important heat seal variable. A general guideline is to use just enough pressure to get by. Some pressure is needed to ensure the film surfaces are in intimate contact for interfacial penetration to occur. Too much pressure, however, may result in squeeze-out of the sealant from the seal area.

In some packaging applications, such as potato chip pouches, an easy opening seal is desired. There are three general methods used to control the seal strength for easy-open or peel–seal performance:

- An interfacial peel–seal mechanism where failure occurs along the sealant–sealant interface or sealant–substrate interface in the case of the lidding structure. This sealing mechanism typically occurs if the seal temperature is low—on the rapidly rising section of the heat seal versus temperature curve that is difficult to control. It is most often used for lidding applications where adhesion is controlled more by wetting considerations rather than polymer interdiffusion.
- A cohesive failure mechanism where a "contaminate" is blended into the polymer to ensure the failure is within the sealant layer during opening. The contaminate is typically an incompatible polymer that forms a weak dispersed phase within the sealant. Polybutene-1 is often used as the contaminate polymer for PE-based sealants.
- A delamination peel–seal mechanism, sometimes known as a "burst" peel where adjusting the sealant layer adhesion to the adjacent layer controls the seal strength. The opening force tears through the sealant layer and the seal fails along the sealant/adjacent layer interface. An example is a cereal liner bag with the structure (HDPE-ionomer).

17.6 Adhesive Polymers

For a coextruded structure to function during manufacture, packaging, distribution, and storage over the product life cycle, there must be sufficient adhesion between the layers. Interlayer adhesion is affected by:

- structure design
- individual layer properties
- process by which it is put together
- manner in which it is used [53].

The structure design and, in particular, the individual layers' compatibility will determine whether specialized adhesive polymers or tie layers are needed [4]. For two polymers to bond to one another, they first must come into intimate contact. In coextrusion, this is achieved as the molten polymer streams are combined in the die or feedblock. For coatings and laminations, this requires good wetting, which is a function of viscosity, polarity, and surface tension. Once good contact is achieved, molecular segments may diffuse across the interface provided:

- Molecules are compatible.
- Polymers are above their glass transition or melting temperature to allow chain segment mobility.
- There is sufficient time and temperature for diffusion to occur.

Diffusion of molecular chains across the interface leads to interfacial strength and adhesion. Even if conditions are not favorable for diffusion, good adhesion can be achieved if specific functional groups are present that lead to a chemical interaction at the interface.

Table 17.7 gives some examples of the adhesion between some typical layer combinations found in coextruded packaging films.

For those cases where adhesion is poor, specialized adhesive polymers or tie resins have been developed [54]. Tie resins are typically a polyolefin or ethylene copolymer matrix resin, with chemical functionality and sometimes a toughener added. The matrix resin is chosen for its compatibility with one bonded layer to promote diffusion. In Fig. 17.5, PE is being adhered to PA. In this example, the matrix resin is typically PE, EVA, or another ethylene copolymer which bonds to the PE layer. Chemical functionality is incorporated into the adhesive polymer via copolymerization, grafting reaction, or alloying. Some common examples of chemical functionality are given in Table 17.8.

For bonding to PA, either acid or anhydride functionality is typically chosen. The acid or anhydride groups react with the PA amine end groups to achieve good adhesion. Finally, various tie resin manufacturers often blend in proprietary modifiers,

Table 17.7 Chart of Adhesion Between Polymers

Polymer	HDPE	PP	PS	PA	EVOH	PVDC	PC
LDPE	G	P	P	P	P	P	P
LLDPE	G	G	P	P	P	P	P
ECO	F	P	X	P	P	G	F
EVA	G	G	G	P	P	G	F
EMA	G	G	X	P	P	G	F
EAA	F	P	P	G	P	P	P
ION (Na)	P	P	P	P	P	P	P
ION (Zn)	P	P	P	G	P	P	P
PE-g-MAH	G	P	P	G	G	P	F

G, good; F, fair; P, poor; X, no data.

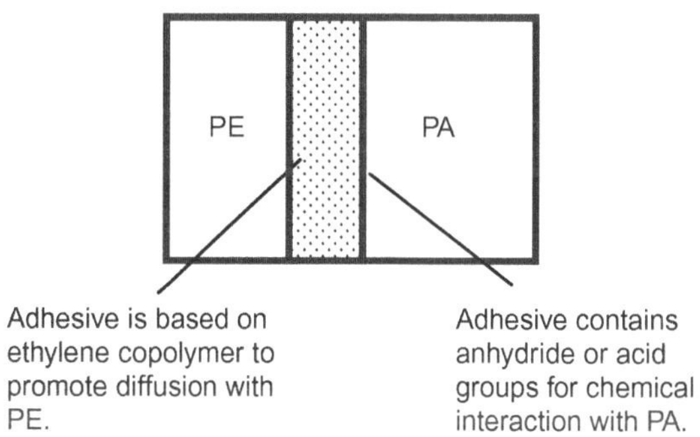

Figure 17.5 Coextrusion adhesive technology.

Table 17.8 Common Functional Groups in Tie Resins

Functionality	Adheres to
Acid	PA, Al foil
Anhydride	PA, EVOH
VA	PVDC, PP, PET
Acrylate	Some inks
Epoxy	PET
Silane	Glass

such as rubber, that impact the tie resin physical properties and its peel strength performance. The most common tie resins used today are anhydride modified polyolefins for bonding to nylon and EVOH in barrier film structures.

When selecting an appropriate tie resin, several factors come into play besides the adhesion performance. The adhesive must comply with appropriate government regulations, such as Food and Drug Administration regulations for food packaging in the United States. It must have the proper flow properties for the given converting process and sufficiently match the flow properties of adjacent layers to avoid flow instabilities during coextrusion. The tie resin may also be called upon to impart other properties, such as moisture barrier, toughness, or clarity. Selecting an appropriate tie resin matrix will often accomplish this. For example, a HDPE-based tie resin will have a lower moisture vapor transmission rate (MVTR) than one based on EVA. Finally, as with any polymer, the product quality and consistency is important for achieving good performance.

Tie resin thickness in the packaging structure can influence adhesion performance, as can coating thickness in extrusion coating [55]. Adhesion is most often measured by pulling the structure apart in a peel strength test. Peel strength is a function of fracture energy, or energy to create new surfaces as the peel front advances. At the advancing edge, the adhesive deforms, creating an energy loss that contributes to the measured fracture energy. Increasing the adhesive thickness may increase the deformation zone, increasing the fracture energy. It may also increase the energy to bend the peel arm during the peel test, although this effect is small for flexible films. In theory, the fracture energy will increase with increasing thickness until a critical value is reached where the deformation zone becomes small compared to the thickness. At this point, the fracture energy plateaus [56]. In many flexible packaging structures, the adhesive thickness is below this critical thickness so that reducing thickness reduces the peel strength performance. There are also practical limits on how low the thickness can be for a given converting process and still ensure adhesive layer continuity. For most processes, 2–3 μm (0.1 mil) is the lower adhesive layer thickness limit.

How the film is fabricated can have a significant impact on adhesion. Typically, less orientation and greater contact time at higher temperatures favors adhesion. Several processing parameters are important for optimizing adhesion performance [54]:

- Adhesion generally increases with increasing processing temperatures. Both diffusion and chemical interaction are favored at higher temperatures.
- Increasing melt contact time generally improves adhesion. Combination technology (feedblock or die design), the order in which layers are brought together and die land length can affect adhesion.
- Increasing line speed generally reduces adhesion. Studies have shown in blown and cast film, as well as coextrusion coating, that peel strength increases with increasing process time, a characteristic parameter related to the time the polymer spends between the die exit and when it freezes off [57–60]. The process time, which is inversely related to line speed, relates to the trade-off between time for the process with the time available for diffusion and chemical reaction to occur as new interfacial area is created and for the imposed stresses to relax.
- Orientation decreases adhesion. Increasing blow up ratio in blown film or increasing draw down ratio in cast film can reduce adhesion. Orientation processes such as double bubble processes can substantially reduce adhesion by increasing stress, reducing thickness, and creating new interfacial area (decreasing bond density).
- Thermoforming can reduce adhesion by some of the same mechanisms as orientation.

Finally, the manner in which the structure is used can impact adhesion [53]. End-use storage conditions (temperature, RH) and the environment the package is subjected to, both physical (eg, cook-in meats, post-packaging sterilization, or pasteurization) and chemical can alter adhesion. The product itself can contain chemical species that migrate to the interface and destroy adhesive bonds. A well-known example is the effect an acid species in orange juice and condiments has on LDPE or ethylene acrylic acid (EAA) adhesion to foil [61–66]. It is critically important to test structures under conditions that closely simulate the end-use before finalizing a packaging structure.

17.7 Applications for Flexible Packaging Film Structures

Packages may be formed in-line by several techniques or may be supplied to the packer as preformed pouches or bags. Packaging may be created by wrapping or shrinking a basic film around a bundle of goods. In-line packaging forming examples include VFFS, HFFS, and thermoform/fill/seal. In VFFS operations, film from a roll is guided through rollers and then shaped by a forming collar into a tube. The film moves in a vertical direction (down) over a filling tube. A vertical seal is made, forming the film into a continuous tube. As the film continues through the machine, a horizontal seal is made, perpendicular to the film machine direction, forming the bottom end-seal of the bag being formed and the top seal of the previously filled bag. The product is dropped into the partially formed bag, advanced to the seal bars and the next bottom and top end-seal is made. The process may operate in a stepwise or continuous manner. One example of a product normally packaged on VFFS equipment is fresh-cut produce. In HFFS operations, the film moves in a horizontal direction during the packaging step. One application that typically uses HFFS equipment is chunk cheese. In thermoform/fill/seal operations, a bottom web is formed, product is added, and the top web, which is normally flat, is sealed to the bottom web. Thermoform/fill/seal packaging is frequently used for bacon and processed meats. Stand-up pouches and other types of packaging may be formed in-line with the filling equipment, or may be fully or partially prefabricated prior to the filling step.

Markets for flexible packaging films have continued to grow in many applications. PE and the various copolymers account for more than 75% of the flexible packaging film. Some major market segments where coextruded film is used include [67]:

- medical packaging
- food packaging
- heavy-duty shipping bags

- stretch wrap
- trash bags
- condiments—OPET/print/LDPE/Al/(EAA–LDPE) and variations
- aseptic packaging OJ—juice boxes—print/LDPE/paperboard/LDPE/Al/(EMAA–LDPE) and variations
- towlettes—OPET/print/LDPE/Al/ION and variations
- condoms—similar structures as towlettes
- laminates for toothpaste tubes—PE-film/(LDPE–EAA)/Al/(EAA/LDPE)/PE-film
- stand-up pouches—OPET/print/adh/m-OPP/adh/(LLDPE–HDPE–LLDPE) and variations.

17.7.1 Medical Packaging

A wide variety of structures are used in the medical packaging area (Table 17.9). The structure requirements include:

- sterilization capability
- microbial barrier
- linear tear properties
- puncture resistance.

Sterilization methods used for medical packaging film include ethylene oxide gas and radiation. The package is typically a forming web and a nonwoven fabric (like TYVEK®) which lets the ethylene oxide in and out of the package. Medical packaging usually does not require oxygen barrier properties so EVOH is normally not required in medical packaging structures. Coextrusion processes are growing in this market. Blown film, cast film, and extrusion coating processes are used in producing flexible packaging structures for medical packaging applications. Films ranging from three to eleven layers are now available. These markets are typically small and require many years to meet the required qualifications.

The forming webs were historically three layer EVA and ionomer films as shown in the syringe package in Fig. 17.6.

Heavy gauge films were sometimes produced by combining a three-layer film to form six layers. Newer film structures have seen more layers being used and incorporating PA for forming and puncture resistance.

Table 17.9 Medical Device Packaging Structures

Structure Design	Layers (%)	Gauge (μm)
3-Layer Structure		
(EVA–ION–EVA)	(20/60/20)	50–400
(ION–EAA–EVA)	(30/30/40)	50–400
Paper/LDPE/Al	Lamination	
5-Layer Structure		
(LLDPE–Tie–PA–Tie–LLDPE)	(40/5/10/5/40)	50–400
(m-LLDPE–Tie–PA–Tie–m-LLDPE)	(40/5/10/5/40)	50–400
(LLDPE–Tie–PA–Tie–PA)	(70/5/10/5/10)	50–400
LDPE/Paper/LDPE/Foil/LDPE	Lamination	
OPET/adh/(LDPE–EVA–ION)	Lamination	
7-Layer Structure		
(LLDPE–Tie–PA–EVOH–PA–Tie–LLDPE)	(30/5/10/10/10/5/30)	50–400
9-Layer Structure:		
(PE–Tie–PA–Tie–PE–Tie–PA–Tie–PE)	(20/l5/10/5/20/5/10/5/20)	50–400
(LLDPE–LDPE–Tie–PA–EVOH–PA–Tie–LDPE–LLDPE)	(10/15/10/10/10//10/10/15/10)	50–400

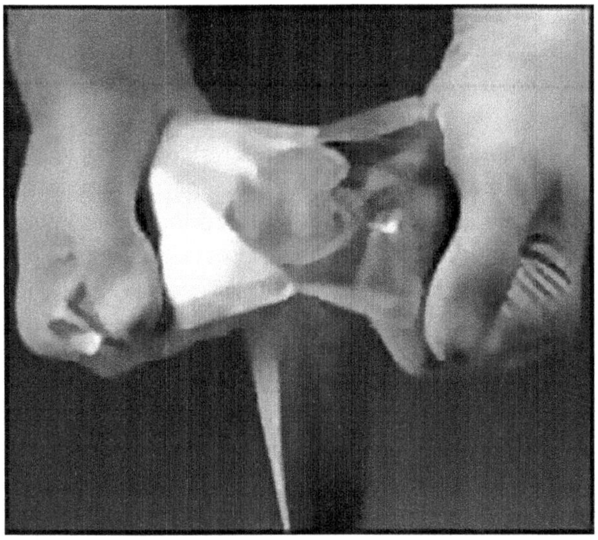

Figure 17.6 Multilayer film used for syringe package.

Consumer and industrial heath care packaging includes:

- medical disposables
- surgical instruments—structures similar to disposables
- resterilization packaging
- syringes and hypodermic needles
  - forming web
  - nonforming web
- sutures
- pharmaceutical
- condoms and towlettes.

Lidding stock is normally produced by extrusion coating and/or lamination processes and combines paper, polymers, and foil to form multilayer structures. Sachets for condom and towelette packaging typically use the structure OPET/print/LDPE/Al/ION and variations, with ionomers used as the sealant for its chemical and flex crack resistance.

17.7.2 Food Packaging

17.7.2.1 Primal Meat Packaging (Shrink)

Packaging primal and subprimal meat requires a package that must:

- provide high shrinkage to fully collapse around irregular shapes
- have excellent optical properties
- shrink at low temperature to prevent product damage
- impart good softness and elasticity
- provide excellent oxygen, moisture, odor, and grease barrier protection
- prevent freezer burn
- facilitate using individual cuts by food preparers
- help reduce purge loss
- extend shelf life
- offer easy disposal
- have good machinability
- have an oxygen transmission rate (OTR) less than 1.0 cc/100 in^2-day-atm (nonfrozen only).

A shrinkable film used in this application is a PVDC barrier film with the sealant layer designed to provide toughness and puncture resistance. These films must be oriented to provide acceptable shrink properties using a double bubble process. Table 17.10 provides some typical film structures used in shrink film for primal and subprimal meat packaging.

Table 17.10 Primal and Subprimal Meat Packaging

5-Layer Structure	Layers (%)	Gauge (μm)
(ULDPE–EVA–PVDC–EVA–ULDPE)	(40/5/10/5/40)	50–120
(m-LLDPE–EVA–PVDC–EVA–m-LLDPE)	(40/5/10/5/40)	50–120
(PA–EVOH–Tie–LLDPE–m-LLDPE)	(30/40/10/10/10)	50–120
(PA–EVOH–Tie–LLDPE–ION)	(30/40/10/10/10)	200–300
OPA/PVDC/(LLDPE–EVA–ION)	(20/30/35/5/10)	50–100

Note: /PVDC/ is an adhesive lamination with PVDC. See appendix for convention used.

17.7.2.2 Processed Meat Packaging

Processed and cook-in meat such as:

- luncheon meat
- ham
- bologna
- salami

are packaged in barrier films that are designed to keep oxygen from entering the package. This extends shelf life and gives the retailer extended product display time. It also allows the consumer to keep the product in their refrigerator, unopened, for some time after purchase. These packages are often printed with eye-catching graphics to increase sales. These films may contain:

- a barrier polymer
- printing surface, such as PET or nylon, that also provides thermal resistance during sealing and helps provide abuse resistance during distribution
- LLDPE or ULDPE toughness layers
- a sealant layer that could be EVA, LLDPE, POP, or ionomer.

The processed meat package comprises a forming film and a backing film. The forming film is thermoformed to the meat product shape. In addition, low oxygen permeability, abuse resistance, and seal integrity are critical to maintaining the proper atmosphere inside the package. Optical properties, such as high gloss and high clarity, are important on the backing film where reverse-printed PET is used to create consumer appeal. Barrier requirements for processed meats range from 0.2 to 1.0 cc/100 in^2-day-atm for OTR and 0.2 to 0.5 g/100 in^2-day for water vapor transmission rates (WVTR). Table 17.11 provides some typical film structures used in processed meat packaging.

17.7.2.3 Poultry/Fish Packaging

Moisture barrier properties are critical for poultry/fish packaging. The packages are normally vacuum packaged (Fig. 17.7) with a good sealant polymer such as EVA, ionomer, or LLDPE (Table 17.12).

Table 17.11 Processed Meat Packaging Film Structures

Product Structure	Layers (%)	Gauge (μm)
Ground Beef		
(PA–Tie–LLDPE)	(20/5/75)	150–200
(PA–EVOH–Tie–LLDPE)	(10/10/5/75)	150–200
(LLDPE–Tie–PA–EVOH–PA–Tie–LLDPE)	(25/10/10/10/10/10/25)	40–150
Barrier Overwrap		
(PA–EVOH–Tie–LLDPE–m-LLDPE)	(10/10/10/40/30)	150–200
(PA–EVOH–Tie–LDPE–LDPE–m-LLDPE)	(15/10/15/10/15/20)	150–200
Chub Films		
PA//PVDC/LLDPE	(20/5/75)	150–200
Foodservice Portion: Steaks/Chops/Roasts		
Forming Web		
(PA–EVOH–Tie–LLDPE)	(15/5/5/75)	150–200
(PA–EVOH–Tie–ION)	(15/5/5/75)	150–200
(PA–Tie–LLDPE)	(20/5/75)	150–200
(PA–ION)	(20/80)	150–200
PA//PVDC/(EVA–ION)	(25/5/10/60)	150–200
(LLDPE–Tie–PA–EVOH–PA–Tie–LLDPE)	(25/10/10/10/10/10/25)	150–200

(Continued)

Table 17.11 Processed Meat Packaging Film Structures—*Cont'd*

Product Structure	Layers (%)	Gauge (μm)
Nonforming Web		
(PA–EVOH–Tie–LLDPE)	(10/10/10/70)	50–80
(PA–EVOH–Tie–ION)	(10/10/10/70)	50–80
OPET//PVDC//LLDPE	(10/5/85)	50–80
OPET//PVDC//ION	(10/5/85)	50–80
(PA–Tie–ION)	(10/10/80)	50–80
(LLDPE–Tie–PA–EVOH–PA–Tie–LLDPE)	(25/10/10/10/10/10/25)	50–80
Skin Packaging Forming Web		
(EVA–Tie–EVOH–Tie–ION)	(35/10/10//10/35)	150–250
Skin Packaging Nonforming Web		
(EVA–Tie–EVOH–Tie–ION)	(35/10/10//10/35)	50–80
Luncheon Meat		
Forming Web		
(PA–EVOH–Tie–LLDPE)	(15/5/5/75)	150–200
(PA–EVOH–Tie–ION)	(15/5/5/75)	150–200
(PA–EVOH–Tie–m-LLDPE)	(15/5/5/75)	150–200
(LLDPE–Tie–PA–EVOH–PA–Tie–LLDPE)	(25/10/10/10/10/10/25)	150–200
Nonforming Web		
OPET//PVDC//(EVA–LLDPE)	(25/5/10/60)	50–100
OPET//adh//(EVOH–Tie–m-LLDPE)	(25/2/5/10/60)	50–100
OPET//adh//(EVOH–Tie–ION)	(25/2/5/10/60)	50–100
(PA–EVOH–Tie–LLDPE)	(25/10/5/25)	50–100
OPA//PVDC//LLDPE	(65/5/25)	50–100
(PA–EVOH–PA–Tie–LLDPE)	(10/10/10/10/60)	50–100
(LLDPE–Tie–PA–EVOH–PA–Tie–LLDPE)	(25/10/10/10/10/10/25)	50–100
Frankfurters		
Forming Web		
PA–EVOH–Tie–LLDPE)	(15/5/5/75)	150–200
(PA–EVOH–Tie–m-LLDPE)	(15/5/5/75)	150–200
(PA–EVOH–Tie–EVA)	(15/5/5/75)	150–200
(PA–EVOH–Tie–ION)	(15/5/5/75)	150–200
(LLDPE–Tie–PA–EVOH–PA–Tie–LLDPE)	(25/10/10/10/10/10/25)	150–200
(PA–Tie–PA–EVOH–PA–Tie–ION)	(15/20/10/5/10/25/15)	150–200
(PA–Tie–PA–EVOH–PA–Tie–LLDPE)	(15/15/10/5/10/25/20)	150–200
(PP–Tie–PA–EVOH–PA–Tie–ION)	(15/20/10/5/10/25/15)	150–200
Nonforming Web		
OPA//PVDC/(EVA–LLDE)	(10/5/25/60)	50–100
OPET//adh/(EVOH–Tie–m-LLDPE)	(15/10/15/60)	50–100

(*Continued*)

Table 17.11 Processed Meat Packaging Film Structures—*Cont'd*

Product Structure	Layers (%)	Gauge (μm)
OPET//PVDC//ION or EVA	(10/5/85)	50–100
(LLDPE–Tie–PA–EVOH–PA–Tie–LLDPE)	(25/10/10/10/10/10/25)	50–100
OPET//adh//(LLDPE–ION)	(25/2/48/25)	50–100
OPET//adh//(LLDPE–Tie–EVOH–Tie–ION)	(15/2/38/10/10/10/15)	50–100
Sausage		
Forming Web		
(LLDPE–LDPE–Tie–EVOH–PA–Tie–EVA)	(30/5/10/10/10/5/30)	150–200
(PA–EVOH–Tie–m-LLDPE)	(75/5/5/15)	150–200
(PA–EVOH–Tie–ION)	(15/5/5/75)	150–200
(LLDPE–Tie–PA–EVOH–PA–Tie–LLDPE)	(25/10/10/10/10/10/25)	150–200
(PA–Tie–PA–EVOH–PA–Tie–ION)	(15/25/10/5/10/20/15)	150–200
(PA–Tie–PA–EVOH–PA–Tie–LLDPE)	(15/25/10/5/10/15/20)	150–200
(PP–Tie–PA–EVOH–PA–Tie–ION)	(15/25/10/5/10/25/15)	150–200
(OPET//adh//(EVOH–Tie–LLDPE)	(15/2/14/10/14/60	50–100
(PA–EVOH–Tie–m-LLDPE)	(15/10/15/60)	50–100
(LLDPE–Tie–PA–EVOH–PA–Tie–LLDPE)	(25/10/10/10/10/10/25)	50–100
Shrink Bags		
(EVA–PVDC–EVA)	(48/5/47)	50–80
(EVA–Tie–EVOH–Tie–EVA)	(38/10/5/10/37)	50–80
(LLDPE–Tie–PA–EVOH–PA–Tie–LLDPE)	(25/10/10/10/10/10/25)	50–80
Ham		
Forming Web		
(PA–EVOH–Tie–LLDPE)	(15/10/15/60)	150–200
(PA–EVOH–Tie–ION)	(15/10/15/60)	150–200
(LLDPE–Tie–PA–EVOH–PA–Tie–LLDPE)	(25/10/10/10/10/10/25)	50–100
Nonforming Web		
PA//PVDC//LLDPE	(25/5/70)	50–80
PA//PVDC/ION	(35/5/70)	50–80
OPET//PVDC//LLDPE	(25/5/70)	50–80
OPET//PVDC//ION	(25/5/70)	50–80
(LLDPE–Tie–PA–EVOH–PA–Tie–LLDPE)	(25/10/10/10/10/10/25)	50–100
Shrink Bags		
(EVA–PVDC–EVA)	(48/5/47)	50–80
(EVA–Tie–EVOH–Tie–EVA)	(38/10/5/10/37)	50–80
(m-LLDPE–Tie–EVOH–Tie–m-LLDPE)	(30/10/10/10/30)	50–80
Bacon		
(PA–EVOH–Tie–ION)	(30/10/10/50)	50–100

(*Continued*)

Table 17.11 Processed Meat Packaging Film Structures—*Cont'd*

Product Structure	Layers (%)	Gauge (μm)
PA/Primer/(PE–ION)	(15/50/35)	50–100
(PA–Tie–PA–EVOH–PA–Tie–ION)	(15/20/10/5/10/20/20)	50–100
Deli Meats		
Shrink Bags		
(EVA–PVDC–EVA)	(48/5/47)	50–80
(EVA–Tie–EVOH–Tie–EVA)	(38/10/5/10/37)	50–80
(m-LLDPE–Tie–EVOH–Tie–m-LLDPE)	(30/10/10/10/30)	50–80
Forming Web		
(PA–EVOH–Tie–LLDPE)	(10/5/5/80)	200–250
(PA–EVOH–Tie–ION)	(10/5/5/80)	200–250
(LLDPE–Tie–PA–EVOH–PA–Tie–LLDPE)	(25/10/10/10/10/10/25)	50–100
Nonforming Web		
LLDPE/adh/(PA–EVOH–Tie–EVA)	(40//10/10/20/30)	50–80
(PA–EVOH–Tie–ION)	(15/10/15/60)	50–80
(LLDPE–Tie–PA–EVOH–PA–Tie–LLDPE)	(25/10/10/10/10/10/25)	50–100
Casings		
OPA//LLDPE	(30//70)	50–80
LLDPE//OPA//PVDC	(70//20//10)	50–80

17.7.2.4 Cereal Box Liners

Cereal box liners (Fig. 17.8) also require good moisture barrier properties to provide good taste and freshness protection.

HDPE polymers are typically used to provide moisture barrier. Sealant polymers such as EVA, ionomer, POP, or blends are used for low-temperature seals, form–fill–seal packaging and easy opening seals. Certain products have additional requirements, such as puncture resistance to keep the product from poking through the packaging film and flavor and aroma barrier for highly flavored cereals. Heat seal initiation temperatures of 90°C and below are commonly required. MVTR less than or equal to 0.1 g/100 in^2-day-atm are often required.

Figure 17.7 Poultry packaging.

Table 17.12 Poultry/Fish Packaging

7-Layers Structure	Layers (%)	Gauge (μm)
(LLDPE–Tie–PA–EVOH–PA–Tie–LLDPE)	(25/10/10/10/10/10/25)	40–150
(LLDPE–Tie–PA–EVOH–PA–Tie–ION)	(25/10/10/10/10/10/25)	40–150

Figure 17.8 Cereal boxes.

Packages requiring aroma or taste barrier properties will contain either nylon or EVOH polymers (Table 17.13).

17.7.2.5 Snack Food Packaging

Potato chips are often packaged in structures that contain metallized films (Table 17.14). Polymer film metallization provides oxygen barrier, moisture barrier, and light barrier. The light barrier is to protect the potato chips from ultraviolet radiation that initializes an oxidation mechanism. Seal strength must be optimized to provide a secure package that can be easily opened by the consumer. Seal integrity and consumer appeal are also critical.

17.7.2.6 Salty Snack Packaging

Salty snacks are frequently high in fat content and may require a package that provides an oxygen barrier in order to prevent the fat in the food from going rancid. They may also require grease resistance to keep the package from leaving an oily spot. Salty snacks may be packaged in barrier films containing foil, a metallized polymer film, or a barrier polymer such as EVOH or PVDC and are gas flushed with nitrogen to maintain a low oxygen concentration inside the package (Table 17.15).

17.7.2.7 Bakery

Moisture barrier is normally the critical property in bakery applications. Polymers used for moisture barrier include LDPE, LLDPE, HDPE, or PP. Typically, EVA polymers are used for sealability and optics. Applications, such as the cake mix pouch, will require aroma, taste, and moisture barrier properties. Nylon is used for taste and aroma barrier. In bread bags, the LLDPE polymer's toughness allows down gauging, while LDPE allows good optics and printability (Table 17.16).

Table 17.13 Cereal Packaging

Market Structure	Layers (%)	Gauge (μm)
Bag-in-Box		
(HDPE–HDPE–EVA + PB)	(45/45/10)	40–60
(HDPE–HDPE–ION)	(45/45/10)	40–60
Bag-in-Box (Peelable Seal)		
(HDPE–HDPE–EVA + ION)	(45/45/10)	40–60
Bag-in-Box (Barrier Bag)		
(HDPE–Tie–EVOH–Tie–EVA + PB)	(60/10/10/10)	40–60
(HDPE–Tie–PA–Tie–EVA + PB)	(60/10/10/10)	40–60

Table 17.14 Snack Food Packaging Films

Structure	Layers (%)	Gauge (μm)
Potato Chips (OTR 2.0, MVTR 0.02)		
OPP//met-OPP	(50/50)	20–60
OPP//LDPE//met-OPP	(25/50/25)	20–60
OPPP-met//(HDPE–EVA)	(15/70/15)	40–80
Tortilla and Corn		
Chips (OTR <2.0, MVTR <0.35)		
OPP//LDPE//OPP	(25/50/25)	20–60
OPP//PVDC//OPP	(50/5/45)	20–60
Pretzels (OTR <2.0, MVTR <0.5)		
OPP//LDPE//OPET	(25/50/25)	30–80
OPP//LDPE//OPP	(25/50/25)	30–80
OPP//PVDC/OPP	(45/5/50)	30–60
Meat Snacks		
OPET//PVDC//LDPE	(50/5/45)	30–60
(PA–Tie–EVOH–Tie–LLDPE)	(10/10/10/10/60)	40–80

Table 17.15 Snack Nuts Packaging

Structure	Layers (%)	Gauge (μm)
OPP//LDPE//OPP	(20/60/20)	40–60
OPP//LDPE//Foil//LDPE	(10/20/5/65)	40–60

17.7.2.8 Cheese Packaging

Most cheese sold in the United States is prepackaged in flexible packaging. The cheese packaging includes:

- individually wrapped processed cheese slices
- chunk cheese
- shredded cheese.

Both chunk cheese and shredded cheese require substantial oxygen barrier to prevent mold growth and spoilage. EVOH, PVDC, or PVOH may provide the oxygen barrier. While EVOH is generally coextruded into the film structure, PVDC or PVOH may be coated on a film via a coating process. Cheese packaging also requires excellent seal integrity and abuse resistance to prevent the controlled atmosphere inside the package from being lost. Cheese packages are often laminations made with reverse-printed outer webs containing PET or nylon for superior graphical presentation. They may also be extrusion-coated structures where the sealant layer has been extrusion coated onto the outer layer. Sealant layers may consist of EVA, an ionomer, or a POP. Low-heat seal initiation temperature (90°C or below) and good seal through contamination performance are required. Processed cheese typically requires films with OTR between 0.6 and1.0 cc/100 in^2-day-atm and WVTR equal to or less than 1.0 g/100 in^2-day (Table 17.17).

An acrylic, PVOH-coated OPP film is also used in cheese packaging in both extrusion and adhesive laminations. Acrylic is coated on one side and PVOH on the other side.

17.7.2.9 Milk Pouches

LLDPE or LDPE/LLDPE blends provide the sealant in both milk powder and liquid pouches (Table 17.18). If oxygen barrier is required for long shelf life, then nylon or EVOH layers are included.

17.7.2.10 Frozen Food

Frozen foods (Fig. 17.9) are packaged in a variety of packaging types.

Table 17.16 Bakery Packaging

Structure	Layers (%)	Gauge (μm)
cPP/PP/cPP	(10/80/10)	30–60
PP/EVA	(80/20)	30–60
(HDPE–EAA–PA–EAA)	(70/10/10/10)	30–70
(LLDPE–PP–LLDPE)	(10/80/10)	30–60
(PA–EVOH–PA–Tie–LLDPE)	(20/10/20/10/30)	100–160

Table 17.17 Cheese Packaging

Structure	Layers (%)	Gauge (μm)
Natural Chuck Cheese Pouches		
OPET/adh/(EVOH–Tie–LLDPE)	(15/5/10/10/55)	40–60
LLDPE//PVDC//OPP//PVDC//LLDPE	(15/10/75)	50–60
OPET//PVDC//LLDPE	(15/10/75)	45–50
EVA//PVDC	(95/5)	50–60
OPA//PVDC//LLDPE	(15/10/75)	45–50
Vacuum Bags for Aging		
OPA//PVDC//EVA	(110/5/85)	40–100
(PA–Tie–LLDPE)	(10/10/80)	40–100
(PA–Tie–EVA)	(10/10/80)	40–100
Shredded Cheese		
PVDC//PA//LDPE	(5/20/75)	80–100
PVDC//PET//LDPE	(5/20/75)	80–100
(PA–EVOH–Tie–LDPE)	(10/10/10/PA)	80–100
Processed Cheese Slices		
OPP//EVA	(50/50)	20
PP/EVA	(20/80)	35–40

Table 17.18 Milk Packaging

Structure	Layers (%)	Gauge (μm)
(LLDPE–Tie–PA–Tie–LLDPE)	(35/10/10/10/35)	40–70
HDPE/(LDPE + LLDPE)	(40/60)	40–80

Examples of frozen foods packaged in flexible packaging include:

- frozen fruits
- vegetables
- French fries
- individually packed quick frozen chicken breasts.

Many frozen foods are packaged in surface-printed PE films. In recent years in the United States, reverse-printed laminations, which may be shaped into stand-up pouches, have taken over the frozen food supermarket aisle. These packages have attractive graphics and may incorporate convenience features such as self-venting for microwave cooking.

Figure 17.9 Frozen food packaging.

Key requirements for frozen food packaging are:

- low-temperature toughness
- modulus
- high hot tack strength
- high seal strength.

Some packages are clear and require good clarity, while others are pigmented and require good gloss. LLDPE, ULDPE, EVA, and POP resins are all commonly used in creating frozen food packaging. Stiffness must be adequate for high-speed packaging and packaging films must have tear and puncture strength high enough to prevent package damage during transportation and storage (Table 17.19).

17.7.2.11 Fresh-Cut Produce

Key performance requirements for fresh-cut produce packaging include proper oxygen and carbon dioxide permeability, seal integrity, machinability, and consumer appeal. Consumer appeal includes both feel and appearance. Feel is generally determined by film thickness and modulus while appearance is governed by print quality and film optical properties, such as clarity, haze, and gloss. In order to extend the shelf life of the produce being packaged, films must provide the proper oxygen permeability that is matched to the packed produce respiration rate. Cut produce respires after harvesting, consuming oxygen and giving off carbon dioxide. By controlling the permeation of gases through the package, the environment inside the package is controlled, respiration is slowed, and shelf life is extended. The bags must have complete seal integrity in order to prevent the unplanned transfer of gases between the bags and the environment. Bags may contain PP, LLDPE, ULDPE, EVA, or POP. OTR requirements vary widely depending on the produce being packaged, but common items range from about 100 cc/100 in^2-day-atm for Caesar salad mixes, to 150–200 cc/100 in^2-day-atm for iceberg salad mixes and 200–350 cc/100 in^2-day-atm for specialty salad mixes such as baby greens and exotic lettuces. Perforation may be used to obtain high transmission rates for some applications.

17.7.2.12 Retortable Pouches

A growing flexible packaging use is in the replacement of metal cans with retortable pouches. These pouches are typically laminations containing biaxially oriented nylon for toughness, foil for oxygen barrier, and a PP sealant film. These pouches may contain items like tuna, pet food, and soup. The food items are held at elevated temperature after packaging, so the packages must remain intact at elevated temperatures. In addition to temperature resistance, toughness, seal strength, and barrier properties are critically important.

Table 17.19 Frozen Food Packaging

Structure	Layers (%)	Gauge (μm)
(EVA–LLDPE–EVA)	(15/70/15)	40–80
PET//Tie//(LDPE–ION)	(15/5/40/40)	40–80
(m-LLDPE–LLDPE–m-LLDPE)	(15/70/15)	40–80
(HDPE–MDPE–EVA)	(15/70/15)	40–80

Table 17.20 Edible Oil Packaging

Structure	Layers (%)	Gauge (μm)
(PA–Tie–EAA)	(10/15/75)	50–150
(EAA–Tie–PA–Tie–EAA)	(35/10/10/10/35)	50–150
(EAA–PA–EAA)	(40/20/40)	50–150

Table 17.21 Bag-in-Box

Structure	Layers (%)	Gauge (μm)
(LLDPE–Tie–PA–Tie–LLDPE)	(35/10/10/10/35)	40–80
(EVA–Tie–PA–Tie–EVA)	(35/10/10/10/35)	40–80
(LLDPE–Tie–EVOH–Tie–LLDPE)	(35/10/10/10/35)	40–80

17.7.2.13 Edible Oil Packaging

The packaging of cooking oil uses nylon to provide oxygen barrier properties. EAA is typically used as the sealant layer. The seal type determines the nylon layer location (Table 17.20).

17.7.2.14 Bag-in-Box

Coextruded films containing oxygen barrier polymers are replacing some metallizing laminates where flex crack resistance is required. LLDPE or EVA polymers are used as sealants (Table 17.21).

Figure 17.10 Rotary pallet stretch cling wrapper.

17.7.3 Industrial/Consumer Films

17.7.3.1 Stretch Wrap

Stretch film, or stretch/cling film, is used to unitize goods for transportation. A thin film is stretched, either by machine or by hand and wrapped around packages to hold the goods together. The film clings to itself and to the pallet, securing the load. In its most common form, stretch/cling film is applied to a stacked pallet using a power prestretch pallet wrapper in an automated operation (Fig. 17.10).

In this operation, the film is stretched, between 100% and 300%, by rollers turning at different speeds and is then applied to a loaded pallet which sits atop a moving turntable. Machine wrap film is typically supplied on rolls that are 20 or 30 in (51 or 76 cm) wide. Hand wrap film is supplied on smaller rolls. Stretch/cling films may be manufactured by either a cast film or blown film process. Most stretch/cling films are coextruded structures containing three to seven layers. Most cast film lines are now being installed with five-layer capability and blown film coextruded films are typically three layers. LLDPE is the primary component in most stretch films. For specialized applications, coextrusions containing minor PP, EVA, POP, m-LLDPE, EPE, or ULDPE layers may be employed (Table 17.22).

For most stretch film structures, a resin with good inherent cling is used on either one or both surface layers. A tackifier, such as poly-isobutylene, can also be added to the structure to provide the desired cling force. Stretch film is used to unitize entire or partial pallets stacked with products such as resin bags, fertilizer bags, consumer goods, and food products during distribution. Most stretch film

Table 17.22 Stretch Cling Pallet Wrap

Structure	Layers (%)	Gauge (μm)
Stretch Cling Film		
(EVA–ULDPE–LLDPE)	(10/80/10)	15–30
(ULDPE–LLDPE–ULDPE)	(10/80/10)	15–30
(ULDPE–LLDPE–m-LLDPE–LLDPE–ULDPE)	(10/25/30/25/10)	15–30
One-Side Cling Film		
(m-LLDPE–LLDPE–MDPE)	(10/80/10)	15–30
(EMA–LLDPE–PP)	(10/80/10)	15–30
(POP–LLDPE–PP)	(10/80/10)	15–30
(ULDPE–LLDPE–LLDPE–e-LLDPE–PP)	(10/20/30/30/10)	10–30

Table 17.23 Heavy Duty Bags

Structure	Layers (%)	Gauge (μm)
(LLDPE–EPE–LLDPE + LDPE)	(20/60/20)	100–200
(EPE–PP–EPE)	(20/60/20)	100–200

is removed by the retailer prior to displaying the packaged items for sale. Stretch films must have:

- good cling
- stretchability
- load retention
- puncture resistance.

17.7.3.2 Heavy-Duty Bags (Shipping Bags)

Heavy-duty shipping sacks are used to transport items such as:

- resin
- salt
- pet food
- fertilizer
- chemicals
- topsoil
- bark mulch
- compressed bales of fiberglass insulation.

When filled they weigh 40 lb (18 kg) or more. Heavy-duty shipping sacks may be supplied as preformed bags or as roll stock, which is formed into bags in a continuous VFFS operation. Special machinery is required to form heavy-duty shipping sacks on VFFS machinery in a high-speed continuous operation. These bags need moderate COF because they must easily pass through the packaging equipment, but stacked bags must not slide off each other. Bags filled with hot products, such as salt, must also withstand the filling temperatures without excessive stretching or dimpling. Film toughness and creep resistance are also important in many heavy-duty shipping sack applications. LLDPE has allowed significant gauge reduction. LDPE is used to reduce creep and improve processability. HDPE and PP are used for stiffness and higher end-use temperature resistance. EVA polymers are used for low-temperature sealability in form–fill–seal applications (Table 17.23).

17.7.3.3 Trash Bags

LLDPE introduction accelerated using coextrusion in trash (or refuse) bags (Table 17.24). Down gauging and using recycled material allows for improved economics and environmental concerns. High-molecular-weight HDPE polymers are also finding increased usage due to further down gauging opportunities. This is the largest coextruded film market segment.

Table 17.24 Trash Bag Coextrusion Structures

Structure	Layers (%)	Gauge (μm)
(LLDPE–LLDPE)	(50/50)	15–70
(LLDPE–RECYCLE–LLDPE)	(33/34/33)	15–70
(LLDPE–HMW–HDPE–LLDPE)	(10/80/10)	15–25

Table 17.25 Grocery Bags

Structure	Layers (%)	Gauge (μm)
(LLDPE–HMW–HDPE–LLDPE)	(10/80/10)	12–20

17.7.3.4 Grocery Sacks (Merchandise Bags)

High-molecular-weight HDPE coextruded with LLDPE provides improved sealability with good down gauging potential (Table 17.25). This film is typically made on high stalk HDPE blown film coextrusion lines.

17.7.3.5 High-Clarity Shrink Film (Oriented)

Oriented, high-clarity shrink film is used to protect and display high-value consumer goods. It is distinguished from regular shrink film by its superior clarity and appearance, as well as increased shrinkage properties and higher stiffness. Goods are packaged by wrapping the film loosely around the goods, sealing the film to make a completely enclosed bag and then shrinking the film in a shrink tunnel or oven. Small holes may be poked in the film before wrapping to allow air to escape while the film is shrinking. As in industrial shrink film, heat causes the polymer molecules to relax, causing the film to return to its original unoriented size and shrink tightly around the packaged goods. Since the polymer molecules in oriented shrink film are highly oriented, high levels of shrinkage may be obtained. Boxed software and stationery products are often wrapped with high-clarity shrink film. Ice cream cartons and other food products are also wrapped in high-clarity shrink film. Optical properties, seal properties, shrinkage, and holding force are key requirements for oriented shrink film. These structures are normally biaxially oriented films of LLDPE and PP (Table 17.26).

Table 17.26 High-Clarity Shrink Film

Structure	Layers (%)	Gauge (μm)
(PP–LLDPE–PP)	(25/50/25)	15–25

17.8 Summary

It is critical for the flexible packaging film producer to understand their market. Even within a given market segment, there are usually several film structures that are used. The film producer must evaluate the alternative film structures to determine the best fit for the market identified and for his production capabilities. Market trends should be analyzed to determine strategy. The market studies will dictate which coextrusion/lamination equipment design will be optimum. Economic evaluation of each market should include the cost incurred in scrap recycle (or disposal). The average production run size and the polymer changes will determine the off-spec film produced. A new consideration for today's market place may also include some thoughts into after-use disposal of the multilayer film. The design and testing of multilayer film, particularly barrier films, will require more manpower and overhead costs than typical monolayer film.

The markets identified for multilayer flexible packaging should value higher performance properties and high-value films. In addition to performance properties, some markets may have other barriers to entry such as qualification cost,

Table 17.27 Conversion Factors

Property	To Convert from:	To:	Multiply by:
Density	g/cc	lb/ft^3	62.43
Pressure	lb/in^2	MPa	0.00689
Output	lb/h	kg/h	0.45359
Output	lb/h	g/s	0.125997
Length	in	mm	25.4
Specific output	lb/h/in-circumference	kg/h/mm-circumference	0.017858
Specific output	lb/h/in-circumference	kg/h/mm-diameter	0.055245
Viscosity	poise	MPa-s	0.1
Viscosity	poise	$lb\text{-}s/in^2$	68947
Stress	lb/in^2	kPa	6.8947
Stress	lb/in^2	$dynes/cm^2$	68947
MVTR	g-mil/100 in^2/day	g-mm/m^2/day	0.394
Permeability	cc-mil/100 in^2/day/atm	cc-mm/mm^2/s/atm	4.56E-06
Permeability	cc-mil/in^2/day/atm	cc-mil/m^2/day/atm	15.5
Weight	lb	kg	0.4536
Peel or seal strength	lb/in	g/in	454
Peel or seal strength	g/in	N/15 mm	0.0059

experience in the market, or lack of business relationships. These considerations must be evaluated. The multilayer coextrusion line design requires knowledge of the specific structures and polymers to be produced. It is recommended that mutual discussions with the equipment supplier and polymer suppliers be made to insure that proper consideration is given to all critical aspects.

New applications continue to be developed for multilayer films and film structures continue to evolve as new market drivers come into play. The structures highlighted in this chapter are intended only as examples and may not represent where packaging is headed in the future. For example, sustainability has recently garnered attention in the marketplace. For packaging, this may mean a number of things, such as down gauging to reduce the carbon footprint or the use of new bio-sourced or biodegradable polymers. No matter what the market driver, the principles developed in this chapter remain true. Combining high-performance polymers and low-cost polymers will expand market opportunities. Understanding how to combine easily the properties of new polymers and knowledge of the market needs and trends will lead to the development of more coextrusion applications.

Table 17.27 gives a list of conversion factors.

References

[1] Smith DJ. Extruder design analysis for typical coextrusion resins. TAPPI coextrusion conference; 1975.

[2] Schrenk WJ, Finch CR. Coextrusion for barrier packaging, SPE RETEC; June 1981; 211.

[3] Schrenk WJ, Veazey EW. Films, multilayer, vol. 7. John Wiley Plastics Encylopedia; 1984. p. 106.

[4] Potts MW. Adhesive resins for coextruded barrier structures. TAPPI PL&C 383; 1987.

[5] Karagiannis A, Mavridis H, Hrymak AN, Vlachopoulos J. Interface determination in multilayer extrusion. TAPPPI PL&C 71; 1987.

[6] Schrenk WJ. Flat die coextrusion of film and sheet by the feedblock method. TAPPI PL&C conference 269; 1984.

[7] Schrenk WJ. Interfacial flow instability in multilayer coextrusion. Polym Eng Sci 1978; 18(8):620.

[8] Soutar AM. Field experience with five layer blown coex lines. TAPPI Coex S/C 73; 1989.

[9] Arvedson MA. Rheological considerations in coextrusion. TAPPI PL&C 513; 1984.

[10] Bode WW. Key considerations in the design of coextrusion blown film dies. TAPPI Coex S/C 49; 1986.

[11] Gates PC. Coextrusion: mechanical and practical considerations in blown film die design. TAPPI PL&C 491; 1987.

[12] Gates PC. Optimizing the coextrusion process. TAPPI Coex S/C 13; 1989.

[13] Gates PC. Operating considerations in coextruded blown film systems. TAPPI Film S/C 29; 1988.

[14] Wright WD. Blown film coextrusion-operation: practicality. TAPPI Coex S/C 29; 1983.

[15] Ossmann H. Blown film coextrusion. TAPPI PL&C 435; 1986.

[16] DeLassus PT. Transport of unusual molecules in polymer films. TAPPI PL&C 445; 1985.

[17] Finch CR. Some effects of melt rheology in coextrusion. TAPPI Coex S/C 109; 1986.

[18] Sacharuk P. How Stationary die/oscillating nip combinations optimize co-extruded film quality. Coex'88 109; 1988.

[19] Schrenk WJ. U.S. Patent 3,884,606; May 20, 1975.

[20] Schrenk WJ, Shastri RK, Ayers RE. U.S. Patent 5,094,788; March 10, 1992.

[21] Schrenk WJ, Shastri RK, Ayers RE. U.S. Patent 5,094,793; March 10, 1992.

[22] Im J, Schrenk W. Coextruded microlayer film and sheet. J Plast Film Sheeting 1988;4:104–15.

[23] Alfey T, Schenk W. U.S. Patent 3,711,176 (optical films); January 16, 1973.

[24] Pointing M, Hiltner A, Baer E. Gradient multilayer films by forced assembly. SPE ANTEC; 2009.

[25] Dooley J, Robacki J, Jenkins S, Lee P, Wrisley R. Producing microlayer blown film structures using layer multiplication and unique die technology. SPE ANTEC; 2011.

[26] Mueller C, Kerns J, Ebeling T, Nazarenko S, Hiltner A, Baer E. Microlayer coextrusion: processing and applications. Polym Process Eng 1997;97:137.

[27] Kerns J, Hsieh A, Hiltner A, Baer E. Mechanical behavior of polymer microlayers. Macromol Symp 1999;147:15–25.

[28] Bernal-Lara TE, Liu RYF, Hiltner A, Baer E. Structure and thermal stability of polyethylene nanolayers. Polymer 2005;46:3043–55.

[29] Oliver G. Performance packaging using layer multiplication. AMI barrier conference, Germany; 2009.

[30] Schirmer HG, Jester R, Medlock GD, Schell T. Nano-layer effects in blown barrier films. SPE international polyolefins conference, Houston, TX; 2010.

[31] Wang H, Keum JK, Hiltner A, Baer E. Confined crystallization of PEO in nanolayered films impacting structure and oxygen permeability. Macromolecules 2000;42:7055–66.

[32] Schirmer HG. U.S. Patent 5,762, 971; June 9, 1998.

[33] Schirmer HG. U.S. Patent 6,413,595; July 2, 2002.

[34] Zumbrunnen ML, Zumbrunnen DA. A new blown film die for controllably forming and extruding micro-layers, polymer blends and composites. SPE ANTEC; 2009.

[35] Djordjevic D. Tailoring films by the coextrusion casting coating. TAPPI Film S/C 1; 1988.

[36] Djordjevic D. Coextrusion—problems and solutions. TAPPI Coex S/C 39; 1989.

[37] Veazey EW. Potential for linear low density polyethylenes in coextrusion film applications. TAPPI Coex S/C 87; 1983.

[38] Daniels CA. Unique performance from specialty resins. TAPPI Film S/C 55; 1985.

[39] Hessenbruch R. Layer combinations, applications, and line concepts for blown barrier films. Specialty plastics conference '88 165; 1988.

[40] Morris BA. The effect of coextrusion on bubble kinematics, temperature distribution and property development in the blown film process. SPE ANTEC; 1998.

[41] Morris BA, Vansant JD. The influence of sealant modulus on the bending stiffness of multilayer films. 1997 TAPPI polymers, laminations and coatings conference; 1997.

[42] Plastics Design Library Staff Permeability and other film properties of plastics and elastomers. William Andrew Publishing/Plastics Design Library; 1995.

[43] Cabedo L, Feijoo J, Lagaron J, Saura J, Gimenez E. Development and characterization

of thermoplastic biodegradable nanocomposites. SPE ANTEC Technical Papers 1300; 2005.

[44] Auras RA, Harte B, Selke S, Hernandez R. Mechanical, physical, and barrier properties of poly(lactide) films. J Plast Film Sheeting 2003;19(2):123–35.

[45] Halle RW. Polymer and processing parameters influencing the heat sealability of polyethylenes. TAPPI polymers, laminations and coatings conference 799; 1989.

[46] Stehling FC, Meka P. Heat sealing of semicrystalline polymer films II. Effect of melting distribution of heat sealing behavior of polyolefins. J Appl Poly Sci 1994;51:105–9.

[47] Qureshi NZ, Stepanov EV, Capaccio G, Hitner A, Baer E. Self-adhesion of polyethylene in the melt. I. Heterogeneous copolymers. Macromolecules 2001;34:1358–64.

[48] Morris BA. Predicting the performance of ionomer films in heat-seal process. TAPPI PLACE conference; 2002.

[49] Van Ness RT. Copolymers for coextrusion, p 241. TAPPI Press Coextrusion Coating and Film Fabrication; 1983.

[50] De Garavilla J. Ionomer acid copolymer, and metallocine polyethylene resins: a comparative assessment of sealant performance. TAPPI J 1995;78(6):191–203.

[51] Qureshi NZ, Rogunova M, Stepanov EV, Capaccio G, Hiltner A, Baer E. Self-adhesion of polyethylene in the melt. 2. Comparison of heterogeneous and homogeneous copolymers. Macromolecules 2001;34(9):3007–17.

[52] Morris BA, Chen JC. The stiffness of ionomers: how it is achieved and its importance to flexible packaging applications. SPE 2003 ANTEC conference, Nashville, TN; May 2003.

[53] Morris BA. Factors influencing adhesion in coextruded structures. TAPPI J 1992;75 (8):107–10.

[54] Guillotte JE, Wright Jr KY. Coextrudable adhesive resins expand coextrusion capabilities. In: Coextrusion coating and film fabrication. Atlanta, GA: TAPPI Press; 1983. p. 231.

[55] Morris BA. Understanding why adhesion in extrusion coating decreases with diminishing coating thickness. J Plast Film Sheeting 2008;24(1):53–85.

[56] Wu S. Polymer interface and adhesion. New York: Marcel Dekker, Inc; 1982.

[57] Morris BA. The effect of processing variables on peel strength in coextruded blown film. SPE ANTEC; 1996.

[58] Morris BA. Peel strength issues in the blown film coextrusion process. TAPPI polymers, laminations and coatings conference; 1996.

[59] Morris BA. Effect of time, temperature and draw down on interlayer peel strength during coextrusion coating, film casting and film blowing. TAPPI PLACE conference, Portsmouth, VA; September 2008.

[60] Morris BA. Correlation of interlayer peel strength to post die processing time in coextrusion blown and cast film. SPE coextrusion conference, Cincinnati, OH; October 2008.

[61] Pieper G. Interactive phenomenon in multilayer flexible packaging. In: Ackermann P, Jagerstad M, Ohlsson T, editors. Foods and packaging materials—chemical interactions. Lund, Sweden: The Royal Society of Chemistry; 1994.

[62] Olasfsson G, Hiildingsson I. Interactions between carboxylic acids and laminated food packaging material. In: Ackermann P, Jagerstad M, Ohlsson T, editors. Foods and packaging materials—chemical interactions. Lund, Sweden: The Royal Society of Chemistry; 1994.

[63] Olafsson G, Jagerstad M, Oste R, Wesslen B. Effects of different organic acids on the adhesion between polyethylene film and aluminum foil. Food Chem 1993;47:227–33.

[64] Olafsson G, Jagerstad M, Oste R, Wesslen B. Delamination of polyethylene and aluminum foil layers of laminated packaging material by acetic acid. J Food Sci 1993;58(1):215.

[65] Olafsson G, Hildingsson I, Bergenstahl B. Transport of oleic and acetic acids from emulsions into low density polyethylene; effects on adhesion with aluminum foil in laminated packaging. J Food Sci 1995;60(2):420.

[66] Schroeder MA, Harte B, Giacin J, Hernandez R. Effect of flavor sorption on adhesive and cohesive bond strengths in laminations. J Plast Film Sheeting 1990;6:232–46.

[67] Keymark Associates North American Flexible Packaging Strategies 2002. Markets, materials & machines, a comprehensive multiple client study; August 20, 1999.

18 Multilayer-Oriented Films

Jürgen Breil
Brückner Maschinenbau GmbH & Co. KG, Siegsdorf, Germany

18.1 Introduction

Chapter 12, "Oriented Film Technology" discussed orientation technology and showed examples of the most widespread film types (biaxially oriented polypropylene (BOPP), biaxially oriented polyethylene terephthalate (BOPET), biaxially oriented polyamide (BOPA), biaxially oriented polystyrene (BOPS), and others). Emphasis was given in particular to the enhancement of the mechanical, optical, and permeation characteristics by means of selective orientation of molecule chains. While monolayer biaxial stretching is most suitable for certain applications, many applications utilize multilayer film structures. The reason is that, as a rule, diverse-specific film requirements are necessary and a single layer film cannot fulfill those requirements. They can only be met with the different combinations possible in a multilayer structure. In a multilayer film, the benefits of different resins are applied in order to attain the required mechanical properties as well as the sliding, seal, and barrier properties. Fig. 18.1 shows a five-layer film structure. Besides meeting the technical requirements, multilayer films provide economic benefits as either recycled or lower priced raw materials are incorporated in the main layers.

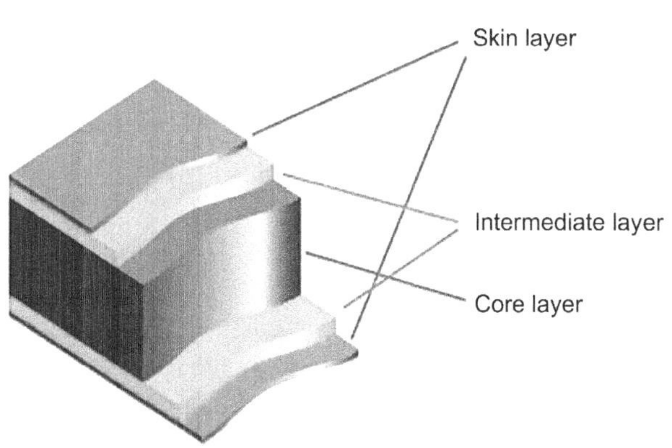

Figure 18.1 A five-layer film structure.

The surface layers essentially serve to define:
- the sliding characteristics—friction value and surface roughness
- sealing properties
- treatment or surface energy for printing and coating adhesion strength
- hot tack
- antistatic activity
- metallizing properties
- optical properties (haze and gloss)
- whiteness
- stiffness
- barrier properties

The application of high barrier layers (eg, ethylene vinyl alcohol (EVOH)) usually requires a tie layer/bonding agent to the base layers, thus resulting in five- or seven-layer structures.

In principle, multilayer configurations can be generated in-line or off-line. In-line processes shall be further explained in this chapter; off-line processes are required as converting steps and therefore shall not be elaborated here.

18.2 Technology for Multilayer-Oriented Films

The in-line process involves three different methods:
- in-line coating
- in-line lamination
- coextrusion

Thereby, the in-line coating-process involves either liquid coating or extrusion coating. During the latter process, an extruder has an extrusion coating device located either in front of the machine direction orienter or in front of the transverse direction orienter so that a melt film is applied to the already prior hardened/reinforced prefilm/base film (Fig. 18.2). However, in view of the fact that this process is rather complex, it is

Multilayer Flexible Packaging. DOI: http://dx.doi.org/10.1016/B978-0-323-37100-1.00018-1

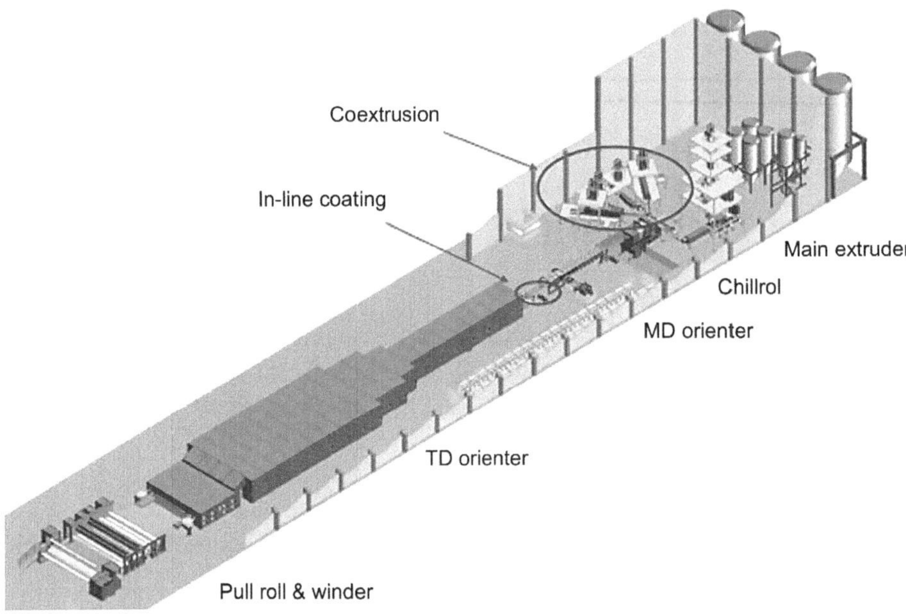

Figure 18.2 Multilayer structures in biaxially orienting lines.

very rarely used. The same applies for in-line lamination, where a second film is laminated prior to stretching. The in-line coating process with liquid coating is, however, widespread. Mainly aqueous solutions are applied with gravure or spray coating on the machine direction–oriented film before the transverse direction orienter; thus the required functional layers are generated on the surface. BOPP and BOPET films can be coated with silicone coatings as release layers. Furthermore, for BOPET films, this process can provide a thin chemical treatment which improves the adhesion properties of printing inks, glue laminates, and metallizing layers.

The coextrusion process represents by far the most widespread method for producing multilayer-oriented films, as this combines all the benefits in terms of the largest flexibility for layer combination and maximum cost-effectiveness/operating efficiency. Fig. 18.3 shows a typical extruder configuration for a five-layer BOPP film production line. The main extruder has a twin-screw additive feeder plus a melt pump and melt filter. This particular configuration has two single-screw and two twin-screw coextruders. The setup can also utilize melt pumps to provide accurate flow of the thin coextruded layers. Twin-screw extruders ensure that the additives, recycling material, and edge trim are well homogenized. Moreover, the specific energy consumption is reduced, a better degasification is achieved and, due to the self-purification, a shorter product switchover time is ensured. In the light of these advantages, over the last 10 years, many film

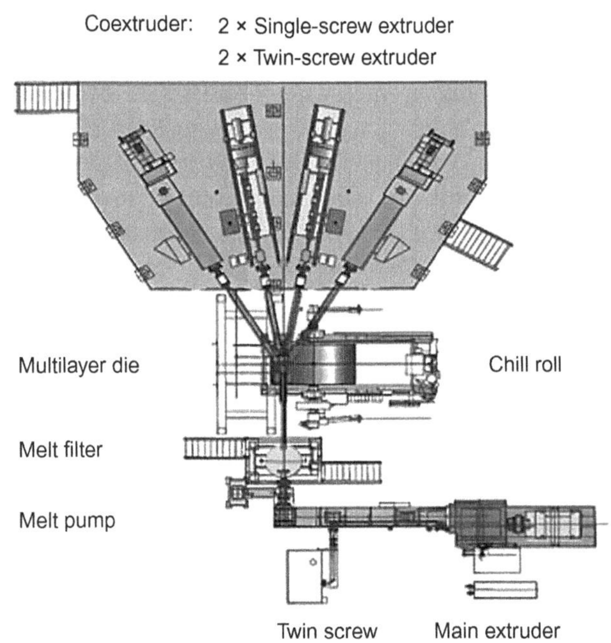

Figure 18.3 A five-layer extruder configuration.

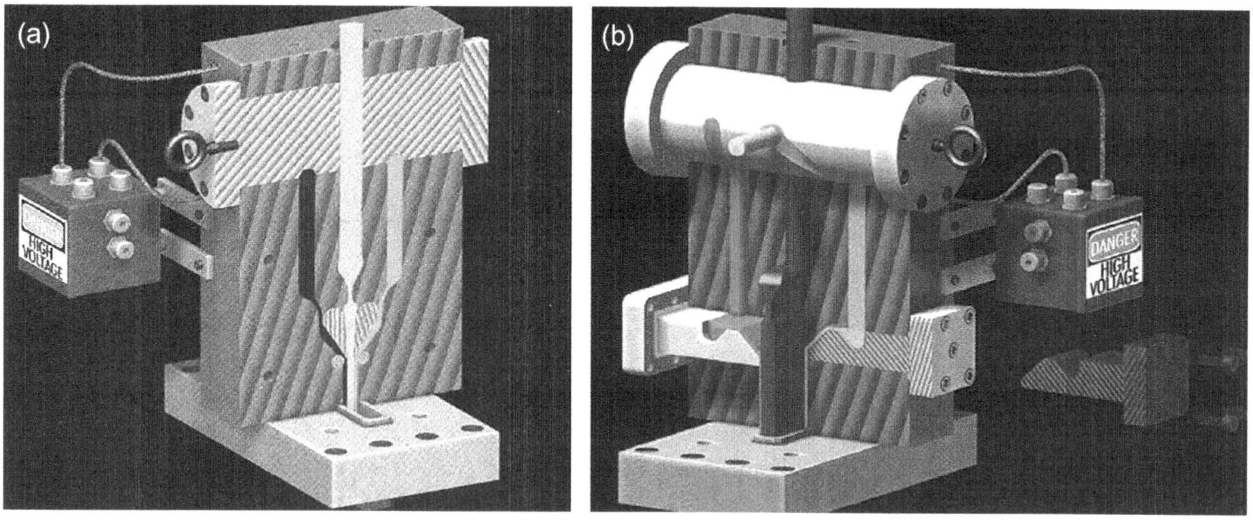

Figure 18.4 (a) Combining plug with adjustable vane adapter; (b) combining plug with laminar adapter.

stretching lines have twin-screw extruders. For BOPET lines, twin-screw extruders, both as main and coextruder, remove the need for complex resin drying.

The melt flow from the main and coextruders is merged in multichannel dies or laminar flow adapters. Focus is placed on the adapter system flexibility. The concern is how many single-layer combinations there are in the adapter system. Fig. 18.4a and b shows the flow pattern for a combining plug with an adjustable vane adaptor and for a combining plug with a laminar flow adaptor.

Multichannel dies represent the most common alternative as one can attain a precisely defined layer distribution over the working width. Single melt flows are merged via coat hanger distribution channels. These channels are specially designed for the resin viscosity and the desired layer width in the extrusion die. Fig. 18.5 shows a cross section of a seven-layer multichannel die in a ABCDCBA configuration. The design criteria are for the film edge trim areas not to have coextrusion layers. This provides an advantage for in-line edge trim recycling, as is commonly practiced with BOPP.

Each polymer that can be oriented has different process parameters. The machine and transverse direction stretch ratios and temperatures have to be carefully selected and adjusted so that the utilizable process windows do not overlap.

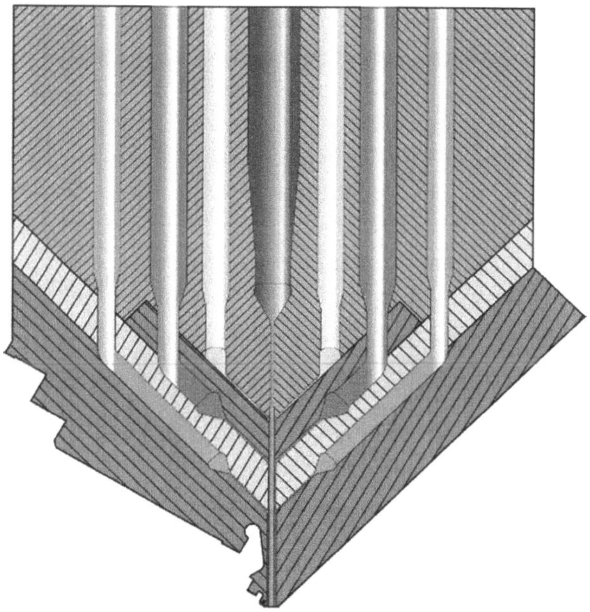

Figure 18.5 Seven-layer coextrusion die.

18.3 Structures

Just as for nonoriented film types, such as cast film and blown film, coextrusion technology is also applied for oriented films, thus yielding a multitude of structures up to seven layers and therefore enhancing the product properties.

Some BOPP film examples and the main applications will be presented along with the beneficial features of multilayer technology.

In the 1970s, three-layer coextrusion technology was adopted for BOPP films to produce heat sealable film for the packaging industry. Polypropylene copolymers and PP terpolymers were applied to the outer layers with thicknesses ranging between 0.5 and 3 μm. These lower melting polymers provided lower heat sealing temperatures in a range that allowed the main PP homopolymer structure to remain geometrically stable. When these heat sealing polymers were developed, it was necessary not only to consider the seal ignition temperature but also to provide adequate hot tack properties, in order to meet the demands made on high-speed packaging lines. Normally, antiblock and slip additives are applied to the outer layers, in order to attain the required sliding properties. Furthermore, special masterbatches are available, which provide either glossy or matte surfaces. In order to attain different skin layers, the line configuration is designed to have one main extruder and two coextruders.

Five-layer film structures allow for more flexibility by implementing intermediate layers between the main layers and skin layers (Fig. 18.6). The following benefits can be obtained:

- use special masterbatches for the main intermediate and skin layers to change optical properties
  - a high opacity film, for example, white opaque films
  - provide high gloss or matte surfaces
- more cost-effective structures without impairing the optical film quality by using higher recycled material and reducing the additives
- increase the seal properties by adding low sealing copolymer onto a comparatively thick PP copolymer layer as an intermediate layer

Fig. 18.7 shows the structure, advantages, applications, and seal strength versus sealing jaw temperature for a five-layer low seal initiation temperature (SIT) BOPP film. A very high seal strength and sealed seam rigidity is achieved with this structure. This characteristic is vitally important for many packaging applications, since both the film and the sealed seam quality provide the necessary product protection. Five-layer technology provides the design flexibility to produce films with excellent properties for various applications (Fig. 18.8).

Fig. 18.9 details a special film produced with a five-layer structure, an ultra-high barrier (UHB) metallized film. In this particular product design, the first surface is a polymer with a high surface energy (PA, EVOH, and others). This polymer requires a tie or an adhesive layer to bond to the PP main layer. The fourth D layer is also PP and is covered with a heat sealable copolymer skin. The metallized properties of this structure, compared to PP homopolymer or PP copolymer surfaces, show a much better adhesion strength and metallized surface uniformity. This results in 50 × better oxygen barrier values than obtained by standard metallized BOPP films.

The UHB-metallized film example shows that when suitable barrier materials are combined, a considerable increase in the barrier properties is possible. Metallized film also provides a light barrier which means that the product inside the package is not visible. There is a demand for transparent barrier films which is being satisfied by clear barrier coatings, as described in chapter "Water- and Solvent-Based Coating Technology," and seven-layer structures.

The seven-layer technology is particularly suitable for such films, since it is possible to attain

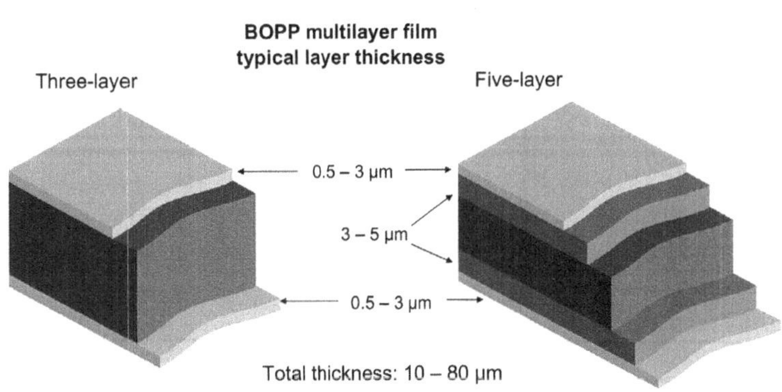

Figure 18.6 Typical layer thickness of a BOPP multilayer film.

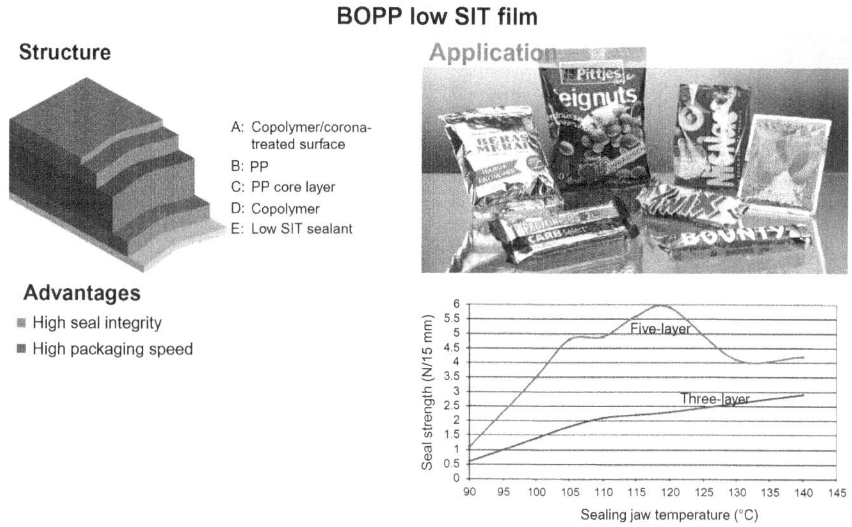

Figure 18.7 Five-layer low seal initiation temperature (*SIT*) BOPP film structure advantages, applications, and seal strength.

Film type category	Thickness (μm)	Examples for end-use application	
Wrap-around labels	35 – 50	White-voided film, both side high gloss, one side treated	
Wrap-around labels	35	White-voided metallized film, high gloss surfaces, very high yield	
Food packaging	35	White-voided metallized film, heat sealable, high protection against light	
Food packaging	30 – 50	White-voided film, both sides heat sealable, high protection against light	
Business cards maps, bags	40 – 80	Synthetic paper	
Paper lamination	15 – 40	Matte film	

Figure 18.8 Five-layer film applications.

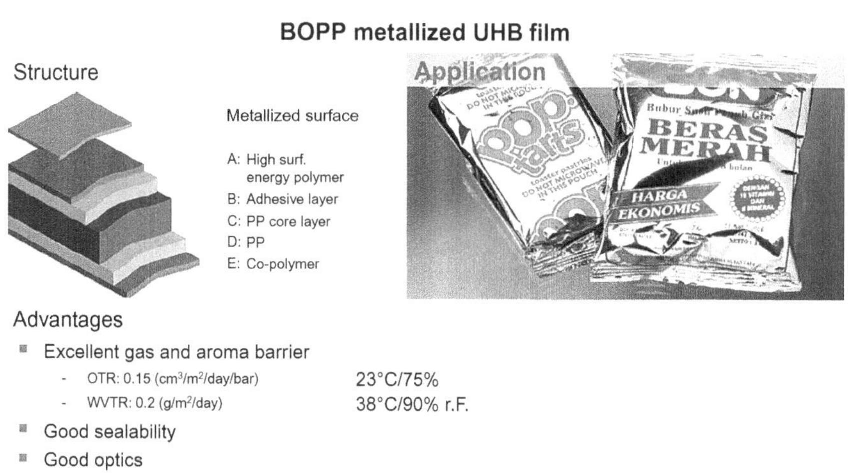

Figure 18.9 Five layer–metallized UHB BOPP film structure and advantages.

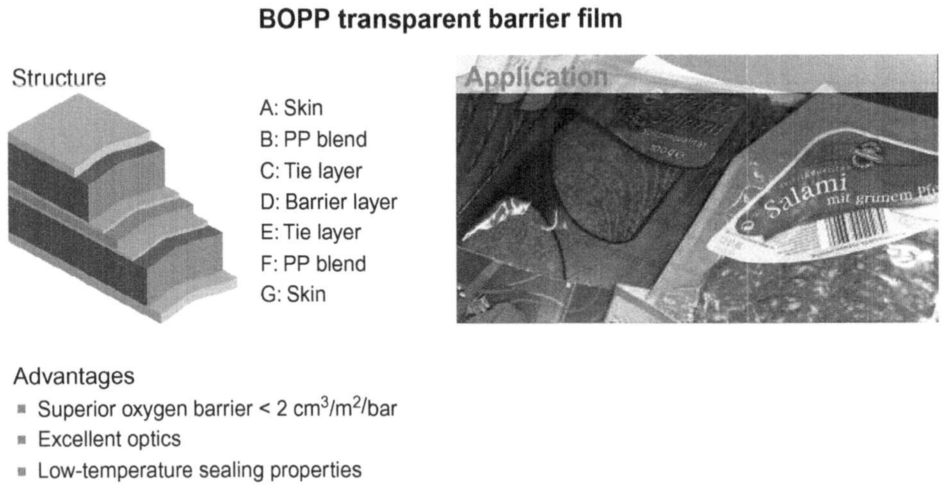

Figure 18.10 A typical seven-layer configuration.

the required barrier values, even without metallizing. In this case, preferentially EVOH as barrier material is applied in the middle of the main layer imbedded between tie layers. Furthermore, such a symmetrical film layer configuration has the advantage that the occurrence of curling becomes less likely. A typical seven-layer configuration is shown in Fig. 18.10. As a rule, such seven-layer structures with the application of various polymers—such as, for example, PP and EVOH—can be manufactured not only with sequential but also with simultaneous orientation technology. However, upon selection of the types of raw material to be used, one has to bear in mind the divergence of the various process requirements for the particular stretching process. The types of available EVOH resin have a distinct correlation: on the one hand, the ethylene content with the stretchability, and, on the other hand, with the barrier values. Generally, it can be said that a higher ethylene proportion implicates a less complicated orientation process, permits higher stretching ratios and, furthermore, the implementation of the sequential orientation process. For the simultaneous orientation process, however, it is possible to stretch all types of EVOH with an ethylene content of 24–47%. Although the barrier properties of the EVOH types are increased, in view of the orientation process, the basic coherence remains that EVOH with a high ethylene content has considerably poorer barrier properties. This is illustrated in Fig. 18.11 showing an evaluation on the oxygen barrier with the use of various EVOH types and orientation with an area-stretching ratio of almost 50.

It can therefore be concluded that the use of EVOH types with an ethylene content of not more than 33% is particularly beneficial, since a good oxygen barrier of less than 2 cm^3/m^2/day/bar can be attained with thin layers of 2 μm thickness. This value can easily compete with other high barrier film types, which are off-line coated.

The examples of multilayer BOPP films demonstrate the wide variability of structures by implementing the coextrusion technology. This applies in general also to the other film types such as BOPET and BOPA.

There is a strong trend to enhance the barrier properties further with a minimum of packaging material, so there is no doubt that coextruded oriented films will continue with stable growth rates.

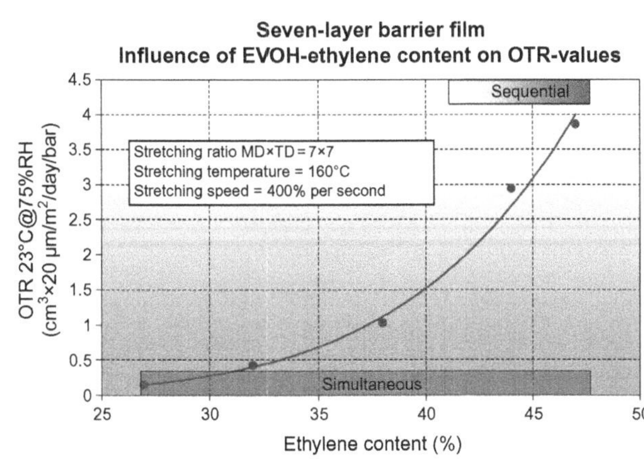

Figure 18.11 Evaluation of the oxygen barrier with the use of various EVOH types.

18.4 Trends for Multilayer-Oriented Films

In 2008 the worldwide production capacity of biaxially stretched films was over 12 million tons. Approximately 66% was BOPP and 25% was BOPET.

As the packaging sector is the majority application for these films, one can analyze flexible film trends within the packaging industry and draw conclusions for future Research and Development challenges.

These trends can be split into the following categories:

- cost efficiency
- consumer requirement
- sustainability

18.4.1 Cost Efficiency

Cost efficiency consists of:

- cost per package
- packaging speed

The prevailing global cost pressure necessitates that packaging experts not only focus their attention on utmost efficiency but also combine minimum material consumption with maximum packaging functionality.

In many cases, biaxially oriented films, especially multilayer films, provide solutions such as downgaging to these challenges.

Also, high mechanical properties and excellent friction characteristics are required to attain high packaging speeds.

Many final packaging films are multilayer laminates produced by combining two or more films. This lamination step strongly impacts the cost. Consequently, intense efforts are being made to combine as many property requirements as possible in one coextruded oriented film. Such endeavors are underway, in particular, combining mechanical properties with barrier, sealing, and surface properties.

This trend will continue and the outcome will be that film stretching lines will become increasingly more complex and diverse. This will reduce, to a certain extent, the cost-intensive subsequent film converting.

18.4.2 Consumer Requirements

Consumer requirements consist of:

- convenience
- product identification
- shelf-life

Wrapping material has to satisfy the consumers' basic needs inasmuch as they should be easy to open and if possible, reclosable. Consequently, there is a demand for multilayer composites with low tear resistance and reclosability. Consumers' requirements to see the package content and have long shelf-life necessitate transparent barrier films. These film characteristics are ideally attained by coextruded oriented films with corresponding barrier layers. This applies in particular to transparent lidding film for products such as sausage and cheese in deep drawn trays that ensure the required shelf-life.

18.4.3 Sustainability

Sustainability consists of:

- energy consumption
- raw material consumption
- renewable sourced materials
- CO_2 footprints

The sustainability conversation has in recent years increasingly influenced the packaging industry and has been driven by large global retailers. In the course of such discussions, all aspects of sustainability have been analyzed and future consequences will increasingly be that energy consumption will be displayed and comparisons drawn. This in turn will result in that complex processes that comprise various production steps will be replaced by single step production processes that provide many end properties, such as coextruded orientation film. This will also support the trend toward replacing aluminum foil, since producing aluminum foil—containing structures is more energy intensive compared to alternative barrier plastic films.

A further trend is the increased renewable sourced raw materials use, such as PLA. Development entails that the property characteristics necessary for wrappings can also be attained with this raw material. In many cases, this is only possible by coextrusion and orientation.

It is to be expected that sustainability will gain importance, as relevant legal requirements are passed, be they on marking or on using certain material.

All these trends will definitely have a positive influence on the extended use of coextruded oriented films in future.

19 Regulatory Aspects of Food Packaging—A Global Matter

Sina Ebnesajjad
FluoroConsultants Group, LLC, Chadds Ford, PA, United States

Globalization and industrialization have led to the consumption of packaged foods around the world. Today, food and packaging are synonymous because of distribution to an ever-growing global population. A variety of plastics in single-layer and multilayer forms are used in the construction of packaging of perishable and nonperishable foods. Applications of plastics in contact with food, in the developed regions, are regulated through a body of laws that are constantly updated. The Food and Drug Administration (FDA) issued the first guidance in the United States in 1949.

Recognition of the impact of packaging material, in particular plastics, on the safety of food stuffs has also led to the promulgation of regulations in the developing countries. These regulatory efforts are at different stages of maturation depending on the country and the region. For example, ASEAN countries have begun an effort to harmonize standards for food-contact material [1]. Australia and New Zealand [2] have established joint packaging regulations.

This chapter presents the readers with an abbreviated discussion of aspects of regulations governing food packaging. A single chapter cannot comprehensively address the regulatory issues relevant to food packaging. Contact with regulatory agencies and study of the complete regulations and most effectively legal advice [3,4] is recommended for anyone who wishes to ascertain the regulatory status of packaging for commercial purposes.

A survey of regulatory issues in the United States and European Union (EU) has been presented in this chapter. To ensure accuracy, in many cases, the exact language and copy published by regulatory agencies have been republished. Otherwise descriptions provided by specialists have been utilized in the presentation. The focus of this chapter is on the materials aspects of packaging without any attention paid to other aspects of foodstuffs such as dietary and nutrition, health, and impurities.

19.1 Introduction

The first FDA advisory about evaluating chemical toxicity in food was issued in 1949. This event followed other milestones in the agency's successful enactments such as improving infant formula content, eliminating botulism in canned foods, and labeling products with possible food allergens. Since then laws have been promulgated to regulate plastic and additives of food packaging and labeling materials; limits in irradiation of packages for terminal sterilization; and functionality of packaging materials.

Extensive food-contact regulations that have been put in place by the US FDA are critical for ensuring the safety of food that is packaged and consumed. As a result, compliance is a key consideration for companies converting and utilizing simple formats or complex multilayer structures for food packaging. Choosing a supplier that follows and uses strong regulatory expertise and strict compliance protocols will help converters and brand owners make the best decisions about selecting and handling materials used in this demanding packaging area [5].

These regulations have been critical for ensuring the safety of the food consumed by people and also to those companies which produce basic plastics and additives, convert the plastics into films, multilayer film structures and containers, package food and market, and sell food packaging. Every player in the value chain has to be aware of the risks of actual or suspected contamination attributed to packaging materials, such as occurrence is certain to attract public censure and potentially further regulatory activity. Examples of public concerns come to light

Multilayer Flexible Packaging. DOI: http://dx.doi.org/10.1016/B978-0-323-37100-1.00019-3

from time to time. One such cautionary episode is the fate of bisphenol A (BPA).

19.1.1 Bisphenol A

The predicament of BPA is instructive in elucidating the impact of actual toxicological data and publicity on governmental action. New studies are conducted on old plastics and additives the results of which can affect their regulatory status. New BPA studies turned up data indicating a review of the incumbent regulatory status of this chemical. These studies received widespread publicity in the media and the Internet which placed immense pressure on governmental agencies and the industry to take immediate action.

BPA is a chemical (or monomer) for the production of polycarbonate plastic and epoxy resins. This monomer has been used safely in manufacturing these plastics for over 40 years and has played an important role in developing numerous consumer products. Examples of BPA applications include can liners, compact disks, impact resistant eyeglass lenses, dental fillings, and food and beverage containers. Polycarbonate has other applications such as light globes, household appliances, components of electrical/electronic devices, automotive parts, telephones, safety and sports helmets, signs, windows, roofing structures, greenhouses, solar and construction glazing, skylights, and others.

Polycarbonate and BPA have undergone extensive safety testing because of its use in food packaging and medical devices. US Environmental Protection Agency (EPA) established an oral reference dose (RfD) of 50 μg/kg/day, or 50 parts per billion (ppb) per day in 1998. This RfD is an "estimate of daily exposure to the human population (including sensitive subgroups) that is likely to be without an appreciable risk of deleterious effects during a lifetime." This RfD is much higher than the actual exposure levels to BPA. The amount of BPA that migrates to food packaged in plastic containers or cans is also extremely low and is much less than this RfD [6].

In August 2008, FDA released a draft report finding that BPA remains safe in food-contact materials. In October 2008, a subcommittee of FDA's science board raised questions about whether FDA's review had adequately considered the most recent scientific information available [7,8]. For example, an article in the September 2008 issue of the *Journal of American Medical Association* (JAMA) reported detrimental effects of BPA on adults [7]. Other studies had reported subtle effects of low BPA doses in laboratory animals [9]. These newer studies have led federal health officials to express some concern about BPA's safety. On January 15, 2010, the FDA issued an interim update on BPA [10] complemented by a public information advisory by the US Department of Health and Human Services. The information indicated the presence of minute amounts of BPA in canned formula, which could increase when heated. It also virtually advised avoidance of any baby bottles or feeding cups made with BPA-containing plastics. There has been some additional recent activity by FDA on BPA. FDA updated its opinion on BPA in March 2012 and also, at the same time, rejected a petition filed by the Natural Resources Defense Council to ban BPA [11]. FDA finalized its BPA ruling in June 2014 indicating BPA can no longer be used in baby bottles, sippy cups, and infant formula packaging. The readers are encouraged to visit the FDA web site www.fda.gov for new rulings and updates on previous ones.

The European Food Safety Authority's (EFSA) Panel on food additives, flavorings, processing aids, and materials in contact with food carried out a comprehensive risk assessment on BPA in 2006 and set a full tolerable daily intake (TDI) of 0.05 mg/kg body weight/day. EFSA received a request in September 2008 from the European Commission (EC) requesting an assessment of the implications to the hazard and risk assessment of BPA based on the JAMA study correlating urinary BPA concentrations to medical disorders in adults [7].

EFSA [12] concluded that this single study [7] does not provide sufficient proof for a causal link between exposure to BPA and the health conditions mentioned in the study, that is, heart disease, diabetes, and elevated liver enzyme activities. EFSA concluded there was no need to revise the TDI as derived by the Panel in 2006.

Shortly after the *JAMA* article, since the beginning of 2009, all major US suppliers of baby bottles and infant feeding cups stop manufacturing products that use BPA for the US market. On May 31, 2011, the EU issued the following statement indicating a ban on BPA: "Baby bottles containing the substance Bisphenol A (BPA) have to be removed from the shelves in stores across the European Union tomorrow, as a ban on the placing on the market and import into the EU of such products enters into force. The ban is foreseen in an EU directive (2011/8/EU) adopted in late January. The industry has been withdrawing voluntarily from the market baby bottles containing BPA. On March 1, the EU had banned the manufacture in the Union of baby bottles containing BPA" [13].

There has been additional activity on BPA in the EU. EFSA made additional statements regarding its investigation of BPA: "In January 2015, EFSA published its latest comprehensive re-evaluation of BPA exposure and toxicity. EFSA's experts concluded that BPA poses no health risk to consumers of any age group (including unborn children, infants and adolescents) at current exposure levels. Although new data and refined methodologies led EFSA's experts to considerably reduce the safe level of BPA from 50 micrograms per kilogram of body weight (bw) per day (μg/kg of bw/day) to 4 μg/kg of bw/day, the highest estimates for dietary exposure and for exposure from a combination of sources (diet, dust, cosmetics and thermal paper) are three to five times lower than the new TDI.

Uncertainties surrounding potential health effects of BPA on the mammary gland, reproductive, metabolic, neurobehavioural and immune systems have been quantified and factored in to the calculation of the TDI. In addition, the TDI is temporary pending the outcome of a long-term study in rats, which will help to reduce these uncertainties" [14].

19.2 Determining the Regulatory Status of Components of a Food-Contact Material in the United States

The overall regulatory status of a food-contact material is dictated by the regulatory status of each individual substance that comprises the article. The individual substance that is reasonably expected to migrate to food because of its intended use in the food-contact material shall be covered by one of the following:

- a regulation listed in Title 21 Code of Federal Regulations
- a prior sanction letter
- meeting the criteria for generally recognized as safe (GRAS) status (including but not limited to a GRAS regulation or GRAS notice)
- a threshold of regulation (TOR) exemption
- or an effective food contact notification (FCN).

Be aware that Section 409(h)(2)(C) of the Federal Food, Drug, and Cosmetic Act (the Act) states that an FCN is effective for the manufacturer, the food contact substance (FCS), and the conditions of use identified in the notification and not effective for a similar or identical substance produced or prepared by a manufacturer other than the manufacturer identified in the notification. FCNs are proprietary to the manufacturer for which the notification is effective; therefore, the FCS must be obtained from that manufacturer.

It is the responsibility of the manufacturer of a food-contact material to ensure that its components comply with the specifications and limitations in all applicable authorizations. When reviewing your composite formulations to determine compliance, consider each authorization to be composed of three parts: the identity of the substance, specifications including purity or physical properties, and limitations on the use.

You may ask the manufacturer for a letter of assurance certifying that a particular product is acceptable for the intended food-contact use. A key question is:

> How do you determine the compliance of the components of your food-contact article with the requirements of the act?

There are a number of ways that a component of your food-contact article may comply with the act. The following list is designed to help the reader make that determination. It allows one to research each component for an appropriate authorization. Here are some possible approaches:

- Consult 21 CFR 174–179 to see if the component is appropriately regulated as an indirect additive for the intended use.
- Consult 21 CFR 182–186 to see if the use of the component is GRAS or consult the list of GRAS notices.
- Consult 21 CFR 181 to see if the use of the component is prior sanctioned
- Consult the listing of TOR exemptions.
- Consult the listing of effective FCS notifications.

Another key question:
What if one or more components of your food-contact article are not covered by any of the categories above, what options do you have?

Submit a TOR exemption request or:
- Submit an FCS notification; or
- Satisfy the criteria for GRAS status.

To find out whether the component is appropriately regulated as an indirect additive for its intended use, consult 21 CFR 174–179 according to the following:
- Adhesives and Components of Coatings (21 CFR Part 175)
- Paper and Paperboard Components (21 CFR Part 176)
- Polymers (21 CFR Part 177)
- Adjuvants, Production Aids, and Sanitizers (21 CFR Part 178)
- Irradiation in the Production, Processing and Handling of Food (21 CFR Part 179)

The requirement for premarket approval in Section 409 of the Federal Food, Drug, and Cosmetic Act (FD&C Act) in 1958 resulted in developing a petition process by which a person could request approval of a food additive for an intended use. The approval resulted in a regulation listed in 21 CFR. Components of a food-packaging material used in compliance with a regulation in 21 CFR (174–179) need no further FDA review. Most regulated indirect food additives can be found in Center for Food Safety and Applied Nutrition's (CFSAN) "Indirect Additive" database [15].

To find out whether the use of the component is listed as GRAS, consult 21 CFR 182–186 according to the following:
- Substances GRAS in food (21 CFR Part 182)
- Substances affirmed as GRAS in food (21 CFR Part 184)
- Substances affirmed as GRAS for use in food packaging (21 CFR Part 186)
- Not all substances which are GRAS are listed in FDA's regulations. FDA has instituted a procedure whereby someone may inform FDA of their own GRAS determination. A list of these GRAS notices, with FDA's response letter to the notifier, is also available at "Summary of all GRAS Notices15."

19.2.1 Food-Contact Formulation Compliance Notification

Occasionally, individuals may wish to verify compliance of the components of a particular food-contact material. In such instances they may submit a notification for an FCS formulation. The purpose of a formulation notification is to verify that components of a food-contact material may legally be used and not to authorize a new FCS. Because these notifications are for the purpose of regulatory status and not safety assessments, notifications for formulations do not require resubmission of the information supporting the safety of the intended use of each FCS in the formulation. A notifier for a formulation need only submit a completed FDA Form 347920 and any additional documentation required to establish that each of the components of the formulation is authorized for its intended use. In cases where the basis for compliance of an individual FCS in a formulation is an effective notification, a notifier of the formulation should establish that he can rely on the notification cited for the intended use of the FCS in the formulation.

19.3 Regulatory Report: FDA's FCS Notification Program

Among the responsibilities of the US FDA is regulation of components of food-contact materials, including packaging. Once known as indirect food additives, FDA now refers to these materials as FCS. In an effort to ensure the safe use of these substances, FDA has established an FCN program within the CFSAN's Office of Food Additive Safety (OFAS). This article describes how to submit an FCS notification to FDA and how the FCN review process works. It also addresses the importance of participating in prenotification consultations, the role of TOR exemptions, and confidentiality concerns [16].

19.3.1 Definitions, History, and Scope

Definitions. Section 201(s) of the FD&C Act defines a "food additive" as any substance the intended use of which results or may reasonably be expected to result, directly or indirectly, in its becoming a component or otherwise affecting the characteristic of any food (including any substance intended for use in producing, manufacturing, packing, processing, preparing, treating, packaging, transporting, or holding food; and including any source of radiation intended for any such use); if such substance is not GRAS or sanctioned prior to 1958. [This is a shortened definition. The full definition, found in the FD&C Act provides for exclusions, as well.] Section 409 of the statute defines an FCS as any substance that is intended for use as a component of materials used in manufacturing, packing, packaging, transporting, or holding food if such use of the substance is not intended to have any technical effect in such food. Common types of FCSs include coatings, plastics, paper, adhesives, as well as colorants, antimicrobials, and antioxidants found in packaging.

History. In 1958, the FD&C Act was amended by adding Section 409, which requires FDA to approve new food additives before they can be used in foods. To meet this requirement FDA established a process by which an individual can petition FDA requesting approval of a food additive. All FCSs that were additives required approval. In 1997, the Food and Drug Administration Modernization Act amended Section 409 of the FD&C Act establishing an FCN process that allows for faster review of FCSs that are food additives. The amendment did not alter the safety standard applied to all food additives and notification for an FCS must contain sufficient scientific information demonstrating that the substance that is the subject of the notification is safe for the intended use [17].

FDA's fiscal year 2000 budget contained an appropriation to fund the FCN program and in October 1999 the OFAS began converting approximately 120 in-house food additive petitions (FAP) and TOR submissions for review under the new amended requirements. [TOR for substances used in food-contact articles can be found in 21 CFR 170.39.] On January 18, 2000, the Agency began accepting new FCNs, and in July 2000 FDA published proposed regulations for the program. The regulations were finalized in May 2002 and codified in 21 CFR Part 170.100–106.

Scope. FCNs are required only for new uses of FCSs that are food additives. Although a notification is not required for an FCS that is GRAS or prior sanctioned for its intended use in contact with food, some companies do choose to notify the Agency in order to clarify the regulatory status of such substances. Manufacturers may also use FCNs to notify FDA of new uses of FCSs that are GRAS or prior sanctioned.

Unlike food additive regulations and TOR exemptions, approvals under the FCN process are proprietary. This is because Section 409(h)(1)(C) of the FD&C Act states that an FCN is effective only for the manufacturer and substance identified in the notification. Thus, any person wishing to rely on an FCN needs to demonstrate that the FCS being marketed has been manufactured or supplied by the manufacturer identified in the FCN and is being used under the conditions that are the subject of the FCN. This may seem a trivial point but it is important since the manufacturer for which the FCN is effective does not need to be the actual manufacturer of the FCS but can also be a supplier who is purchasing the substance and distributing it for the intended use. If a notifier is not careful in defining the entities for which the FCN is effective, the proprietary approval received in a successful FCN may not be applied to the intended company.

19.3.2 The Notification Process

Anyone may submit an FCN to FDA. (Although the FD&C Act refers to manufacturers or suppliers submitting a notification, FDA interprets the term supplier broadly to include persons who could previously have requested authorization under the FAP process. This ensures that no one was unfairly excluded by the transition to the FCN process.) The FCN process is initiated when a submission, prepared by the notifier, arrives at the OFAS. A notification control assistant logs the submission and assigns it a notification number. The submission is then distributed to a review team for assignment. The review team consists of a consumer safety officer, a chemist, a toxicologist, and an environmental scientist. During the initial month of the review period, FDA does not ordinarily contact the notifier; however, the notifier is free to contact the consumer safety officer in charge of its notification.

A "phase one" review meeting is held within the first 3 weeks after receipt of the FCN to ensure that the basic data and informational elements are present and that the submission meets the administrative requirements set forth in the FD&C Act and FDA's regulations. If the submission is complete it is accepted and FDA sends the notifier an acknowledgment letter. The acknowledgment letter initiates formal communication between FDA and the notifier and gives the notifier an opportunity to comment on the Agency's understanding of the identity and intended use of the FCS that is the subject of the notification. Notifiers who disagree with FDA's description of the FCS should respond as quickly as possible to this acknowledgment letter. The acknowledgment letter also establishes the date of receipt of the notification, which indicates when the 120-day FCN review period began, and that the review team is entering into "phase two" review.

During "phase two" the team evaluates the safety of the FCS as it is proposed for use. If there are no concerns during "phase two" review, the FCN automatically becomes effective on the 120-day date and the consumer safety officer sends a letter to the notifier confirming the effective date. Information about the notification is then added to the Inventory of Effective Premarket Notifications for FCSs posted on CFSAN's web page [18].

If after "phase one" review the notification is deemed "not acceptable" and FDA must request additional information, the date of receipt of the additional information may be considered the date of receipt of the complete notification and the 120-day review period would begin on that date. This determination is based on the nature of the additional information received, that is, whether it is substantive or nonsubstantive in nature. Since the time for review of FCNs is short, requests for and receipt of additional information must be handled in a timely manner as one of several scenarios may play out this point. If the requested information is not received within approximately 2 weeks and the notifier does not withdraw the submission, FDA may deem the notification incomplete and a nonacceptance letter may be issued to the notifier indicating termination of FDA's review. If FDA does receive the additional information within 2 weeks but still deems the notification to be incomplete the notifier may be given the option of withdrawing the submission without prejudice before receiving a nonacceptance letter. (About one in four FCNs is withdrawn because of deficiencies.) If the information adequately completes the submission it is accepted, an acknowledgment letter is sent, and "phase two" review begins. Provided that there are no safety concerns during "phase two" review, a final letter is sent to the notifier confirming the FCN effective date and the notification is added to the Inventory of Effective Premarket Notifications for FCSs posted on the Internet.

What happens if there are safety concerns during the "phase two" review? If safety concerns emerge, the notifier is given the option of withdrawing the submission without prejudice to a future submission. If the notifier does not withdraw the notification, an objection letter is sent to the notifier.

Under 21 CFR 170.100(c) FDA may also issue a nonacceptance letter for the following:

- FCS uses that increase the cumulative estimated daily intake (EDI) of the FCS to a level equal to or greater than 3 mg per person per day for a substance that is not a biocide (eg, it is intended to exert microbial toxicity); and 0.5 mg per person per day for a biocide; and/or
- where a bioassay exists on the FCS, FDA has not reviewed the bioassay, and the bioassay is not clearly negative for carcinogenic effects.

Threshold of regulation. Some FCNs are for uses that qualify for review under FDA's "TOR" process. The TOR process is an abbreviated review process for uses of indirect food additives that result in exposures below 1.5 mg per person per day. The TOR process can result in a faster approval; however, that approval is not proprietary [19].

Confidentiality. A common concern of many notifiers is the availability of information submitted in an FCN. The FD&C Act prohibits FDA from disclosing information in an FCN during the 120-day review period. In addition, FDA has issued regulations establishing that information in FCNs withdrawn before FDA issues an objection letter or before the FCN becomes effective will not be disclosed. FDA also considers information in an FCN that is "not accepted" to be confidential commercial information protected from disclosure.

Some notifiers seek to protect information by including it in a food additive master file. Such information is protected only if it is not used as the basis of an FDA safety decision or if it is not otherwise publicly available. In these scenarios the information in the food master file is treated as if it was in the FCN. In short, information in an FCN is protected from disclosure during review and if the FCN is withdrawn or if a nonacceptance letter is issued, but is not protected from disclosure after an FCN becomes effective or if FDA objects to the FCN, unless it is trade secret or confidential information.

19.3.3 Increasing the Odds of Success

FDA encourages notifiers to participate in a prenotification consultation to facilitate the development of a complete FCN submission. Prenotification consultations are interactions between the OFAS and industry prior to submission of an FCN, an FAP, or a TOR exemption request [20]. They are particularly advisable in situations where current guidance is not completely applicable. There are three circumstances in which FDA specifically recommends a prenotification consultation. The first specific circumstance is when the notifier believes that an FCN is not sufficient and an FAP is required for the use of an FCS. Such a meeting may be used to verify that a petition is required and that an appropriate level of information is supplied in the petition. The second specific circumstance for which FDA recommends a prenotification consultation is when different interpretations of available data would result in different conclusions, which in turn could affect whether it is appropriate to submit a notification or a petition, for example, if different interpretations of data in a bioassay could change the conclusion regarding the likely carcinogenicity of the substance. The third specific circumstance for which a prenotification consultation is recommended is when there are uncertainties about how specific data may be interpreted and those uncertainties are of such magnitude that they may affect the outcome of the overall safety determination, for example, if the EDI is close enough to the acceptable daily intake that different choices for the no effect level may cause the resulting acceptable daily intake to be larger or smaller than the EDI. A prenotification consultation request may be sent via e-mail, facsimile, or letter, and should be addressed to the notification control assistant.

19.3.4 FCS Formulations

Occasionally, individuals may wish to verify compliance of the components of a particular food-contact material. In such instances they may submit a notification for an FCS formulation. Such notifications are only meant to verify that components of a food-contact material may legally be used and not to authorize a new FCS. Notifications for formulations do not require resubmission of the information supporting the safety of the intended use of each FCS in the formulation. A notifier for a formulation need only submit a completed FDA Form 3479 and any additional documentation required to establish that each of the components of the formulation is authorized for its intended use. In cases where the basis for compliance of an individual FCS in a formulation is an effective notification, a notifier of the formulation should establish that he can rely on the notification cited and that such notification is effective for the intended use in the formulation.

When checking compliance of components of a food-contact material, remember that substances that are reasonably expected to migrate to food because of their intended use in contact with food should be one of the following: (1) approved indirect food additives used in compliance with 21 CFR Parts 174–179; (2) GRAS for use in or on contact with food; (3) substances whose use in contact with food is the subject of a prior sanction issued by FDA or US department of Agriculture (USDA) before 1958 [21] used under a TOR exemption by FDA [22] since October of 1999, the subject of an effective FCN. If the substance does not fall into one of these categories, the FCN processes or the TOR processes are the preferred methods of providing for the safe use of the substance.

Individuals wishing to determine if FDA has a regulation for a specific food additive can view a list of food additives and their corresponding regulations at www.cfsan.fda.gov/~dms/opa-indt.html. The individual regulations for food additives and GRAS substances listed for use in the United States are located in Title 21 CFR Parts 182–186 and can be accessed through CFSAN's web site. However, not all substances which are GRAS are listed in FDA's regulations. FDA has instituted a procedure whereby companies may inform FDA of their own GRAS determination (GRAS notification procedure). A list of these GRAS notices, with FDA's response letter to the notifier, can be found at www.fda.gov/Food/IngredientsPackagingLabeling/GRAS/ucm083022.htm.

It is important to remember that the use of any chemicals to fabricate food-contact articles is done with the understanding that it is the responsibility of the manufacturer to ensure that any resultant food-contact articles comply with the specifications and limitations in all applicable regulations. When reviewing composite formulations to determine compliance, consider each regulation or notification to be composed of three parts:

- The identity of the substance
- Specifications, including purity or physical properties
- Limitations on the conditions of use

In order for your products to be suitable for use in contact with food in the United States, each chemical component must comply with all three criteria.

19.4 Preservation of Foods by Irradiation

Food-borne diseases are a source of concern for every person who consumes food. These are the illnesses contracted from eating contaminated foods or beverages. The illnesses include food-borne intoxications and infections and are often incorrectly referred to as food poisoning [23]. There are more than 250 different food-borne diseases caused by viruses, bacteria, parasites, toxins, metals, and prions. According to the US Center for Disease Control, 31 pathogens are known to cause 20% of the food-borne illnesses, with 80% being caused by unknown agents [24].

Concern with food safety dates back thousands of years. Many techniques have been devised over the centuries including drying, pickling, salting, sugaring, cold storage, smoking, canning, and blanching. Every one of these techniques has limitations such a preservation of freshness, applicability, scalability, and cost. The population growth and the demand for food grown in distant locations have resulted in a complex global distribution system increasing the challenge of keeping foods safe. Ionizing irradiation has been recognized to be a convenient technique for food preservation for over a century.

Present-day food processors prefer that food be prepackaged in the final packaging form before irradiation to prevent recontamination and to expedite shipment to market after irradiation. Prepackage food could potentially be contaminated by the by-products formed if the packaging materials are degraded by irradiation. This may lead to a safety concern and, therefore, testing of packaging materials after exposure to irradiation is an integral part of the premarket safety assessment of packaging materials irradiated in contact with food [25]. A partial preventive measure is to *limit* the radiation used to treat the packaged food (Table 19.1).

Table 19.1 FDA Food Radiation Specifications [25]

Food Type	Purpose	Radiation Dose
Fresh, nonheated-processed pork	Control of *Trichinella spiralis*	0.3 kGy minimum to 1 kGy maximum
Fresh food	Growth and maturation inhibition	1 kGy maximum
Foods	Arthropod disinfection	10 kGy maximum
Dry or dehydrated enzyme preparations	Microbial disinfection	1 kGy maximum
Dry or dehydrated spices/ seasonings	Microbial disinfection	30 kGy maximum
Fresh or frozen, uncooked poultry products	Pathogen Control	3 kGy Maximum
Frozen packaged meats (solely NASA)	Sterilization	44 kGy minimum
Refrigerated, uncooked meat products	Pathogen control	4.5 kGy maximum
Frozen uncooked meat products	Pathogen control	7 kGy maximum
Fresh shell eggs	Control of *Salmonella*	3.0 kGy maximum
Seeds for sprouting		8.0 kGy maximum
Fresh or frozen molluscan shellfish (per FDA 2005)	Control of *Vibrio* species and other food-borne pathogens	5.5 kGy maximum

19.4.1 FDA Regulations for Treatment of Foods with Radiation [26]

There are different reasons to irradiate foodstuff. Many foods are irradiated to kill bacteria and parasites that pose health risks to humans. This makes the food safer and is comparable to pasteurizing milk. Irradiation at lower dose levels also extends shelf life and can be used to control insects. Food is also irradiated to destroy bacteria, fungi, or parasites that cause human disease or cause food to spoil. Irradiation kills harmful bacteria such as *Escherichia coli* O157:H7, *Salmonella*, *Listeria*, *Campylobacter*, and *Vibrio*, which are main contributors to the illnesses and deaths from food-borne poisoning [27,28]. Moreover, *parasites* such as *Cryptosporidium* sp., *Cyclospora* sp., *Toxoplasma gondii*, and *Trichinella* are eliminated.

When used in this manner, irradiation is comparable to pasteurizing milk, in that the product is left fresh, but is much safer. Irradiation also extends the food shelf life by retarding maturation in vegetables and reducing spoilage organisms that can even grow under refrigeration. Irradiated strawberries can last weeks in the refrigerator without developing mold. Irradiation can also be used in place of fumigants and other quarantine procedures to allow fruits and vegetables to be imported or exported without risking the introduction of harmful insects to the receiving country [27].

The primary regulation that covers irradiation in the production, processing, and handling of food is 21 CFR 179. It is divided into subparts and sections summarized in Table 19.2. Subpart B of 21 CFR 179 describes the type of radiation and its sources; they include gamma ray, electron beam, and X-ray. The general provisions for food irradiation are described. Subpart B also describes other radiation methods such as radiofrequency radiation, ultraviolet, and pulsed light. Subpart C 21 CFR 179 describes packaging materials for irradiated foods. The regulations listed in Part 179 are due to the approved FAPs via the codified process of 21 CFR 171 [29]. The actual content of each section in Table 19.2 have been described in this chapter.

Table 19.2 Title 21Part 179 Irradiation in the Production, Processing, and Handling of Food

Regulation Section	Regulation Coverage
Subpart B— Radiation and Radiation Sources	
179.21	Sources of radiation used for inspection of food, for inspection of packaged food, and for controlling food processing
179.25	General provisions for food irradiation
179.26	Ionizing radiation for the treatment of food
179.30	Radiofrequency radiation for the heating of food, including microwave frequencies
179.39	Ultraviolet radiation for the processing and treatment of food
179.41	Pulsed light for the treatment of food
Subpart C— Packaging Materials for Irradiated Foods	
179.45	Packaging materials for use during the irradiation of prepackaged foods

19.4.2 Title 21 CFR 179—Subpart B: Radiation and Radiation Sources

Sec. 179.21 Sources of radiation used for inspection of food, for inspection of packaged food, and for controlling food processing.

Sources of radiation for the purposes of inspection of foods, for inspection of packaged food, and for controlling food processing may be safely used under the following conditions:

1. The radiation source is one of the following:
    a. X-ray tubes producing X-radiation from operation of the tube source at a voltage of 500 kV peak or lower.
    b. Sealed units producing radiations at energy levels of not more than 2.2 million electron volts (MeV) from one of the following isotopes: americium-241, cesium-137, cobalt-60, iodine-125, krypton-85, radium-226, and strontium-90.
    c. Sealed units producing neutron radiation from the isotope californium-252 (CAS Reg. No. 13981-17-4) to measure moisture in food.
    d. Machine sources producing X-radiation at energies no greater than 10 MeV.
    e. Monoenergetic neutron sources producing neutrons at energies not less than 1 MeV but no greater than 14 MeV.
2. To assure safe use of these radiation sources:
    a. The label of the sources shall bear, in addition to the other information required by the Act:
        i. Appropriate and accurate information identifying the source of radiation.
        ii. The maximum energy of radiation emitted by X-ray tube sources.
        iii. The maximum energy of X-radiation emitted by machine source.
        iv. The minimum and maximum energy of radiation emitted by neutron source.
    b. The label or accompanying labeling shall bear:
        i. Adequate directions for installation and use.
        ii. A statement that no food shall be exposed to radiation sources listed in paragraph (a) (1) and (2) of this section so as to receive an absorbed dose in excess of 10 grays (Gy).
        iii. A statement that no food shall be exposed to a radiation source listed in paragraph (a)(3) of this section so as to receive an absorbed dose in excess of 2 mGy.
        iv. A statement that no food shall be exposed to a radiation source listed in paragraph (a)(4) of this section so as to receive a dose in excess of 0.5 Gy.
        v. A statement that no food shall be exposed to a radiation source listed in paragraph (a)(5) of this section so as to receive a dose in excess of 0.01 Gy.

[42 FR 14635, Mar. 15, 1977, as amended at 48 FR 46022, Oct. 11, 1983; 61 FR 14246, Apr. 1, 1996; 64 FR 69191, Dec. 10, 1999; 66 FR 18539, Apr. 10, 2001; 69 FR 76404, Dec. 21, 2004]

Sec. 179.25 General provisions for food irradiation

For the purposes of 179.26, current good manufacturing practice is defined to include the following restrictions:

a. Any firm that treats foods with ionizing radiation shall comply with the requirements of part 110 of this chapter and other applicable regulations.
b. Food treated with ionizing radiation shall receive the minimum radiation dose reasonably required to accomplish its intended technical effect and not more than the maximum dose specified by the applicable regulation for that use.
c. Packaging materials subjected to irradiation incidental to the radiation treatment and processing of prepackaged food shall be in compliance with 179.45, shall be the subject of an exemption for such use under 170.39 of this chapter, or shall be the subject of an effective premarket notification for an FCS for such use submitted under 170.100 of this chapter.
d. Radiation treatment of food shall conform to a scheduled process. A scheduled process for food irradiation is a written procedure that ensures that the radiation dose range selected by the food irradiation processor is adequate under commercial processing conditions (including atmosphere and temperature) for the radiation to achieve its intended effect on a specific product and in a specific facility. A food irradiation processor shall operate with a scheduled process established by qualified persons having expert knowledge in radiation processing requirements of food and specific for that food and for that irradiation processor's treatment facility.
e. A food irradiation processor shall maintain records as specified in this section for a period of time that exceeds the shelf life of the irradiated food product by 1 year, up to a maximum of 3 years, whichever period is

shorter, and shall make these records available for inspection and copy by authorized employees of the FDA. Such records shall include the food treated, lot identification, scheduled process, evidence of compliance with the scheduled process, ionizing energy source, source calibration, dosimetry, dose distribution in the product, and the date of irradiation.

[51 FR 13399, Apr. 18, 1986, as amended at 67 FR 9585, Mar. 4, 2002; 67 FR 35731, May 21, 2002]

Sec. 179.26 Ionizing radiation for the treatment of food

Ionizing radiation for treatment of foods may be safely used under the following conditions:

1. Energy sources. Ionizing radiation is limited to:
   a. Gamma rays from sealed units of the radionuclides cobalt-60 or cesium-137.
   b. Electrons generated from machine sources at energies not to exceed 10 MeV.
   c. X-rays generated from machine sources at energies not to exceed 5 MeV, except as permitted by paragraph (a)(4) of this section.
   d. X-rays generated from machine sources using tantalum or gold as the target material and using energies not to exceed 7.5 MeV.
2. Limitations are listed in Table 19.3.
3. Labeling.
   a. The label and labeling of retail packages of foods irradiated in conformance with paragraph (b) of this section shall bear the following logo:

      along with either the statement "Treated with radiation" or the statement "Treated by irradiation" in addition to information required by other regulations. The logo shall be placed prominently and conspicuously in conjunction with the required statement. The radiation disclosure statement is not required to be more prominent than the declaration of ingredients required under 101.4 of this chapter. As used in this provision, the term "radiation disclosure statement" means the written statement that discloses that a food has been intentionally subject to irradiation.
   b. For irradiated foods not in package form, the required logo and phrase "Treated with radiation" or "Treated by irradiation" shall be displayed to the purchaser with either (i) the labeling of the bulk container plainly in view or (ii) a counter sign, card, or other appropriate device bearing the information that the product has been treated with radiation. As an alternative, each item of food may be individually labeled. In either case, the information must be prominently and conspicuously displayed to purchasers. The labeling requirement applies only to a food that has been irradiated, not to a food that merely contains an irradiated ingredient but that has not itself been irradiated.
   c. For a food, any portion of which is irradiated in conformance with paragraph (b) of this section, the label and labeling and invoices or bills of lading shall bear either the statement "Treated with radiation—do not irradiate again" or the statement "Treated by irradiation—do not irradiate again" when shipped to a food manufacturer or processor for further processing, labeling, or packing.

[51 FR 13399, Apr. 18, 1986, as amended at 53 FR 12757, Apr. 18, 1988; 53 FR 53209, Dec. 30, 1988; 54 FR 32335, Aug. 7, 1989; 55 FR 14415, Apr. 18, 1990; 55 FR 18544, May 2, 1990; 60 FR 12670, Mar. 8, 1995; 62 FR 64121, Dec. 3, 1997; 63 FR 43876, Aug. 17, 1998; 65 FR 45282, July 21, 2000; 65 FR 64607, Oct. 30, 2000; 69 FR 76846, Dec. 23, 2004; 70 FR 48072, Aug. 16, 2005; 73 FR 49603, Aug. 22, 2008]

Sec. 179.30 Radiofrequency radiation for the heating of food, including microwave frequencies.

Radiofrequency radiation, including microwave frequencies, may be safely used for heating food under the following conditions:

- The radiation source consists of electronic equipment producing radio waves with specific

Table 19.3 Food Permitted to Be Irradiated and the Allowed Limitations

Use	Limitations
1. For control of *Trichinella spiralis* in pork carcasses or fresh, nonheat-processed cuts of pork carcasses.	Minimum dose 0.3 kGy (30 kilorad (krad)); maximum dose not to exceed 1 kGy (100 krad).
2. For growth and maturation inhibition of fresh foods.	Not to exceed 1 kGy (100 krad).
3. For disinfestation of arthropod pests in food.	Do.
4. For microbial disinfection of dry or dehydrated enzyme preparations (including immobilized enzymes).	Not to exceed 10 kGy (1 megarad (Mrad)).
5. For microbial disinfection of the following dry or dehydrated aromatic vegetable substances when used as ingredients in small amounts solely for flavoring or aroma: culinary herbs, seeds, spices, vegetable seasonings that are used to impart flavor but that are not either represented as, or appear to be, a vegetable that is eaten for its own sake, and blends of these aromatic vegetable substances. Turmeric and paprika may also be irradiated when they are to be used as color additives. The blends may contain sodium chloride and minor amounts of dry food ingredients ordinarily used in such blends.	Not to exceed 30 kGy (3 Mrad).
6. For control of food-borne pathogens in fresh or frozen, uncooked poultry products that are: (1) whole carcasses or disjointed portions of such carcasses that are "ready-to-cook poultry" within the meaning of 9 CFR 381.1(b) (44), or (2) mechanically separated poultry product (a finely comminuted ingredient produced by the mechanical deboning of poultry carcasses or parts of carcasses).	Not to exceed 3 kGy (300 krad); any packaging used shall not exclude oxygen.
7. For the sterilization of frozen, packaged meats used solely in the National Aeronautics and Space Administration space flight programs	Minimum dose 44 kGy (4.4 Mrad). Packaging materials used need not comply with 179.25(c) provided that their use is otherwise permitted by applicable regulations in parts 174 through 186 of this chapter.
8. For control of food-borne pathogens in, and extension of the shelf life of, refrigerated or frozen, uncooked products that are meat within the meaning of 9 CFR 301.2 (rr), meat by-products within the meaning of 9 CFR 301.2 (tt), or meat food products within the meaning of 9 CFR 301.2 (uu), with or without nonfluid seasoning, that are otherwise composed solely of intact or ground meat, meat by-products, or both meat and meat by-products.	Not to exceed 4.5 kGy maximum for refrigerated products; not to exceed 7.0 kGy maximum for frozen products.
9. For control of *Salmonella* in fresh shell eggs.	Not to exceed 3.0 kGy.
10. For control of microbial pathogens on seeds for.	Not to exceed 8.0 kGy.

frequencies for this purpose authorized by the Federal Communications Commission.

- The radiation is used or intended for use in the production of heat in food wherever heat is necessary and effective in the treatment or processing of food.

Sec. 179.39 Ultraviolet radiation for the processing and treatment of food.

Ultraviolet radiation for the processing and treatment of food may be safely used under the following conditions:

- The radiation sources consist of low-pressure mercury lamps emitting 90% of the emission at a wavelength of 253.7 nm (2537 Å).
- The ultraviolet radiation is used or intended for use as follows in Table 19.4.

Sec. 179.41 Pulsed light for the treatment of food

Pulsed light may be safely used for treatment of foods under the following conditions:

- The radiation sources consist of xenon flashlamps designed to emit broadband radiation consisting of wavelengths covering the range of 200–1100 nm, and operated so that the pulse duration is no longer than 2 ms.
- The treatment is used for surface microorganism control.
- Foods treated with pulsed light shall receive the minimum treatment reasonably required to accomplish the intended technical effect.
- The total cumulative treatment shall not exceed 12.0 J/cm^2.

[61 FR 42383, Aug. 15, 1996]

19.4.3 Title 21 CFR 179—Subpart C: Packaging Materials for Irradiated Foods

Sec. 179.45 Packaging materials for use during the irradiation of prepackaged foods

The packaging materials identified in this section may be safely subjected to irradiation incidental to the radiation treatment and processing of prepackaged foods, subject to the provisions of this section and to the requirement that no induced radioactivity is detectable in the packaging material itself:

1. The radiation of the food itself shall comply with regulations in this part.
2. The following packaging materials may be subjected to a dose of radiation, not to exceed 10 kGy, unless otherwise indicated, incidental to the use of gamma, electron beam, or X-radiation in the radiation treatment of prepackaged foods:
   a. Nitrocellulose-coated or vinylidene chloride copolymer-coated cellophane complying with 177.1200 of this chapter.
   b. Glassine paper complying with 176.170 of this chapter.
   c. Wax-coated paperboard complying with 176.170 of this chapter.
   d. Polyolefin film prepared from one or more of the basic olefin polymers complying with 177.1520 of this chapter. The finished film may contain:
      i. Adjuvant substances used in compliance with 178.3740 and 181.22 through 181.30 of this chapter, sodium citrate, sodium lauryl sulfate, polyvinyl

Table 19.4 Ultraviolet Radiation for Treatment of Food

Irradiated Food	Limitations	Use
Food and food products	Without ozone production: high fat content food irradiated in vacuum or in an inert atmosphere; Intensity of radiation, 1 W (of 2537 Angstrom radiation) per 5–10 ft^2	Surface microorganism control.
Potable water	Without ozone production; coefficient of absorption, 0.19/cm or less; flow rate, 100 gal/h/W of 2537 Angstrom radiation; water depth, 1 cm or less; lamp-operating temperature, 36–46°C.	Sterilization of water used in food production.
Juice products	Turbulent flow through tubes with a minimum Reynolds number of 2200.	Reduction of human pathogens and other microorganisms.

chloride, and materials as listed in paragraph (d)(2)(i) of this section.

ii. Coatings comprising a vinylidene chloride copolymer containing a minimum of 85% vinylidene chloride with one or more of the following comonomers: acrylic acid, acrylonitrile, itaconic acid, methyl acrylate, and methyl methacrylate.

e. Kraft paper prepared from unbleached sulfate pulp to which rosin, complying with 178.3870 of this chapter, and alum may be added. The kraft paper is used only as a container for flour and is irradiated with a dose not exceeding 500 Gy.

f. Polyethylene terephthalate (PET) film prepared from the basic polymer as described in 177.1630(e)(4)(i) and (ii) of this chapter. The finished film may contain:
   i. Adjuvant substances used in compliance with 178.3740 and 181.22 through 181.30 of this chapter, sodium citrate, sodium lauryl sulfate, polyvinyl chloride, and materials as listed in paragraph (d)(2)(i) of this section.
   ii. Coatings comprising a vinylidene chloride copolymer containing a minimum of 85% vinylidene chloride with one or more of the following comonomers: acrylic acid, acrylonitrile, itaconic acid, methyl acrylate, and methyl methacrylate.
   iii. Coatings consisting of polyethylene conforming to 177.1520 of this chapter.

g. Polystyrene film prepared from styrene basic polymer. The finished film may contain adjuvant substances used in compliance with 178.3740 and 181.22 through 181.30 of this chapter.

h. Rubber hydrochloride film prepared from rubber hydrochloride basic polymer having a chlorine content of 30–32 weight percent and having a maximum extractable fraction of 2 weight percent when extracted within hexane at reflux temperature for 2 h. The finished film may contain adjuvant substances used in compliance with 178.3740 and 181.22 through 181.30 of this chapter.

i. Vinylidene chloride–vinyl chloride copolymer film prepared from vinylidene chloride–vinyl chloride basic copolymers containing not less than 70 weight percent of vinylidene chloride and having a viscosity of 0.50–1.50 centipoises as determined by ASTM method D729-81, "Standard Specification for Vinylidene Chloride Molding Compounds," which is incorporated by reference. Copies may be obtained from the American Society for Testing Materials, 100 Barr Harbor Dr., West Conshohocken, Philadelphia, PA 19428-2959, or may be examined at the National Archives and Records Administration (NARA). For information on the availability of this material at NARA, call 202-741-6030, or go to: http://www.archives.gov/federal_register/code_of_federal_regulations/ibr_locations.html. The finished film may contain adjuvant substances used in compliance with 178.3740 and 181.22 through 181.30 of this chapter.

j. Nylon 11 conforming to 177.1500 of this chapter.

3. Ethylene–vinyl acetate copolymers complying with 177.1350 of this chapter. The ethylene–vinyl acetate packaging materials may be subjected to a dose of radiation, not to exceed 30 kGy (3 Mrad), incidental to the use of gamma, electron beam, or X-radiation in the radiation treatment of packaged foods.

4. The following packaging materials may be subjected to a dose of radiation, not to exceed 60 kGy incidental to the use of gamma, electron beam, or X-radiation in the radiation processing of prepackaged foods:
   a. Vegetable parchments, consisting of a cellulose material made from waterleaf paper (unsized) treated with concentrated sulfuric acid, neutralized, and thoroughly washed with distilled water.
   b. Films prepared from basic polymers and with or without adjuvants, as follows:
      i. Polyethylene film prepared from the basic polymer as described in 177.1520 (a) of this chapter. The finished film may contain one or more added substances.

19.5 Regulatory Aspects of Recycled Plastics—US FDA View

There is a great deal of interest in recycling plastics in today's society. The drivers include economic, environmental, and sustainability issues. Continuous increase in the cost of petroleum provides economic incentive for the reuse of plastic parts and packaging materials beyond a threshold oil price. One of the largest applications of plastics is food packaging where there is opportunity for the consumption of recycled plastics.

FDA becomes involved when industry collects used polymeric materials (usually food containers) and proposes to recycle these materials to make new food containers. FDA's main safety concerns with the use of recycled plastic materials in food-contact articles are (From here forward the regulation text is republished from FDA's web site www.fda.gov.):

1. that contaminants from the postconsumer material may appear in the final food-contact product made from the recycled material;
2. that recycled postconsumer material not regulated for food-contact use may be incorporated into food-contact packaging;
3. that adjuvants in the recycled plastic may not comply with the regulations for food-contact use.

19.5.1 Introduction

Currently, to address these concerns, FDA [30] considers each proposed use of recycled plastic on a case-by-case basis and issues informal advice as to whether the recycling process is expected to produce plastic suitable for food-contact applications. FDA has prepared a document entitled Guidance for Industry—*Use of Recycled Plastics in Food Packaging: Chemistry Considerations* that is aimed to assist manufacturers of food packaging in evaluating processes for recycling plastic into food packaging. This guidance document represents the FDA's current thinking on this topic. This Guide is provided in the next section.

If a manufacturer would like FDA to consider the use of recycled plastic for a food-contact application, the following information should to be submitted:

- A complete description of the recycling process, including a description of the source of the recyclable plastic and a description of any source controls in place intended to ensure that only plastic that initially complied with the applicable regulations is recycled. Also, a description of any steps that are taken to ensure that the recyclable plastic is not contaminated at some point, either before collection for recycling or during the recycling process.
- The results of any tests performed to show that the recycling process removes possible contaminants. For use of the recycled material as a substitute for plastic made from virgin materials, it would be necessary either to show that there has been no possibility of contamination with substances other than food or to demonstrate, through surrogate contaminant testing and, if appropriate, additional migration testing, that your recycling process successfully removes possible contaminants. However, surrogate contaminant testing is no longer considered necessary to demonstrate that postconsumer recycled (PCR) PET or polyethylene naphthalate (PEN) produced by a tertiary recycling process is suitable for food-contact use. Because FDA has determined that tertiary recycling processes produce PCR-PET or PEN of suitable purity for food-contact use, the Agency no longer sees a need to evaluate tertiary recycling processes for PET or PEN or to issue individual opinion letters for them.
- A description of the proposed conditions of use of the plastic (eg, information on intended temperature of use, type of food with which the plastic will come into contact, the duration of the contact, and whether the food-contact plastic will be for repeated or single-use applications).

19.5.2 Use of Recycled Plastics in Food Packaging: Chemistry Considerations [30]

19.5.2.1 Introduction

The purpose of this document is to highlight the chemistry issues that FDA recommends that a manufacturer of recycled plastic consider during the manufacturer's evaluation of a recycling process for producing material suitable for food-contact applications. This document supersedes the December 1992 "Points to Consider for the Use of Recycled Plastics in Food Packaging: Chemistry Considerations." The possibility that chemical contaminants in plastic materials intended for recycling may remain in the recycled material and could migrate into the food the material contacts is one of the major considerations for the safe use of recycled plastics for food-contact applications. Other aspects of plastics recycling, such as microbial contamination and structural integrity of the recycled plastic, are also important, but are not discussed in this document.

FDA's guidance documents, including this guidance, do not establish legally enforceable responsibilities. Instead, guidances describe the Agency's current thinking on a topic and should be viewed only as recommendations, unless specific regulatory or statutory requirements are cited. The use of the word should in Agency guidances means that something is suggested or recommended, but not required.

19.5.2.2 Background

Historically, glass, steel, aluminum, and paper have been recycled for food-contact use. Postconsumer use contamination has not been a major concern with glass and metals. These materials are generally impervious to contaminants and are readily cleaned at the temperatures used in their recycling. In addition, pulp from reclaimed fiber in paper and paperboard may be used for food-contact articles provided it meets the criteria in Title 21 of the Code of Federal Regulations, Section 176.260 (Pulp from reclaimed fiber).

Manufacturers of food-contact articles made from recycled plastic are responsible for ensuring that, like virgin material, recycled material is of suitable purity for its intended use and will meet all existing specifications for the virgin material. These requirements, which are described in 21 CFR, Parts 174 through 179, serve as the framework for the testing protocol and evaluation procedures outlined in this guidance document. In particular, Section 174.5 (General provisions applicable to indirect food additives), subparagraph (a)(2) states, "Any substance used as a component of articles that contact food shall be of a purity suitable for its intended use."

Several general methodologies exist by which plastic packaging can be recycled, and each presents distinct issues regarding the contaminant residues that may be present in postconsumer material. The agency presents below a preliminary discussion of the basic types of recycling and identifies specific concerns associated with each type. This guidance then describes a recommended approach for estimating the maximum level of a chemical contaminant in the recycled material that would result in an EDI that does not exceed 1.5 μg/person/day (0.5 ppb dietary concentration (DC)), the level that FDA would generally consider to be of negligible risk for a contaminant migrating from recycled plastic. Finally, the guidance recommends a protocol for developing chemistry data that would be useful for evaluating the adequacy of a recycling process to remove chemical contaminants. FDA notes that the testing protocol and evaluation procedures recommended in this guidance may change as new knowledge is acquired.

The following changes to the testing protocol and evaluation procedures that were previously recommended in the December 1992 "Points to Consider for the Use of Recycled Plastics in Food Packaging: Chemistry Considerations" are included in this document:

Lowering from 1 to 0.5 ppb the DC that FDA would generally consider to be of negligible risk for a contaminant migrating from recycled plastic. These DCs correspond to EDIs of 3 and 1.5 μg/person/day, respectively.

Increasing the number of recommended options for surrogate contaminants for use in evaluating a recycling process.

Eliminating the recommendation to include a heavy metal contaminant in the surrogate testing of recycling processes for polyethylene terephthalate (PET).

Providing recommendations to address secondary recycling of plastics for cases in which plastic containers from nonfood-contact applications (those that originally contained, eg, household cleaners, soaps, shampoos, or motor oil) are included in the postconsumer feedstock.

Eliminating all data recommendations for tertiary recycling processes for PET and polyethylene naphthalate (PEN).

Recommending the use of 0.05 as the default consumption factor (CF) for any plastic recycled for food-contact use.

Although not required by law or regulation, recyclers of plastics intended for the manufacture of food-contact articles are invited to submit information on their recycling process to FDA for evaluation and comment. Please send submissions to the OFAS at the address given on the cover of this guidance.

19.5.2.3 Recycling Processes

In 1991, the EPA introduced a nomenclature that describes the three distinct approaches to the recycling of plastic-packaging materials. Primary recycling (1°) refers to the use of preconsumer industrial scrap and salvage to form new packaging, a common practice in industry. Secondary recycling (2°) refers to the physical reprocessing (eg, grinding and melting) and reformation of postconsumer plastic-packaging materials. Tertiary recycling (3°) involves subjecting postconsumer plastic packaging to chemical treatment whereby its components are isolated and reprocessed for use in manufacture.

EPA considers "recycling" to be the processing of waste to make new articles. Because bottles intended for reuse are not made to be discarded and become waste, reuse is not considered recycling by EPA. Rather, reuse is regarded simply as one form of source reduction, that is, minimizing the amount of material entering the environment. In simple reuse, the package remains intact and is reused in its original form. In secondary and tertiary recycling, the original package is destroyed and new packaging is formed from the remains. This guidance focuses on the recycling of plastic-packaging materials, and reuse will not be discussed further.

19.5.2.3.1 Preconsumer Scrap: Primary Recycling

Primary recycling is the recycling of industrial scrap produced during the manufacture of food-contact articles and is not expected to pose a hazard to the consumer. The recycling of this scrap ("home scrap" as defined by the EPA in 56 FR 49992, Oct. 2, 1991) is acceptable, provided good manufacturing practices are followed. If the home scrap is collected from several different manufacturers, however, FDA recommends that the recycler consider whether the level and type of adjuvants in the recycled plastic would comply with existing approvals.

19.5.2.3.2 Physical Reprocessing: Secondary Recycling

Physical reprocessing involves grinding, melting, and reforming plastic-packaging material. The basic polymer is not altered during the process. Prior to melting and reforming, the ground, flaked, or pelletized polymer is generally washed to remove contaminants. The size of the polymer flakes or pellets can influence the effectiveness of the washing. Smaller particles provide a greater surface area for enhancing the effectiveness of the wash. Different polymers may also undergo different reforming conditions, such as different processing temperatures, the use of vacuum stripping, or other procedures, that could influence contaminant levels. In some cases, during the grinding or melting phases, the reprocessed material may be blended with virgin polymer.

Recyclers must be able to demonstrate that contaminant levels in the reformed plastic have been reduced to sufficiently low levels to ensure that the resulting packaging is of a purity suitable for its intended use. To produce a polymer with the desired qualities, however, additional antioxidants, processing aids, or other adjuvants may need to be added to the recycled polymer. The type and total amount of these additives must comply with existing authorizations, and any adjuvants already in the plastic may not react during the recycling process to form substances whose safety has not been evaluated by the FDA. Use in the recycled polymer of a new additive or an amount of an approved additive in excess of what is currently authorized for the virgin polymer would require an FCN or FAP [31]. This guidance has been prepared by the Division of Food Contact Notifications in the CFSAN at the US FDA.

A secondary recycling process presents some unique challenges that might cause it to be inappropriate for the production of food-contact articles, particularly if the recycler had little or no control over the waste stream entering the recycling facility (eg, commingling of food-contact and nonfood-contact materials). Where effective source control or sorting procedures can be established, however, the potential for postconsumer food-contact materials to be recycled together with other postconsumer plastics will be minimized or eliminated. Nevertheless, even if all the incoming postconsumer polymers were comprised of food-contact materials, limitations on food type or conditions of use could be compromised in the finished recycled product. For example, an additive approved for use only in contact with aqueous food or only for refrigerated use could be incorporated into packaging intended for high-temperature use with fatty foods. The resulting food-contact article would not comply with existing approvals. This concern may be mitigated by development of sorting procedures that result in reprocessing of only a single characteristic container, for example, a PET soda bottle.

Given the foregoing, FDA recommends that secondary recyclers address these concerns by, for example, implementing controls on the source of the postconsumer polymer, adequate sorting procedures for the incoming postconsumer material, use limitations on the finished recycled packaging (such as use at room temperature or below), or food-type restrictions (such as dry or aqueous foods only). In any submissions to FDA regarding 2° recycling processes, a discussion of these types of actions would be very helpful in FDA's evaluation of the processes.

19.5.2.3.3 Chemical Reprocessing: Tertiary Recycling

The primary goal of tertiary recycling is the regeneration of purified starting materials. Chemical reprocessing may involve depolymerization of the used packaging material with subsequent regeneration and purification of resulting monomers (or oligomers). The monomers are then repolymerized and the regenerated or reconstituted polymer is formed into new packaging. Regenerated monomer, polymer, or both may be blended with virgin materials. The regeneration process may involve a variety of monomer/polymer purification steps in addition to washings, such as distillation, crystallization, and additional chemical reaction.

19.5.2.4 Exposure to Chemical Contaminants

The FDA believes that acute consumer exposure to chemical contaminants from food containers produced from plastic that has been processed by 2° or 3° recycling will be extremely low because of the low concentrations of contaminant residues in the recycled polymers (see below). It is possible, however, that traces of a toxic substance could be carried through a 2° or 3° recycling process, become a part of the packaging, and migrate into food in contact with the packaging. Although subsequent recycling of the packaging will result in dilution of the toxin, a very low steady-state concentration of certain toxins could conceivably develop in the recycled material over the long term. Therefore, there is a potential for a consumer to be exposed to low concentrations of a particular toxin over a long period of timc. In order to develop a recommendation for the maximum acceptable level(s) of residual contaminants in the recycled material, FDA has considered the question of risk in a probabilistic way rather than on a compound-by-compound basis.

To recommend a maximum acceptable level for chemical contaminants in recycled food-contact articles that can form the basis of good manufacturing practice with respect to recycled material, FDA has determined the residual concentration of a contaminant that corresponds to an acceptable upper limit of dietary exposure. Using the scientific analysis supporting the TOR approach to evaluating indirect food additives as a basis (see 21 CFR 170.39), FDA believes that EDIs of contaminants from recycled food-contact articles on the order of 1.5 μg/person/day (0.5 ppb DC) or less are generally of negligible risk. The following exercise illustrates the calculation of the maximum acceptable level in the plastic of a contaminant in PET that would result in an EDI of no more than 1.5 μg /person/day.

In the case of PET, combining its density of 1.4 g/cm^3 with an assumed container thickness of 0.50 mm (~0.02 in) gives a package with a mass-to-surface area ratio of 70 mg/cm^2 (450 mg/in^2). FDA further assumed the following: individuals consume 3 kg of food per day, 10 g of food contacts one square inch of container, a consumption factor (CF) of 0.05 for recycled PET [FDA will assume a default CF of 0.05 for any recycled polymer. Previously, FDA used a CF for both virgin and recycled PET of 0.05. Based on recent market data that demonstrated that, since 1990, virgin PET has captured a dramatically larger share of the food-packaging market, FDA increased the CF for virgin PET to 0. Such an increase in market share has not been observed for recycled PET, so FDA uses a CF for recycled PET of 0.05. Since PET is recycled into food containers at a higher rate than any other polymer, it can be assumed that the CF for any other recycled polymer will not exceed 0.05.], and a food-type distribution factor (fT) of 1.0 for all food types (see "Preparation of Food Contact Notifications and Food Additive Petitions for Food Contact Substances: Chemistry Recommendations"). The relationships among EDI,

DC, and the CF, fT, and migration level from package to food are as follows:

$$\mathrm{DC} = \mathrm{CF}\cdot\langle M\rangle = \mathrm{CF}\cdot\sum_{i=1}^{4}(M\cdot f_{\mathrm{T}})_i$$
$$\mathrm{EDI} = \mathrm{DC}\cdot 3\ \mathrm{kg\ food/person/day}$$

where M is the concentration of migrant in a food simulant, i, where i represents the four simulated food types: aqueous, acidic, alcoholic, and fatty foods. Use of the parameters noted above leads to:

$$\mathrm{DC} = 0.5\ \mathrm{ppb} = 0.05\ \langle M\rangle = 0.05(M)(1.0)$$

and

$$\begin{aligned}\langle M\rangle &= (0.5\times 10^{-9}\ \mathrm{g\ contaminant/g\ food})\div(0.05)\\ &= 1\times 10^{-8}\ \mathrm{g\ contaminant/g\ food}\end{aligned}$$

Then,

$$\begin{aligned}&(450\times 10^{-3}\ \mathrm{g\ packaging/in^2})\div(10\ \mathrm{g\ food/in^2})\\ &\quad = 0.045\ \mathrm{g\ packaging/g\ food}\\ &(1\times 10^{-8}\ \mathrm{g\ contaminant/g\ food})\\ &\quad \div(0.045\ \mathrm{g\ packaging/g\ food})\\ &\quad = 2.2\times 10^{-7}\ \mathrm{g\ contaminant/g\ packaging}\end{aligned}$$

or 220 μg/kg of contaminant in the packaging material. In other words, if a contaminant were present at 220 μg/kg in the PET container made from the recycled material and if 100% migration of the contaminant into food were assumed (a conservative assumption for room-temperature applications of a high barrier material like PET), the DC of the contaminant would be 0.5 ppb (EDI: 1.5 μg/person/day).

The maximum acceptable level of a residual contaminant in a polymer (Table 19.5) that corresponds to an EDI equal to 1.5 μg/person/day will depend on the polymer density, polymer thickness, and CF. Table 19.5 reflects residue levels in several polymers that would result in an EDI of a contaminant of 1.5 μg/person/day. These calculations were done using a CF of 0.05 for each recycled polymer [FDA will assume a default CF of 0.05 for any recycled polymer. Previously, FDA used a CF for both virgin and recycled PET of 0.05. Based on recent market data that demonstrated that, since 1990, virgin PET has captured a dramatically larger share of the food-packaging market, FDA increased the CF for virgin PET to 0. Such an increase in market share has not been observed for recycled PET, so FDA uses a CF for recycled PET of 0.05. Since PET is recycled into food containers at a higher rate than any other polymer, it can be assumed that the CF for any other recycled polymer will not exceed 0.05.], a *container wall thickness of 0.50 mm (~0.02 in)*, and the conservative assumptions that all food types are used with each polymer and that the finished article will consist of 100% recycled polymer.

Table 19.5 Maximum Acceptable Level of Residual Contaminant in a Polymer

Recycled Polymer	Density (g/cm^3)	Maximum Residue
PET	1.4	220 μg/kg
Polystyrene	1.05	300 μg/kg
PVC	1.58	200 μg/kg
Polyolefins	0.965	320 μg/kg

Thus, to achieve EDIs below 1.5 μg/person/day for recycled containers of 0.50-mm thickness, individual chemical contaminants should not be present at levels greater than those given above. It must be emphasized that the calculated levels depend on the thickness of the packaging—the thicker the packaging, the lower the maximum residue levels must be to meet the 1.5 μg/person/day EDI limit. If a specialized use for a recycled polymer can be documented, it may be possible to estimate a lower CF for use in calculating a maximum acceptable contaminant level. Finally, in cases in which recycled polymer is expected to be blended with virgin polymer, and thus, contaminants in the recycled polymer are diluted with virgin polymer, the maximum acceptable contaminant level calculated using the agency's recommended approach set out above may be divided by the fraction of recycled polymer in the blend.

19.5.2.5 Surrogate Contaminant Testing

How may the ability of a 2° or 3° recycling process to remove contaminants from plastic containers or packaging that has been subjected to consumer misuse or abuse (eg, through storage of pesticides or automotive chemicals) be demonstrated? FDA recommends simulating consumer misuse by exposing virgin polymer (either in container form or as flake) to selected surrogate contaminants and then running the exposed or "challenged" polymer through the recycling process. Subsequent analysis of the processed polymer for the surrogate contaminants would provide a means to evaluate the efficacy of the recycling process.

19.5.2.5.1 Choice of Surrogates

The FDA recommends that recyclers use materials that have a variety of chemical and physical properties to simulate consumer misuse. In particular, FDA recommends that the surrogate contaminants represent "common" materials accessible to the consumer and include a volatile polar organic substance, a volatile nonpolar organic substance, a nonvolatile polar organic substance, a nonvolatile nonpolar organic substance, and a heavy metal salt (except for PET, see below). Examples of recommended surrogates are given below; FDA believes that one surrogate per category is sufficient for the testing.

Volatile Polar	*Nonvolatile Polar*
Chloroform	Benzophenone
Chlorobenzene	Methyl salicylate
1,1,1-Trichloroethane	
Diethyl ketone	*Nonvolatile Nonpolar*
	Tetracosane
Volatile Nonpolar	Lindane
Toluene	Methyl stearate
	Phenylcyclohexane
Heavy Metal	1-Phenyldecane
Copper(II) 2-ethylhexanoate	2,4,6-Trichloroanisole

Chloroform and toluene are components of cleaning solvents; benzophenone is a suitable substitute for nonvolatile polar pesticides such as Diazinon; and tetracosane is a good representative for the long-chain hydrocarbons that comprise motor oil. A heavy metal salt such as copper(II) 2-ethylhexanoate, a substitute for the toxic salts commonly used in herbicides, would complete the range of properties noted. [Although FDA previously recommended testing with a polymer-specific surrogate, eg, ortho-cresol, which is known to significantly swell PET, such data are no longer considered necessary because (1) the range of possible contaminant properties are already covered by the five surrogate categories selected, and (2) a consumer's storage of a polymer-specific solvent in a bottle would significantly degrade the bottle to the extent that it would be rejected during the sorting process.] OFAS is available to discuss the use of surrogates other than those listed in the previous table.

In the case of PET, FDA does not recommend including a heavy metal salt in surrogate testing. In the PET recycling submissions reviewed over the past decade, migration of the heavy metal surrogate has never been detected in food simulants. FDA reviewed data for a number of heavy metal surrogates, including the metal salts calcium monomethylarsonate (an herbicide for crabgrass), copper acetoarsenite (Paris green), cadmium acetate, zinc stearate, and copper(II) 2-ethylhexanoate. The data show that, unlike small organic molecules, metal salts do not sorb as readily into PET and also that the salts are more easily washed out of PET, probably because they simply adsorb to the PET surface. In one case, the heavy metal surrogate was incorporated into the resin by blending and still was not detected in migration experiments. FDA believes that the metal-containing substances to which consumers have access are primarily in the salt form, and even if this were not the case, it is unlikely that nonionic organometallic species would behave any differently than the organic compounds represented by the four general surrogate categories.

To date, surrogate testing data for recycling processes for polymers other than PET are insufficient to support general conclusions about the behavior of heavy metals in those polymers. Therefore, FDA continues to recommend the use of a heavy metal surrogate in the testing of recycling processes for polymers other than PET.

19.5.2.5.2 Contamination of the Plastic

In order to test the recycling process, FDA recommends the following approach.

First, containers made of the virgin plastic of interest are contaminated or "challenged" by filling them with the surrogate contaminants, either "neat" or in "at use" concentrations, using a solvent such as hexane as a diluent. An alternative approach that would reduce the amount of potentially hazardous wastes would be to soak several kilograms of flaked virgin plastic of the type actually used in the recycling process in the selected contaminants at either "neat" or "at use" concentrations. A mixture, or "cocktail," of the contaminants could be used so long as the components of the "cocktail" do not react with each other. Agency recommendations for minimum concentrations of surrogates for a "cocktail" are given in Table 19.6.

Once the bottles are filled or after the contaminants are thoroughly mixed with the flakes, the bottles or flakes should be stored sealed for 2 weeks at 40°C with periodic agitation. After the contaminants are drained and the bottles or flakes are rinsed, the concentration of each surrogate should be determined in the polymer. The challenged polymer should then be subjected to the proposed recycling process, and regenerated components or packaging material formed from the reprocessed polymer should be analyzed for residual contaminants. This approach represents a worst-case scenario, that is, all material entering the recycling stream is assumed to be contaminated.

Table 19.6 Examples of Minimum Concentrations of Contaminants in a Surrogate Cocktail

Contaminant	Concentration
Chloroform (volatile polar)	10% v/v[a]
Toluene (volatile nonpolar)	10% v/v
Benzophenone (nonvolatile polar)	1% v/v
Tetracosane or lindane (nonvolatile nonpolar)	1% w/w[b]
Copper(II) 2-ethylhexanoate (heavy metal)	1% w/w
Balance:	
2-Propanol (as solvent for Cu(II) 2-ethylhexanoate)	10% v/v
Hexane or heptane (as overall solvent for cocktail)	68% v/v

[a]*v/v, volume of contaminant per unit volume of entire cocktail.*
[b]*w/w, mass of surrogate per unit mass of entire cocktail.*

Testing protocols may be submitted to OFAS for comment before any contamination studies are done. FDA recommends that all analyses be validated as discussed in the "Preparation of Food Contact Notifications and Food Additive Petitions for Food Contact Substances: Chemistry Recommendations."

19.5.2.5.3 Other Considerations

If a proposed recycling process cannot be shown to remove contaminants to maximum acceptable levels under the scenario discussed above, then additional factors or limitations on use could justify a conclusion that the recycled package will not introduce contaminants into the diet at unacceptable levels. The following additional factors/limitations may result in an acceptable upper limit of dietary exposure: the use of a recycled/virgin blend, source controls, restricted uses, the fraction of contaminant that migrates into food or a food simulant, or the use of an effective barrier. FDA recommends that the effect of measured or mathematically modeled factors be supported by adequate documentation (eg, studies on a specific source control program, studies on the actual extent of contaminated material entering the recycling stream, or information that demonstrates that the recycled polymer is separated from food contact by an effective barrier).

In cases where the postconsumer plastic feedstock is intended to consist of food containers only (ie, containers that were used to package nonfood substances such as household cleaners are intentionally excluded), FDA would consider data submitted by recyclers that show the extent of contaminated material entering the recycling stream as a result of consumer misuse in order to demonstrate or allow a prediction of the actual incidence of chemical contamination of recycled articles. FDA believes that, due to the incidental nature of potential consumer misuse of a food container and subsequent introduction of that container into the recycling stream, this information (if properly validated) can be factored into the exposure calculations to obtain more realistic values.

19.5.2.6 Plastic Containers from Nonfood-Contact Applications as Feedstock

The 1992 "Points to Consider" was developed to address incidental contamination of food containers by consumers, following the original intended use of the containers. Containers with nonfood contents (eg, household cleaners, furniture polish, shampoos, soaps, pesticides, or motor oil) were purposely not addressed.

The amount of custom PET containers (ie, containers other than soda bottles that are used to package specialty foods as well as nonfood substances) collected via curbside programs has increased dramatically over the past several years Of 280,000 tons of custom PET containers generated in 1993, 3.6% was recycled. Of 820,000 tons generated in 1999, 9.8% was recycled (see Refs. [32,33] under the "Other plastic containers" entries for PET) and is predicted to continue [34,35]. In addition, the use of PET in nonfood containers is increasing [36,37]. These trends indicate that the contamination introduced into postconsumer PET feedstock via nonfood containers, as well as the probability that up to 100% of the postconsumer plastic feedstock might consist of nonfood containers, is likely to increase. Therefore, FDA recommends that worst-case contamination assumptions be made for nonfood containers entering the recycling stream (see below).

The FDA believes that the following two questions are relevant in determining the appropriateness of recycling nonfood containers into new food containers by two processes:

- Are the nonfood container plastic, its adjuvants, and the adjuvant levels in the plastic currently authorized for use in contact with food? FDA believes that it is highly unlikely that two recycling processes will remove unapproved adjuvants (such as colorants or antioxidants) because these substances are incorporated into the polymer matrix during manufacture.
- Can the levels of chemical contaminants introduced into the postconsumer feedstock by nonfood containers, which can sorb relatively large amounts of chemicals from their contents over very long shelf lives, be sufficiently reduced so that finished recycled containers are suitable for food-contact use? FDA believes that it is likely that 2° recycling processes could remove residual amounts of nonfood substances such as household cleaning solutions from nonfood containers because these substances are simply sorbed into the container's surface.

These questions have been addressed for PET:

- The FDA has received information from the plastics industry that verifies that all PET resins used to manufacture containers in the United States are authorized for food-contact use, that is, food-grade PET is used to manufacture both food and nonfood containers [38].
- The FDA has used a mathematical model, based on Fick's law of diffusion, to predict the amount of a contaminant (represented by the surrogate contaminants described above) that will sorb into a PET bottle during a period of 1 year at 25°C, the shelf life and use temperature of a typical nonfood substance packaged in PET (see Appendix 1). Because the model tends to overpredict sorption (see Appendix 1), FDA's experimental sorption results and data from the literature have also been used to calculate the sorption of several surrogates into PET under these use conditions [39,40]. The PET sorption values for several surrogates are summarized in Table 19.7.

FDA recommends that recyclers who wish to include nonfood PET containers in their feedstock establish that the concentrations of the surrogates in challenged PET flake, prior to its being run through their recycling process, are greater than or equal to the sorption values given in Table 19.7. FDA does not recommend that this criterion be included in surrogate testing for a recycling process that uses only food containers as feedstock—simply exposing virgin flake or intact bottles to the surrogate cocktail for 2 weeks at 40°C is sufficient to model incidental misuse of containers by consumers.

In order to obtain the initial concentrations given in Table 19.7, FDA recommends exposing PET flake rather than intact bottles to the surrogate solutions for at least 14 days at 40°C. The use of intact PET bottles in surrogate testing potentially can result in at least an order of magnitude less sorption of the surrogate contaminants than the values given in Table 19.2, due to the lower surface area of a PET bottle compared to an equal mass of flake [41]. FDA also recommends that a surrogate cocktail contain, at a minimum, the concentrations of contaminants given in Table 19.1. Finally, FDA recommends that recyclers consider using higher concentrations of volatile surrogates such as chloroform in the cocktail to offset losses that could occur before the flake is analyzed.

FDA recommends that, in cases in which the starting concentrations are found to be lower than the values in Table 19.7, recyclers correct for the shortfall by multiplying the surrogate testing results by a simple factor. For example, if the actual starting concentration of toluene were 500 mg/kg, then the factor would be 1.6 ((780 mg/kg)/(500 mg/kg) = 1.6).

Table 19.7 Sorption of Surrogate Contaminants into PET After 365 Days at 25°C

Surrogate	Sorption Value (mg/kg)	Reference
Volatile, Polar		
Chloroform	4860	[39] (modeled value)
Chlorobenzene	1080	[40][a]
1,1,1-Trichloroethane	1050	[40][a]
Diethyl ketone	4860	Assumed to be the same as chloroform, based on similar molecular weights
Volatile, Nonpolar		
Toluene	780	[39]
Nonvolatile, Polar		
Benzophenone	49	[39]
Methyl salicylate	200	[39]
Nonvolatile, Nonpolar		
Tetracosane	154	[39] (modeled value)
Lindane	750	[39]
Methyl stearate	150	Assumed to be the same as tetracosane, based on preliminary FDA experimental results
Phenylcyclohexane	390	[40][a]
1-Phenyldecane	170	[40][a]
2,4,6-Trichloroanisole	1100	Based on value for lindane with molecular weight correction

[a]*These values were measured at 40°C.*

FDA recommends against analyzing actual batches of postconsumer plastics for potential chemical contaminants in order to refine exposure calculations because these analyses provide only a snapshot of the composition of curbside-collected plastic containers, and there is no guarantee that the composition of nonfood containers in curbside recycling programs will remain the same over time.

Recyclers who have already had their PET recycling processes evaluated by FDA should not assume based on the conclusions in this guidancc document that their letters from FDA apply to the use of nonfood PET containers as feedstock. If these recyclers desire FDA's opinion on their use of nonfood PET containers, they may ask FDA to reevaluate their surrogate testing data and issue a separate letter.

19.5.2.7 The Use of an Effective Barrier

The use of two or three recycled material as a nonfood-contact layer of a multilayer food package is a potential application for recycled plastics. FDA believes that this use would not present a concern about potential contaminant migration into food as long as the recycled polymer is separated from the food by an effective barrier made from virgin polymer or other appropriate material, for example, an aluminum film. Based on experimental and mathematically modeled diffusion data obtained by the OFAS laboratory (and others) for three-layered coextruded PET films in which the center layer contained surrogate contaminants and the outer layers were comprised of virgin material [42], FDA has determined that virgin PET is an effective barrier to contaminants that could potentially migrate from a recycled plastic inner layer under the following conditions:

- at a thickness $\geq 25\ \mu m$ (~ 0.001 in) at room temperature and below;
- at a thickness $\geq 50\ \mu m$ (~ 0.002 in) at higher temperatures, including use as a dual-ovenable container for cooking food at 150°C for 30 min, provided that only food containers are used in the feedstock to manufacture the recycled layer.

In these cases, the presence of a virgin PET layer ensures that migration of a contaminant to food will result in an EDI no greater than 1.5 μg/person/day. The calculation assumes a CF of 0.05 for recycled plastic packaging. [FDA will assume a default CF of 0.05 for any recycled polymer. Previously, FDA used a CF for both virgin and recycled PET of 0.05. Based on recent market data that demonstrated that, since 1990, virgin PET has captured a dramatically larger share of the food-packaging market, FDA increased the CF for virgin PET to 0. Such an increase in market share has not been observed for recycled PET, so FDA uses a CF for recycled PET of 0.05. Since PET is recycled into food containers at a higher rate than any other polymer, it can be assumed that the CF for any other recycled polymer will not exceed 0.05.] [Note: Although an EDI ≤ 1.5 μg/person/day for a contaminant is generally of negligible safety concern, a substance intentionally used as a component of a food-contact article is still subject to the food additive definition and might require FDA premarket approval via an FCN (see Preparation of Food Contact Notifications: Administrative) or a TOR submission (see Submitting Requests Under 21 CFR 170.39 TOR of Substances Used in Food Contact Articles 3), even if the EDI of the substance is ≤ 1.5 μg/person/day.]

To demonstrate that a given thickness of a virgin polymer functions as an effective barrier to the migration of contaminants, FDA recommends that the recyclers subject intentionally contaminated polymer to the recycling process and incorporate the recycled polymer into a nonfood-contact layer of a finished article, using virgin polymer as the barrier layer. FDA recommends that migration studies be performed with food simulants as described in the "Preparation of Food Contact Notifications and Food Additive Petitions for Food Contact Substances: Chemistry Recommendations." If migration studies show that the EDI of individual contaminants would not exceed 1.5 μg/person/day, FDA would consider the virgin layer of the specified thickness to be an effective barrier to contaminants migrating from the nonfood-contact layer of recycled material. If data from studies other than migration experiments establish that a given thickness of a particular polymer is sufficiently impermeable under anticipated time/temperature use conditions, those data could serve to replace migration experiments.

19.5.2.8 Elimination of Data Recommendations for 3° Recycling Processes for PET and PEN

Based on a comprehensive review of all surrogate testing data submitted over the past decade for 3° recycling processes for PET and PEN, FDA concludes that 3° recycling of PET or PEN by methanolysis or glycolysis results in the production of monomers or oligomers that are readily purified to produce a finished polymer that is suitable for food-contact use. Both 3° processes will clean the polyester sufficiently to allow it to be considered of suitable purity, even assuming 100% migration of residual surrogate to food. This is a significant difference from the surrogate testing of 2° recycling processes. Secondary recycling processes often produce PET that is insufficiently cleaned to withstand 100% migration calculations for the residual surrogates. Under these circumstances, FDA recommends additional migration tests to demonstrate that the finished PET meets the 1.5 μg/person/day EDI limit.

Based on a determination that 3° recycling processes produce PET or PEN of suitable purity for food-contact use, FDA no longer recommends that such recyclers submit data for agency evaluation. Because 3° processes for polymers other than PET and PEN have not been the subject of FDA reviews, however, recyclers who wish to engage in 3° recycling of polymers other than PET and PEN are encouraged to submit data for evaluation.

19.5.2.9 Appendix 1: Model of the Sorption of Surrogate Contaminants into Plastic

FDA recommends using the following equation from Ref. [43] to model the sorption, at a fixed temperature, of any substance (including surrogate contaminants) into plastic:

$$\frac{M_t}{M_\infty} = (1+\alpha)\left[1-\exp\left(\frac{T}{\alpha^2}\right)\mathrm{erfc}\left(\frac{T}{\alpha^2}\right)^{1/2}\right] \quad (19.1)$$

$$T = \frac{Dt}{l^2} \quad (19.2)$$

In this equation, M_t and M_∞ are, respectively, the sorption (g surrogate/g plastic) at time t and the sorption at equilibrium (or "infinite" time), α is the ratio of the volume of the surrogate solution to the volume of the plastic, D is the diffusion coefficient (cm^2/s) of the surrogate in the plastic at a given temperature, t is the time in seconds, and l is the thickness of the plastic (cm). The full thickness is used for a single-sided experiment (ie, a plastic bottle filled with surrogate solution) while half the thickness is used for a double-sided experiment (ie, a plastic strip soaked in a surrogate solution).

In order to solve Eq. (19.1) for M_t, a value for M_∞ is needed. In the absence of experimentally determined values, FDA recommends using the following equation derived from Ref. [43] to calculate M_∞:

$$M_\infty = \frac{a}{w} C_0 \frac{1}{1+\alpha} \quad (19.3)$$

In Eq. (19.3) a is the volume of the surrogate solution (mL), w is the mass of the polymer (g), and C_0 is the starting concentration of the surrogate in the solution (g/mL). FDA's version of this equation differs from Crank's in that the mass of the polymer is included in the denominator to obtain M_∞ in the units g surrogate/g plastic.

The following parameters were used for a typical nonfood PET bottle: a 1-L capacity, a mass of 38.26 g, a density of 1.37 g/cm^3, and a wall thickness of 0.03 cm. The bottle's mass was divided by its density to obtain the volume of PET in contact with the surrogate solution. A 10% w/w concentration of each surrogate in solution was assumed to represent the maximum concentration of any given component of a nonfood substance packaged in PET. [A search of a database of ingredients in household products (www.happi.com) shows that liquid detergents, shampoos, and hand cleaners might contain up to 30% long-chain fatty acids; however, these compounds are not expected to sorb into PET to any greater extent than lower molecular weight compounds from 10% solutions.] The densities of the surrogate and the remainder of the surrogate solution (assumed to be 1 g/cm^3) were used to calculate C_0. The diffusion coefficients in Table 19.8 were used for several surrogates in PET.

Table 19.8 Diffusion Coefficients for Several Surrogates in PET

Surrogate	D (cm^2/s), 25°C	Reference
Chloroform	9.1×10^{-14}	Calculated with Piringer model [45]
Toluene	4.3×10^{-15}	[44]
Benzophenone	4.8×10^{-14}	Calculated with Piringer model [45]
Lindane	1.6×10^{-14}	Calculated with Piringer model [45]
Tetracosane	1×10^{-16}	[44]

In order to validate the model, the modeled results were compared with experimental measurements of toluene sorption into a PET strip (ie, a two-sided experiment) made by Ref. [40] after 15 and 40 days at 40°C. Sadler's D value of 3.92×10^{-14} cm^2/s for toluene at 40°C, which has been well characterized, was used in the model [44]. Specifications for the PET strip and the surrogate solution as given in the Demertzis article were also used, the results of which are summarized in Table 19.9.

The excellent agreement of the modeled results with experimental measurements indicates that the model adequately predicts surrogate sorption into PET over time. However, the model tends to overpredict sorption when experimentally determined D or M_∞ values are not available and semiempirical

Table 19.9 Results of Model Validation for Sorption of Toluene

Time (days) at 40°C	Modeled Toluene Sorption from 16.7% (w/w) Solution	Experimentally Measured Toluene Sorption from 16.7% (w/w) Solution [40]
15	3.1 mg/dm^2 of PET (750 mg/kg in PET)	3.5 mg/dm^2 of PET
40	5.1 mg/dm^2 of PET (1230 mg/kg in PET)	6.0 mg/dm^2 of PET

or theoretical values are used. The Piringer model, an empirical correlation based on the molecular weight of the migrant, was used to calculate D for chloroform, benzophenone, and lindane for input to the sorption model. Comparisons of modeled D values with experimental measurements indicate that the Piringer model can overestimate D by several orders of magnitude [45]. This discrepancy has been attributed to failure of the model to account for partitioning of the migrant between the solvent and polymer phases. Similarly, Eq. (19.3) will generally overpredict M_∞ because it does not take into account partitioning of the migrant between the solvent and polymer phases [43]. Because the sorption model depends on D and M_∞, overestimates of these values will result in overestimates of sorption. Modeled results for a typical bottle are given in Table 19.10.

Table 19.10 Modeled Sorption of Surrogates into 1-L PET Bottles Filled with 10% w/w Surrogate Solutions After 365 Days at 25°C

Surrogate	Modeled Sorption into PET (mg/kg)
Chloroform	4860
Toluene	1000
Benzophenone	3390
Lindane	1920
Tetracosane	15

FDA believes that the modeled sorption value for chloroform is probably reasonable because there is very little partitioning of chloroform between the solvent and polymer phases. However, the measured sorption values for benzophenone and lindane have shown the modeled values to be significant overestimates due to the need to rely on semiempirical or theoretical D and M_∞ values [39].

19.6 EU Legislation on Food-Contact Plastics

The situation in EU is more complex than in United States because of the existing national laws predating the formation of EU. The EC began working on replacing the Plastics Directive with a regulation that would be directly applicable in all of the EU Member States in 2004. This new regulation was enacted and published in the Journal of the European Union on January 15, 2011. The new regulation consolidates all previous directives and regulations on plastic food-contact materials and implements some significant changes. This regulation, previously referred to as the Plastics Implementing Measure, and now commonly referred to as the Plastics Regulation, went into effect on February 4, 2011. As of May 1, 2011, the Plastics Regulation replaced the Plastics Directive (2002/72/EC) and its amendments, as well as various directives on compliance (migration) testing. The Plastics Regulation became directly applicable in all of the EU Member States as of May 1, 2011 [46].

19.6.1 EU Regulation No. 10/2011 on Plastic Materials Intended to Come into Contact with Food

This section presents an abridged version of the EU Regulation No. 10/2011. The readers can refer to the original document for complete information [47].

19.6.2 Consolidating Paragraphs

1. Regulation (EC) No. 1935/2004 lays down the general principles for eliminating the differences between the laws of the Member States as regards food-contact materials. Article 5(1) of that Regulation provides for the adoption of specific measures for groups of materials and articles and describes in detail the procedure

for the authorization of substances at EU level when a specific measure provides for a list of authorized substances.

2. This Regulation is a specific measure within the meaning of Article 5(1) of Regulation (EC) No. 1935/2004. This Regulation should establish the specific rules for plastic materials and articles to be applied for their safe use and repeal Commission Directive 2002/72/EC of August 6, 2002, on plastic materials and articles intended to come into contact with foodstuffs.
3. Directive 2002/72/EC sets out basic rules for the manufacture of plastic materials and articles. The Directive has been substantially amended six times. For reasons of clarity, the text should be consolidated and redundant and obsolete parts removed.
4. In the past, Directive 2002/72/EC and its amendments have been transposed into national legislation without any major adaptation. For transposition into national law, usually a time period of 12 months is necessary. In case of amending the lists of monomers and additives in order to authorize new substances, this transposition time leads to a retardation of the authorization and thus slows down innovation. Therefore, it seems appropriate to adopt rules on plastic materials and articles in the form of a Regulation directly applicable in all Member States.
5. Directive 2002/72/EC applies to materials and articles purely made of plastics and to plastic gaskets in lids. In the past, these were the main use of plastics on the market. However, in recent years, besides materials and articles purely made of plastics, plastics are also used in combination with other materials in the so-called multimaterial multilayers. Rules on the use of vinyl chloride monomer laid down in Council Directive 78/142/EEC of January 30, 1978, on the approximation of the laws of the Member States relating to materials and articles which contain vinyl chloride monomer and are intended to come into contact with foodstuffs. OJ L 44, 15.2.1978, p. 15 already applies to all plastics. Therefore, it seems appropriate to extend the scope of this Regulation to plastic layers in multimaterial multilayers.
6. Plastic materials and articles may be composed of different layers of plastics held together by adhesives. Plastic materials and articles may also be printed or coated with an organic or inorganic coating. Printed or coated plastic materials and articles as well as those held together by adhesives should be within the scope of the Regulation. Adhesives, coatings, and printing inks are not necessarily composed of the same substances as plastics. Regulation (EC) No. 1935/2004 foresees that for adhesives, coatings, and printing inks, specific measures can be adopted. Therefore, plastic materials and articles that are printed, coated, or held together by adhesives should be allowed to contain in the printing, coating, or adhesive layer other substances than those authorized at EU level for plastics. Those layers may be subject to other EU or national rules.
7. Plastics as well as ion exchange resins, rubbers, and silicones are macromolecular substances obtained by polymerization processes. Regulation (EC) No. 1935/2004 foresees that for ion exchange resins, rubbers, and silicones, specific measures can be adopted. As those materials are composed of different substances than plastics and have different physicochemical properties, specific rules for them need to apply and it should be made clear that they are not within the scope of this Regulation.
8. Plastics are made of monomers and other starting substances, which are chemically reacted to a macromolecular structure, the polymer, which forms the main structural component of the plastics. To the polymer additives are added to achieve defined technological effects. The polymer as such is an inert high-molecular-weight structure. As substances with a molecular weight above 1000 Da usually cannot be absorbed in the body, the potential health risk from the polymer itself is minimal. Potential health risk may occur from non- or incompletely reacted monomers or other starting substances or from low-molecular-weight additives which are transferred into food via migration from the plastic food-contact material. Therefore, monomers, other starting substances, and additives should be risk assessed and

authorized before their use in the manufacture of plastic materials and articles.

9. The risk assessment of a substance to be performed by the European Food Safety Authority (hereinafter the Authority) should cover the substance itself, relevant impurities and foreseeable reaction and degradation products in the intended use. The risk assessment should cover the potential migration under worst foreseeable conditions of use and the toxicity. Based on the risk assessment, the authorization should if necessary set out specifications for the substance and restrictions of use, quantitative restrictions, or migration limits to ensure the safety of the final material or article.
10. No rules have yet been set out at EU level for the risk assessment and use of colorants in plastics. Therefore, their use should remain subject to national law. That situation should be reassessed at a later stage.
11. Solvents used in the manufacture of plastics to create a suitable reaction environment are expected to be removed in the manufacturing process as they are usually volatile. No rules have yet been set out at EU level for the risk assessment and use of solvents in the manufacture of plastics. Therefore, their use should remain subject to national law. That situation should be reassessed at a later stage.
12. Plastics can also be made of synthetic or natural occurring macromolecular structures which are chemically reacted with other starting substances to create a modified macromolecule. Synthetic macromolecules used are often intermediate structures which are not fully polymerized. Potential health risk may occur from the migration of non- or incompletely reacted other starting substances used to modify the macromolecule or an incompletely reacted macromolecule. Therefore, the other starting substances as well as the macromolecules used in the manufacture of modified macromolecules should be risk assessed and authorized before their use in the manufacture of plastic materials and articles.
13. Plastics can also be made by microorganisms that create macromolecular structures out of starting substances by fermentation processes. The macromolecule is then either released to a medium or extracted. Potential health risk may occur from the migration of non- or incompletely reacted starting substances, intermediates, or by-products of the fermentation process. In this case, the final product should be risk assessed and authorized before its use in the manufacture of plastic materials and articles.
14. Directive 2002/72/EC contains different lists for monomers or other starting substances and for additives authorized for the manufacture of plastic materials and articles. For monomers, other starting substances, and additives, the Union list is now complete; this means that only substances authorized at EU level may be used. Therefore, a separation of monomers or other starting substances and of additives in separate lists due to their authorization status is no longer necessary. As certain substances can be used both as monomer or other starting substances and as additive for reasons of clarity, they should be published in one list of authorized substances indicating the authorized function.
15. Polymers can not only be used as main structural component of plastics but also as additives achieving defined technological effects in the plastic. If such a polymeric additive is identical to a polymer that can form the main structural component of a plastic material, the risk from polymeric additive can be regarded as evaluated if the monomers have already been evaluated and authorized. In such a case it should not be necessary to authorize the polymeric additive, but it could be used on the basis of the authorization of its monomers and other starting substances. If such a polymeric additive is not identical to a polymer that can form the main structural component of a plastic material, then the risk of the polymeric additive cannot be regarded as evaluated by evaluation of the monomers. In such a case, the polymeric additive should be risk assessed as regards its low-molecular-weight fraction below 1000 Da and authorized before its use in the manufacture of plastic materials and articles.

16. In the past no clear differentiation has been made between additives that have a function in the final polymer and polymer production aids (PPA) that only exhibit a function in the manufacturing process and are not intended to be present in the final article. Some substances acting as PPA had already been included in the incomplete list of additives in the past. These PPA should remain in the Union list of authorized substances. However, it should be made clear that the use of other PPA will remain possible, subject to national law. That situation should be reassessed at a later stage.
17. The Union list contains substances authorized to be used in the manufacture of plastics. Substances such as acids, alcohols, and phenols can also occur in the form of salts. As the salts usually are transformed in the stomach to acid, alcohol, or phenol, the use of salts with cations that have undergone a safety evaluation should in principle be authorized together with the acid, alcohol, or phenol. In certain cases, where the safety assessment indicates concerns on the use of the free acids, only the salts should be authorized by indicating in the list the name as "... acid(s), salts."
18. Substances used in the manufacture of plastic materials or articles may contain impurities originating from their manufacturing or extraction process. These impurities are nonintentionally added together with the substance in the manufacture of the plastic material (nonintentionally added substance—NIAS). As far as they are relevant for the risk assessment, the main impurities of a substance should be considered and if necessary be included in the specifications of a substance. However it is not possible to list and consider all impurities in the authorization. Therefore, they may be present in the material or article but not included in the Union list.
19. In the manufacture of polymers, substances are used to initiate the polymerization reaction such as catalysts and to control the polymerization reaction such as chain transfer, chain extending, or chain stop reagents. These aids to polymerization are used in minute amounts and are not intended to remain in the final polymer. Therefore, they should at this point of time not be subject to the authorization procedure at EU level. Any potential health risk in the final material or article arising from their use should be assessed by the manufacturer in accordance with internationally recognized scientific principles on risk assessment.
20. During the manufacture and use of plastic materials and articles, reaction and degradation products can be formed. These reaction and degradation products are nonintentionally present in the plastic material (NIAS). As far as they are relevant for the risk assessment, the main reaction and degradation products of the intended application of a substance should be considered and included in the restrictions of the substance. However it is not possible to list and consider all reaction and degradation products in the authorization. Therefore, they should not be listed as single entries in the Union list. Any potential health risk in the final material or article arising from reaction and degradation products should be assessed by the manufacturer in accordance with internationally recognized scientific principles on risk assessment.
21. Prior to the establishment of the Union list of additives, other additives than those authorized at EU level could be used in the manufacture of plastics. For those additives, which were permitted in the Member States, the time limit for the submission of data for their safety evaluation by the Authority with a view to their inclusion in the Union list expired on December 31, 2006. Additives for which a valid application was submitted within this time limit were listed in a provisional list. For certain additives on the provisional list, a decision on their authorization at EU level has not yet been taken. For those additives, it should be possible to continue to be used in accordance with national law until their evaluation is completed and a decision is taken on their inclusion in the Union list.
22. When an additive included in the provisional list is inserted in the Union list or when it is decided not to include it in the Union list, that additive should be removed from the provisional list of additives.

23. New technologies engineer substances in particle size that exhibit chemical and physical properties that significantly differ from those at a larger scale, for example, nanoparticles. These different properties may lead to different toxicological properties and therefore these substances should be assessed on a case-by-case basis by the Authority as regards their risk until more information is known about such new technology. Therefore, it should be made clear that authorizations which are based on the risk assessment of the conventional particle size of a substance do not cover engineered nanoparticles.
24. Based on the risk assessment, the authorization should if necessary set out specific migration limits to ensure the safety of the final material or article. If an additive that is authorized for the manufacture of plastic materials and articles is at the same time authorized as food additive or flavoring substance, it should be ensured that the release of the substance does not change the composition of the food in an unacceptable way. Therefore, the release of such a dual-use additive or flavoring should not exhibit a technological function on the food unless such a function is intended and the food-contact material complies with the requirements on active food-contact materials set out in Regulation (EC) No. 1935/2004 and Commission Regulation (EC) No. 450/2009 of May 29, 2009, on active and intelligent materials and articles intended to come into contact with food. The requirements of Regulations (EC) No. 1333/2008 of the European Parliament and of the Council of December 16, 2008, on food additives(2) or (EC) No. 1334/2008 of the European Parliament and of the Council of December 16, 2008, on flavorings and certain food ingredients with flavoring properties for use in and on foods and amending Council Regulation (EEC) No. 1601/91, Regulations (EC) No. 2232/96, and (EC) No, 110/2008, and Directive 2000/13/EC(3) should be respected where applicable.
25. According to Article 3(1)(b) of Regulation (EC) No. 1935/2004, the release of substances from food-contact materials and articles should not bring about unacceptable changes in the composition of the food. According to good manufacturing practice, it is feasible to manufacture plastic materials in such a way that they are not releasing more than 10 mg of substances per 1 dm^2 of surface area of the plastic material. If the risk assessment of an individual substance is not indicating a lower level, this level should be set as a generic limit for the inertness of a plastic material, the overall migration limit. In order to achieve comparable results in the verification of compliance with the overall migration limit, testing should be performed under standardized test conditions including testing time, temperature, and test medium (food simulant) representing worst foreseeable conditions of use of the plastic material or article.
26. The overall migration limit of 10 mg per 1 dm^2 results for a cubic packaging containing 1 kg of food to a migration of 60 mg per kg food. For small packaging where the surface to volume ratio is higher, the resulting migration into food is higher. For infants and small children which have a higher consumption of food per kilogram bodyweight than adults and do not yet have a diversified nutrition, special provisions should be set in order to limit the intake of substances migrating from food-contact materials. In order to allow also for small volume packaging the same protection as for high volume packaging, the overall migration limit for food-contact materials that are dedicated for packaging foods for infants and small children should be linked to the limit in food and not to the surface area of the packaging.
27. In recent years plastic food-contact materials are being developed that do not only consist of one plastic but combine up to 15 different plastic layers to attain optimum functionality and protection of the food, while reducing packaging waste. In such a plastic multilayer material or article, layers may be separated from the food by a functional barrier. This barrier is a layer within food-contact materials or articles preventing the migration of substances from behind that barrier into the food. Behind a functional barrier, non-authorized substances may be used,

provided below a given detection limit. Taking into account foods for infants and other particularly susceptible persons, as well as the large analytical tolerance of the migration analysis, a maximum level of 0.01 mg/kg in food should be established for the migration of a nonauthorized substance through a functional barrier. Substances that are mutagenic, carcinogenic, or toxic to reproduction should not be used in food-contact materials or articles without previous authorization and should therefore not be covered by the functional barrier concept. New technologies that engineer substances in particle size that exhibit chemical and physical properties that significantly differ from those at a larger scale, for example, nanoparticles, should be assessed on a case-by-case basis as regards their risk until more information is known about such new technology. Therefore, they should not be covered by the functional barrier concept.

28. In recent years, food-contact materials and articles are being developed that consist of a combination of several materials to achieve optimum functionality and protection of the food while reducing packaging waste. In these multimaterial multilayer materials and articles, plastic layers should comply with the same compositional requirements as plastic layers which are not combined with other materials. For plastic layers in a multimaterial multilayer, which are separated from the food by a functional barrier, the functional barrier concept should apply. As other materials are combined with the plastic layers and for these other materials, specific measures are not yet adopted at EU level and it is not yet possible to set out requirements for the final multimaterial multilayer materials and articles. Therefore, specific migration limits and the overall migration limit should not be applicable except for vinyl chloride monomer for which such a restriction is already in place. In the absence of a specific measure at EU level covering the whole multimaterial multilayer material or article, Member States may maintain or adopt national provisions for these materials and articles provided they comply with the rules of the Treaty.
29. Article 16(1) of Regulation (EC) No. 1935/2004 provides that materials and articles covered by specific measures be accompanied by a written declaration of compliance stating that they comply with the rules applicable to them. To strengthen the coordination and responsibility of the suppliers at each stage of manufacture, including that of the starting substances, the responsible persons should document the compliance with the relevant rules in a declaration of compliance which is made available to their customers.
30. Coatings, printing inks, and adhesives are not yet covered by a specific EU legislation and therefore not subject to the requirement of a declaration of compliance. However, for coatings, printing inks, and adhesives to be used in plastic materials and articles, adequate information should be provided to the manufacturer of the final plastic article that would enable him to ensure compliance for substances for which migration limits have been established in this Regulation.
31. Article 17(1) of Regulation (EC) No. 178/2002 of the European Parliament and of the Council of January 28, 2002, laying down the general principles and requirements of food law, establishing the European Food Safety Authority, and laying down procedures in matters of food safety requires the food business operator to verify that foods are compliant with the rules applicable to them. To this end and subject to the requirement of confidentiality, food business operators should be given access to the relevant information to enable them to ensure that the migration from the materials and articles to food complies with the specifications and restrictions laid down in food legislation.
32. At each stage of manufacture, supporting documentation, substantiating the declaration of compliance, should be kept available for the enforcement authorities. Such demonstration of compliance may be based on migration testing. As migration testing is complex, costly, and time consuming, it should be admissible that compliance can be demonstrated also by calculations, including modeling, other analysis, and scientific

evidence or reasoning if these render results which are at least as severe as the migration testing. Test results should be regarded as valid as long as formulations and processing conditions remain constant as part of a quality assurance system.

33. When testing articles not yet in contact with food, for certain articles, such as films or lids, it is often not feasible to determine the surface area that is in contact with a defined volume of food. For these articles, specific rules should be set out for verification of compliance.
34. The setting of migration limits takes into account a conventional assumption that 1 kg of food is consumed daily by a person of 60 kg bodyweight and that the food is packaged in a cubic container of 6 dm^2 surface area releasing the substance. For very small and very large containers, the real surface area to volume of packaged food is varying a lot from the conventional assumption. Therefore, their surface area should be normalized before comparing testing results with migration limits. These rules should be reviewed when new data on food packaging uses become available.
35. The specific migration limit is a maximum permitted amount of a substance in food. This limit should ensure that the food-contact material does not pose a risk to health. It should be ensured by the manufacturer that materials and articles not yet in contact with food will respect these limits when brought into contact with food under the worst foreseeable contact conditions. Therefore, compliance of materials and articles not yet in contact with food should be assessed and the rules for this testing should be set out.
36. Food is a complex matrix and therefore the analysis of migrating substances in food may pose analytical difficulties. Therefore, test media should be assigned that simulate the transfer of substances from the plastic material into food. They should represent the major physicochemical properties exhibited by food. When using food simulants, standard testing time and temperature should reproduce, as far as possible, the migration which may occur from the article into the food.
37. For determining the appropriate food simulant for certain foods, the chemical composition and the physical properties of the food should be taken into account. Research results are available for certain representative foods comparing migration into food with migration into food simulants. On the basis of the results, food simulants should be assigned. In particular, for fat-containing foods, the result obtained with food simulant may in certain cases significantly overestimate migration into food. In these cases, it should be foreseen that the result in food simulant is corrected by a reduction factor.
38. The exposure to substances migrating from food-contact materials was based on the conventional assumption that a person consumes daily 1 kg of food. However, a person ingests at most 200 g of fat on a daily basis. For lipophilic substances that only migrate into fat, this should be taken into consideration. Therefore, a correction of the specific migration by a correction factor applicable to lipophilic substances in accordance with the opinion of the Scientific Committee on Food (SCF) and the opinion of the Authority should be foreseen.
39. Official control should establish testing strategies which allow the enforcement authorities to perform controls.
40. Basic rules on migration testing should be set out in this Regulation. As migration testing is a very complex issue, these basic rules can, however, not cover all foreseeable cases and details necessary for performing the testing. Therefore, an EU guidance document should be established, dealing with more detailed aspects of the implementation of the basic migration testing rules.
41. The updated rules on food simulants and migration testing provided by this Regulation will supersede those in Directive 78/142/EEC and the Annex to Council Directive 82/711/EEC of October 18, 1982, laying down the basic rules necessary for testing migration of the constituents of plastic materials and articles intended to come into contact with foodstuffs.
42. Substances present in the plastic but not listed in Annex I to this Regulation have not necessarily been risk assessed as they had not been subject to an authorization procedure.

Compliance with Article 3 of Regulation (EC) No. 1935/2004 for these substances should be assessed by the relevant business operator in accordance with internationally recognized scientific principles taking into account exposure from food-contact materials and other sources.

43. Recently additional monomers, other starting substances, and additives have received a favorable scientific evaluation by the Authority and should now be added to the Union list.
44. As new substances are added to the Union list, the Regulation should apply as soon as possible to allow for manufacturers to adapt to technical progress and allow for innovation.
45. Certain migration testing rules should be updated in view of new scientific knowledge. Enforcement authorities and industry need to adapt their current testing regime to these updated rules. To allow for this adaptation, it seems appropriate that the updated rules only apply 2 years after the adoption of the Regulation.
46. Business operators are currently basing their declaration of compliance on supporting documentation following the requirements set out in Directive 2002/72/EC. Declaration of compliance need, in principle, only to be updated when substantial changes in the production bring about changes in the migration or when new scientific data are available. In order to limit the burden to business operators, materials which have been lawfully placed on the market based on the requirements set out in Directive 2002/72/EC should be able to be placed on the market with a declaration of compliance based on supporting documentation in accordance with Directive 2002/72/EC until 5 years after the adoption of the Regulation.
47. Analytical methods for testing migration and residual content of vinyl chloride monomer as described in Commission Directives 80/766/EEC of July 8, 1980, laying down the Community method of analysis for the official control of the vinyl chloride monomer level in materials and articles which are intended to come into contact with foodstuffs and 81/432/EEC of April 29, 1981, laying down the Community method of analysis for the official control of vinyl chloride released by materials and articles into foodstuffs are outdated. Analytical methods should comply with the criteria set out in Article 11 of Regulation (EC) No. 882/2004 of the European Parliament and of the Council on official controls performed to ensure the verification of compliance with feed and food law, animal health, and animal welfare rules. Therefore, Directives 80/766/EEC and 81/432/EEC should be repealed.
48. The measures provided for in this Regulation are in accordance with the opinion of the Standing Committee on the Food Chain and Animal Health.

19.6.3 Chapter I—General Provisions

19.6.3.1.1 Article 1: Subject Matter

1. This Regulation is a specific measure within the meaning of Article 5 of Regulation (EC) No. 1935/2004.
2. This Regulation establishes specific requirements for the manufacture and marketing of plastic materials and articles:
   a. intended to come into contact with food; or
   b. already in contact with food; or
   c. which can reasonably be expected to come into contact with food.

19.6.3.1.2 Article 2: Scope

1. This Regulation shall apply to materials and articles which are placed on the EU market and fall under the following categories:
   a. materials and articles and parts thereof consisting exclusively of plastics;
   b. plastic multilayer materials and articles held together by adhesives or by other means;
   c. materials and articles referred to in points (a) or (b) that are printed and/or covered by a coating;
   d. plastic layers or plastic coatings, forming gaskets in caps and closures, that together with those caps and closures compose a set of two or more layers of different types of materials;
   e. plastic layers in multimaterial multilayer materials and articles.
2. This Regulation shall not apply to the following materials and articles which are placed on

the EU market and are intended to be covered by other specific measures:
   a. ion exchange resins;
   b. rubber;
   c. silicones.
3. This Regulation shall be without prejudice to the EU or national provisions applicable to printing inks, adhesives, or coatings.

19.6.3.1.3 Article 3: Definitions

For the purpose of this Regulation, the following definitions shall apply:

1. "plastic materials and articles" means:
   a. materials and articles referred to in points (a), (b), and (c) of Article 2(1); and
   b. plastic layers referred to in Article 2(1) (d) and (e);
2. "plastic" means polymer to which additives or other substances may have been added, which is capable of functioning as a main structural component of final materials and articles;
3. "polymer" means any macromolecular substance obtained by:
   a. a polymerization process such as polyaddition or polycondensation, or by any other similar process of monomers and other starting substances; or
   b. chemical modification of natural or synthetic macromolecules; or
   c. microbial fermentation;
4. "plastic multilayer" means a material or article composed of two or more layers of plastic;
5. "multimaterial multilayer" means a material or article composed of two or more layers of different types of materials, at least one of them a plastic layer;
6. "monomer or other starting substance" means:
   a. a substance undergoing any type of polymerization process to manufacture polymers; or
   b. a natural or synthetic macromolecular substance used in the manufacture of modified macromolecules; or
   c. a substance used to modify existing natural or synthetic macromolecules;
7. "additive" means a substance which is intentionally added to plastics to achieve a physical or chemical effect during processing of the plastic or in the final material or article; it is intended to be present in the final material or article;
8. "polymer production aid" means any substance used to provide a suitable medium for polymer or plastic manufacturing; it may be present but is neither intended to be present in the final materials or articles nor has a physical or chemical effect in the final material or article;
9. "nonintentionally added substance" means an impurity in the substances used or a reaction intermediate formed during the production process or a decomposition or reaction product;
10. "aid to polymerization" means a substance which initiates polymerization and/or controls the formation of the macromolecular structure;
11. "overall migration limit" (OML) means the maximum permitted amount of nonvolatile substances released from a material or article into food simulants;
12. "food simulant" means a test medium imitating food; in its behavior, the food simulant mimics migration from food-contact materials;
13. "specific migration limit" (SML) means the maximum permitted amount of a given substance released from a material or article into food or food simulants;
14. "total specific migration limit" (SML(T)) means the maximum permitted sum of particular substances released in food or food simulants expressed as total of moiety of the substances indicated;
15. "functional barrier" means a barrier consisting of one or more layers of any type of material which ensures that the final material or article complies with Article 3 of Regulation (EC) No. 1935/2004 and with the provisions of this Regulation;
16. "nonfatty food" means a food for which in migration testing only food simulants other than food simulants D1 or D2 are laid down in Table 19.2 of Annex V to this Regulation;
17. "restriction" means limitation of use of a substance or migration limit or limit of content of the substance in the material or article;
18. "specification" means composition of a substance, purity criteria for a substance,

physicochemical characteristics of a substance, details concerning the manufacturing process of a substance, or further information concerning the expression of migration limits.

19.6.3.1.4 Article 4: Placing on the Market of Plastic Materials and Articles

Plastic materials and articles may only be placed on the market if they:

a. comply with the relevant requirements set out in Article 3 of Regulation (EC) No. 1935/2004 under intended and foreseeable use; and
b. comply with the labeling requirements set out in Article 15 of Regulation (EC) No. 1935/2004; and
c. comply with the traceability requirements set out in Article 17 of Regulation (EC) No. 1935/2004; and
d. are manufactured according to good manufacturing practice as set out in Commission Regulation (EC) No. 2023/2006; and
e. comply with the compositional and declaration requirements set out in Chapters II, III, and IV of this Regulation.

19.6.4 Chapter II: Compositional Requirements

19.6.4.1 Section 1: Authorized Substances

19.6.4.1.1 Article 5: Union List of Authorized Substances

1. Only the substances included in the Union list of authorized substances (hereinafter referred to as the Union list) set out in Annex I may be intentionally used in the manufacture of plastic layers in plastic materials and articles.
2. The Union list shall contain:
   a. monomers or other starting substances;
   b. additives excluding colorants;
   c. polymer production aids excluding solvents;
   d. macromolecules obtained from microbial fermentation.
3. The Union list may be amended in accordance with the procedure established by Articles 8 to 12 of Regulation (EC) No. 1935/2004.

19.6.4.1.2 Article 6: Derogations for Substances Not Included in the Union List

1. By way of derogation from Article 5, substances other than those included in the Union list may be used as polymer production aids in the manufacture of plastic layers in plastic materials and articles subject to national law.
2. By way of derogation from Article 5, colorants and solvents may be used in the manufacture of plastic layers in plastic materials and articles subject to national law.
3. The following substances not included in the Union list are authorized subject to the rules set out in Articles 8, 9, 10, 11, and 12:
   a. salts (including double salts and acid salts) of aluminum, ammonium, barium, calcium, cobalt, copper, iron, lithium, magnesium, manganese, potassium, sodium, and zinc of authorized acids, phenols, or alcohols;
   b. mixtures obtained by mixing authorized substances without a chemical reaction of the components;
   c. when used as additives, natural or synthetic polymeric substances of a molecular weight of at least 1000 Da, except macromolecules obtained from microbial fermentation, complying with the requirements of this Regulation, if they are capable of functioning as the main structural component of final materials or articles;
   d. when used as monomer or other starting substance, prepolymers and natural or synthetic macromolecular substances, as well as their mixtures, except macromolecules obtained from microbial fermentation, if the monomers or starting substances required to synthesize them are included in the Union list.
4. The following substances not included in the Union list maybe present in the plastic layers of plastic materials or articles:
   a. unintentionally added substances;
   b. aids to polymerization.
5. By derogation from Article 5, additives not included in the Union list may continue to be used subject to national law after January 1, 2010, until a decision is taken to include or not to include them in the Union list provided they are included in the provisional list referred to in Article 7.

19.6.4.1.3 Article 7: Establishment and Management of the Provisional List

1. The provisional list of additives that are under evaluation by the European Food Safety Authority (hereinafter referred to as the Authority) that was made public by the Commission in 2008 shall be regularly updated.
2. An additive shall be removed from the provisional list:
   a. when it is included in the Union list set out in Annex I; or
   b. when a decision is taken by the Commission not to include it in the Union list; or
   c. if during the examination of the data, the Authority calls for supplementary information and that information is not submitted within the time limits specified by the Authority.

19.6.4.2 Section 2: General Requirements, Restrictions, and Specifications

19.6.4.2.1 Article 8: General Requirements on Substances

Substances used in the manufacture of plastic layers in plastic materials and articles shall be of a technical quality and a purity suitable for the intended and foreseeable use of the materials or articles. The composition shall be known to the manufacturer of the substance and made available to the competent authorities on request.

19.6.4.2.2 Article 9: Specific Requirements on Substances

1. Substances used in the manufacture of plastic layers in plastic materials and articles shall be subject to the following restrictions and specifications:
   a. the specific migration limit set out in Article 11;
   b. the overall migration limit set out in Article 12;
   c. the restrictions and specifications set out in column 10 of Table 1 of point 1 of Annex I;
   d. the detailed specifications set out in point 4 of Annex I.
2. Substances in nanoform shall only be used if explicitly authorized and mentioned in the specifications in Annex I.

19.6.4.2.3 Article 10: General Restrictions on Plastic Materials and Articles

General restrictions related to plastic materials and articles are laid down in Annex II.

19.6.4.2.4 Article 11: Specific Migration Limits

1. Plastic materials and articles shall not transfer their constituents to foods in quantities exceeding the specific migration limits (SML) set out in Annex I. Those specific migration limits (SML) are expressed in mg of substance per kg of food (mg/kg).
2. For substances for which no specific migration limit or other restrictions are provided in Annex I, a generic specific migration limit of 60 mg/kg shall apply.
3. By derogation from paragraphs 1 and 2, additives which are also authorized as food additives by Regulation (EC) No. 1333/2008 or as flavorings by Regulation (EC) No. 1334/2008 shall not migrate into foods in quantities having a technical effect in the final foods and shall not:
   a. exceed the restrictions provided for in Regulation (EC) No. 1333/2008 or in Regulation (EC) No. 1334/2008 or in Annex I to this Regulation for foods for which their use is authorized as food additive or flavoring substances; or
   b. exceed the restrictions set out in Annex I to this Regulation in foods for which their use is not authorized as food additive or flavoring substances.

19.6.4.2.5 Article 12: Overall Migration Limit

1. Plastic materials and articles shall not transfer their constituents to food simulants in quantities exceeding 10 mg of total constituents released per dm^2 of food-contact surface (mg/dm^2).
2. By derogation from paragraph 1, plastic materials and articles intended to be brought into contact with food intended for infants and young children, as defined by Commission Directives 2006/141/EC(1), shall not transfer their constituents to food simulants in quantities exceeding 60 mg of total of constituents released per kg of food simulant.

19.6.5 Chapter III: Specific Provisions for Certain Materials and Articles

19.6.5.1.1 Article 13: Plastic Multilayer Materials and Articles

1. In a plastic multilayer material or article, the composition of each plastic layer shall comply with this Regulation.
2. By derogation from paragraph 1, a plastic layer which is not in direct contact with food and is separated from the food by a functional barrier, may:
   a. not comply with the restrictions and specifications set out in this Regulation except for vinyl chloride monomer as provided in Annex I; and/or
   b. be manufactured with substances not listed in the Union list or in the provisional list.
3. The migration of the substances under paragraph 2(b) into food or food simulant shall not be detectable measured with statistical certainty by a method of analysis set out in Article 11 of Regulation (EC) No. 882/2004 with a limit of detection of 0.01 mg/kg. That limit shall always be expressed as concentration in foods or food simulants. That limit shall apply to a group of compounds, if they are structurally and toxicologically related, in particular isomers or compounds with the same relevant functional group, and shall include possible setoff transfer.
4. The substances not listed in the Union list or provisional list referred to in paragraph 2(b) shall not belong to either of the following categories:
   a. substances classified as "mutagenic," "carcinogenic," or "toxic to reproduction" in accordance with the criteria set out in Sections 3.5, 3.6, and 3.7 of Annex I to Regulation (EC) No. 1272/2008 of the European Parliament and the Council;
   b. substances in nanoform.
5. The final plastic multilayer material or article shall comply with the specific migration limits set out in Article 11 and the overall migration limit set out in Article 12 of this Regulation.

19.6.5.1.2 Article 14: Multimaterial Multilayer Materials and Articles

1. In a multimaterial multilayer material or article, the composition of each plastic layer shall comply with this Regulation.
2. By derogation from paragraph 1, in a multimaterial multilayer material or article, a plastic layer which is not in direct contact with food and is separated from the food by a functional barrier may be manufactured with substances not listed in the Union list or the provisional list.
3. The substances not listed in the Union list or provisional list referred to in paragraph 2 shall not belong to either of the following categories:
   a. substances classified as "mutagenic," "carcinogenic," or "toxic to reproduction" in accordance with the criteria set out in sections 3.5, 3.6, and 3.7 of Annex I to Regulation (EC) No. 1272/2008;
   b. substances in nanoform.
4. By derogation from paragraph 1, Articles 11 and 12 of this Regulation do not apply to plastic layers in multimaterial multilayer materials and articles.
5. The plastic layers in a multimaterial multilayer material or article shall always comply with the restrictions for vinyl chloride monomer laid down in Annex I to this Regulation.
6. In a multimaterial multilayer material or article, specific and overall migration limits for plastic layers and for the final material or article may be established by national law.

19.6.6 Chapter IV: Declaration of Compliance and Documentation

19.6.6.1.1 Article 15: Declaration of Compliance

1. At the marketing stages other than at the retail stage, a written declaration in accordance with Article 16 of Regulation (EC) No. 1935/2004 shall be available for plastic materials and articles, products from intermediate stages of their manufacturing as well as for the substances intended for the manufacturing of those materials and articles.

2. The written declaration referred to in paragraph 1 shall be issued by the business operator and shall contain the information laid down in Annex IV.
3. The written declaration shall permit an easy identification of the materials, articles, or products from intermediate stages of manufacture or substances for which it is issued. It shall be renewed when substantial changes in the composition or production occur that bring about changes in the migration from the materials or articles or when new scientific data becomes available.

19.6.6.1.2 Article 16: Supporting Documents

1. Appropriate documentation to demonstrate that the materials and articles, products from intermediate stages of their manufacturing, as well as the substances intended for the manufacturing of those materials and articles comply with the requirements of this Regulation shall be made available by the business operator to the national competent authorities on request.
2. That documentation shall contain the conditions and results of testing, calculations, including modeling, other analysis, and evidence on the safety or reasoning demonstrating compliance. Rules for experimental demonstration of compliance are set out in Chapter V.

19.6.7 Chapter V: Compliance

19.6.7.1.1 Article 17: Expression of Migration Test Results

1. To check the compliance, the specific migration values shall be expressed in mg/kg applying the real surface to volume ratio in actual or foreseen use.
2. By derogation from paragraph 1 for:
   a. containers and other articles, containing or intended to contain, less than 500 mL or g or more than 10 L;
   b. materials and articles for which, due to their form it is impracticable to estimate the relationship between the surface area of such materials or articles and the quantity of food in contact therewith;
   c. sheets and films that are not yet in contact with food;
   d. sheets and films containing less than 500 mL or g or more than 10 L, the value of migration shall be expressed in mg/kg applying a surface to volume ratio of 6 dm^2 per kg of food. This paragraph does not apply to plastic materials and articles intended to be brought into contact with or already in contact with food for infants and young children, as defined by Directives 2006/141/EC and 2006/125/EC.
3. By derogation from paragraph 1, for caps, gaskets, stoppers, and similar sealing articles, the specific migration value shall be expressed in:
   a. mg/kg using the actual content of the container for which the closure is intended or in mg/dm^2 applying the total contact surface of sealing article and sealed container if the intended use of the article is known, while taking into account the provisions of paragraph 2;
   b. mg/article if the intended use of the article is unknown.
4. For caps, gaskets, stoppers, and similar sealing articles, the overall migration value shall be expressed in:
   a. mg/dm^2 applying the total contact surface of sealing article and sealed container if the intended use of the article is known;
   b. mg/article if the intended use of the article is unknown.

19.6.7.1.2 Article 18: Rules for Assessing Compliance with Migration Limits

1. For materials and articles already in contact with food verification of compliance with specific migration limits shall be carried out in accordance with the rules set out in Chapter 1 of Annex V.
2. For materials and articles not yet in contact with food verification of compliance with specific migration limits shall be carried out in food or in food simulants set out in Annex III in accordance with the rules set out in Chapter 2, Section 2.1 of Annex V.
3. For materials and articles not yet in contact with food screening of compliance with the specific migration limit can be performed by applying screening approaches in accordance with the rules set out in Chapter 2, Section 2.2 of Annex V. If a material or article fails to comply with the migration limits in the

screening approach, a conclusion of noncompliance has to be confirmed by verification of compliance in accordance with paragraph 2.

4. For materials and articles not yet in contact with food verification of compliance with the overall migration limit shall be carried out in food simulants A, B, C, D1, and D2 as set out in Annex III in accordance with the rules set out in Chapter 3, Section 3.1 of Annex V.
5. For materials and articles not yet in contact with food, screening of compliance with the overall migration limit can be performed by applying screening approaches in accordance with the rules set out in Chapter 3, Section 3.4 of Annex V. If a material or article fails to comply with the migration limit in the screening approach, a conclusion of noncompliance has to be confirmed by verification of compliance in accordance with paragraph 4.
6. The results of specific migration testing obtained in food shall prevail over the results obtained in food simulant. The results of specific migration testing obtained in food simulant shall prevail over the results obtained by screening approaches.
7. Before comparing specific and overall migration test results with the migration limits, the correction factors in Chapter 4 of Annex V shall be applied in accordance with the rules set out therein.

19.6.7.1.3 Article 19: Assessment of Substances Not Included in the Union List

Compliance with Article 3 of Regulation (EC) No. 1935/2004 of substances referred to in Articles 6(1), 6(2), 6(4), 6(5), and 14(2) of this Regulation which are not covered by an inclusion in Annex I to this Regulation shall be assessed in accordance with internationally recognized scientific principles on risk assessment.

19.6.8 Chapter VI: Final Provisions

19.6.8.1.1 Article 20: Amendments of EU Acts

The Annex to Council Directive 85/572/EEC is replaced by the following:

> *The food simulants to be used for testing migration of constituents of plastic materials and articles intended to come into contact with a single food or specific groups of foods are set out in point 3 of Annex III to Commission Regulation (EU)No 10/2011.*

19.6.8.1.2 Article 21: Repeal of EU Acts

Directives 80/766/EEC, 81/432/EEC, and 2002/72/EC are hereby repealed with effect from May 1, 2011.

References to the repealed Directives shall be construed as references to this Regulation and shall be read in accordance with the correlation tables in Annex VI.

19.6.8.1.3 Article 22: Transitional Provisions

1. Until December 31, 2012, the supporting documents referred to in Article 16 shall be based on the basic rules for overall and specific migration testing set out in the Annex to Directive 82/711/EEC.
2. As from January 1, 2013, the supporting documents referred to in Article 16 for materials, articles, and substances placed on the market until December 31, 2015, may be based on:
   a. the rules for migration testing set out in Article 18 of this Regulation; or
   b. the basic rules for overall and specific migration testing set out in the Annex to Directive 82/711/EEC.
3. As from January 1, 2016, the supporting documents referred to in Article 16 shall be based on the rules for migration testing set out in Article 18, without prejudice to paragraph 2 of this Article.
4. Until December 31, 2015, additives used in glass fiber sizing for glass fiber reinforced plastics which are not listed in Annex I have to comply with the risk assessment provisions set out in Article 19.
5. Materials and articles that have been lawfully placed on the market before May 1, 2011, may be placed on the market until December 31, 2012.

19.6.8.1.4 Article 23: Entry into Force and Application

This Regulation shall enter into force on the 20th day following its publication in the Official Journal of the European Union. It shall apply from May 1, 2011. The provision of Article 5 as regards the use of additives, others than plasticizers, shall apply for plastic layers or plastic coatings in caps and closures referred to in Article 2(1)(d), as from December 31, 2015. The provision of Article 5 as regards the use of additives used in glass fiber sizing for glass fiber reinforced plastics shall apply from December 31, 2015. The provisions of Articles 18(2), 18(4), and 20 shall apply from December 31, 2012.

19.7 EU Legislation for Recycled Plastics

EC has promulgated regulations governing the use of used and scrapped plastics. The entire legislation, Commission Regulation (EC) No. 282/2008 of March 27, 2008, can be found in the Official Journal of the European Union, March 27, 2008 [48].

A helpful questions and answer document published by the Commission accompanies the recycled plastic regulations. The exact document has been provided in the next section.

19.7.1 Questions and Answers on Recycled Plastics in Food-Contact Materials

The EC adopted (in 2008) a Regulation [49] on recycled plastic that will ensure that food packaging made of recycled plastics is safe. The Regulation sets the conditions under which the manufacturers of food-contact materials can use recycled plastics and promotes the objectives of the strategies on the sustainable use of natural resources and the prevention and recycling of waste. At the same time, the regulation is expected to achieve a high level of consumer protection. Here are some more details about the issue:

Question: What are food-contact materials?

Food-contact materials are all materials and articles coming into contact with foods. These include packaging materials but also cutlery, dishes, processing machines, containers, etc. The term also includes materials and articles, which are in contact with water intended for human consumption but it does not cover fixed public or private water supply equipment.

Question: What has been the legal status for the use of recycled plastic in food-contact materials in the EU market?

Recycled plastics in food contact have been covered by the general requirements on food-contact materials laid down in Regulation (EC) No. 1935/2004. Member States have adopted different national measures for recycled plastics. Some Member States prohibit the use of recycled plastic in food contact, some have established an authorization procedure, others have issued recommendations, and there is also a group of EU countries that does not have any national rules in place.

Question: Why is the Commission introducing a Regulation for recycled plastics in food contact?

The main objective of this Regulation is to open a market opportunity for recycling of plastics for food contact. A single market will be created for the use of recycled plastics in food contact by laying down criteria, which ensure the safety of the food-contact material without posing unnecessary burdens on the manufacturers and at the same time allowing for a competitive position as compared to virgin plastics and other materials.

Question: What are the main rules for recycled plastics in the Regulation?

The proposed Regulation seeks to create a clear centralized authorization system, which will allow a unified approach for authorization of food-contact materials throughout the EU. The European Food Safety Authority (EFSA) will be responsible for carrying out the risk assessment on the processes used for recycling of plastic intended for food contact, while the Commission will manage the dossiers of each applicant, putting forward a proposal for the authorization of recycling processes, which are found to be safe. Individual recycling processes are based on specific know-how and technology, therefore the regulation provides for individual authorizations of each process. The authorization holders are responsible for the process being implemented in all recycling premises according to the authorization that has been granted. They have to notify to Member States the premises that apply in the recycling process. Member States are responsible for the control of the recycling premises in their territory.

Question: What are the requirements for a recycling process to be authorized for use in the manufacture of food-contact materials in the EU?

For a recycling process to be authorized, it needs to be demonstrated that an adequate control system is in place ensuring that the plastic being recycled originates only from plastic materials and articles that have been manufactured in accordance with Community legislation on plastic food-contact materials and articles. It must be demonstrated that the process is able either to reduce any contamination in the plastic from previous use to a concentration that does not pose a risk to human health or to rule out a contamination of the plastic. Furthermore, it needs to be demonstrated that the quality of the recycled plastic is characterized and controlled by a quality assurance system. Authorization will only be granted for those applications for which safe use can be demonstrated.

Question: What are the labeling requirements for the use of recycled plastics?

Labeling of recycled plastics used in food contact is voluntary. However, if a manufacturer wants to label the use of recycled plastics, he shall follow the rules of ISO 14021:1999 or equivalent.

Question: Will the legislation affect innovation in the sector?

The aim of the Regulation is to create a level playing field within the EU for virgin and recycled plastics and to encourage innovation in the area of recycled and sustainable packaging.

Question: How will the Regulation affect food-contact materials containing recycled plastics from Third Countries?

The Regulation also covers recycled plastics from Third Countries. Also, these can only be used if the recycling process is authorized. The Commission has to notify about premises in Third Countries that use the authorized recycling processes. A level playing field is established for products from Third Countries and those originating from within the EU.

Question: Can Member States ban/approve recycled plastics for food contact independently of the EU?

As from the establishment of the first list of authorized recycling processes, Member States cannot ban or approve independently of the EU recycled plastics for food contact.

Acknowledgment

A previous version of this chapter was published in Plastic Films in Food Packaging—Materials, Technology, and Applications, edited by S. Ebnesajjad, Elsevier, December 2012.

References

[1] The 11th Meeting ACCSQ—PFPWG in Manila, Philippines, http://foodcontact.dss.go.th; July 2010.

[2] www.foodstandards.gov.au/code/proposals/Documents/P1034-Packaging-CFS-SD2.pdf; 2015.

[3] Catherine RN. www.packaginglaw.com; 2012.

[4] Pamela LL. New EU Legislation on Food Contact Plastics Enacted, Significant changes on horizon, Keller and Heckman, LLP, www.packaginglaw.com/3125_.shtml; March 2011.

[5] Understanding FDA Food Packaging Regulations, White Paper, DOW Corp, Form No. 768-23201-0212 SMG; February 2012.

[6] Saulo AA, Bisphenol A. Food safety and technology; May 2008.

[7] Lang IA, Galloway TS, Scarlett A, et al. Association of urinary bisphenol A concentration with medical disorders and laboratory abnormalities in adults. JAMA 2008;300(11):1303–10.

[8] Stahlhut RW, van Wijngaarden E, Dye TD, Cook S, Swan SH. Concentrations of urinary phthalate metabolites are associated with increased waist circumference and insulin resistance in adult U.S. males. Environ Health Perspect 2007;115(6):876–82 pmid:

[9] Richter CA, Birnbaum LS, Farabollini F, et al. In vivo effects of bisphenol A in laboratory rodent studies. Reprod Toxicol 2007;24(2):199–224.

[10] www.fda.gov/NewsEvents/PublicHealthFocus/ucm064437.htm; 2012.

[11] www.cnn.com/2012/03/30/health/bpa-ban-denial/ index.html; 2015.

[12] STATEMENT OF EFSA on a study associating bisphenol A with medical disorders, European Food Safety Authority J 2008;838:1–3.

[13] http://europa.eu/rapid/pressReleasesAction.do?reference = IP/11/664&format = HTML; 2015.

[14] http://www.efsa.europa.eu/en/topics/topic/bisphenol.htm; 2015.

[15] www.fda.gov/food/ingredientspackaginglabeling/packagingfcs/indirectadditives/default.htm; 2015.

[16] www.fda.gov/food/ingredientspackaginglabeling/packagingfcs/ucm064161.htm; 2015, Reprinted from: Shanklin AP, Sánchez ER, Series edited by Cianci S, Food Safety Magazine; October/November 2005.
[17] Guidance for Submitting Petitions and Notifications, www.fda.gov/Food/Guidance Regulation/GuidanceDocumentsRegulatory Information/IngredientsAdditivesGRAS Packaging/default.htm; 2015.
[18] Inventory of Effective Premarket Notifications for Food Contact Substances, www.accessdata.fda.gov/scripts/fdcc/?set = fcn; 2015.
[19] Submitting Requests Under 21 CFR 170.39 Threshold of Regulation for Substances Used in Food-Contact Articles, www.accessdata.fda.gov/scripts/fdcc/?set = fcn; 2015.
[20] Pre-petition Consultations for Food Additives and Color Additives, www.accessdata.fda.gov/scripts/fdcc/?set = fcn; 2015.
[21] Prior Sanctioned Food Ingredients, http://www.accessdata.fda.gov/scripts/cdrh/cfdocs/cfcfr/CFRSearch.cfm?CFRPart = 181; 2015 and Inventory of Effective Premarket Notifications for Food Contact Substances, www.accessdata.fda.gov/scripts/fdcc/?set = fcn; 2015.
[22] Threshold of Regulation Exemptions, www.fda.gov/Food/IngredientsPackagingLabeling/PackagingFCS/ThresholdRegulationExemptions/default.htm; or Submitting Requests Under 21 CFR 170.39 Threshold of Regulation For Substances Used in Food-Contact Articles, www.fda.gov/Food/IngredientsPackaging Labeling/PackagingFCS/ThresholdRegulation Exemptions/default.htm; 2015.
[23] McKeen LW. The effect of sterilization methods on plastics and elastomers. 2nd ed. Oxford, UK: Plastics Design Library, Elsevier; 2012.
[24] Scallan E, Hoekstra RM, et al. Foodborne illness acquired in the United States—major pathogens US Centers for Disease Control and Prevention (www.CDC.gov) Emerg Infect Dis 2011;17(1).
[25] Han JH, editor. Packaging for foods treated with ionizing radiation, packaging for non thermal processing of food. Blackwell Publishing; 2007.
[26] www.accessdata.fda.gov/scripts/cdrh/cfdocs/cfcfr/CFRSearch.cfm?CFRPart = 179; 2015.
[27] Feed Irradiation. Food Irradiation Processing Alliance, an affiliate of the International Irradiation Association, http://foodirradiation.org; 2015.
[28] Sommers CH, Fan X. Food irradiation research and technology. IFT Press, Blackwell Publishing; 2006.
[29] www.fda.gov/Food/IngredientsPackaging Labeling/IrradiatedFood Packaging/ucm081399.htm; 2015.
[30] www.fda.gov/Food/IngredientsPackaging Labeling/PackagingFCS/RecycledPlastics/default.htm; 2015.
[31] Food Ingredients and Packaging Approval and Notifications Programs, www.fda.gov/Food/IngredientsPackagingLabeling/default.htm; 2015.
[32] Environmental Protection Agency (EPA). Characterization of Municipal Solid Waste in the United States: 1994 Update, EPA/530-S-94-042, Washington, DC 20460, Table 19.7; November 1994.
[33] EPA. Municipal Solid Waste in the United States: 1999 Final Report, Washington, DC 20460, Table 19.7, http://www.epa.gov/epaoswer/ non-hw/muncpl/msw99.htm#links4; 1999.
[34] Plastics Market Analysis. Plastics Recycling Update; April 1996. P. 1.
[35] Adding Custom PET Bottles Helps Increase Plastics Recycling. Waste age's recycling times; April 16, 1996. p. 7.
[36] PET Finds Growing Use in Non-Food Containers. Modern plastics; April 1997. p. 60–65.
[37] Schumacher JW, Petesch M, Sakuma Y, Leder A, Jaeckel M. Polyethylene terephthalate (PET) solid-state resins, chemical economics handbook—SRI international, plastics and resins; July 1997. p. 580.1180 H, 580.1180 W-X, and 580.1181 C-I.
[38] Phoenix Technologies, L.P. RPET source control submission data, FDA submission CTS# 51519. This submission is accessible via a Freedom of Information Act request; June 18, 1998.
[39] Begley TH, McNeal TP, Biles JE, Paquette KE. Evaluating the potential for recycling all PET bottles into new food packaging. Food Addit Contam 2002;19(1):135–43.

[40] Demertzis PG, Johansson F, Lievens C, Franz R. Development of a quick inertness test procedure for multi-use PET containers—sorption behavior of bottle wall strips. Packaging Technol Sci 1997;10:45–58.

[41] Komolprasert V, Lawson A. Residual contaminants in recycled poly(ethylene terephthalate): effects of washing and drying. Plastics, rubber and paper recycling: a pragmatic approach, Ch. 35. ACS Symposium Series vol. 609; 1995. p. 435–44.

[42] Piringer O, Franz R, Huber M, Begley TH, McNeal TP. Migration from food packaging containing a functional barrier: mathematical and experimental evaluation. J Agric Food Chem 1998;46(4):1532–8.

[43] Crank J. The mathematics of diffusion. 2nd ed. Oxford: Clarendon Press; 1975. p. 56–9 [equations No. 4.43, 4.41]

[44] Sadler G, Pierce D, Lawson A, Suvannunt D, Senthil V. Evaluating organic compound migration in poly(ethylene terephthalate): a simple test with implications for polymer recycling. Food Addit Contam 1996;13(8):979–89.

[45] Baner A, Brandsch J, Franz R, Piringer O. The application of a predictive migration model for evaluating the compliance of plastic materials with European Food Regulations. Food Addit Contam 1996;13(5):587–601.

[46] Pamela LL. New EU legislation on food contact plastics enacted; significant changes on horizon, Keller and Heckman, LLP, www.packaginglaw.com/3125_.shtml; March 2011.

[47] The Official Journal of European Union, http://eur-lex.europa.eu/LexUriServ/LexUriServ.do?uri = OJ:L:2011:012:0001:0089:EN:PDF; January 15, 2011.

[48] http://eur-lex.europa.eu/oj/direct-access.html; 2015.

[49] Journal of the European Union, http://eur-lex.europa.eu/legal-content/en/TXT/?uri = OJ:L:2008:081:TOC, 2015; March 27, 2008.

Appendix: Writing Guide for Packaging Films and Other Multilayer Structures

Scott B. Marks
E.I. du Pont de Nemours & Company, Wilmington, DE, United States

A wide variety of multilayer structures are used in the flexible and rigid packaging industry, which are put together by an assortment of techniques including: extrusion lamination, adhesive lamination, extrusion coating, as well as coextrusion blown film, cast film, sheet and blow molding.

Currently, there does not exist a standard method for communicating structures in a written format between companies (and for the most part, even within companies). The guidelines described below are derived from formats used within DuPont. While no one method is best, we have found this guide to be useful in that it presents a systematic way to not only communicate the components of the structure but also the technology used to put it together. By following the guidelines, one can better understand whether, for example, a structure is manufactured using either coextrusion or tandem extrusion coating. Such concise writing can improve communication. Thus we offer these guidelines to the industry in the hope that they will increase communication effectiveness and understanding between all persons in our business.

The format guidelines allow for use in handwritten communication or for typing using a keyboard, which is especially important for the electronic communication methods in use today.

One commonly misunderstood character is the dash or minus sign. There are actually three different dashes available in most word processors. They are:

- The minus sign − (a short length dash)
- An En dash – (a medium length dash)
  {With the "Num Lock" on, hold down the Alt key and type 0150 on the keypad.}, or
  {With the "Num Lock" on, hold down the Ctrl key and type "-" on the keypad.}
- An Em dash — (a long length dash)
  {With the "Num Lock" on, hold down the Alt key and type 0151 on the keypad.}, or
  {With the "Num Lock" on, hold down the Ctrl &Alt keys and type "-" on the keypad.}

In Microsoft Word, they are also shown on the Special Character tab of the symbol window.

In this guide the dash is called out for one symbol and is meant to have a space before and after the dash to make the code more legible. One can also use the En dash to replace the dash with the before and after spaces.

Note: The information contained herein is only a guide. The structures listed herein are a mixture of actual and potential applications, but users should not rely upon it absolutely for specific applications since performance properties will vary with processing and specific end-use conditions. It is given and accepted at user's risk and confirmation of its validity and suitability in particular cases should be obtained independently. We make no guarantees of results and assume no obligation or liability in connection with its advice. This publication is not to be taken as a license to operate under, or recommendation to infringe, any patents (Tables A.1–A.14).

Table A.1 Key Definitions

Primer	A bonding agent that is applied in a liquid form to a solid surface. It is intended to promote the adhesion of another liquid or molten material that will subsequently be applied on top of the primer. Typically, it is dried, prior to having the subsequent material applied to it. *(This is usually associated with extrusion coating.)* In some parts of the world these are referred to as anchor coatings (*AC*). Adhesion generally increases after curing time.
Adhesive	A bonding agent that is applied in a liquid or paste form to one or more solid surfaces. It is intended to promote adhesion of the solid surfaces to one another. The adhesive is typically dried, and then the solid surfaces are brought together with heat and/or pressure to active the adhesive. *(This usually is associated with water based, solvent based, or solventless dry lamination.)* Adhesion generally increases after curing time. An adhesive can be abbreviated as "ADH," or "adh."

Table A.2 Key Abbreviations

VM	A vacuum metallization process for applying a thin metal layer to a solid surface, such as OPET and OPP films. This can also be applied to paper, printed surfaces, and other materials. (*VM* can also be written, "*met.*") The metal is usually aluminum, but can also be other metals such as copper, gold, tin, etc. (If it is not aluminum, please specify in your communication.)
vm	Vacuum metallized
met.	Metallized
SiO_x	A silicon oxide coating that is typically applied to a film such as OPET. This usually transparent coating is applied in a vacuum deposition process.
O.L.	An overlacquer that is applied over ink to add scratch protection and gloss.
W.O.	White opaque. This is a white pigmented material with higher opacity.
Tie	Generically indicates a coextrudable adhesive resin. This is used in structure notations when the resin brand or grade is unknown. If the tie material base resin is known then please indicate: for example, LLDPE tie, EVA tie, PP tie, etc. If the brand is known, please indicate, such as DuPont Bynel®, or Mitsui Admer®. If the grade number is known, please indicate.
MB	A "masterbatch" or concentrate. This could be for pigments such as TiO_2 (white MB) or for slip and antiblock concentrates. For example: white MB, slip MB, etc.

Table A.3 Key Symbols

/	The boundary between a solid interface and a liquid or molten material being applied to it. For example: *OPET/ink/primer/33 μm EMAA* is an oriented polyester film, printed with one or more inks, and dried. A primer is applied to the printed surface, dried, then extrusion coated with 33 μm of an EMAA resin.
( )	A set of parentheses. These indicate that the materials within are being coextruded. Look for a short dash " – " sign with spaces or an En dash "–" between materials.
" – "	Indicates the boundary between two molten materials being coextruded together. They should be enclosed in a set of parentheses, "( )." Use a space on each side of the dash or En dash. For example: *(1 mil LDPE–2 mil EAA)* is a two-layer coextruded film of low-density polyethylene and EAA.
[ ]	A set of straight-sided brackets. These indicate that the materials within are being blended. If known, percentages by *weight ratio* for the materials being blended should be indicated. Look for a plus sign between the materials.
"+"	Indicates two materials being blended together. They should be enclosed in a set of straight-sided brackets, "[ ]." Use a space on each side the plus. For example: *([75% Nylon 6 + 25% Amorphous PA] – tie – LLDPE)* is a blend of Nylon 6 and Amorphous Nylon, in the *weight ratios* indicated, is the outer layer of a three-layer coextrusion. The middle layer is an unknown coextrudable adhesive resin, and the inner layer is an LLDPE.
//	Indicates the boundary between two solid surfaces, one is a film, and the other is a film that is often coated with an adhesive. (Typical for adhesive lamination.) For example: *OPET/adhesive//25 μm ionomer film* an oriented polyester film that has an adhesive applied to it and dried. An ionomer film is then combined, typically with a heated nip roll to activate the adhesive.
"//TL or BL//"	Indicates the boundary between two warm solid layers that are joined together by only heat and pressure. Examples would be: "*//TL//*" Thermally laminated films. Two films (or one film and one foil), combined typically between rubber rolls, following preheating of one or both films. "*//BL//*" Blown films that are intentionally blocked in the collapsing nip. A "*B*locked *L*amination". In addition, surround the structure with braces or curly brackets, "{ }", to indicate a combined web. See examples in the following pages.
< or >	Indicates direction to which a surface treatment has been applied, such as; flame, corona, ozone, plasma. For example: *Paper< flame/LDPE/foil/LDPE.*

Table A.4 Materials (Generic Representations) for Webs/Films/Substrates

OPET or BOPET	Biaxially oriented polyester film. (Occasionally called BOPET.) Using "OPET" will prevent confusion with extruded PET.
VM-OPET	Vacuum-metallized OPET film Metallization on the left
OPET-VM	Vacuum-metallized OPET film Metallization on the right
OPET-SiO*x*	OPET film with silicon oxide on the right side of the film
SiO*x*-OPET	OPET film with silicon oxide on the left side of the film
OPET-PVdC	Oriented polyester film with a PVdC coating on the right
PVdC-OPET	Oriented polyester film with a PVdC coating on the left
PVdC-OPET-PVdC	Oriented polyester film, two side coated with PVdC
OPP or BOPP	Biaxially oriented polypropylene film
VM-OPP	Oriented polypropylene film, with metallization on the left side
OPP-VM	Oriented polypropylene film, with metallization on the right side
OPP-A	Oriented polypropylene film, with an acrylic coating on the right side
A-OPP	Oriented polypropylene film, with an acrylic coating on the left side
CPP	Cast polypropylene film
VM-CPP	Cast polypropylene film with metallization on the left side
CPP-VM	Cast polypropylene film with metallization on the right side
C-Nylon	Cast nylon film, also written as *Cast PA*, for cast polyamide film
O-Nylon	Uniaxially oriented nylon film
BONy	Biaxially oriented nylon film
BOPA	Biaxially oriented polyamide (nylon) film
K-Nylon	Biaxially oriented nylon film, with PVdC on one side
BONy-PVdC	"K-Nylon," but written to specify PVdC on the right side
Alu	Aluminum (Aluminium), can also be written as "*AL*"
Foil	Usually thin aluminum. Can be other metals, if not aluminum, specify. Examples: tin foil, gold foil. (In Europe the word "foil" is used to indicate any thin film, and can cause communication issues globally.)
Cello	Plain, uncoated cellophane
K-Cello	Two-side PVdC-coated cellophane
NC-Cello	Two-side nitrocellulose-coated cellophane
Paper	Lightweight paper-based webs. Please specify when possible: Kraft, Bleached Kraft, SBS, clay coated, etc.
Board	Heavy weight paper-based webs. Please specify when possible.
Nonwoven	Nonwoven webs. If the brand is unknown, use nonwoven HDPE, nonwoven PP, and nonwoven PET. Use brand name if known; such as DuPont Tyvek®, PGI Reemay®, PGI Typar®.
Woven Fabric	Fabric weaves. For example: "tight weave"; woven PP, woven HDPE, woven PP.
Scrim	Woven mesh type fabrics with an "open weave"; for example: Scrim PP, Scrim PET, Scrim HDPE.

Table A.5 Thickness or Gauge Representations

#	"pounds per ream" (a ream being 3000 square feet)
Mils	English measurement of 0.001 or 1/1000 of an inch.
Inches	English unit for thick layers, such as in bottles and sheet applications
μm	Metric measurement (micron), 0.001 of a millimeter (it is common in this industry to also just write the "μ" symbol when referring to microns or micrometers, instead of the full "μm".)
mm	Millimeters, metric unit used for thick layers, such as in bottle and sheet applications
gsm	grams per square meter
pt.	"point", which is a unit of thickness for paperboard. (1 pt. approximately = 1 mil)
ga	"gauge" of a film in some systems. e.g. 100 ga = 1 mil = 25.4 μm: 48 ga = 12 μm = 0.5 mils

Table A.6 Adhesives and Primers

PUR	Polyurethane; *PUR adhesive or PUR primer*, can be solvent or water based
PEI	Polyethylene-imine; *PEI primer*, usually a water-based primer
EAA	Ethylene acrylic acid; *EAA primer* (water based)
EAC	Ethylene acrylate; *EAC primer* (water based)
Other primers	(there are several others such as organic titanates, etc.)

Table A.7 Surface Treatments

Corona	Electron discharge excitement of atmospheric air over the film surface.
Plasma	Electron discharge excitement of a gas other than standard air over the film surface. Examples would include: nitrogen, argon, helium, neon, etc. (as well as gas mixtures.)
Flame	Surface treatment with burning gases. The gas mixture could be either an oxidizing flame, reducing flame, or stoichiometric flame.
Ozone	Molten web oxidation by exposing the polymer to a dry ozone flow created off-line and pumped to the molten web area, through a tube with small outlet holes along its length. Occasionally seen on extrusion coating lines.

Table A.8 Extrudable Resins (Generic Names for When Brand/Grade is Unknown)

HDPE	High-density polyethylene
HMW-HDPE	High molecular weight HDPE
MDPE	Medium-density polyethylene
LDPE	Low-density polyethylene
LLDPE	Linear low-density polyethylene
VLDPE	Very low-density (linear) polyethylene (also- VLLDPE)
ULDPE	Ultra low-density polyethylene (also- ULLDPE)
mPE	Metallocene polyethylene, generic
mLMDPE	Metallocene linear medium-density polyethylene
mLLDPE	Metallocene linear low-density polyethylene
mVLDPE	Metallocene very low (linear) density polyethylene (sometimes referred to as a "plastomer"), such as ExxonMobil Exact®, Dow Engage®, Dow Affinity®, LG Chem Lucene™
PP	Polypropylene (generic indication when minimal information is known)
CoPP	Copolymer polypropylene
HoPP	Homopolymer polypropylene
ACR	Acid copolymer resin, generic name for EAA and EMAA resins
EAA	Ethylene acrylic acid copolymer, such as DuPont Nucrel®, Dow Primacor®, and ExxonMobil Escor®
EMAA	Ethylene methacrylic acid copolymer, such as DuPont Nucrel®
Ionomer	Ionomer copolymer resins, such as DuPont Surlyn®
Acrylate	Acrylate copolymers, such as Arkema Lotryl®, DuPont Elvaloy AC®, ExxonMobil Optema®, Westlake EMAC®
EBA	Ethylene butyl acrylate
EEA	Ethylene ethyl acrylate
EMA	Ethylene methyl acrylate
EMMA	Ethylene methyl methacrylate
EiBA	Ethylene *iso*-butyl acrylate
EnBA	Ethylene normal-butyl acrylate
EVA	Ethylene vinyl acetate; such as DuPont Elvax®, Celanese Ateva®, LyondellBasell Ultrathene®, ExxonMobil Escorene Ultra®
PS	Polystyrene
EPS	Expanded (or foamed) polystyrene
HIPS	High impact polystyrene
GPPS	General purpose polystyrene
PVC	Polyvinyl chloride
PVdC	Polyvinylidene chloride. Most commonly seen as a coating on a film, but there are also extrudable grades, such as Dow Saran®, and Solvay Ixan®
PA	Polyamide, commonly called *Nylon*
Nylon	Polyamide, when unspecified, quite often is *Nylon 6*
Nylon 6	Polymer from caprolactam (also written "PA 6")

(*Continued*)

Table A.8 Extrudable Resins (Generic Names for When Brand/Grade is Unknown)—*Cont'd*

Nylon 6.6	Polymer from adipic acid & hexamethylenediamine (also written "PA 6,6" or "PA 6:6" or "PA 66")
Nylon 6/6.6	Copolymer of 6 and 6.6 types of polyamide (also written "*PA 6/6.6*" or *PA 6:66*)
Nylon (other)	other nylon polymers: Nylon 11, Nylon 12, Nylon 6/12, etc...
Amorphous PA	Amorphous nylon; DuPont Selar PA®, or EMS Grivory®
MXD6	Crystalline nylon; Mitsubishi Nylon-MXD6
EVOH	Ethylene vinyl alcohol; Kuraray Eval®, Nippon-Gohsei Soarnol®, Arkema/CCPC Evasin®
PAN	Polyacrylontrile, sometimes written as "an"
SAN	Styrene acrylonitrile copolymer
AN-MA	Acrylonitrile methyl acrylate copolymer (such as Ineos Barex®, which was being decommissioned end of 2014)
ABS	Acrylonitrile butadiene styrene copolymer
LCP	Liquid crystal polymer
COC	Cyclic olefin copolymer, such as TAP Topas®, or Mitsui Apel®
PUR	Polyurethane, extrudable type when seen in a coextrusion, for example
SBC	Styrene butadiene copolymer (such as Chevron Philips K-Resin®)
PET	Extruded polyester. Could be monolayer or in a coextrusion
PET ext.ctg.	Extrusion coating of polyester
APET	Amorphous polyester
CPET	Crystalline polyester
PETG	Polyester copolymer with glycol
PEN	Polyethylene napthalate
PLA	Polylactic acid
PGA	Polyglycolic acid
PHA	Polyhydroxyalkanote

Table A.9 Terpolymers

EVACO	Terpolymer of ethylene, vinyl acetate, and carbon monoxide, such as DuPont Elvaloy®
EnBACO	Terpolymer of ethylene, normal-butyl acrylate, and carbon monoxide, such as DuPont Elvaloy®
EiBAMAA	Terpolymer of ethylene, isobutyl acrylate, and methacrylic acid
Other terpolymers	(There are various other terpolymers in the industry based using a variety of monomers...)

Table A.10 Grafted Resins

EVA-gMAh	Ethylene vinyl acetate with a graft of maleic anhydride
LLDPE-gMAh	LLDPE with a graft of maleic anhydride
Other grafted polymers	(There are various other grafted resins in the industry based on HDPE, PP, LDPE, EPR, EPDM, PS, etc.)

Table A.11 Others

Peel Seal or Easy Peel	Generic name for a peelable sealant resin of unknown brand/grade. If brand is known, please indicate, such as DuPont Appeel®, Mitsui-DuPont CMPS®, Yasuhara Hirodyne®, or Toyo Topco®
H.S. lacquer	Generic name for a "heat seal lacquer," usually applied by gravure
Hot Melt	Generic name for a hot melt adhesive applied as a sealant

Table A.12 Trademarks

Arkema	Evatane, Lotader, Lotryl, Orevac, Orgalloy, Pebax, Rilsan, and Evasin (in conjunction with Chang Chun Petrochem (Taiwan)
Avintiv	Reemay, Typar
Celanese	Ateva
Chevron Phillips Chemical:	K-Resin, Marlex, MarFlex
Bemis Company, Inc.:	EZ Peel
Dow Chemical:	Affinity, Amplify, Attane, Dowlex, Elite, Engage, Flexomer, Infuse, Nordel, Primacor, Saran, Tuflin, Versify
DuPont-Teijin Films	Mylar, Melinex
E.I. du Pont de Nemours & Co.	Appeel, Booster, Bynel, Conpol, Elvaloy, Elvaloy AC, Elvax, Entira, Fusabond, Hytrel, Nucrel, Selar, Surlyn, Tyvek, Zytel
EMS Group:	Grivory, Grilon
LyondellBasell Chemical	Ultrathene, Petrothene, Plexar, Pro-Fax, Moplen, Adflex, Adstif, Adsyl, Toppyl
Evonik	Vestamid
ExxonMobil Chemical:	Enable, Escor, Escorene, Exact, Exceed, Exxelor, Optema, Vistalon, Vistamaxx
Ineos:	Barex
Kuraray Chemical:	Eval
Mitsui Chemical:	Apel, Admer
Mitsubishi Chemical:	Modic
Mitsui-DuPont Chemical:	CMPS
Nippon Gohsei	Soarnol
Phillips 66	Copylene
Solvay Chemical:	Ixan (resin), Diofan (dispersion)
Topas Advanced Polymers:	Topas
Toyo Chemical:	Topco
Westlake Chemical:	Hifor, Elevate, EMAC, EBAC, Mxsten, Tymax
Yasuhara Chemical:	Hirodyne, Hirotac

Table A.13 Structure Examples (Structures Should Be Written From "Outside to Inside," With Regards to the Order of Layers, Starting From the Left)

OPET/ink/adhesive//CPP
An adhesive lamination of cast polypropylene film to reverse printed polyester film. An unknown adhesive is applied to the printed polyester film surface.
OPET/primer/CoPP
An extrusion coating of copolymer polypropylene on to an unprinted polyester film. There is an unknown primer applied to the OPET.
48 ga OPET/ink/PEI primer/28# LDPE
A reverse printed 48-gauge OPET film that is primed with a PEI primer, and then extrusion coated with 28 pounds per ream of low density polyethylene.
15 μm BONy/ink/PUR adh//88 μm LDPE
An adhesive lamination of an LDPE film to a reverse printed biaxially oriented nylon film. The PUR adhesive is applied to the reverse printed nylon film.
0.5 mils OPET-PVdC/ink/primer/0.8 mils LDPE/1.5 mils Peel Seal
A 0.5-mils OPET film with PVdC coating is reverse printed on the PVdC side. The printed side is then primed and extrusion coated with 0.8-mils LDPE. A 1.5-mils Easy Peel or Peel Seal extrusion coating resin is applied onto the LDPE. This is done in either tandem or a two-pass extrusion coating.
12 μm OPET/primer/15 μm LDPE/6 μm alu/(11 μm EMAA – 25 μm Easy Peel)
A 12-μm OPET film is primed and extrusion laminated with 15 μm of LDPE to 6 μm aluminum foil. The bare aluminum side is then coextrusion coated with a two-layer coating that is 11 μm of EMAA and 25 μm of Easy Peel resin.
OPET/primer/(LDPE – [LDPE + white MB]- LDPE)/foil/EMAA
An OPET film which is primed, and then has foil extrusion laminated to it using a three-layer coextrusion, with the center layer containing a white masterbatch. The foil is then extrusion coated on the other side with EMAA.
20 μm OPP/ink/adhesive//7 μm alu/33 gsm EAA
A 20-μm OPP film that is reverse printed, then adhesively laminated to 7 μm aluminum foil. The adhesive is applied to the printed surface, not the aluminum. The laminate is then extrusion coated with 33 gsm of EAA.
20 μm OPP-PVdC/low temp PUR primer/35 μm EVA
PVdC coated OPP extrusion coated with an 18% EVA. A special low activation temperature PUR primer is used to allow for direct extrusion coating of an EVA at 235°C onto a primed film.
BONy/ink/solventless adh//(LLDPE – LDPE – Ionomer)
A solventless adhesive lamination of a reverse printed biaxially oriented nylon film to a three-layer coextruded film. The adhesive is applied to the printed surface of the nylon film.
(Nylon 6:66 – Tie A – Nylon 6 – Tie B – Ionomer)
A five-layer coextruded film. You will need to specify if this is made by a cast film or blown film process. In a straight coextrusion, this is helpful to know for physical property issues.
OPP< corona/solvent adhesive//([LLDPE + LDPE] – tie - Nylon 6 – tie – LLDPE - LDPE – Ionomer)
An OPP film that is corona treated in-line, then has a solventless adhesive applied to its surface. This film is then combined with a seven-layer coex film, likely with a typical heated pressure nip.
O.L./ink/40# Paper< flame/0.7 mils LDPE/0.35 mils alu/1.8 mils Ionomer
A surface printed and overlacquered paper is flame treated, and extrusion laminated using LDPE to aluminum. The other side of the aluminum is then extrusion coated with the Ionomer. You will need to specify if this is done in a two-pass operation on a single station extrusion coating line, or one pass on a tandem extrusion coating line.

(*Continued*)

Table A.13 Structure Examples (Structures Should Be Written From "Outside to Inside," With Regards to the Order of Layers, Starting From the Left)—*Cont'd*

O.L./ink/Paper/EMAA/alu/[Ionomer + Slip MB]
A surface printed and overlacquered paper that is extrusion laminated to aluminum with EMAA. The other side of the aluminum is then extrusion coated with a blend of Ionomer and masterbatch.
(CoPP – tie A – Nylon 6 – EVOH – Nylon 6 – tie B – Ionomer)
A seven-layer coextruded film. The tie layer resins are not known, but it is known that there are two different grades being used in the one structure. You will need to specify whether it is made by a cast film or blown film process.
(Nylon 6 – tie – Nylon 6 – tie – LLDPE – mPE – Ionomer)
A seven-layer coextruded film. The tie layers are believed to be the same, so are not labeled differently. A mPE is used to adhere the Ionomer to the LDPE. An upgrade would be to specify the mPE grade or density and the Ionomer grade. You will need to specify whether it is made by a cast film or blown film process.
(15 μm Nylon 6 – 5 μm LLDPE tie – 45 μm LLDPE)
A three-layer coextruded film. As each layer thickness is known, they are specified in the structure. While the tie resin grade is not known, it is known to be based on LLDPE. You will need to specify whether it is a cast film or blown film.
(20 μm [Nylon 6 + Amorphous PA] – 4 μm tie – 40 μm LLDPE)
A three-layer coex film. The outer layer blend ratio is unknown. The layer thickness for each layer is specified. You will have to specify if this is made by a blown or cast film process.
([80% Nylon 6 + 20% Amorphous PA] – tie – EVOH – tie – LLDPE)
A five-layer coextruded film. The outer layer is an 80% Nylon 6 plus 20% amorphous nylon blend. You will need to specify if the film is made by a blown or cast process.
OPET/ink /adh//SiOx-OPET/primer/(LDPE – Ionomer)
A reverse printed OPET is adhesively laminated to a silicon oxide coated OPET. The adhesive is applied to the printed surface of the outer OPET. Then the plain side of the silicon oxide coated OPET is primed and coextrusion coated with a two-layer LDPE and Ionomer coating.
OPET/ink/primer/met/adhesive//(Nylon - tie - LLDPE)
A reverse printed OPET is primed on the ink surface, then metallized. This film is then adhesively laminated to a three-layer coex film. The adhesive is applied to the metallized surface.
OPET/ink/adhesive//OPET-VM/(LDPE – Ionomer)
A reverse printed OPET is adhesively laminated to a metallized OPET. The adhesive is applied to the ink surface. The metallized layer is facing to the inside of the package structure, not the outside. The metallized surface is coex-extrusion coated with a two-layer LDPE and Ionomer coating.
OPP/ink/primer/LDPE/primer/VM-CPP
A reverse printed OPP that is extrusion laminated using LDPE to a metallized cast polypropylene film. Primers are applied to the printed OPP, and the metallized CPP.
OPP/ink/primer/LDPE/VM-OPET//adhesive/CPP
A reverse printed OPP that is extrusion laminated using LDPE to a metallized OPET. This is then adhesively laminated to a cast polypropylene film. The adhesive is applied to the CPP film.
OPET/ink/primer/LDPE/VM-OPET/adhesive//CPP
A reverse printed OPET that is extrusion laminated using LDPE to a metallized OPET. This is then adhesively laminated to a cast polypropylene film. The adhesive is applied to the OPET film.

(*Continued*)

Table A.13 Structure Examples (Structures Should Be Written From "Outside to Inside," With Regards to the Order of Layers, Starting From the Left)—*Cont'd*

12 μm OPET/met/adh//15 μm BONy/adhesive//corona> (20 μm LLDPE– 10 μm LDPE–45 μm EVA)
An OPET that has been metallized, then adhesively laminated to BONy film. The other side of the nylon film is then adhesively laminated to a coex film which has been in-line corona treated (likely re-treated to burn off slip agents in the film).
OPET/ink/primer/LDPE/met-OPET< corona/primer/LDPE/LLDPE film
A reverse printed OPET film that is coated with an unknown primer, and then extrusion laminated to metallized PET film using LDPE. The plain side of the PET film is then corona treated and primed then extrusion laminated to LLDPE film using LDPE.
20 μm OPP/ink/primer/15 μm LDPE/13 μm met-OPET< corona/primer/12 μm LDPE/12 μm EVA
A reverse printed OPP that is extrusion laminated to a metallized PET film. The plain side of the PET film is extrusion coated with LDPE, and the LDPE is extrusion coated with EVA.
(LLDPE – W.O. LLDPE – LDPE) film/EAA/alu/EAA/(LDPE – LLDPE) film
An extrusion lamination with EAA of a three-layer film to one side of the aluminum. The outer film is white, but the pigment is only in the middle layer. The other side of the aluminum has a two-layer film extrusion laminated to it with an EAA.
50 μm Blue EAA//TL//alu//TL//50 μm EAA
A thermal lamination of blue pigmented EAA film to aluminum on one side. Another film of clear EAA is thermally laminated to the other side of the alu. (A cable shielding type structure.)
Green EAA/200 μm alu/EAA
An extrusion coating of green pigmented EAA is applied to one side of the 200 μm aluminum. A clear EAA is extrusion coated to the other side of the aluminum. (A cable shielding type structure.)
(LDPE – EAA)//TL//150 μm alu//TL//(EAA – LDPE)
150 μm aluminum that has coex films thermally laminated to each side. (A cable shielding type structure.)
{ (mLLDPE – LDPE – tie – Nylon – EVOH – Nylon – EVA tie)//BL//(EVA tie – Nylon – EVOH – Nylon – tie – LDPE – mLLDPE) }
A 13-layer structure that is made by intentionally blocking a coex blown film in the main collapsing nip. The film from the die is actually seven layers. In the main nip, the bubble is blocked on purpose. The nip rolls may be heated to help this. The tower height is usually very short so that the film is still quite warm entering the nips. The tie resin at the blocking interface is known to be an EVA based tie resin, but the exact grade is not known.
Cello/ink/primer/LDPE/alu/EMAA
Cellophane that is reverse printed, primed, and extrusion laminated to aluminum with LDPE. The other side of the aluminum is then extrusion coated with EMAA.
K-Cello/ink/primer/LDPE/alu/Ionomer
A PVdC coated cellophane (two side coated), that is reverse printed, primed, and extrusion laminated to aluminum with LDPE. The other side of the aluminum is then extrusion coated with an ionomer.
O.L./Ink/50 gsm Paper/LDPE/7 μm Alu/35 μm EAA (made on a tandem extrusion line)
Paper that is surface printed and overlacquered. It is extrusion laminated with LDPE to aluminum. The other side of the aluminum is extrusion coated with EAA. This is specified as being processed by tandem extrusion lamination instead of two-pass coating.
LDPE/ink/18 pt. Board/LDPE/alu/(EAA- LDPE)
A surface printed 18 Point Paperboard that is surface extrusion coated with LDPE, and then extrusion laminated with LDPE to aluminum. The other side of the aluminum is then coextrusion coated with a combination of LDPE and EAA.

(*Continued*)

Table A.13 Structure Examples (Structures Should Be Written From "Outside to Inside," With Regards to the Order of Layers, Starting From the Left)—*Cont'd*

(850 μm HDPE – 55 μm HDPE tie – 55 μm Nylon 6)
A three-layer coextrusion, which by looking at the thickness, likely is a coex blow molded bottle. This layer thickness is often used for barrier bottles, such as for agricultural chemicals, where nylon is often used as the inner layer. Please indicate the application when drafting out structures such as this to prevent confusion, as this could also be a sheeting application.
(725 μmHDPE – 12 μm regrind – 55 μm HDPE tie – 55 μm Nylon 6)
Similar to above structure for an agchem bottle, except with a layer of regrind material between the outer HDPE and the tie layer.
(225 μm PP – PP tie – 35 μm EVOH – PP tie – 175 μm PP)
A coextrusion that by the thickness is likely a sheet material. The structure may seem to indicate that it might be used to thermoform into a cup or tray. A description of the actual application for the structure should be included in any communication.
OPET/primer/LDPE/sealant//TL//leather
An oriented polyester film that is primed and extrusion coated with LDPE and then the sealant resin. This structure is then thermally laminated to leather. (The application may be for an emblem or other fabric decoration that requires surface protection.)
Ink/30 μm alu<flame/(EMAA – Easy Peel)
A surface printed aluminum foil that is in-line flame treated, and then coextrusion coated with a two-layer coextrusion. The EMAA would be for high foil adhesion. The "easy peel" is a sealant to various cup stock materials.

Table A.14 Sample Structures for Practice Descriptions

K-Nylon< corona/adhesive//(mLLDPE – 12% EVA – peel seal resin)
BONy-PVdC/adhesive//(mLLDPE – mLLDPE – EMAA)
BONy-PVdC/primer/(LDPE – LDPE – EMAA)
OPET/ink/adh/alu/adhesive//(LLDPE – LLDPE – EMAA)
PET-VM/adhesive//BONy/adhesive//corona> (LLDPE – LDPE – EVA)
OPP/ink/adh//met-OPET/primer/(LDPE – Ionomer)
OPET-PVdC/ink/adhesive/(LDPE – EMAA – Ionomer)
OPP/ink<corona/(tie resin-[LDPE+ adhesion modifier]> corona/met-OPET<corona/primer/LDPE

Index

Note: Page numbers followed by "*f*" and "*t*" refer to figures and tables, respectively.

G

H

I

K

L

M

R

S

T

Edwards Brothers Malloy
Ann Arbor MI. USA
April 12, 2016